Geophysical Monograph Series

Geophysical Monograph Series

180 **Arctic Sea Ice Decline: Observations, Projections, Mechanisms, and Implications** *Eric T. DeWeaver, Cecilia M. Bitz, and L.-Bruno Tremblay (Eds.)*
181 **Midlatitude Ionospheric Dynamics and Disturbances** *Paul M. Kintner, Jr., Anthea J. Coster, Tim Fuller-Rowell, Anthony J. Mannucci, Michael Mendillo, and Roderick Heelis (Eds.)*
182 **The Stromboli Volcano: An Integrated Study of the 2002–2003 Eruption** *Sonia Calvari, Salvatore Inguaggiato, Giuseppe Puglisi, Maurizio Ripepe, and Mauro Rosi (Eds.)*
183 **Carbon Sequestration and Its Role in the Global Carbon Cycle** *Brian J. McPherson and Eric T. Sundquist (Eds.)*
184 **Carbon Cycling in Northern Peatlands** *Andrew J. Baird, Lisa R. Belyea, Xavier Comas, A. S. Reeve, and Lee D. Slater (Eds.)*
185 **Indian Ocean Biogeochemical Processes and Ecological Variability** *Jerry D. Wiggert, Raleigh R. Hood, S. Wajih A. Naqvi, Kenneth H. Brink, and Sharon L. Smith (Eds.)*
186 **Amazonia and Global Change** *Michael Keller, Mercedes Bustamante, John Gash, and Pedro Silva Dias (Eds.)*
187 **Surface Ocean–Lower Atmosphere Processes** *Corinne Le Quèrè and Eric S. Saltzman (Eds.)*
188 **Diversity of Hydrothermal Systems on Slow Spreading Ocean Ridges** *Peter A. Rona, Colin W. Devey, Jérôme Dyment, and Bramley J. Murton (Eds.)*
189 **Climate Dynamics: Why Does Climate Vary?** *De-Zheng Sun and Frank Bryan (Eds.)*
190 **The Stratosphere: Dynamics, Transport, and Chemistry** *L. M. Polvani, A. H. Sobel, and D. W. Waugh (Eds.)*
191 **Rainfall: State of the Science** *Firat Y. Testik and Mekonnen Gebremichael (Eds.)*
192 **Antarctic Subglacial Aquatic Environments** *Martin J. Siegert, Mahlon C. Kennicut II, and Robert A. Bindschadler (Eds.)*
193 **Abrupt Climate Change: Mechanisms, Patterns, and Impacts** *Harunur Rashid, Leonid Polyak, and Ellen Mosley-Thompson (Eds.)*
194 **Stream Restoration in Dynamic Fluvial Systems: Scientific Approaches, Analyses, and Tools** *Andrew Simon, Sean J. Bennett, and Janine M. Castro (Eds.)*
195 **Monitoring and Modeling the Deepwater Horizon Oil Spill: A Record-Breaking Enterprise** *Yonggang Liu, Amy MacFadyen, Zhen-Gang Ji, and Robert H. Weisberg (Eds.)*
196 **Extreme Events and Natural Hazards: The Complexity Perspective** *A. Surjalal Sharma, Armin Bunde, Vijay P. Dimri, and Daniel N. Baker (Eds.)*
197 **Auroral Phenomenology and Magnetospheric Processes: Earth and Other Planets** *Andreas Keiling, Eric Donovan, Fran Bagenal, and Tomas Karlsson (Eds.)*
198 **Climates, Landscapes, and Civilizations** *Liviu Giosan, Dorian Q. Fuller, Kathleen Nicoll, Rowan K. Flad, and Peter D. Clift (Eds.)*
199 **Dynamics of the Earth's Radiation Belts and Inner Magnetosphere** *Danny Summers, Ian R. Mann, Daniel N. Baker, and Michael Schulz (Eds.)*
200 **Lagrangian Modeling of the Atmosphere** *John Lin (Ed.)*
201 **Modeling the Ionosphere-Thermosphere** *Jospeh D. Huba, Robert W. Schunk, and George V. Khazanov (Eds.)*
202 **The Mediterranean Sea: Temporal Variability and Spatial Patterns** *Gian Luca Eusebi Borzelli, Miroslav Gacic, Piero Lionello, and Paola Malanotte-Rizzoli (Eds.)*
203 **Future Earth – Advancing Civic Understanding of the Anthropocene** *Diana Dalbotten, Gillian Roehrig, and Patrick Hamilton (Eds.)*
204 **The Galápagos: A Natural Laboratory for the Earth Sciences** *Karen S. Harpp, Eric Mittelstaedt, Noémi d'Ozouville, and David W. Graham (Eds.)*
205 **Modeling Atmospheric and Oceanic Flows: Insightsfrom Laboratory Experiments and Numerical Simulations** *Thomas von Larcher and Paul D. Williams (Eds.)*
206 **Remote Sensing of the Terrestrial Water Cycle** *Venkat Lakshmi (Ed.)*
207 **Magnetotails in the Solar System** *Andreas Keiling, Caitríona Jackman, and Peter Delamere (Eds.)*
208 **Hawaiian Volcanoes: From Source to Surface** *Rebecca Carey, Valerie Cayol, Michael Poland, and Dominique Weis (Eds.)*
209 **Sea Ice: Physics, Mechanics, and Remote Sensing** *Mohammed Shokr and Nirmal Sinha (Eds.)*
210 **Fluid Dynamics in Complex Fractured-Porous Systems** *Boris Faybishenko, Sally M. Benson, and John E. Gale (Eds.)*
211 **Subduction Dynamics: From Mantle Flow to Mega Disasters** *Gabriele Morra, David A. Yuen, Scott King, Sang Mook Lee, and Seth Stein (Eds.)*
212 **The Early Earth: Accretion and Differentiation** *James Badro and Michael Walter (Eds.)*
213 **Global Vegetation Dynamics: Concepts and Applications in the MC1 Model** *Dominique Bachelet and David Turner (Eds.)*
214 **Extreme Events: Observations, Modeling and Economics** *Mario Chavez, Michael Ghil, and Jaime Urrutia-Fucugauchi (Eds.)*
215 **Auroral Dynamics and Space Weather** *Yongliang Zhang and Larry Paxton (Eds.)*
216 **Low-Frequency Waves in Space Plasmas** *Andreas Keiling, Dong-Hun Lee, and Valery Nakariakov (Eds.)*
217 **Deep Earth: Physics and Chemistry of the Lower Mantle and Core** *Hidenori Terasaki and Rebecca A. Fischer (Eds.)*
218 **Integrated Imaging of the Earth: Theory and Applications** *Max Moorkamp, Peter G. Lelievre, Niklas Linde, and Amir Khan (Eds.)*
219 **Plate Boundaries and Natural Hazards** *Joao Duarte and Wouter Schellart (Eds.)*
220 **Ionospheric Space Weather: Longitude and Hemispheric Dependences and Lower Atmosphere Forcing Timothy Fuller-Rowell**, *Endawoke Yizengaw, Patricia H. Doherty, and Sunanda Basu (Eds.)*
221 **Terrestrial Water Cycle and Climate Change Natural and Human-Induced Impacts** *Qiuhong Tang and Taikan Oki (Eds.)*
222 **Magnetosphere-Ionosphere Coupling in the Solar System** *Charles R. Chappell, Robert W. Schunk, Peter M. Banks, James L. Burch, and Richard M. Thorne (Eds.)*
223 **Natural Hazard Uncertainty Assessment: Modeling and Decision Support** *Karin Riley, Peter Webley, and Matthew Thompson (Eds.)*
224 **Hydrodynamics of Time-Periodic Groundwater Flow: Diffusion Waves in Porous Media** *Joe S. Depner and Todd C. Rasmussen (Auth.)*
225 **Active Global Seismology** *Ibrahim Cemen and Yucel Yilmaz (Eds.)*
226 **Climate Extremes** *Simon Wang (Ed.)*
227 **Fault Zone Dynamic Processes** *Marion Thomas (Ed.)*
228 **Flood Damage Survey and Assessment: New Insights from Research and Practice** *Daniela Molinari, Scira Menoni, and Francesco Ballio (Eds.)*
229 **Water-Energy-Food Nexus – Principles and Practices** *P. Abdul Salam, Sangam Shrestha, Vishnu Prasad Pandey, and Anil K Anal (Eds.)*
230 **Dawn–Dusk Asymmetries in Planetary Plasma Environments** *Stein Haaland, Andrei Rounov, and Colin Forsyth (Eds.)*
231 **Bioenergy and Land Use Change** *Zhangcai Qin, Umakant Mishra, and Astley Hastings (Eds.)*
232 **Microstructural Geochronology: Planetary Records Down to Atom Scale** *Desmond Moser, Fernando Corfu, James Darling, Steven Reddy, and Kimberly Tait (Eds.)*
233 **Global Flood Hazard: Applications in Modeling, Mapping and Forecasting** *Guy Schumann, Paul D. Bates, Giuseppe T. Aronica, and Heiko Apel (Eds.)*
234 **Pre-Earthquake Processes: A Multidisciplinary Approach to Earthquake Prediction Studies** *Dimitar Ouzounov, Sergey Pulinets, Katsumi Hattori, and Patrick Taylor (Eds.)*
235 **Electric Currents in Geospace and Beyond** *Andreas Keiling, Octav Marghitu, and Michael Wheatland (Eds.)*

Geophysical Monograph 236

Quantifying Uncertainty in Subsurface Systems

Céline Scheidt
Lewis Li
Jef Caers

This Work is a co-publication of the American Geophysical Union and John Wiley and Sons, Inc.

AGU 100
ADVANCING EARTH AND SPACE SCIENCE

WILEY

This Work is a co-publication between the American Geophysical Union and John Wiley & Sons, Inc.

This edition first published 2018 by John Wiley & Sons, Inc., 111 River Street, Hoboken, NJ 07030, USA and the American Geophysical Union, 2000 Florida Avenue, N.W., Washington, D.C. 20009

Published under the aegis of the AGU Publications Committee

Brooks Hanson, Executive Vice President, Science
Lisa Tauxe, Chair, Publications Committee

For details about the American Geophysical Union visit us at www.agu.org.

Wiley Global Headquarters

111 River Street, Hoboken, NJ 07030, USA

For details of our global editorial offices, customer services, and more information about Wiley products visit us at www.wiley.com.

Library of Congress Cataloging-in-Publication data is available.

ISBN: 978-1-119-32583-3

Cover design: Wiley
Cover image: (Ganges Delta) © NASA, USGS EROS Data Center Satellite Systems Branch; (Graph) © Jef Karel Caers, Professor, School of Earth, Energy & Environmental Sciences, Stanford University

Set in 10/12pt Times by SPi Global, Pondicherry, India

10 9 8 7 6 5 4 3 2 1

CONTENTS

Preface vii
Authors xi

1. The Earth Resources Challenge 1
2. Decision Making Under Uncertainty 29
3. Data Science for Uncertainty Quantification 45
4. Sensitivity Analysis 107
5. Bayesianism 129
6. Geological Priors and Inversion 155
7. Bayesian Evidential Learning 193
8. Quantifying Uncertainty in Subsurface Systems 217
9. Software and Implementation 263
10. Outlook 267

Index 273

PREFACE

"I think that when we know that we actually do live in uncertainty, then we ought to admit it; it is of great value to realize that we do not know the answers to different questions. This attitude of mind – this attitude of uncertainty – is vital to the scientist, and it is this attitude of mind which the student must first acquire"

Richard P. Feynman, Noble Laureate in Physics, 1965

This book offers five substantial case studies on decision making under uncertainty for subsurface systems. The strategies and workflows designed for these case studies are based on a Bayesian philosophy, tuned specifically to the particularities of the subsurface realm. Models are large and complex; data are heterogeneous in nature; decisions need to address conflicting objectives; the subsurface medium is created by geological processes that are not always well understood; and expertise of a large variety of scientific and engineering disciplines need to be synthesized.

There is no doubt that we live in an uncertain time. With growing population, resources such as energy, materials, water, and food will become increasingly critical in their exploitation. The subsurface offers many such resources, important to the survival of humankind. Drinking water from groundwater systems is gaining in importance, as aquifers are natural purifiers and can store large volumes. However, the groundwater system is fragile, subject to contamination from agriculture practices and industries. Before renewables become the dominant energy sources, oil and gas will remain a significant resource in the next few decades. Geothermal energy both deep (power) and shallow (heating) can contribute substantially to alleviating reliance on fossil fuels. Mining minerals used for batteries will aid in addressing intermittency of certain renewables, but mining practices will need to address environmental concerns.

Companies and governmental entities involved in the extraction of these resources face considerable financial risk because of the difficulty in accessing the poorly understood subsurface and the cost of engineering facilities. Decisions regarding exploration methods, drilling, extraction methods, and data-gathering campaigns often need to balance conflicting objectives: resource versus environmental impact, risk versus return. This can be truly addressed only if one accepts uncertainty as integral part of the decision game. A decision based on a deterministic answer when uncertainty is prevailing is simply a poor decision, regardless of the outcome. Decisions and uncertainty are part of one puzzle; one does not come before the other.

Uncertainty on key decision variables such as volumes, rates of extraction, time of extraction, spatiotemporal variation on fluid movements needs to be quantified. Uncertainty quantification, in this book shortened to UQ, requires a complex balancing of several fields of expertise such as geological sciences, geophysics, data science, computer science, and decision analysis. We gladly admit that we do not have a single best solution to UQ. The aim of this book is to provide the reader with a principled approach, meaning a set of actions motivated by a mathematical philosophy based on axioms, definitions, and algorithms that are well understood, repeatable, and reproducible, as well as a software to reproduce the results of this book. We consider uncertainty not simply to be some posterior analysis but a synthesized discipline steeped in scientific ideas that are still evolving. Ten chapters provide insight into our way of thinking on UQ.

Chapter 1 introduces the five case studies: an oil reservoir in Libya, a groundwater system in Denmark, a geothermal source for heating buildings in Belgium, a contaminated aquifer system in Colorado, and an unconventional hydrocarbon resource in Texas. In each case study, we introduce the formulation of the decision problem, the types of data used, and the complexity of the modeling problem. Common to all these cases is that the decision problem involves simple questions: Where do we drill? How much is there? How do we extract? What data to gather? The models involved on the other hand are complex and high dimensional, the forward simulators time-consuming. The case studies set the stage.

Chapter 2 introduces the reader to some basic notions in decision analysis. Decision analysis is a science, with its own axioms, definitions, and heuristics. Properly formulating the decision problem, defining the key decision variables, the data used to quantify these, and the objectives of the decision maker are integral to such decision analysis. Value of information is introduced as a formal framework to assess the value of data before acquiring it.

Chapter 3 provides an overview of the various data science methods that are relevant to UQ problems in the subsurface. Representing the subsurface requires a high-dimensional model parametrization. To make UQ problems manageable, some form of dimension reduction is needed. In addition, we focus on several methods of regression such as Gaussian process regression and CART (classification and regression trees) that are useful for statistical learning and development of statistical proxy models. Monte Carlo is covered extensively as this is instrumental to UQ. Methods such as importance sampling and sequential importance resampling are discussed. Lastly, we present the extension of Monte Carlo to Markov chain Monte Carlo and bootstrap; both are methods to address uncertainty and confidence.

Chapter 4 is dedicated to sensitivity analysis (SA). Although SA could be part of Chapter 3, because of its significance to UQ, we dedicate a single chapter to it. Our emphasis will be on global SA and more specifically Monte Carlo-based SA since this family of methods (Sobol', regionalized sensitivity analysis, CART) provides key insight into understanding what model variables most impact data and prediction variables.

Chapter 5 introduces the philosophy behind Bayesian methods: Bayesianism. We provide a historical context to why Bayes has become one of the leading paradigms to UQ, having evolved from other paradigms such as induction, deduction, and falsification. The most important contribution of Thomas Bayes is the notion of the prior distribution. This notion is critical to UQ in the subsurface, simply because of the poorly understood geological medium that drives uncertainty. The chapter, therefore, ends with a discussion on the nature of prior distributions in the geosciences, how one can think about them and how they can be established from physical, rather than statistical principles.

Chapter 6 then extends on Chapter 5 by discussion on the role of prior distribution in inverse problems. We provide a brief overview of both deterministic and stochastic inversion. The emphasis lies on how quantification of geological heterogeneity (e.g., using geostatistics) can be used as prior models to solve inverse problems, within a Bayesian framework.

Chapter 7 is perhaps the most novel technical contribution of this book. This chapter covers a collection of methods termed Bayesian evidential learning (BEL). Previous chapters indicated that one of the major challenges in UQ is model realism (geological) as well as deal with large computing times in forward models related to data and prediction responses. In this chapter, we present several methods of statistical learning, where Monte Carlo is used to generate a training set of data and prediction variables. This Monte Carlo approach requires the specification of a prior distribution on the model variables. We show how learning the multivariate distribution of data and prediction variables allows for predictions based on data without complex model inversions.

Chapter 8 presents various strategies addressing the decision problem of the various case studies introduced in Chapter 1. The aim is not to provide the best possible method but to outline choices in methods and strategies in combination to solve real-world problems. These strategies rely on materials presented in Chapters 2–7.

Chapter 9 provides a discussion of the various software components that are necessary for the implementation of the different UQ strategies presented in the book. We discuss some of the challenges faced when using existing software packages as well as provide an overview of the companion code for this book.

Chapter 10 concludes this book by means of seven questions that formulate important challenges that when addressed may move the field of UQ forward in impactful ways.

We want to thank several people who made important contributions to this book, directly and indirectly. This book would not have been possible without the continued support of the Stanford Center for Reservoir Forecasting. The unrestricted funding provided over the last 30 years has aided us in working on case studies as well as fundamental research that focuses on synthesis in addition to many technical contributions in geostatistics, geophysics, data science, and others. We would also like to thank our esteemed colleagues at Stanford University and elsewhere, who have been involved in many years of discussion around this topic. In particular, we would like to thank Tapan Mukerji (Energy Resources Engineering & Geophysics), who has been instrumental in educating us on decision analysis as well as on the geophysical aspects of this book. Kate Maher (Earth System Science) provided important insights into the modeling of the case study on uranium contamination. We thank the members of the Ensemble project funded by the Swiss government, led by Philippe Renard (University of Neuchatel), Niklas Linde (University of Lausanne), Peter Huggenberger (University of Basel), Ivan Lunati (University of Lausanne), Grégoire Mariethoz (University of Lausanne), and David Ginsbourger (University of Bern). To our knowledge, this was one of the first large-scale governmental project involving both research and education for quantifying uncertainty in the subsurface. We would also like to thank Troels Vilhelmsen (University of Aarhus) for the short but intensive collaboration on the Danish groundwater case. We welcome the data provided by Wintershall (Michael Peter Suess) and Andarko (Carla Da Silva). The Belgian case was done with Thomas Hermans (University of Gent), when he was postdoctoral researcher at Stanford. Discussions with Fréderic Nguyen (University of Liege) were also instrumental for that case study. We would also like to thank Emanuel Huber (University of Basel) for the construction of the hydrological (DNAPL) test case used in Chapters 3 and 4 during his postdoc at Stanford.

PhD students also have been integral part of this work, at Stanford and elsewhere. In particular, we would like to thank Addy Satija, Orhun Aydin, Ognjen Grujic, Guang Yang, Jihoon Park, Markus Zechner, and Adrian Barfod (University of Aarhus).

The thumbtack game on decision analysis was introduced to us by Reidar Bratvold (University of Stavanger). Early reviews on Chapter 5 (Bayesianism) by Klaus Mosegaard (University of Kopenhagen), Don Dodge (Retired, San Francisco), and Matthew Casey (The Urban School, San Francisco) were instrumental to the writing and clarity of the chapter. We also thank Darryl Fenwick (Kappa Engineering) for early reviews of Chapters 4 and 6 and for many fruitful discussions. We are very grateful to the 10 anonymous reviewers and the Wiley editors for their critical comments.

We hope you enjoy our work.

Céline Scheidt
Lewis Li
Jef Caers

AUTHORS

Céline Scheidt
Senior Research Engineer
Departments of Energy Resources Engineering
Stanford University, Stanford, CA, USA

Lewis Li
Doctoral Student
Departments of Energy Resources Engineering
Stanford University, Stanford, CA, USA

Jef Caers
Professor of Geological Sciences
Director, Stanford Center for Earth Resources Forecasting
Stanford University, Stanford, CA, USA

CONTRIBUTORS

Ognjen Grujic
Department of Energy Resources Engineering,
Stanford University, Stanford, CA, USA

Thomas Hermans
University of Liege, Liege, Belgium

Kate Maher
Department of Geological Sciences, Stanford University,
Stanford, CA, USA

Jihoon Park
Department of Energy Resources Engineering,
Stanford University, Stanford, CA, USA

Carla Da Silva
Anadarko, The Woodlands, TX, USA

Troels Norvin Vilhelmsen
Department of Geoscience, Aarhus University, Aarhus,
Denmark

Guang Yang
Department of Energy Resources Engineering,
Stanford University, Stanford, CA, USA

1

The Earth Resources Challenge

Co-Authored by: Troels Norvin Vilhelmsen[1], Kate Maher[2], Carla Da Silva[3], Thomas Hermans[4], Ognjen Grujic[5], Jihoon Park[5], and Guang Yang[5]

1.1. WHEN CHALLENGES BRING OPPORTUNITIES

Humanity is facing considerable challenges in the 21st century. Population is predicted to grow well into this century and saturate between 9 and 10 billion somewhere in the later part. This growth has led to climate change (see the latest IPCC reports), has impacted the environment, and has affected ecosystems locally and globally around the planet. Virtually no region exists where humans have had no footprint of some kind [*Sanderson et al.*, 2002]; we now basically "own" the ecosystem, and we are not always a good Shepard. An increasing population will require an increasing amount of resources, such as energy, food, and water. In an ideal scenario, we would transform the current situation of unsustainable carbon-emitting energy sources, polluting agricultural practices and contaminating and over-exploiting drinking water resources, into a more sustainable and environmentally friendly future. Regardless of what is done (or not), this will not be an overnight transformation. For example, natural gas, a green-house gas (either as methane or burned into CO_2), is often called the blue energy toward a green future. But its production from shales (with vast amounts of gas and oil reserves, 7500 Tcf of gas, 400 billion barrels of oil, US Energy Information, December 2014) has been questioned for its effect on the environment from gas leaks [*Howarth et al.*, 2014] and the unsolved problem of dealing with the waste water it generates. Injecting water into kilometer-deep wells has caused significant earthquakes [Whitaker, 2016], and risks to contamination of the groundwater system are considerable [*Osborn et al.*, 2011].

Challenges bring opportunities. The Earth is rich in resources, and humanity has been creative and resourceful in using the Earth to advance science and technology. Batteries offer promising energy storage devices that can be connected to intermittent energy sources such as wind and solar. Battery technology will likely develop further from a better understanding of Earth materials. The Earth provides a naturally emitting heat source that can be used for energy creation or heating of buildings. In this book, we will contribute to exploration and exploitation of geological resources. The most common of such resources are briefly described in the following:

1. *Fossil fuels* will remain an important energy source for the next several decades. Burning fossil fuels is not a sustainable practice. Hence, the focus will be on the transformation of this energy, least impacting the environment as possible. An optimal exploitation, by minimizing drilling, will require a better understanding of the risk associated with the exploration and production. Every mistake (drilling and spilling) made by an oil company has an impact on the environment, direct or indirect. Even if fossil fuels will be in the picture for a while, ideally we will develop these resources as efficient as possible, minimally impacting the environment.

2. *Heat* can be used to generate steam, drive turbines, and produce energy (high enthalpy heat systems). However, the exploitation of geothermal systems is costly and not always successful. Injecting water into

[1]Department of Geoscience, Aarhus University, Aarhus, Denmark
[2]Department of Geological Sciences, Stanford University, Stanford, CA, USA
[3]Anadarko, The Woodlands, TX, USA
[4]University of Liege, Liege, Belgium
[5]Department of Energy Resources Engineering, Stanford University, Stanford, CA, USA

Quantifying Uncertainty in Subsurface Systems, Geophysical Monograph 236,
First Edition. Céline Scheidt, Lewis Li, and Jef Caers.

kilometer-deep wells may end up causing earthquakes [Glanz, 2009]. Reducing this risk is essential to a successful future for geothermal energy. In a low enthalpy system, the shallow subsurface can be used as a heat exchanger, for example through groundwater, to heat buildings. The design of such systems is dependent on how efficient heat can be exchanged with groundwater that sits in a heterogeneous system, and the design is often subject to a natural gradient.

3. *Groundwater* is likely to grow as a resource for drinking water. As supply of drinking water, this resource is however in competition with food (agriculture) and energy (e.g., from shales). Additionally, the groundwater system is subject to increased stresses such as from over-pumping and contamination.

4. *Minerals resources* are exploited for a large variety of reasons. For example, the use of Cu/Fe in infrastructure, Cd/Li/Co/Ni for batteries, rare earth elements for amplifiers in fiber-optic data transmission or mobile devices, to name just a few. An increase in the demand will require the development of mining practices that have minimal effect on the environment, such as properly dealing with waste as well as avoiding groundwater contamination.

5. *Storage* of fluids such as natural gas, CO_2, or water (aquifer storage and recovery) in the subsurface is an increasing practice. The porous subsurface medium acts as a permanent or temporary storage of resources. However, risks of contamination or loss need to be properly understood.

The geological resource challenge will require developing basic fields of science, applied science and engineering, economic decision models, as well as creating a better understanding regarding human behavioral aspects. The ultimate aim here is to "predict" what will happen, and based on such prediction what are best practices in terms of optimal exploitation, maximizing sustainability, and minimizing of impact on the environment. The following are the several areas that require research: (i) fundamental science, (ii) predictive models, (iii) data science, and (iv) economic and human behavior models.

Fundamental science. Consider, for example, the management of groundwater system. The shallow subsurface can be seen as a biogeochemical system where biological, chemical agents interact with the soils or rock within which water resides. The basic reactions of these agents may not yet be fully understood nor does the flow of water when such interactions take place. To understand this better, we will further need to develop such understanding based on laboratory experiments and first principles. Additionally, the flow in such systems depends on the spatial variability of the various rock properties. Often water resides in a sedimentary system. A better understanding of the processes that created such systems will aid in predicting such flow. However, the flow of particles in a viscous fluid, which leads to deposition and erosion and ultimately stratigraphy, is fundamentally not well understood; hence, the basic science around this topic needs to be further developed. A common issue is that basic science is conducted in laboratories at a relatively small scale; hence, the question of upscaling to application scales remains, equally, a fundamental research challenge.

Predictive models. Fundamental science or the understanding of process alone does not result in a prediction or an improvement into what people decide in practice. Predictions require predictive models. These could be a set of partial differential equations, reactions, phase diagrams, and computer codes developed from basic understanding. In our groundwater example, we may develop codes for predictive modeling of reactive transport in porous media. Such codes require specification of initial and boundary conditions, geochemical reaction rates, biogeochemistry, porous media properties, and so on. Given one such specification, the evolution of the system can then be predicted at various space-time scales.

Data science. Predictive models alone do not make meaningful predictions in practical settings. Usually, site-specific data are gathered to aid such predictions. In the groundwater case, this may consist of geophysical data, pumping data, tracer data, geochemical analysis, and so on. The aim is often to integrate predictive models with data, generally denoted as inversion. The challenge around this inversion is that no single model predicts the data; hence, uncertainty about the future evolution of the system exists. Because of the growing complexity of the kind of data we gather and the kind of models we develop, an increased need exists in developing data scientific methods that handle such complexities fully.

Economic decision models and social behavior. The prediction of evolution of geological resource systems cannot be done without the "human context." Humans will make decision on the exploitation of geological resources and their behavior may or may not be rational. Rational decision making is part of decision science, and modeling behavior (rational or not) is part of game theory. Next to the human aspects, there is a need for global understanding of the effect of the evolution of technology on geological resources. For example, how will the continued evolution affect the economy of mineral resources? How will any policy change in terms of rights to groundwater resources change the exploitation of such resources?

In this book, we focus mostly on making predictions as input to decision models. Hence, we focus on development of data scientific tools for uncertainty quantification in geological resources systems. However, at the same time, we are mindful about the fact that we do not yet have a fundamental understanding of some of the basic science. This is important because after all UQ is about quantifying lack of understanding. We are also mindful about the fact the current predictive models only approximate any physical/chemical reality in the sense that these are based on (still) limited understanding of process. In the subsurface, this is quite prevalent. We do not know exactly how the subsurface system, consisting of solids and fluids, was created and how solids and fluids interact (together with the biological system) under imposed stresses or changes. Most of our predictive models are upscaled versions of an actual physical reality. Last, we are also mindful that our predictions are part of a larger decision model and that such decision models themselves are only approximate representation of actual human behavior.

Hence, we will not provide an exact answer to all these questions and solve the world's problems! In that sense, the book is contributing to sketching paths forward in this highly multidisciplinary science. This book is part of an evolution in the science of predictions, with a particular application to the geological resources challenge. The best way to illustrate this is with real field case studies on the above-mentioned resources, how predictive models are used, how data come into the picture, and how the decision model affects our approach to using such predictive models in actual practical cases, with actual messy data. Chapter 1 introduces these cases and thereby sets the stage.

1.2. PRODUCTION PLANNING AND DEVELOPMENT FOR AN OIL FIELD IN LIBYA

1.2.1. Reservoir Management from Discovery to Abandonment

Uncertainty quantification in petroleum systems has a long history and perhaps one of the first real-world applications of such quantification, at least for the subsurface. This is partly due to the inherent large financial risk (sometime billions of dollars) involved in decision making about exploration and production. Consider simply that the construction of a single offshore platform may cost several billion dollars and may not pay back return if uncertainty/risk is poorly understood, or if estimates are too optimistic. Uncertainty quantification is (and perhaps should be) an integral part of decision making in such systems.

Modern reservoir management aims at building complex geological models of the subsurface and running computationally demanding models of multiphase flow that simulates the combined movement of fluids in the subsurface under induced changes, such as from production by enhancing the recovery by injection of water, CO_2, polymers, or foams. In particular, for complex systems and costly operations, numerical models are used to make prediction and run numerical optimizations since simple analytical solution can only provide very rough estimates and cannot be used for individual well-planning or for assessing the worth of certain data acquisition methods.

Reservoir management is not a static task. First, the decision to use certain modeling and forecasting tools depends on what stage of the reservoir life one is dealing with, which is typically divided into (i) exploration, (ii) appraisal, (iii) early production, (iv) late production, and (v) abandonment. Additionally, several types of reservoir systems exist. Offshore reservoirs may occur in shallow to very deep water (1500–5000 ft of water column) and are found on many sedimentary margins in the world (e.g., West Africa, Gulf of Mexico, Brazil). To produce such reservoirs, and generate return on investments, wells need to be produced at a high rate (as much as 20,000 BBL/day). Often wells are clustered from a single platform. Exploration consists of shooting 2D seismic lines, from which 2D images of the subsurface are produced. A few exploration wells may be drilled to confirm a target or confirm the extent of target zone. From seismic alone it may not be certain that a sand is oil-filled or brine-filled. With interesting targets identified, 3D seismic surveys are acquired to get a better understanding of the oil/gas trap in terms of the structure, the reservoir properties, and distribution of fluids (e.g., contacts between gas/oil, oil/water). Traps are usually 1–10 km in magnitude aerially and 10–100s of feet vertically. The combination of additional exploration wells together with seismic data allows for the assessment of the amount of petroleum product (volume) available and how easy it is to recover the reservoir, and how such recovery will play out over time: the recovery factor (over time).

Because of the lack of sufficient data, any estimate of volume or recovery at the appraisal stage is subject to considerable uncertainty. For example, a reservoir volume (at surface conditions, meaning accounting for volume changes due to extraction to atmospheric conditions) is determined as

$$\text{Volume} = \text{area} \times \text{thickness} \times \text{porosity} \times \text{oil saturation} \times \text{formation volume factor} \quad (1.1)$$

However, this simple expression ignores the (unknown) complexity in the reservoir structure (e.g., presence of faults). Each of the above factors is subject to uncertainty. Typically, a simple Monte Carlo analysis is performed to determine uncertainty on the reservoir volume. This requires stating probability distributions for each variable, often taken as independent, and often simply guessed by the modeler. However, such analysis assumes a rather simple setting such as shown in Figure 1.1 (left). Because only few wells are drilled, the reservoir may look fairly simple from the data point of view. The combination of a limited number of wells (samples) with the low-resolution seismic (at least much lower than what can be observed in wells) may obfuscate the presence of

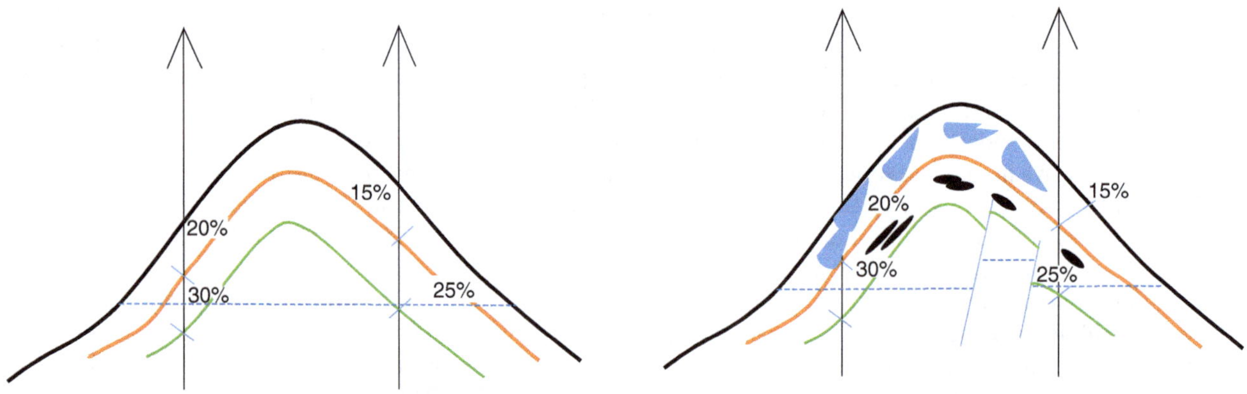

Figure 1.1 Idealized vs. real setting in estimating original oil in place.

complexity that affects volume, such as geological heterogeneity (the reservoir is not a homogenous sand but has a considerable non-reservoir shale portion), presence of faults not detectable on seismic, or presence of different fluid contacts as shown in Figure 1.1 (right). This requires then a careful assessment of the uncertainty of each variable involved.

While offshore reservoirs are produced from a limited set of wells (10–50), onshore systems allow for much more extensive drilling (100–1000). Next to the conventional reservoir systems (those produced in similar ways as the offshore ones and in similar geological settings), a shift has occurred to unconventional systems. Such systems usually consist of shales, which were considered previously to be "unproducible," but have become part of oil/gas production due to the advent of hydraulic fracturing (HF). Thus, starting in 2005, a massive development of unconventional shale resources throughout North America has interrupted both the domestic and the international markets. From a technical perspective, development of shale reservoirs is challenging and is subject to a substantial learning curve. To produce value, shale operators often experiment with new technologies, while also testing applicability of the best practices established in other plays. Traditional reservoir modeling methods and Monte Carlo analysis (see next) become more difficult in these cases, simply because the processes whereby rock breaks, gas/oil released and produced at the surface are much less understood and require in addition to traditional fields of reservoir science knowledge about the joint geomechanical and fluid flow processes in such systems. As a result, and because of fast development of shale plays (e.g., one company reporting drilling more than 500/year of "shale" wells), a more data centric approach to modeling and uncertainty quantification is taken. This data scientific approach relies on using production of existing wells, in combination with the production and geological parameters to directly model and forecast new wells or estimate how long a producing well will decline (hydraulic fractured wells typically start with a peak followed by a gradual decline). In Section 1.6, we will present these types of systems. Here we limit ourselves to conventional reservoir systems.

1.2.2. Reservoir Modeling

In the presence of considerable subsurface complexity, volume or recovery factor assessment becomes impossible without explicitly modeling the various reservoir elements and all the associated uncertainties. Reservoirs requiring expensive drilling are therefore now routinely assessed by means of computer (reservoir) models, whether for volume estimate, recovery factor estimates, placement of wells, or operations of existing wells. Such models are complex, because the reservoir structure is complex. The following are the various modeling elements that need to be tackled.

1. *Reservoir charge*. No oil reservoir exists without migration of hydrocarbon "cooked" from a source rock and trapped in a sealing structure. To assess this, oil companies build basin and petroleum system models to assess the uncertainty and risk associated with finding hydrocarbons in a potential trap. This requires modeling evolution of the sedimentary basins, the source rock, burial history, heat flow, and timing of kerogen migration, all of which are subject to considerable uncertainty.

2. *Reservoir structure*, consisting of faults and layers. These are determined from wells and seismic, and these may be very uncertain in cases with complex faulting (cases are known to contain up to 1000 faults), or due to difficult and subjective interpretation from seismic. In addition, the seismic image itself (the data on which interpretation are done) is uncertain. Structures are usually modeled as surfaces (2D elements). Their modeling requires accounting of tectonic history, informing the age relationships between faults, and several rules of interaction between the structural elements (see Chapter 6).

3. *The reservoir petrophysical properties*. The most important are porosity (volume) and permeability (flow). However, because of the requirement to invert and model

seismic data (3D or 4D), other properties and their spatial distribution are required such as lithology, velocity (p-wave, s-wave), impedance, density, compressibility, Young's modulus, Poisson coefficient, and so on. First, the spatial distribution of these properties depends on the kind of depositional system present (e.g., fluvial, deltaic), which may itself be uncertain, with few wells drilled. The depositional system will control the spatial distribution of lithologies/facies (e.g., sand, shale, dolomite), which in turn controls the distribution of petrophysical properties, as different lithologies display different petrophysical characteristics. In addition, all (or most) petrophysical properties are (co)-related, simply because of the physical laws quantifying them. Rock physics is a field of science that aims to understand these relationships, based on laboratory experiments, and then apply them to understand the observed seismic signals in terms of rock and fluid properties. These relationships are uncertain because (i) the scale of laboratory experiments and ideal conditions are different from reservoir conditions and (ii) the amount of reservoir (core) samples that can be obtained to verify these relationships are limited. This has led to the development of the field of statistical rock physics [*Avseth et al.*, 2005; *Mavko et al.*, 2009].

4. *Reservoir fluid properties.* A reservoir usually contains three types of fluids: gas, oil, and brine (water), usually layered in that order because of density difference. The (initial) spatial distribution of these fluids may, however, not be homogeneous depending on temperature, pressure, geological heterogeneity, and migration history (oil matures from a source rock, traveling toward a trap). Reservoir production will initially lead to a pressure decline (primary production), then to injection of other fluids (e.g., water, gas, polymers, foams) into the reservoir. Hence, to understand all these processes, one needs to understand the interaction and movement of these various fluids under changing pressure, volume, and temperature conditions. This requires knowing the various thermodynamic properties of complex hydrocarbon chains and their phase changes. These are typically referred to as the PVT (pressure–volume–temperature) properties. The following are some basic properties involved that are crucial (to name just a few):

- Formation volume factor: The ratio of a phase volume (water, oil, gas) at reservoir conditions, relative to the volume of a surface phase (water, oil, or gas).
- Solution gas–oil ratio: The amount of surface gas that can be dissolved in a stock tank oil when brought to a specific pressure and temperature.
- API specific gravity: A common measure of oil specific gravity.
- Bubble-point pressure: The pressure when gas bubbles dissolve from the oil phase.

In a reservoir system, several fluids move jointly through the porous systems (multiphase flow). A common way to represent this is through relative permeability and capillary functions. These functions determine how one fluid moves under given saturation of another fluid. However, they in turn depend on the nature of the rock (the lithology) and the pore fabric system, which is uncertain, both in characteristics (which mineral assemblages occur) and in spatial distribution. Limited samplings (cores) are used in laboratory experiments to determine all these properties.

Building a reservoir model, namely representing structure and rock and fluid properties, requires a complex set of software tools and data. Because of the limited resolution of such models, the limited understanding of reservoir processes, and the limited amount of data, such models are subject to considerable uncertainty. The modern approach is to build several (hundreds) of alternative reservoir models, which comes with its own set of challenges, in terms of both computation and storage. In addition, any prediction of flow and saturation changes (including the data that inform such changes such as 4D seismic and production data) requires running numerical implementation of multiphase flow, which depending on the kind of physics/chemistry represented (compressibility, gravity, compositional, reactive) may take hours to sometimes days.

1.2.3. The Challenge of Addressing Uncertainty

As production of oil/gas takes place in increasingly complex and financially risky situations, the traditional simple models of reservoir decline are gradually replaced by more comprehensive modeling of reservoir systems to understand better uncertainty in predictions made from such models. Based on the above description, Table 1.1 lists the various modeling components, subject to uncertainty, and the data involved in determining their uncertainty.

Despite the complexity in modeling, the target variables of such exercise are quite straightforward. In all, one can distinguish four categories of such prediction variables.

1. *Volumes.* How much target fluid is present? (a scalar)
2. *Recovery.* How much can be recovered over time under ideal conditions? (a time series)
3. *Wells.* Where should wells be placed and in what sequence? What strategy of drilling should be followed? Injectors/producer? Method of enhanced recovery? These are simply locations of wells and the time they will be drilled (a vector), and whether they are injecting or producing.
4. *Well controls.* How should wells produce? More complex wells are drilled, such as horizontal wells, that can be choked at certain points and their rates controlled in that fashion.

The primordial question is not necessarily the quantification of uncertainty of all the reservoir variables in Table 1.1 but of a decision-making process involving any of the target variables in question, which are

Table 1.1 Overview of the various modeling components, fields of study, and data sources for UQ and decision making in conventional oil/gas reservoirs.

Type	Class	Uncertain variable	Field of study	Main data
Charge	Basin	Deposition, erosion	Basin and petroleum system modeling, geochemistry	Wells seismic core/ log oil samples
	Source rock	Organic content; heat flow		
	Migration	Timing of kerogen transformation		
Structural	Faults	Amount		
		Location	Structural geology	Wells
		Slip/throw		
		Transmissibility	Geomechanics	3D seismic
		Fractures		
		Fault network hierarchy	Rock mechanics	Well tests
	Horizons	Depth variation	Stratigraphy	Wells
		Layer thickness variation		3D seismic
	Contacts	WOC	Hydrostatics	Pressure data
		GOC		
Petrophysical	Reservoir	Porosity	Sedimentary geology Carbonate geology	Core/log Seismic Production data
		Permeability		
		Lithology		
		Depositional system		
	Seismic	Velocity (P/S)	Seismic processing Rock physics	Seismic Core/logs
		Density		
		Impedance (P/S)		
	Geo-mechanics	Poisson modulus	Geomechanics	Cores
		Young's modulus		
Fluid	Fluid	PVT	Thermodynamics	Lab samples
		Relative permeability	Multiphase flow	Core experiments
		Capillary pressure		

uncertain due to various reservoir uncertainties. Is the 2D seismic data warranting drilling exploration wells? Is there enough volume and sufficient recovery to go ahead with reservoir development? Which wells and where do we drill to optimize reservoir performance? To further constrain reservoir development, is there value in acquiring 4D seismic data and how? As such, there is a need to quantify uncertainty with these particular questions in mind.

1.2.4. The Libya Case

1.2.4.1. Geological Setting. To illustrate the various challenges in decision making under uncertainty for a realistic reservoir system, we consider a reservoir in the Sirte Basin in north central Libya. This system contains 1.7% of the world's proven oil reserves according to *Thomas* [1995]. Its geological setting as described by *Ahlbrandt et al.* [2005] considers the area to have been structurally weakened due to alternating periods of uplift and subsidence originating in the Late Precambrian period, commencing with the Pan-African orogeny that consolidated several proto-continental fragments into an early Gondwanaland. Rifting is considered to have commenced in the Early Cretaceous period, peaked in the Late Cretaceous period, and ended in the early Cenozoic. The Late Cretaceous rifting event is characterized by formation of a sequence of northwest-trending horsts (raised fault blocks bounded by normal faults) and grabens (depressed fault blocks bounded by normal faults) that step progressively downward to the east. These horsts and grabens extend from onshore areas northward into a complex offshore terrene that includes the Ionian Sea abyssal plain to the northeast [*Fiduk*, 2009]. This structural complexity has important ramifications to reservoir development.

The N-97 field under consideration is located in the Western Hameimat Trough of the Sirte Basin (see Figure 1.2).

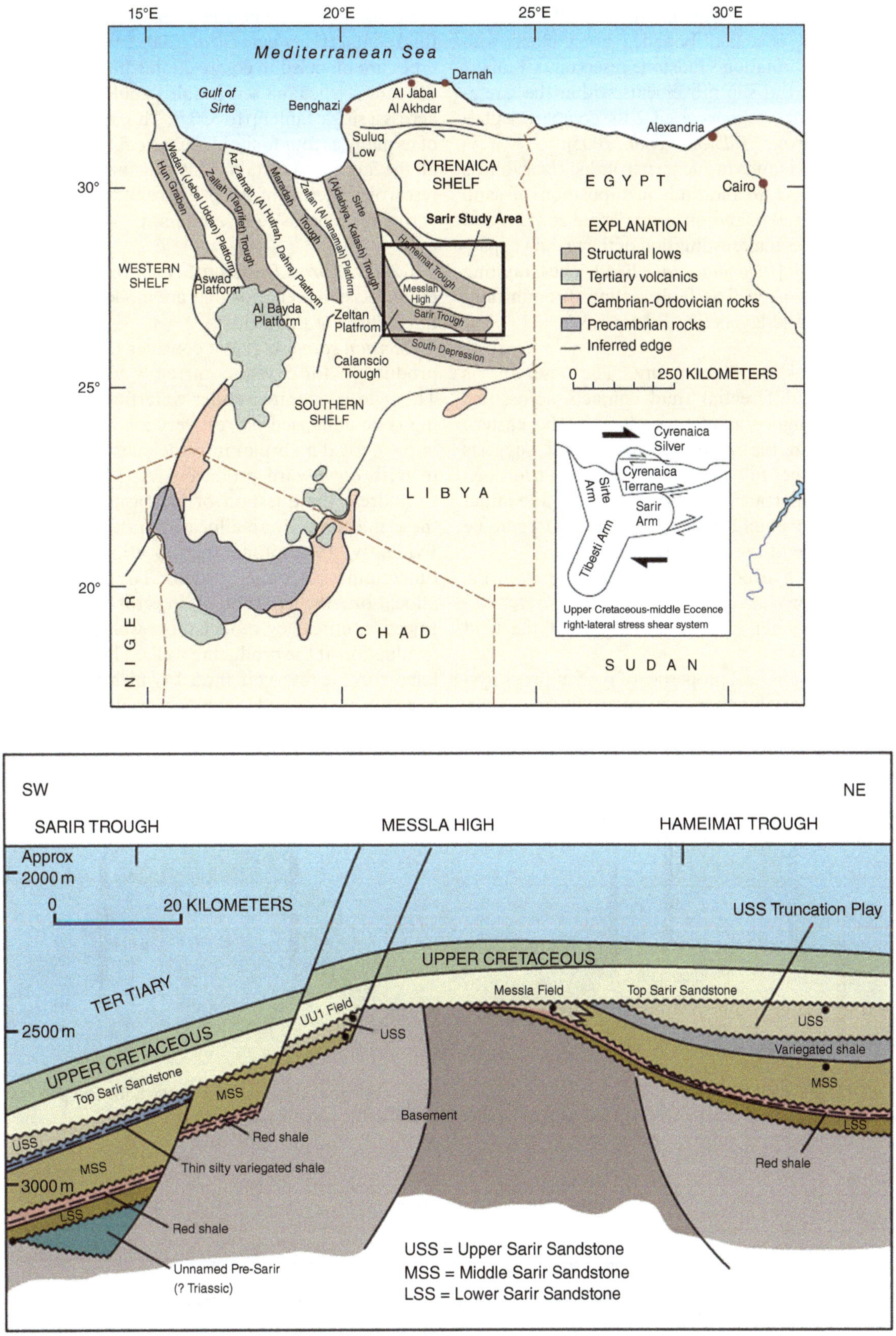

Figure 1.2 Structural elements of Sirte Basin. Schematic, structural cross-section from the Sarir Trough showing hydrocarbons in the Sarir Sandstone [*Ambrose*, 2000; *Ahlbrandt et al.*, 2005].

The reservoir under consideration, the WintersHall Concession C97-I, is a fault-bounded horst block with the Upper Sarir Formation sandstone reservoir. Complex interactions of dextral slip movements within the Cretaceous–Paleocene rift system have led to the compartmentalization of the reservoir [*Ahlbrandt et al.*, 2005].

Fluid flow across faults in such heterolithic reservoirs is particularly sensitive to the fault juxtaposition of sand layers. But the variable and uncertain shale content and diagenetic processes make estimation of the sealing capacity of faults difficult [*Bellmann et al.*, 2009]. Thus, faulting impacts fluid flow as well as fault sealing through fault juxtaposition of sand layers (see Figure 1.3).

1.2.4.2. Sources of Uncertainty. The reservoir is characterized by differential fluid contacts across the compartments. Higher aquifer pressure in the eastern compartment than the western compartment suggests the presence of either fully sealing faults or low transmissibility faults compartmentalization. However, the initial oil pressure is in equilibrium. Such behavior can be modeled using one of the two mechanisms:

1. a differential hydrodynamic aquifer drive from the east to the west, or
2. a perched aquifer in the eastern part of the field (see Figure 1.2).

By studying the physical properties of the fault-rock system such as pore-size distribution, permeability and capillary curves, the presence of only a single fault was falsified since that would not be able to explain the difference in the fluid contacts [*Bellmann et al.*, 2009]. When fault seal properties are modeled in conjunction with fault displacement, the cata-clastic fault seal is able to hold oil column heights across a single fault up to 350 ft. This indicates the presence of as many as four faults in the system. The displacement of all the faults is uncertain. This structural uncertainty in the reservoir in terms of the presence of faults and fluid flow across them needs to be addressed.

1.2.4.3. Three Decision Scenarios. Figure 1.4 shows three decision scenarios that are modeled to occur during the lifetime of this field.

Decision scenario 1. We consider the field has been in production for 5 years, currently with five producers. The field is operated under waterflooding. Waterflooding is an enhanced oil recovery method that consists of injecting water (brine) into the subsurface via injectors to push oil toward producers. At 800 days, one needs to address the question of increasing the efficiency of these injectors, by re-allocating rate between injectors. Evidently, the optimal re-allocation depends on the (uncertain) reservoir system. To determine this re-allocation, the concept of injector efficiency is used. Injection efficiency models how well each injector aids production at the producing wells. This measure is calculated from a reservoir model (which is uncertain). The question is simple: How much needs to be re-allocated and where?

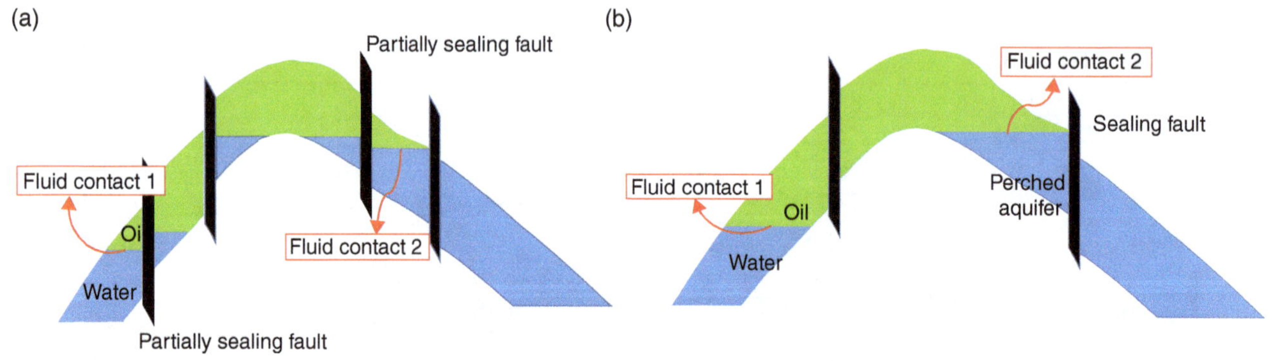

Figure 1.3 (a) Differential hydrodynamic trapping mechanism leading to different levels in fluid contact. (b) The perched aquifer explained as the reason. Contact levels depend on the number of faults in the system.

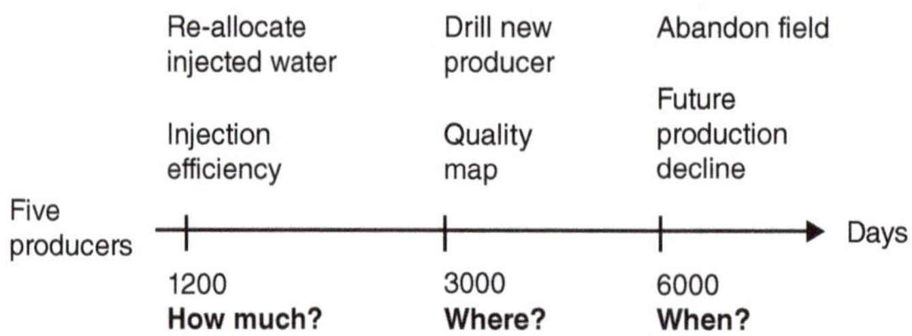

Figure 1.4 Three decision scenarios with three decision variables: injector efficiency, quality map, and production decline.

Decision scenario 2. At some point, optimizing just injectors will not cut it and new producing wells will need to be drilled, which comes at considerable cost. These wells should tap into un-swept areas of the reservoir system, for example, where the oil saturation is high. To do so, one often constructs "quality maps" [*da Cruz et al.*, 2004], for example, maps of high oil saturations. These maps can then be used to suggest locations where this new well can be drilled. The question here is again straightforward: Where to drill a new producer?

Decision scenario 3. At the final stages of a reservoir life, production will need to be stopped when the field production falls below economic levels of current operating situations. This will depend on how fast production declines, which itself depends on the (uncertain) reservoir system. Companies need to plan for such phase, that is, determine when this will happen, to allocate the proper resources required for decommissioning. The question is again simple: What date to stop production?

The point made here is that the engineering of subsurface systems such as oil reservoir involves a larger number of fields expertise, expensive data, and possibly complex modeling, yet the question stated in these scenarios involve a simple answer: how much, where, when?

1.3. DECISION MAKING UNDER UNCERTAINTY FOR GROUNDWATER MANAGEMENT IN DENMARK

1.3.1. Groundwater Management Challenges under Global Change

Global change, in terms of climate, energy needs, population, and agriculture, will put considerable stress on freshwater supplies (IPCC reports, [*Green et al.*, 2011; *Oelkers et al.*, 2011; *Srinivasan et al.*, 2012; *Kløve et al.*, 2014]). Increasingly, the shift from freshwater resources toward groundwater resources put more emphasis on the proper management of such resources [*Famiglietti*, 2014]. Currently, groundwater represents the largest resources of freshwater accounting for one third of freshwater use globally [*Siebert et al.*, 2010; *Gleeson et al.*, 2015]. Lack of proper management where users maximize their own benefit at the detriment of the common good has led to problems of depletion and contamination, affecting ecosystems and human health, due to decreased water quality [*Balakrishnan et al.*, 2003; *Wada et al.*, 2010].

Solutions are sought to this tremendous challenge both in academia and in wider society. This requires a multidisciplinary approach involving often fragments of fields of science and expertise as diverse as climate science, land-use change, economic development, policy, decision science, optimization, eco-hydrology, hydrology, hydrogeology, geology, geophysics, geostatistics, multiphase flow, integrated modeling, and many more. Any assessment of the impact of policy and planning, change in groundwater use or allocation, will increasingly rely on integrated quantitative modeling and simulation based on understanding of the various processes involved, whether through economic, environmental, or subsurface modeling. Regardless of the complexity and sophistication of modeling, there is increased need for acquiring higher quality data for groundwater management. Computer models are only useful in simulating reality if such models are constrained by data informing that reality. Unfortunately, the acquisition of rigorous, systematic, high quality, and diverse data sources, as done in the petroleum industry, has not reached the same status in groundwater management, partly because such resources were often considered cheap or freely available. Data are needed both to map aquifers spatially (e.g., using geophysics) and to assess land use/land-use change (remote sensing), precipitation (remote sensing), hydraulic heads (wells), aquifer properties (pump tests), and heterogeneity (geological studies). It is likely that with an increased focus on the freshwater supply such lack of data and lack of constraints in computer modeling and prediction will gradually dwindle.

Quantitative groundwater management will play an increasing role on policy and decision making at various scales. Understanding the nature of the scale and the magnitude of the decision involved is important in deciding what quantitative tools should be used. For example, in modeling transboundary conflict [*Blomquist and Ingram*, 2003; *Chermak et al.*, 2005; *Alker*, 2008; *Tujchneider et al.*, 2013], it is unlikely that modeling of any local heterogeneity will have the largest impact because such problems are dominated by large-scale (read averaged) groundwater movement or changes and would rather benefit from an integrated hydro-economic, legal, and institutional approach [*Harou and Lund*, 2008; *Harou et al.*, 2009; *Maneta et al.*, 2009; *Khan*, 2010]. A smaller-scale modeling effort would be at the river or watershed scale where groundwater and surface water are managed as a single resource, by a single entity or decision maker, possibly accounting for impact on ecosystem, or land use [*Feyen and Gorelick*, 2004, 2005]. The impact of data acquisition and integrated modeling can be highly effective for resource management in particular in areas that are highly dependent on groundwater (such as the Danish case). In this context, there will be an increased need for making informed predictions, as well as optimization under uncertainty. Various sources of uncertainty present themselves in all modeling parameters, whether economical or geoscientific due to a lack of data and lack of full understanding of all processes, and their interactions.

In this book, we focus on the subsurface components of this problem with an eye on decision making under the various sources of subsurface uncertainty. Such uncertainty cannot be divorced from the larger framework of other uncertainties, decision variables or constraints, such as climate, environmental, logistical, and economic constraints, policy instruments, or water right structures. Subsurface groundwater management over the longer term, and possibly at larger scales, will be impacted by all these variables. Here we consider smaller-scale modeling (e.g., watershed) possibly over a shorter-term time span (e.g., years instead of decades).

Within this context, often, a simulation–optimization approach is advocated [*Gorelick*, 1983; *Reed et al.*, 2013; *Singh*, 2014a, 2014b] where two types of problems are integrated: (i) engineering design, focusing on minimizing cost and maximizing extraction under certain constraints and (ii) hydro-economics to model the interface between hydrology and human behavior to evaluate the impact of policy. Such models require integrating the optimization method with integrated surface–subsurface models. The use of optimization methods under uncertainty (similar to reservoir engineering) is not within the scope of this book, although the methods developed can be readily plugged into such framework. Instead, we focus on smaller-scale engineering type, groundwater management decision analysis for a specific case, namely groundwater management in the country of Denmark.

Table 1.2 Danish government's 10-point program from 1994.

Danish government's 10-point program (1994)
Pesticides dangerous to health and environment shall be removed from the market
Pesticide tax – the consumption of pesticides shall be halved
Nitrate pollution shall be halved before 2000
Organic farming shall be encouraged
Protection of areas of special interest for drinking water
New Soil Contamination Act – waste deposits shall be cleaned up
Increased afforestation and restoration of nature to protect groundwater
Strengthening of the EU achievements
Increased control of groundwater and drinking water quality
Dialogue with the farmers and their organisations

Source: http://www.geus.dk/program-areas/water/denmark/case_groundwaterprotection_print.pdf.

This structure has been changed since 1994. Denmark no longer has 14 counties but 5 regions. The regions are not directly involved in the groundwater protection, which now has been moved to state level, and the local management is controlled by municipalities.

1.3.2. The Danish Case

1.3.2.1. Overview. Groundwater management in Denmark is used as a backdrop to illustrate and present methods for decision analysis, uncertainty quantification, and their inherent challenges, as applied to aquifers. The Danish case is quite unique but perhaps also foretelling of the future of managing such resources through careful and dedicated top-down policy making, rigorous use of scientific tools, and most importantly investment in a rich and heterogeneous source of subsurface data to make management less of a guessing game.

Freshwater supply in Denmark is based on high-quality groundwater, thereby mitigating the need for expensive purification [*Thomsen et al.*, 2004; *Jørgensen and Stockmarr*, 2009]. However, increasing pollution and sea-level changes (and hence seawater intrusion) have increased stresses on this important resource of Danish society. As a result, the Danish government approved a ten-point plan (see Table 1.2) to improve groundwater protection, of which one subarea consisted in drawing up a water-resources protection plan. The government delegated that 14 county councils be responsible for water-resources planning based on dense spatial mapping (using geophysics) and hydrogeological modeling as the basis for such protection. This high-level government policy therefore trickled down into mandates for local, site-specific, groundwater protection, a strategy and ensuing action plan (decision making) by local councils at the river/watershed level.

The widespread availability of high-quality groundwater limits extensive pipeline construction. It was also recognized that some areas are more vulnerable to contamination from industry and agriculture than others; that despite extensive drilling, the aquifer heterogeneity and its impact on pumping could not be simply deduced or modeled from wells only. Hence, a more data-rich, modeling-intensive approach is required for proper management and to meet the goals in the government action plan. In that context, it was also established that simple drinking-well protection models based on multilevel radial protection zones ignored the impact of geological heterogeneity on how contaminants reach wells [*Sonnenborg et al.*, 2015]. This is particularly relevant in Denmark where the shallow subsurface is largely dominated by the presence of "buried valleys." Buried valleys are mainly thought to be formed below the ice by erosion into the substratum caused by pressurized meltwater flow [*Jørgensen and Sandersen*, 2006]. Typically formed close to and perpendicular to the ice margin, these valleys often end abruptly, their cross-sections are typically U-shaped and can occur at a depth of up to 350 m. While the valleys are formed as isolated structures, they often show cross-cutting relationships. Often younger valleys are eroded into the fill of older valleys, where these deposits are easily erodible than the surroundings. A complex network of

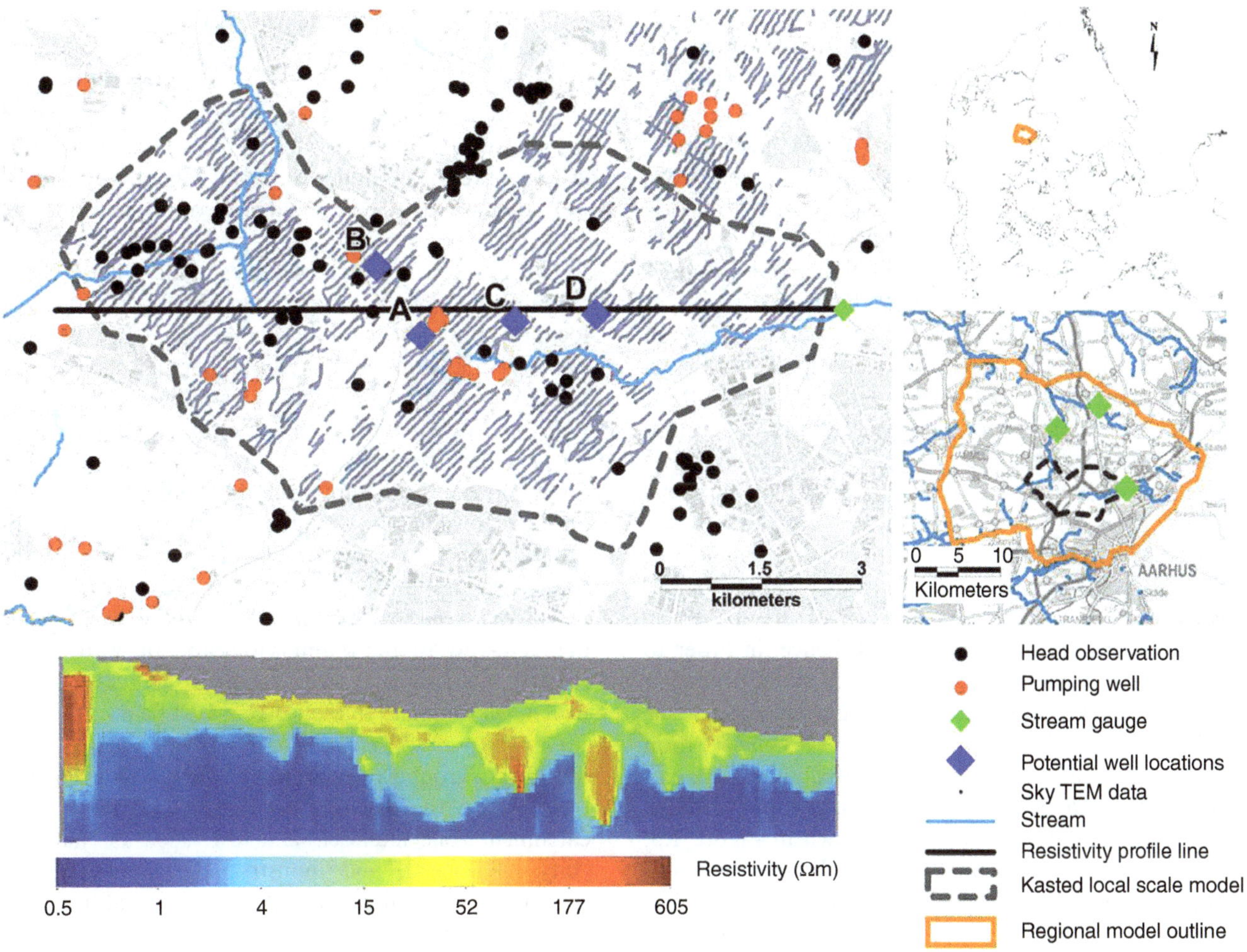

Figure 1.5 Location of the decision problem near the city of Kasted. The blue diamonds are the four alternative well locations (A, B, C, and D) in the decision problem. The grey lines are locations with SkyTEM data. Bottom: vertical profile of inverted SkyTEM data showing buried valleys.

cross-cutting valleys creates significant heterogeneity in the subsurface that influence groundwater recharge and flow. About half of the valleys are filled with hydraulic conductive sand, the rest filled with clayey deposits, but valleys with combined sand/clay infill are also quite common [*Sandersen and Jørgensen*, 2003]. Some of the valleys act as groundwater reservoirs, while others constitute barriers for groundwater flow/protection, making groundwater management unlikely to be reliable without any modeling or based on simple basic assumptions.

This geological phenomenon cannot be comprehensively modeled from boreholes only as such "point" information does not allow for an accurate mapping of the subsurface, leading to considerable uncertainty and risk in establishing protection zones.

Understanding the heterogeneity caused by the buried valleys depositional system is therefore critical to assessing aquifer vulnerability. Such valleys act as underground "rivers," but such structure may themselves contain or act as flow-barriers created by the presence of clay [*Refsgaard et al.*, 2010; *Hoyer et al.*, 2015]. The complex intertwining of sand and clay makes such assessment difficult, and also because the majority of buried valleys are not recognizable from the terrain. In such depositional system, clay serves not only as a purifier, sand as a conduit, of water but also as a contaminant. This requires a comprehensive modeling of the various physical, chemical, and biological processes that take place in the heterogeneous subsurface. For that reason, a large geophysical data acquisition campaign was initiated, in particular through the use of various transient electro-magnetic (TEM) surveys [*Møller et al.*, 2009] (see Figure 1.5). Such geophysical surveys provide a more detailed insight into the geological heterogeneity but their use does not necessarily result in a perfectly accurate map of the subsurface, due to limited resolution of the data source (similar to the limited resolution of seismic data in reservoir modeling), limitations in data coverage, and the subjectivity of interpretations made from such data [*Jørgensen et al.*, 2013].

1.3.2.2. A Specific Decision Problem. Aquifer management requires dealing with conflicting objectives, uncertain predictions, limited data, and decision making within such context. At the local level, the decision to drill wells for drinking water extraction requires balancing the need for using resources versus the impact of extraction on the environment. In Denmark, the benefit of using aquifers for drinking water supply has to be weighed against the risk of (i) affecting streamflow, (ii) affecting wetland restoration, and (iii) risk of contamination from agriculture. These factors are related to EU regulations in which the Water Framework Directive is based.

We consider an area in Denmark, near the small town of Kasted, that requires considering such careful balancing act (see Figure 1.5). It has been observed that an extraction area is affecting wetlands; hence, in order to restore wetlands closer to their original state, a portion of the current groundwater abstraction will need to be re-allocated to a different area. Based on consideration of existing wells, current water distribution system, accessibility, and geological knowledge, four locations are proposed, A, B, C, and D, as shown in Figure 1.5. Jointly, the local municipality council and the water supply company must now decide on one of these locations. Evidently, we need to justify that the new location can indeed make up for the reduction in abstraction from the current well field, but also that this would not have any adverse effect on the environment, which would defeat the purpose of this re-allocation. We will treat this problem within a formal decision analytic framework using state-of-the-art groundwater modeling, sensitivity analysis, and uncertainty quantification.

Chapter 2 will introduce a formal decision analysis framework requiring stating objectives and using such objectives to compare stated alternatives on which decisions are based. This requires a formal statement of (i) what the alternatives are; no decision is better than the choice made from the stated alternatives, (ii) the objectives under which alternatives will be evaluated, typically in the form of "maximize this," "minimize that," and (iii) a quantitative measure of how well each alternative achieves the stated objectives (termed the "attribute"). Because of the existence of multiple competing objectives in this case, some statements of preferences are needed. In a decision analysis framework, these preferences are stated as value function, which transform preference to a common scale (e.g., 0–100). More details will be discussed in Chapter 2, more specifically, the means of weighting the various conflicting objectives. Formally, we have constructed the following definitions:

1. *Alternatives*: the four locations/zones of pumping wells we are considering, assuming the well rates are fixed and known (corresponding to 20% of the abstraction at the existing well field). We could also consider several well rates.
2. *Objectives*: four objectives are stated:
   - *minimize drawdown extraction*: preferably, the new location should bear the burden of the 20% extraction due to re-allocation and anything more is an additional plus. A large drawdown indicates poor aquifer conditions, and hence needs to be minimized.
   - *maximize streamflow reduction potential*: depends on the flow in the stream given the existing abstraction, and the flow on the stream if we move 20% of the groundwater abstraction from the existing wells to the new well at any of the four locations.
   - *maximize increased groundwater outflow to wetlands*: due to re-allocation, the aim is to restore the water table, thereby increasing the outflow of groundwater to the wetlands proximate to the existing well field.
   - *minimize risk of contamination of drinking water*: the abstracted groundwater from the new well originates from within the so-called well catchment zone. This catchment zone intersects land use, such as "nature," "city," "farmland," and "industry." We aim to maximize the part of the well catchment that is located in nature and minimize that part of the catchment located within the category "industry" and "farmland." The city is considered as neutral.

The four target variables are calculated from a groundwater model, but because this model is uncertain, so are the payoffs associated with each target. This groundwater model has the following uncertain parameters (model components):

1. Uncertainty in the lithology distribution
2. Uncertainty in the hydraulic conductivity
3. Uncertainty on the boundary conditions
4. Uncertainty on the aquifer recharge
5. Uncertainty related to streams: connection with the aquifer (conductance) and digital elevation model (DEM) model used to define their elevation

To constrain this uncertainty, several data sources are available.

Conceptual geological understanding of buried valleys. The availability of dense borehole data in conjunction with high-quality geophysical data allows for a better understanding of the nature of the depositional system. Based on the large amount of studies in Denmark and neighboring areas [*Sandersen and Jørgensen*, 2003; *Sandersen et al.*, 2009; *Høyer et al.*, 2015], a conceptual model has been drawn (Figure 1.6), conveying the

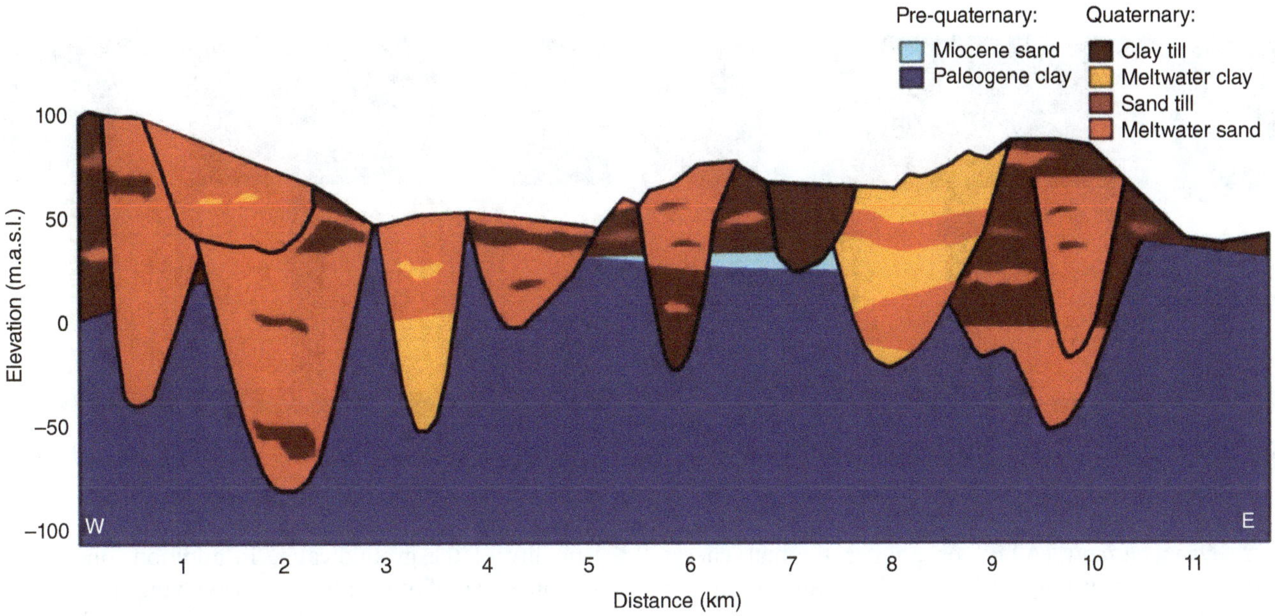

Figure 1.6 Conceptual geological model (a sketch) of the buried valet deposits. Valleys are with different lithologies. Hence, hydraulic properties cross-cut each other [*Hoyer et al.*, 2015].

interpretation of the lithological architecture created by subsequent glaciation periods.

Hydraulic head observations. A Danish national well database, JUPITER [*Møller et al.*, 2009], can be queried for measurements in the area of study. These measurements vary in quality, either because of how they are measured, type, and age of the borehole or because of the difference in coordinate and datum recording. A total of 364 head data were used in the study.

Stream discharge measurements were available from three gauging stations. Two stations had time series spanning approximately 20 years, while the third station had a span of 3 years.

Borehole data. The study area holds approximately 3000 boreholes with lithological logs of which the majority of boreholes are relatively shallow in depth (<50 m). Borehole information consists of lithology variation with depth. This data is also of different quality and based on metadata (drill-type, age) it is grouped into four quality groups.

Geophysical data. One of the defining features of the Danish groundwater management case is the availability of a rich and high-quality set of direct current (DC) and TEM geophysical data (see Figure 1.5). DC methods typically resolve the very shallow subsurface, while TEM methods resolve resistivity contrasts at greater depths. The TEM data were collected either through a port of numerous ground-based campaigns or through two campaigns (in 2003 and 2014) using the SkyTEM system [*Sørensen and Auken*, 2004] with the main purpose to delineate important buried valley structures, serving as aquifers. Altogether, geophysical data collected in the area span 30 years, and 50 individual surveys, and they have all been stored in the national Danish geophysical database GERDA [*Møller et al.*, 2009].

The question now is simple: What is the best location to re-allocate drinking water, A, B, C, or D?

1.4. MONITORING SHALLOW GEOTHERMAL SYSTEMS IN BELGIUM

1.4.1. The Use of Low-Enthalpy Geothermal Systems

Low-enthalpy geothermal systems are increasingly used for climatization (heating/cooling) of buildings, in an effort to reduce the carbon footprint of this type of energy use. It is estimated [*Bayer et al.*, 2012] that the potential reduction of CO_2 emission reduction is around 30% compared to conventional systems. The main idea is the utilization of the subsurface, whether rocks, soils, saturated, or unsaturated, as a heat source or heat sink (cooling). To make this work in practice, two types of systems are used [*Stauffer et al.*, 2013] (see Figure 1.7).

1. *Closed systems* (BTES or borehole thermal energy storage): a series of vertical or horizontal pipes, often plastics, are installed in the subsurface. Fluids such as antifreeze solutions are circulated in the pipes to exchange heat with the subsurface. The system can be used for warming in winter and cooling in summer. Such systems are often installed in low-permeability soils, mitigating the risk of leakage of pipes.

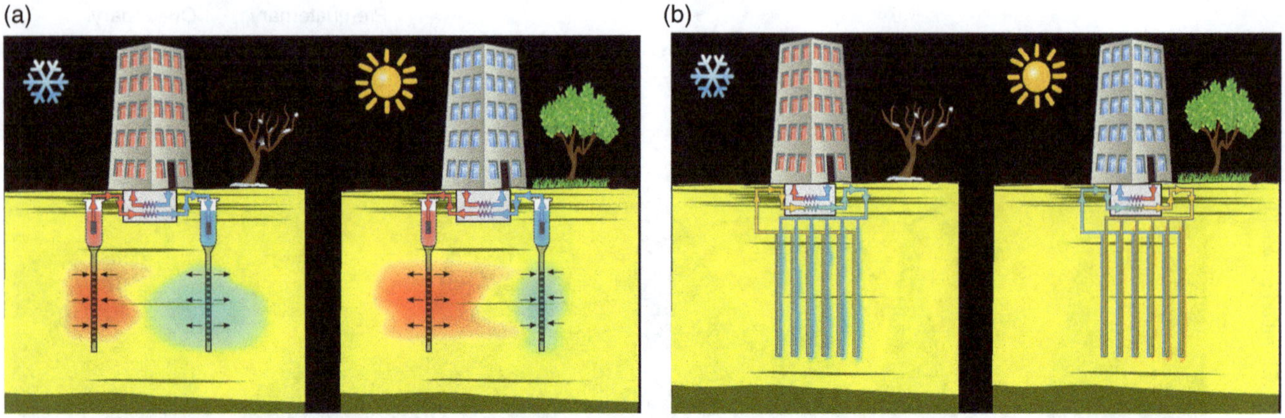

Figure 1.7 (a) An open (ATES) and (b) closed (BTES) shallow geothermal systems [*Bonte*, 2013].

2. *Open systems* (ATES or aquifer thermal energy storage): using drilling and boreholes, water is directly circulated between a production and an injection well through a heat exchanger (also called groundwater heat pump). Evidently, this requires a high-permeability subsurface. Heat stored in summer can theoretically be used in winter. However, because the open system is more sensitive to the ambient subsurface, its design needs to be done more carefully than a closed system. There is a risk that, if the system operates suboptimal, the energy stored cannot be fully recovered (e.g., in case of hydraulic gradient).

While the idea is straightforward, the practical implementation raises a number of important questions. Next to the evident question on how to design the system, questions related to the impact of such thermal perturbation on the subsurface system need to be addressed. These impacts are multifold:

1. *Hydrological*. Changes in temperature induces a heat flux, which may affect areas further away from wells (the thermally affected zone). Catchment areas of existing drinking water wells may be affected, which in turn may impact flow and hence such change increases the risk for unwanted (and unforeseen) contamination or cross-aquifer flow.

2. *Thermal*. A long-term warming or cooling may occur. This may cause interference with other uses of groundwater. In addition, it may affect the performance of the system because of possible freezing or short-circuiting the heat exchange. This thermal impact needs to be considered jointly with other long-term sources of thermal changes such as climate change and urbanization.

3. *Chemical*. Rainwater is filtrated in the subsurface and such a process produces fresh drinking water, leading to a specific vertical groundwater stratification with shallow oxidized, nitrate-rich groundwater and reduced iron-rich deeper water. ATES can introduce a mixing that affects the quality of the groundwater. In addition, one needs to be concerned of other effects such as change in reaction kinetics, organic matter oxidation, and mineral solubility. Urban areas are already vulnerable to contamination from various pollution sources and chemical changes may further enhance that effect.

4. *Microbial*. The groundwater system is an ecosystem (consisting of bacteria, fungi, pathogens, and nutrients). Any temperature changes may affect this system, and hence affect the balance of this ecosystem, possibly leading to changes in water quality. In addition, microbial changes may lead to clogging of this system, which is particularly relevant near boreholes.

Since exploitation of the subsurface for heat will add an additional stress to a system already subject to stresses from other sources, such as drinking water extraction, contaminants, and geotechnical construction, it is likely that new policies and regulations will need to address the shared use of this resource. Such regulations are likely to include monitoring (perhaps in the same sense as required for CO_2 sequestration) to mitigate risk or reduce the impact of the thermal footprint. Next we discuss the design of such monitoring system and what affect the unknown subsurface properties have on that design. Then, we introduce a specific case of data acquired in an aquifer in Belgium.

1.4.2. Monitoring by Means of Geophysical Surveys

1.4.2.1. Why Geophysics? The design as well as monitoring of the shallow geothermal system, like many other subsurface applications, require a multidisciplinary approach, involving several fields such as geology, hydrogeology, physics, chemistry, hydraulics engineering design, and economics. For example, *Blum et al.* [2011] showed (based on systems in Germany) that subsurface characteristics are insufficiently considered for a proper design. Characterization of heat flow, temperature changes, and its effect on the ambient environment requires characterizing geological heterogeneity, combined fluid and thermal properties, as well as geochemical characteristics (to study impact on subsurface chemistry). Early models relied mostly on analytical equations. However, such approaches ignore the complexity of the subsurface and the observed (see later) heterogeneity of temperature and temperature changes in the subsurface, leading to inadequate design. The more modern approach relies on creating groundwater models and modeling combined fluid flow and heat transport using numerical simulators. Next to traditional tests such as borehole flowmeter tests, slug tests, hydraulic pumping tests, and tracers, two field experiments are used to constrain the thermal parameters required for such simulators: the thermal response test (TRT) and the thermal tracer test (TTT). These are used to characterize thermal diffusivities and hydraulic and thermal conductivities required for simulations. These values can be obtained both from field and from laboratory data. However, both TRT and TTT are borehole centric tests. For example, with a TRT one circulates a hot fluid and continuously measures temperature changes of the fluid. TTT involved two wells and works like a tracer but now for heat. Such experiments can be short or long term (short = hours, long = months). In the short-term experiments, heated or cooled water is injected as a tracer, and temperature changes are measured in a nearby observation well. To derive the required properties, one can either use analytical equations (relying on simplifying assumptions) or build numerical models and solve inverse problems. There are several problems that arise when limiting oneself to only these types of test. First, they provide only information near the well (TRT) or between well locations. Second, geological heterogeneity makes direct interpretation difficult for such tests and hence inverse modeling becomes tedious.

New techniques are therefore needed to more directly and more effectively monitor the spatial and temporal distributions of temperature in the system which could lead to (i) better design the geothermal system and the monitoring network, (ii) prevent any thermal feedback/recycling, and (iii) image and control the thermal affected zone [*Hermans et al.*, 2014]. Here we focus on the use of a specific method, namely electrical resistivity tomography (ERT) and its time-lapse variety to characterize temperature and its changes under shallow geothermal exploitation and monitoring.

1.4.2.2. ERT and Time-Lapse ERT. ERT is a method that images the bulk electrical resistivity distribution of the subsurface (Figure 1.8). Electrical resistivity depends on several properties of relevance for shallow geothermal systems: (i) clay mineral content, (ii) water saturation and salinity, (iii) porosity, and (iv) temperature. As with any geophysical technique, the target physical property (temperature here) needs to be untangled from other influences. Consequently, because of geological heterogeneity, this becomes more difficult to achieve and requires knowledge of such heterogeneity as well as the various rock physics relations between the properties involved.

Practically, electrical currents are injected between two current electrodes, either on the surface or in the borehole. Then, the resulting potential difference is measured simultaneously between two different (potential) electrodes. Because the current is known (a control), the ratio between the measured difference of electrical potentials equals the electrical resistance, as follows directly from Ohm's law. This process is repeated along one or several profiles using many quadrupoles to acquire 2D or 3D datasets. The acquired values of electrical resistance

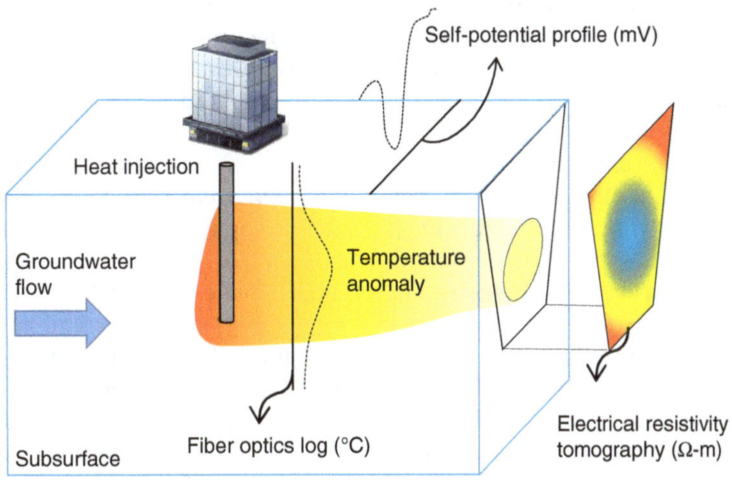

Figure 1.8 The use of electrical resistivity tomography in the design of shallow geothermal systems. From *Hermans et al.* [2014].

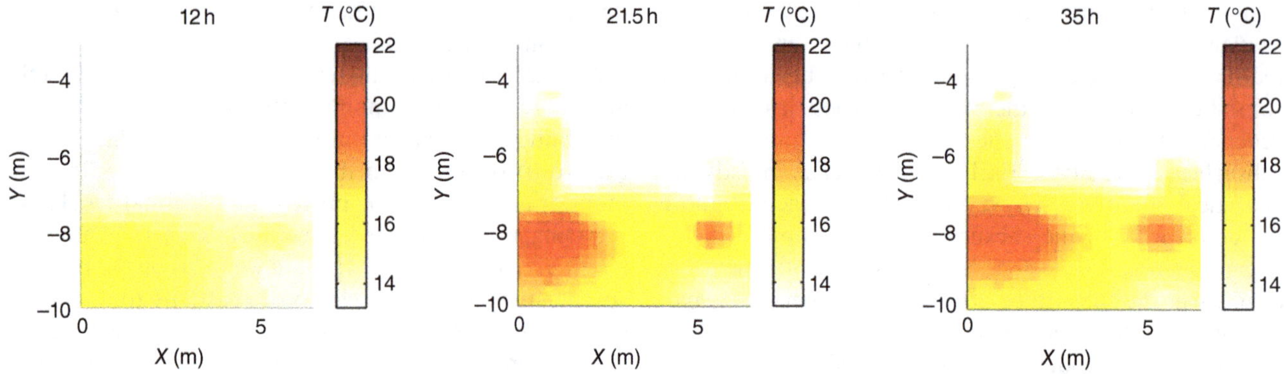

Figure 1.9 Example of temperature monitoring during a heat-tracing experiment from ERT. Modified after *Hermans et al.* [2015].

measured at each (quadrupole) location needs to be inverted into an electrical resistivity distribution that can then be linked to a target physical property (e.g., temperature).

Monitoring is a time-varying study; hence, instead of taking one snapshot (in time), ERT imaging can be repeated to detect changes. Inversion into electrical resistivity is repeated and compared with the base survey. Similar to the static inversion, changes in electrical resistivity can be related to changes in target physical properties such as temperature changes. One of the advantage of time-lapse ERT as applied to geothermal monitoring is that temperature is the dominant change; hence, time-lapse ERT becomes easier to interpret in terms of temperature, as other effects are mostly constant. As an example, *Hermans et al.* [2015] monitored a heat-tracing experiment with cross-borehole ERT. Assuming no changes in chemistry and the absence of clayey minerals, *Hermans et al.* were able to image from ERT changes in temperature as low as 1.2°C with a resolution of a few tenths of degree Celcius (Figure 1.9).

1.4.2.3. Issues. Despite the straightforward advantage of ERT and its time-lapse variety, several challenges occur because of non-ideal conditions in the subsurface and in performing such surveys.

Smoothing. As with any geophysical technique, ERT data provides only a smooth view of the physical properties of the subsurface. As a result, any inversion of such data is non-unique (see Chapter 6 on inverse modeling). However, most current approaches rely on some smooth inversion (using regularization terms, see Chapter 6). The lack of proper representation of actual subsurface variability has led to poor recovery of mass-balance in tracing experiments [*Singha and Gorelick*, 2005; *Muller et al.*, 2010] and over- or underestimation of the physical properties due to over-smoothing of the geophysical image [*Vanderborght et al.*, 2005; *Hermans et al.*, 2015]. Additionally, to convert electrical resistivity changes to temperature changes, one needs to rely on petrophysical relationships established in small-scale laboratory experiments that become difficult to apply (without error) to the larger-scale inversions. Such approaches will work in relatively homogeneous deposits but lose their applicability in more heterogeneous systems. In Chapter 6, we show how standard regularization methods do not lead to an adequate quantification of the uncertainty in the obtained temperature changes. Such an uncertainty is needed for risk quantification in the design of the system.

Noise. Noise in ERT measurements is composed of a random and a systematic component. The latter may be correlated in time. Random error arises from variations in the contact between the electrodes and the ground [*Slater et al.*, 2000]. Systematic errors are related to the data acquisition, hence any problems with electrode placement (e.g., misplaced, disconnected). Time-lapse geophysical measurements are subject to the repeatability issues, namely that exact same conditions and configurations need to occur over time, which is rarely the case. One way to address noise is to make use of the so-called reciprocal measurements, which involves reversing the current and potential electrodes. Under ideal, non-noise conditions, this should result in identical readings. It is often observed that the error obtained by means of reciprocal measurement increases with resistance.

1.4.2.4. Field Case. We consider a specific field case where geophysical data is used to assess the potential for a geothermal heat exchanger for building heating. The aim is to assess whether an alluvial aquifer allows storing thermal energy and restore it at a later stage. The aim is therefore to predict heat storage capacity of the system undergoing an injection and pumping cycle. Here we study one such cycle of injecting hot water for 30 days, then extracting for 30 days. In other words, the target is to predict the change in temperate during extraction. This quantifies the efficiency of the recovery and aids in the design of the heat pump. The target for prediction is a simple function of time: $\Delta T(t)$.

The study site is located in the alluvial aquifer of the Meuse River in Hermalle-sous-Argenteau, Belgium, consisting of a 10 m thick zone. The water level is located at 3.2 m depth. Based on borehole logs, the saturated deposits can be divided into two distinct layers. The upper layer, between 3 and about 7 m depth, is composed of gravel in a sandy matrix. The bottom layer is composed of coarse clean gravel. The bedrock composed of low permeability carboniferous shale and sandstones lies at 10 m depth and constitutes the basement of the alluvial aquifer. The reader can refer to *Klepikova et al.* [2016] and *Hermans et al.* [2015] for more details on the site.

For 30 days, heated water is continuously injected into a well at a rate of 3 m^3/h. The temperature of the injected water is 10°C above the background temperature. At the end of the 30-day period, water is extracted from the well at the same rate of 3 m^3/h. The change in temperature of the pumped water compared to the initial value in the aquifer (i.e., before injection) is recorded for another 30-day period. The thermal energy recovery can be estimated using the temperature of the extracted water. The simulation is limited in time to avoid considering changing boundary conditions with time. Similarly, a rate of 3 m^3/h allows to reduce the size of the investigated zone.

In Chapter 8, we will address the following questions:

1. In the design of this specific system, which model parameter most impacts the prediction of $\Delta T(t)$. Many uncertainties exist as discussed earlier; hence, we need to focus on those that matter.

2. What data can be used to narrow those model uncertainties that most impact the decision variable, here $\Delta T(t)$. Will it be useful to use ERT data?

3. If we would decide to go ahead with acquiring ERT data, how would the uncertainty on $\Delta T(t)$ be reduced?

4. Given that we are working in such a complex system, is there a way to test our various modeling assumptions (meaning aiming to falsify them, see Chapter 5)?

1.5. DESIGNING STRATEGIES FOR URANIUM REMEDIATION IN THE UNITED STATES

1.5.1. Global Environmental Challenges

Similar to the need to protect and manage groundwater resources in the next century, stresses on the environment, due to anthropogenic activities, will continue to grow. For example, the United States government, under the Department of Energy (DOE), formulated the explicitly stated goals of developing sustainable solutions to such environmental challenges, based on a predictive understanding of environmental systems. Some of the stated goals are as follows: (i) synthesize new process knowledge and innovative computational methods that advance next generation, integrated models of the human–Earth system; (ii) develop, test, and simulate process-level understanding of atmospheric systems and terrestrial ecosystems, extending from bedrock to the top of the vegetative canopy; (iii) advance fundamental understanding of coupled biogeochemical processes in complex subsurface environments to enable systems-level prediction and control; and (iv) identify and address science gaps that limit translation of fundamental science into solutions for the most pressing environmental challenges (http://science.energy.gov/ber/research/cesd/).

As in the groundwater case, there is a need to understand and study watersheds as complex hydrobiogeochemical systems, in particular how such systems respond to contaminant loading. This systems approach is similar to the "reservoir system" or the "groundwater system." However, now an additional set of processes may complicate matters. For example, an understanding of the complex processes and interactions that occur from bedrock to the top of the canopy is necessary [*Brantley et al.*, 2007]. This requires modeling and understanding processes from the molecular to a planet-wide scale of perturbations or changes. The latter has led to the building of mechanistic (numerical models) reactive transport models. Such models incorporate knowledge of microbial processes, speciation, and interactions of inorganic elements with microbes, and how these processes act on different time and space scales [*Li et al.*, 2017].

These broad scientific questions are often addressed within specific decision-making frameworks and policy outcomes in mind. Hence, there is a need to integrate these various processes within a single framework, to provide guidance on what data should be collected to adhere to regulations, to design any remediation strategy, and to ultimately monitor and verify the effects of such remediation.

1.5.2. Remediation: Decision Making Under Uncertainty

On a more local scale, there will be an increased need to remediate contaminated soils or groundwater that may pose a significant risk to human health. Decision making for contaminant remediation may well be more complex than for petroleum or (uncontaminated) groundwater systems. Such a decision making varies highly by country, regions, or even state (see Table 1.3). For example, at an EPA Superfund site near Davis, California, three lead agencies oversee the cleanup of chemical spills: the Central Valley Regional Water Quality Control Board (RWQB), the California Department of Toxic Substances

Table 1.3 Example of stakeholders in the decision context of contamination remediation.

Facility owner	Regulatory agencies	Local/county agencies
• Achieve regulatory compliance • Utilize risk-based techniques • Minimize/eliminate disruption of operations • Minimize costs • Reduce long-term treatment and liabilities	• Protect human health and the environment, including groundwater resources • Protect groundwater resources • Achieve regulatory compliance • Eliminate off-site impacts to receptors • Involve stakeholders • Maintain reasonable schedule • Obtain reimbursement for oversight costs	• Optimize zoning • Maximize tax revenues • Accelerate schedule • Protect human health and the environment • Maximize quality of life • Protect groundwater resources

Source: Adapted from "A decision-making framework for cleanup of sites impacted with light non-aqueous phase liquids (LNAPL)" (2005))

Control, and the US EPA (United Stated Environmental Protection Agency). Often environmental consultants are hired by responsible parties (here UC Davis and the DOE), leaving the decisions on which remedial options to consider. Recommendations are made to the lead agencies for implementation and final decision making.

Within this context, the US EPA guidelines are outlined in a document entitled "A decision-making framework for cleanup of sites impacted with light non-aqueous phase liquids (LNAPL)" (2005). Although specific to LNAPL, the document is only a guide (not a policy or legal requirement), aiming to provide practicable and reasonable approaches for management of petroleum hydrocarbons in the subsurface (and hence is specific to contamination from petroleum product around refineries, pipelines, etc.). Although not based on any formal decision analysis (such as in Chapter 3), the document outlines the complex interaction of stake holders (oil industry, local communities, government agencies), formulation of high-level goals and objectives, the implementation of remediation strategies, modeling of potential exposure pathways, and data acquisition. This decision process involves different parties with different competing objectives. The treatment of these competing objectives within a formal decision-making process will be discussed in Chapter 2.

1.5.3. Remediation: Data and Modeling

Targeted data collection is critical to the evaluation of several remediation options and the selection of the most appropriate alternative for a given site, as many alternatives may present themselves such as "clean up the site to pristine conditions," and "clean only the most impacted portions and contain the remainder of the contamination on site." A decision will ultimately consist of choosing among these alternatives (and accompanied remediation methods) based on the stated objectives. Similar to the groundwater case, data and modeling will have value as long as they inform the "payoffs" for each alternative. If all aspects of the subsurface are clearly informed, the effect of remediation would be perfectly known, then a decision is simply made from the highest payoff (or lowest cost). However, because of the various uncertainties involved, decisions will need to be made under uncertainty.

The importance of prediction of contaminant distribution in space and time through numerical modeling has long been acknowledged [*Wagner and Gorelick*, 1987; *Morgan et al.*, 1993; *Andričević and Cvetković*, 1996; *James and Oldenburg*, 1997; *Maxwell et al.*, 1999]. Such a prediction is especially important toward designing and evaluating remediation strategies. For example, in one mercury contamination remediation case study, various risk assessment/prediction tools are developed to evaluate options of active management such as capping and dredging, or passive natural attenuation [*Wang et al.*, 2004]. As monitoring costs are very expensive and such monitoring data provide only a short-term inference on future events within a limited spatial area, numerical modeling is needed to provide meaningful long-term predictions. Behavior of contaminant plumes, both conservative and reactive, has been studied extensively both at the lab-scale and at the field-scale experiments, to assist developing better numerical modeling tools that provide these predictions [*Lovley et al.*, 1991; *Yabusaki et al.*, 2007; *Li et al.*, 2011; *Williams et al.*, 2011]. However, uncertainties are naturally associated with numerical modeling. Such uncertainties in models of contaminant transport come from spatially variable hydraulic properties, physical and chemical descriptions or the initial and boundary conditions, knowledge of the contaminant source, and importantly the rates and mechanisms associated with the physical and biochemical processes.

As decisions made in contaminant remediation depend on long-term predictions within a small tolerance range, efforts have been made across research areas, such as hydrogeology, geology, geostatistics, geophysics, biogeochemistry, and numerical modeling, to create better models that improve accuracy and reduce the uncertainties of predictions [*Cirpka and Kitanidis*, 2000; *Sassen et al.*,

2012; *Steefel et al.*, 2015]. More recently, there have been rising concerns over remediation solutions to real-world, polluted groundwater systems. Such contaminations were caused by destructive human activities during the 20th century [*Wang et al.*, 2004; *Yabusaki et al.*, 2007; *Chen et al.*, 2012].

1.5.4. Uranium Contamination in the United States

In this book, we will be concerned with groundwater contamination resulting from the extraction and processing of uranium ore during the Cold War era that poses environmental risk across hundreds of sites in the United States, particularly within the upper Colorado River Basin [*Palmisano and Hazen*, 2003]. A US DOE site near Rifle, Colorado, contains a contaminated floodplain that has been the focus of detailed investigation [as reviewed in *Anderson et al.*, 2003; *Williams et al.*, 2009, 2011; *Orozco et al.*, 2011]. The site once contained a milling facility for ores rich in uranium and other redox sensitive metals (e.g., vanadium, selenium, and arsenic). After removal of the contaminated overburden, low but persistent levels of contamination within subsurface sediments still affect groundwater quality and flow directly to the Colorado River. Elevated concentrations of contaminants can be harmful to young-of-year fish that use the backwater channels as habitat during late summer. The Rifle site is also within the above described wider context of building models to quantify how land use and climate change affect subsurface carbon fluxes and transformations, flow paths, subsurface microbial communities, and ultimately the biogeochemical behavior of a watershed. The wealth of data collected at this site provides a testbed for developing such models, testing hypotheses, generating predictive uncertainty, and ultimately quantitative prediction of short- and long-term evolution of this biogeochemical system [*Williams et al.*, 2011; *Zachara et al.*, 2013; see also *Williams et al.*, 2013].

Acetate injection has been evaluated at the Rifle pilot site to examine the effectiveness of in situ bio-remediation [*Yabusaki et al.*, 2007; *Li et al.*, 2011; *Williams et al.*, 2011]. The acetate amendment stimulates the indigenous dissimilatory iron reducing microorganisms to catalyze the reduction of U(VI) in groundwater to insoluble U(IV) [*Lovley et al.*, 1991] and offers a cost-effective, in situ remediation solution.

1.5.5. Assessing Remediation Efficacy

To study the efficacy of acetate injection as a remediation strategy, four field bio-stimulation experiments have been conducted at the US DOE's Integrated Field Research Challenge site in Rifle, Colorado, as shown in Figure 1.10. Previous experiments have shown that acetate injection is capable of immobilizing uranium [*Li et al.*, 2010; *Williams et al.*, 2011]. Figure 1.11 shows

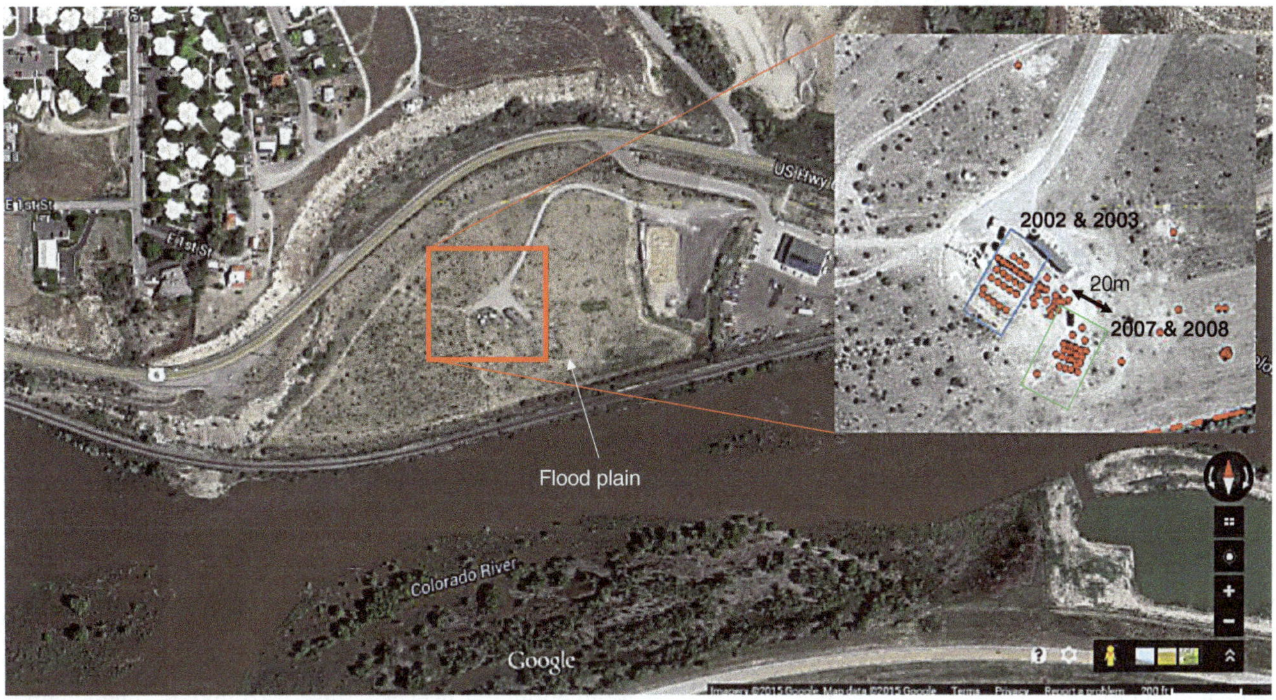

Figure 1.10 Rifle site with locations of the four uranium bioremediation experiments conducted in the years 2002, 2003, 2007, and 2008. The areal extent of 2007 and 2008 experiments are approximately 20 m × 20 m. The terrain map is obtained from Google Earth. Location of the wells (indicated by the red dots) for different years are referred from *Yabusaki et al.* [2007].

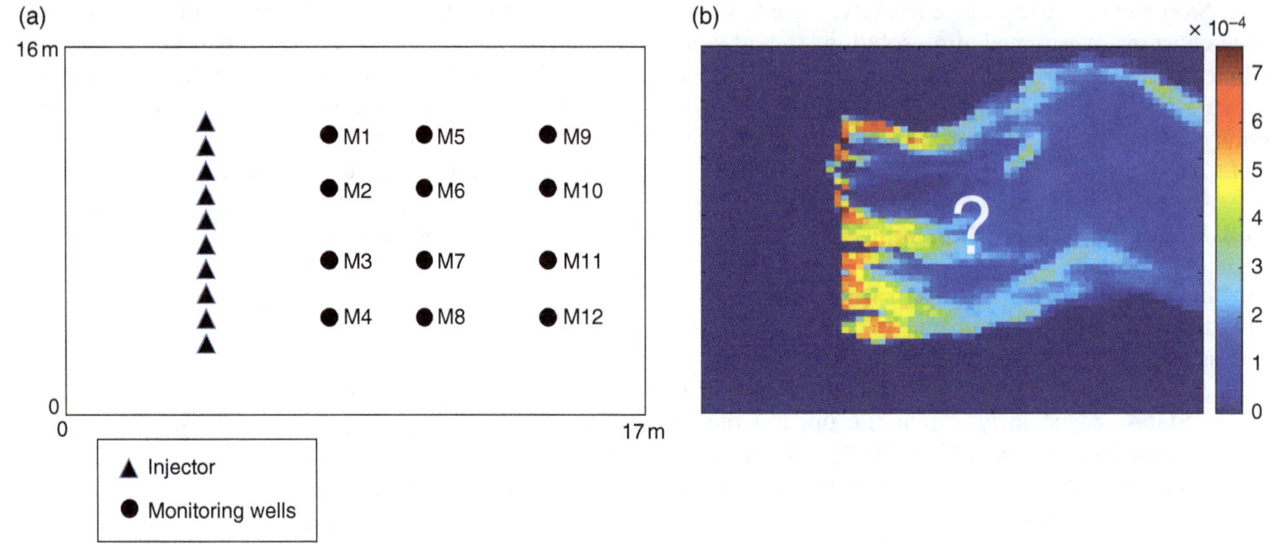

Figure 1.11 The 2007 Winchester experimental setup (a). The aim is to predict the volume and spatial distribution of immobilized uranium. One simulation result for immobilized uranium is shown (b). M1–M12 are wells where tracers or concentrations are monitored.

the setup of the 2007 Winchester experiment, which is the data we will be using in this book. Acetate mixed with conservative tracer is injected into a set of injector wells. Concentrations of tracers as well as acetate, sulfate, and UO_2^{2+} are measured at observation wells. The goal of this study is to predict the volume of immobilized uranium, since this will indicate the efficacy of the injection experiment.

The uncertainties associated with predicting the extent of uranium immobilization are substantial. Similar to the case of shallow geothermal monitoring, the amount of data (even at these kinds of sites) remains limited (although geophysical surveys have also been conducted at these sites) [*Williams et al.*, 2009; *Orozco et al.*, 2011]. We distinguish three groups of uncertainties:

1. *Geological.* This pertains to the uncertain hydraulic conductivity and porosity, their statistical properties and spatial variability.

2. *Biogeochemical.* This pertains to the various geochemical reactions taking place upon acetate injection, in particular the kinetics of such reactions, as well as the initial concentrations and volumes and surface areas of iron-bearing minerals.

3. *Hydrological.* This pertains to the various boundary conditions such as hydraulic gradients, recharge, and so on.

Therefore, the questions are as follows:

1. Which of all these uncertainties impacts most the remediation efficacy?

2. Having this knowledge, how much can we predict long-term efficacy from short-term tracer monitoring?

1.6. DEVELOPING SHALE PLAYS IN NORTH AMERICA

1.6.1. Introduction

A new and vast source of energy, organically rich shale rocks, has changed the global energy landscape in the last decade (see Figure 1.12). Development of such resources is very complex and relies on substantial amount of drilling (operators drill hundreds of wells per year). Decision questions regarding shale systems are not very different from those in conventional systems: Where to drill? How to fracture the rock? What production to expect? However, the well-established approaches developed for conventional reservoirs are not readily applicable to shales, mainly due to the rapid development of these plays. Unconventional wells are drilled at a rate of several wells per week, while one comprehensive prediction and uncertainty quantification could take anywhere from a few weeks to several months. This peculiar nature of unconventional reservoirs calls for the development of new, rapid and comprehensive data analysis and uncertainty quantification methods. In that sense, the problems described in this section are unique and the methods different from those introduced in the previous applications fields. Here statistical and machine learning methods appeared to be more attractive because of their rapid learning and prediction. However, this learning is challenging, involving spatial, temporal, and multivariate elements of high degrees of complexities.

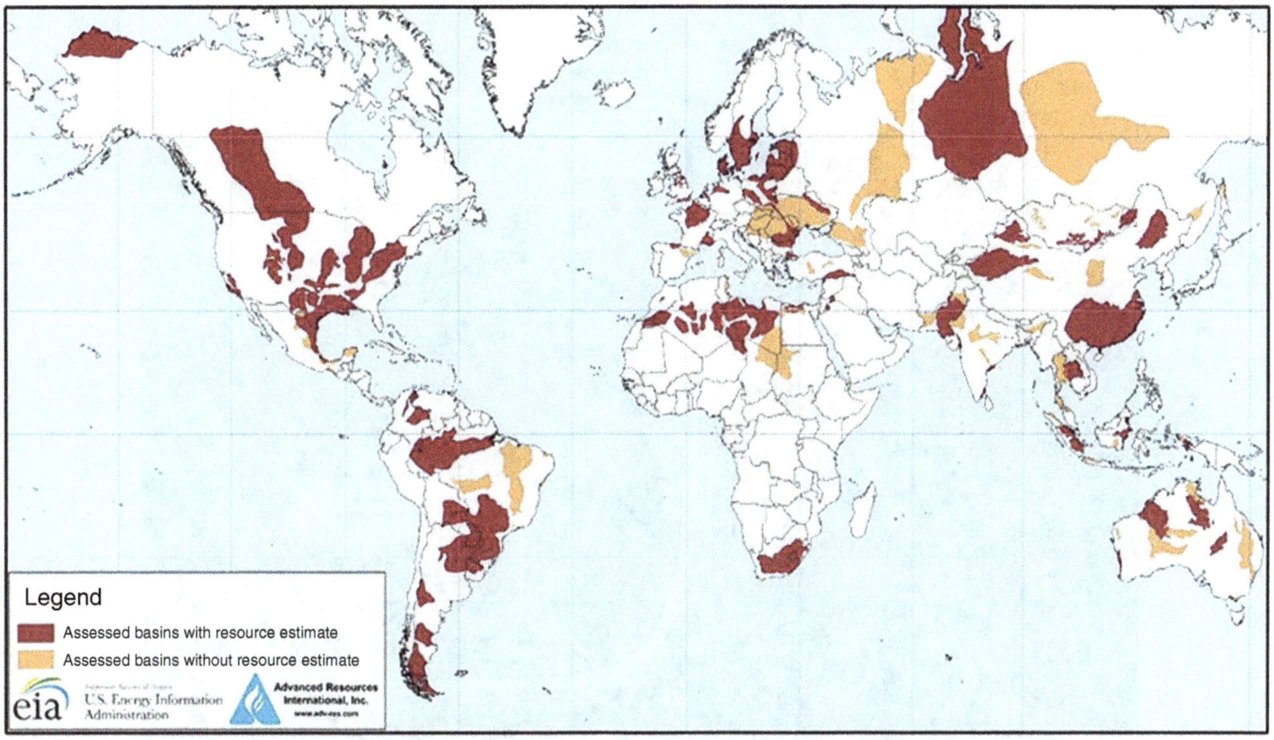

Figure 1.12 Overview of world shale resources (oil or gas). Image taken from http://www.eia.gov.

1.6.2. What are Shales Reservoirs and How are They Produced?

In oceans, a large amount of organic matter from microorganisms and planktons falls to the seabed and mixes with silt, clay, and other materials already present, forming an organic source. Sediment inflow from rivers brings clastic material that is simply deposited on top of the organic source, further burying sediments into the subsurface. Over the course of millions of years, such organically rich and finely grained mixture turns into a specific type of rock, shale, and ends up buried at very large depths. Large depth means large overburden, which imposes high pressures on the organically rich shale, hence increasing its overall temperature. When the temperature of the shale exceeds 120°C the organic matter starts to "cook." Hydrocarbon molecules start forming from the organic matter already contained within the rock. When the volume of hydrocarbons in the rock becomes critical, low density and buoyant forces push hydrocarbons toward the shallower zones of the subsurface (toward lower pressure) in a process called "migration." Normally, hydrocarbons would migrate all the way to the surface (seeping holes); however, they often end up trapped in highly porous sandstones forming hydrocarbon reservoirs. These hydrocarbon reservoirs are also known as the conventional reservoirs, while the organically rich shale that generated the hydrocarbons is commonly referred to as the "source rock."

The amount of hydrocarbons contained in conventional reservoirs is only a small portion of the oil that the source rocks originally generated. Source rocks still contain a large amount of immobile hydrocarbons and as such represent a potentially large energy resource. Every shale rock is different, and the way in which it bounds with the hydrocarbons is also different. This bounding is a result of complex interplay of the rock and fluid compositions and complex physical interactions. Some shales are capable of chemically absorbing gas (sorption), while others are not, sometimes the viscosity of oil is high, and sometimes it is low. Some shales are very brittle with dense networks of natural fractures, while some others are very ductile with almost no natural fractures. What all shales have in common is the fact that they are all almost impermeable and highly organically rich rocks.

Early efforts to produce shale reservoirs through vertical drilling and completion have mostly resulted in failure, due the low permeability of shales. However, with the advent of hydraulic fracturing (HF) and in some cases usage of explosives, production of commercial quantities of hydrocarbons at specific shale plays was possible. The most notable ones are the Big Sandy gas field in Eastern Kentucky, North part of the Marcellus shale in the state of New York where some wells were drilled in the early

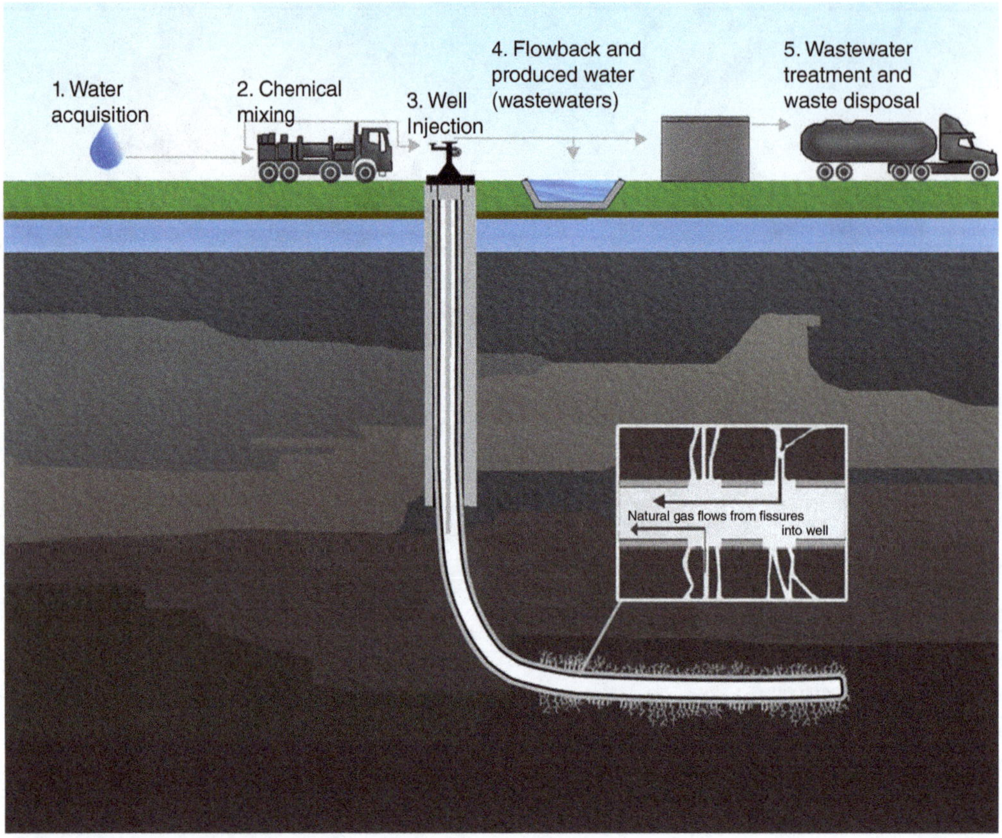

Figure 1.13 Overview of shale HF operations. Image taken from Wikimedia.

1800s (indeed almost 200 years ago), and New Albany shale. These sporadic successes were later attributed to the very well developed networks of natural fractures (producing high permeability flow paths). Real, organized, large-scale effort to tap into shales did not occur until the first big oil crisis in the late 1970s when US DOE initiated the game-changing Eastern Shales Gas Project (ESGS). This project is the largest research project ever taken on shale reservoirs whose successful result is best reflected on US independence on foreign oil at present. ESGS identified that horizontal drilling technology with multistage fracturing is a technique capable of unlocking the potential of organically rich shales. The idea is simple, try to maximize the contact between the well and the rock by producing as many as possible artificial flow paths/fractures.

Today, operators drill long horizontal wells (several thousands of feet long) and conduct massive HF jobs with anywhere between 10 and 40 man-made hydraulic fractures (commonly referred to as "stages") (see Figure 1.13). HF is a complicated and expensive procedure with many different parameters whose complex interplay with the geology determines the quality of the produced hydraulic fractures and ultimately affects the hydrocarbon production. Table 1.4 provides just a few of the many engineering and geological parameters involved. (The abbreviations given in the last column of the table will be used in Chapter 8.) Obviously, optimization of such parameters achieves significant cost reductions, hence maximizes profit. Given that every shale is different, best fracturing practices identified in one shale play do not necessarily translate directly as the most optimal to other shale plays. Therefore, every shale play data are analyzed independently with the aim to understand production, interplay between HF and geology, and ultimately use such understanding to produce some forecasting models. All this in an effort to answer the simple business questions: where to drill, how to complete, and what to expect?

Analysis of data from shale production is not a trivial endeavor. First, the input data (covariates) are very high dimensional (see Table 1.4), making standard statistical techniques difficult to apply. Second, production data from different wells comprise of time series, but of different time intervals, depending on how long the well has been in production. In Chapter 8, we will consider two real field cases, one from the Barnett shale with thousands of wells and one from newly developed system with only 172 hydraulically fractured horizontal wells.

Table 1.4 Overview of some of the parameters involved in designing unconventional shale operations.

	Type of uncertainty	Parameter	Unit	Abbreviation
TARGET	Production	Oil rates	stb/day	Function of time
		Gas rates	stb/day	Function of time
		Water rates	stb/day	Function of time
INPUT (Covariates)	Completions	Number of completion stages	#	CMP STAGES STIMULATED
		Total amount of injected fluid	gal	CMP TOTAL FLUID PUMPED GAL
		Total amount of injected proppant	lbs	CMP TOTAL PROPPANT USED
		Stimulated lateral length	ft	CMP STIMULATED LATERAL LENGTH
		Total amount of slick water	bbl	CMP AMT SLICKWATER BBL
		Total amount of injected x-link fluid	bbl	CMP AMT CROSSLINK BBL
		Completion stage interval	ft	CompStageInterval
		Total amount of linear fluid	bbl	CMP AMT LINEAR BBL
	Geographical	X location	ft	GeolX Rel
		Y location	ft	GeolY Rel
		Z location (depth)	ft	GeolZ
	PVT	Oil API gravity	api units	GeolAPIGrav
	Petrophysical	Total organic content (TOC)	%	PetroTOC
		Clay content	%	PetroVClay
		Water saturation	%	PetroSwt
		Porosity	%	PetroPor
		Total amount of quartz	%	PetroVQtz
		Amount of pyrite	%	PetroPyr

1.6.3. Shale Development Using Data Science

Examples of data centric modeling for shales are in *Mohaghegh et al.* [2011] who utilized artificial intelligence to data mine and forecast production in unconventional reservoirs (see also *Bhattacharya and Nikolaou* [2013]). Most of these methods predict scalar values, such as a rate at a given time. However, decision variables in shale systems are rates or volumes of produced hydrocarbons as they vary over time. Therefore, understanding shales and identifying value-creating practices with data-driven techniques require proper handling of production time series. This is often challenging since production time series come as noisy, discrete observations of production rates over time. In addition, any data scientific method will need to account for the large number of variables involved as well as the spatial heterogeneity of the shale play itself, leading to spatial variation of production, even if they would be produced under the exact same engineering conditions.

Shale management from exploration and production comes with a large series of problems and questions. Here we will focus on those that pertain to the use of data science to predict and quantify uncertainty to what happens when the play is in production. As more wells are drilled and produced, more data become available about geological parameters, completion parameters, and production decline.

The following are the questions we will be most interested in:

1. Which geological and completion parameters most impact production? This a question of sensitivity and it is needed to make the high-dimensional problem manageable before developing any prediction or UQ methods.

2. How to predict and quantify uncertainty on production decline in a new well for given geological and completion parameters? This question requires building a statistical relationship between several covariates and an uncertain function.

3. How many wells need to be in production before a statistical model can confidently estimate production decline

in a new location? With too few data, data scientific method will fail to produce meaningful prediction because uncertainty is too large.

1.7. SYNTHESIS: DATA-MODEL-PREDICTION-DECISION

While the various applications of prediction and UQ are quite diverse, there are various common elements that are useful in summarizing. To do this, let us consider the following statements, including some additional comments.

In engineering the subsurface, uncertainty quantification is only relevant within a decision framework.

This book is about applied science, not pure science. Hence, in such application there is a "utility," or at a minimum "use-inspired" part of the scientific process of UQ. In the various applications, we saw how, ultimately, a decision is what is needed:

1. case 1: how much to re-allocate, where to drill new wells, when to stop production
2. case 2: choose between four alternative well fields
3. case 3: design of the geothermal system by deciding on the type of heat pump
4. case 4: deciding to perform acetate injection and if so, how to inject
5. case 5: deciding where to drill wells, how to complete wells

If for some reason, any UQ does not affect the decision made, simply because a deterministic model leads to a "good" decision, then no UQ is needed. It is therefore important to consider the decisions as an integral part of any UQ; otherwise, one may endlessly model to quantify uncertainty, then only to discover such exercise has negligible impact. This concept will be treated in Chapters 2 and 4 using methods of decision analysis and sensitivities involved in such decisions.

Decisions are made based on key prediction variables that are often simple quantities. Rarely are decisions made directly on complex models.

Rarely do modelers look at hundreds of complex models and decide on that basis. Key prediction variables in decision problems are often simple quantities, certainly simpler than the models on which they are based:

1. case 1: an injector efficiency (scalar), a quality map (map), a rate decline (time series)
2. case 2: recharge area (map), wetlands (rate), river (rate), contamination (map)
3. case 3: heat variation in the subsurface (space-time variable)
4. case 4: volume (scalar) and spatial distribution (map) of precipitated uranium
5. case 5: location (two parameters) or completion (about 10–20 parameters)

The fact that key prediction variables are much simpler (of lower dimension) than models (much higher dimension) can be exploited in a fit-for-purpose (sometimes also termed top-down) modeling approach. It is difficult to reduce model dimension, but it is easier to reduce dimensions in the key prediction variables. This idea will be exploited in Chapter 4 in terms of quantifying sensitivity of models and model variables on prediction variables. It will also be exploited in Chapter 7 to avoid difficult model inversion by directly focusing on the posterior distribution of key prediction variables.

Uncertainty quantification without data (and only models) is meaningless within an engineering–type, decision-making context.

Models alone cannot make accurate predictions. They can be used to understand sensitivity of model variables on prediction variables or data variables. They may provide a way to optimize data acquisition campaigns. But ultimately, if a prediction needs to be made and decisions are to be based on them, in a quantitative fashion, then field measurements are needed. The oil industry has long invested in measurements for the simple reason that they pay back tremendously in terms of management and decision making in reservoirs. The environmental sector has lagged in gathering quality measurements simply because of cost. However, if the goal is to gain a "predictive understanding" of environmental systems and to attain quantitative decision making, then gathering more data will be a prerequisite. As such, the introduction of geophysical data as presented in cases 2, 3, and 4 has gained increased attraction. Like the role of models, data are only useful if it alter decisions, not necessarily only because it inform better models or predictions. This will be treated in Chapter 2 as a "value of information" problem.

REFERENCES

Ahlbrandt, T. S., U S Geological Survey Bulletin F, and G. A. Norton (2005), The Sirte Basin province of Libya – Sirte-Zelten total petroleum system. *U.S. Geological Survey Bulletin 2202–F*: 29.

Alker, M. (2008), *The Nubian Sandstone Aquifer System A Case Study for the Research Project "Transboundary Groundwater Management in Africa"*. German Development Institute, 273 pp.

Ambrose, G. (2000), The geology and hydrocarbon habitat of the Sarir Sandstone, SE Sirt Basin, Libya, *J. Pet. Geol.*, *23*(2), 165–191, doi:10.1111/j.1747-5457.2000.tb00489.x.

Anderson, R. T., H. A. Vrionis, I. Ortiz-Bernad, C. T. Resch, P. E. Long, R. Dayvault, K. Karp, S. Marutzky, D. R. Metzler, A. Peacock, D. C. White, M. Lowe, and D. R. Lovley (2003), Stimulating the in situ activity of geobacter species to remove uranium from the groundwater of a uranium-contaminated aquifer, *Appl. Environ. Microbiol.*, *69*(10), 5884–5891, doi:10.1128/AEM.69.10.5884-5891.2003.

Andričević, R., and V. Cvetković (1996), Evaluation of risk from contaminants migrating by groundwater, *Water Resour. Res.*, *32*(3), 611–621, doi:10.1029/95WR03530.

Avseth, P., T. Mukerji, and G. Mavko (2005), *Quantitative Seismic Interpretation: Applying Rock Physics to Reduce Interpretation Risk*, vol. *53*, Cambridge University Press, Cambridge, doi:10.1017/CBO9781107415324.004.

Balakrishnan, S., A. Roy, M. G. Ierpetritou, G. P. Flach, and P. G. Georgopoulos (2003), Uncertainty reduction and characterization for complex environmental fate and transport models: An empirical Bayesian framework incorporating the stochastic response surface method. *Water Resour. Res.*, *39*(12), 13, doi:10.1029/2002WR001810.

Bayer, P., D. Saner, S. Bolay, L. Rybach, and P. Blum (2012), Greenhouse gas emission savings of ground source heat pump systems in Europe: A review, *Renew. Sustain. Energy Rev.*, *16*(2), 1256–1267, doi:10.1016/j.rser.2011.09.027.

Bellmann, L. H., W. Kouwe, and G. Yielding (2009), Fault Seal Analysis in the N-97 Oil Field in Wintershall Concession C97-I, Libya, *2nd EAGE International Conference on Fault and Top Seals-From Pore to Basin Scale 2009*, Montpellier, France, September 21–24, 2009.

Bhattacharya, S., and M. Nikolaou (2013), Analysis of production history for unconventional gas reservoirs with statistical methods, *SPE J.*, *18*(5), 878–896, doi:10.2118/147658-PA.

Blomquist, W., and H. M. Ingram (2003), Boundaries seen and unseen: Resolving transboundary groundwater problems, *Water Int.*, *28*(2), 162–169, doi:10.1080/02508060308691681.

Blum, P., G. Campillo, and T. Kölbel (2011), Techno-economic and spatial analysis of vertical ground source heat pump systems in Germany, *Energy*, *36*(5), 3002–3011, doi:10.1016/j.energy.2011.02.044.

Bonte, M. (2013), *Impacts of Shallow Geothermal Energy on Groundwater Quality: A Hydrochemical and Geomicrobial Study of the Effects of Ground Source Heat Pumps and Aquifer Thermal Energy Storage*, Vrije Universiteit, Amsterdam, The Netherlands, pp. 178.

Brantley, S. L., M. B. Goldhaber, and K. Vala Ragnarsdottir (2007), Crossing disciplines and scales to understand the critical zone, *Elements*, *3*(5), 307–314, doi:10.2113/gselements.3.5.307.

Chen, X., H. Murakami, M. S. Hahn, G. E. Hammond, M. L. Rockhold, J. M. Zachara, and Y. Rubin (2012), Three-dimensional Bayesian geostatistical aquifer characterization at the Hanford 300 area using tracer test data, *Water Resour. Res.*, *48*(6), 1–20, doi:10.1029/2011WR010675.

Chermak, J. M., R. H. Patrick, and D. S. Brookshire (2005), Economics of transboundary aquifer management, *Ground Water*, *43*, 731–736, doi:10.1111/j.1745-6584.2005.00070.x.

Cirpka, O. A., and P. K. Kitanidis (2000), An advective-dispersive stream tube approach for the transfer of conservative-tracer data to reactive transport, *Water Resour. Res.*, *36*(5), 1209–1220, doi:10.1029/1999WR900355.

da Cruz, P., R. N. Horne, and C. V. Deutsch (2004). The quality map: A tool for reservoir uncertainty quantification and decision making, *SPE Reservoir Eval. Eng.*, *7*(1), 6–14, doi:10.2118/87642-PA.

Famiglietti, J. S. (2014), The global groundwater crisis, *Nat. Clim. Change*, *4*(11), 945–948, doi:10.1038/nclimate2425.

Feyen, L., and S. M. Gorelick (2004), Reliable groundwater management in hydroecologically sensitive areas, *Water Resour. Res.*, *40*, 1–14, doi:10.1029/2003WR003003.

Feyen, L., and S. M. Gorelick (2005), Framework to evaluate the worth of hydraulic conductivity data for optimal groundwater resources management in ecologically aensitive areas, *Water Resour. Res.*, *41*(3), 1–13, doi:10.1029/2003WR002901.

Fiduk, J. C. (2009), Evaporites, petroleum exploration, and the cenozoic evolution of the libyan shelf margin, central north africa, *Mar. Pet. Geol.*, *26*(8), 1513–1527, doi:10.1016/j.marpetgeo.2009.04.006.

Glanz, J. (2009), Quake threat leads Swiss to close geothermal project. *New York Times*, 10 December.

Gleeson, T., K. M. Befus, S. Jasechko, E. Luijendijk, and M. B. Cardenas (2015), The global volume and distribution of modern groundwater, *Nat. Geosci.*, *9*, 161–167, doi:10.1038/ngeo2590.

Gorelick, S. M. (1983), A review of distributed parameter groundwater management modeling methods, *Water Resour. Res.*, *19*(2), 305–319, doi:10.1029/WR019i002p00305.

Green, T. R., M. Taniguchi, H. Kooi, J. J. Gurdak, D. M. Allen, K. M. Hiscock, H. Treidel, and A. Aureli (2011), Beneath the surface of global change: Impacts of climate change on groundwater, *J. Hydrol.*, *405*(3–4), 532–560, doi:10.1016/j.jhydrol.2011.05.002.

Harou, J. J., and J. R. Lund (2008), Ending groundwater overdraft in hydrologic-economic systems, *Hydrogeol. J.*, *16*(6), 1039–1055, doi:10.1007/s10040-008-0300-7.

Harou, J. J., M. Pulido-Velazquez, D. E. Rosenberg, J. Medellín-Azuara, J. R. Lund, and R. E. Howitt (2009), Hydro-economic models: Concepts, design, applications, and future prospects, *J. Hydr.*, *375*(3), 627–643, doi:10.1016/j.jhydrol.2009.06.037.

Hermans, T., F. Nguyen, T. Robert, and A. Revil (2014), Geophysical methods for monitoring temperature changes in shallow low enthalpy geothermal systems, *Energies*, *7*(8), 5083–5118, doi:10.3390/en7085083.

Hermans, T., S. Wildemeersch, P. Jamin, P. Orban, S. Brouyère, A. Dassargues, and F. Nguyen (2015), Quantitative temperature monitoring of a heat tracing experiment using cross-borehole ERT, *Geothermics*, *53*, 14–26, doi:10.1016/j.geothermics.2014.03.013.

Howarth, R.W., R. Santoro, and A. Ingraffea (2011), Methane and the greenhouse-gas footprint of natural gas from shale formations, *Climatic Change*, *106*(4), 679.

Høyer, A. S., F. Jørgensen, N. Foged, X. He, and A. V. Christiansen (2015), Three-dimensional geological modelling of AEM resistivity data: A comparison of three methods. *J. Appl. Geophys.*, *115*, 65–78. doi:10.1016/j.jappgeo.2015.02.005.

Hoyer, A. S., F. Jorgensen, P. B. E. Sandersen, A. Viezzoli, and I. Moller (2015), 3D geological modelling of a complex buried-valley network delineated from borehole and AEM data, *J. Appl. Geophys.*, *122*, 94–102, doi:10.1016/j.jappgeo.2015.09.004.

James, A. L., and C. M. Oldenburg (1997), Linear and monte carlo uncertainty analysis for subsurface contaminant transport simulation, *Water Resour. Res.*, *33*(11), 2495–2508, doi:10.1029/97WR01925.

Jørgensen, F., and P. B. E. Sandersen (2006), Buried and open tunnel valleys in denmark-erosion beneath multiple ice sheets, *Quat. Sci. Rev.*, *25*(11), 1339–1363, doi:10.1016/j.quascirev.2005.11.006.

Jørgensen, L. F., and J. Stockmarr (2009), Groundwater monitoring in denmark: Characteristics, perspectives and comparison with other countries, *Hydrogeol. J.*, *17*(4), 827–842, doi:10.1007/s10040-008-0398-7.

Jørgensen, F., R. Møller, L. Nebel, N. P. Jensen, A. V. Christiansen, and P. B. E. Sandersen (2013), A method for cognitive 3D geological voxel modelling of AEM data, *Bull. Eng. Geol. Environ.*, *72*(3–4), 421–432, doi:10.1007/s10064-013-0487-2.

Khan, S. (2010), A regional hydrologic-economic evaluation to devise environmentally sustainable rice farming systems in southern Murray Darling Basin, Australia, *Paddy Water Environ.*, *8*(1), 41–50, doi:10.1007/s10333-009-0172-z.

Klepikova, M., S. Wildemeersch, T. Hermans, P. Jamin, P. Orban, F. Nguyen, S. Brouyère, and A. Dassargues (2016), Heat tracer test in an alluvial aquifer: Field experiment and inverse modelling, *J. Hydrol.*, *540*, 812–823, doi:10.1016/j.jhydrol.2016.06.066.

Kløve, B., P. Ala-Aho, G. Bertrand, J. J. Gurdak, H. Kupfersberger, J. K. Værner, T. Muotka, H. Mykrä, E. Preda, P. Rossi, C. Bertacchi Uvo, E. Velasco, and M. Pulido-Velazquez (2014), Climate change impacts on groundwater and dependent ecosystems, *J. Hydrol.*, *518*, 250–266, doi:10.1016/j.jhydrol.2013.06.037.

Li, L., C. I. Steefel, M. B. Kowalsky, A. Englert, and S. S. Hubbard (2010), Effects of physical and geochemical heterogeneities on mineral transformation and biomass accumulation during biostimulation experiments at Rifle, Colorado, *J. Contam. Hydrol.*, *112*(1–4), 45–63, doi:10.1016/j.jconhyd.2009.10.006.

Li, L., N. Gawande, M. B. Kowalsky, C. I. Steefel, and S. S. Hubbard (2011), Physicochemical heterogeneity controls on uranium bioreduction rates at the field scale, *Environ. Sci. Technol.*, *45*(23), 9959–9966, doi:10.1021/es201111y.

Li, L., K. Maher, A. Navarre-Sitchler, J. Druhan, C. Meile, C. Lawrence, J. Moore, J. Perdrial, P. Sullivan, A. Thompson, L. Jin, E. W. Bolton, S. L. Brantley, W. E. Dietrich, K. Ulrich Mayer, C. I. Steefel, A. Valocchi, J. Zachara, B. Kocar, J. Mcintosh, B. M. Tutolo, M. Kumar, E. Sonnenthal, C. Bao, and J. Beisman (2017), Expanding the role of reactive transport models in critical zone processes, *Earth-Sci. Rev.*, *165*, 280–301.

Lovley, D. R., E. J. P. Phillips, Y. A. Gorby, and E. R. Landa (1991), Microbial reduction of uranium, *Nature*, *350*(6317), 413–416, doi:10.1038/350413a0.

Maneta, M. P., M. D. O. Torres, W. W. Wallender, S. Vosti, R. Howitt, L. Rodrigues, L. H. Bassoi, and S. Panday (2009), A spatially distributed hydroeconomic model to assess the effects of drought on land use, farm profits, and agricultural employment, *Water Resour. Res.*, *45*(11), W11412, doi:10.1029/2008WR007534.

Mavko, G., T. Mukerji, and J. Dvorkin (2009), *The Rock Physics Handbook*, Second Edition, Cambridge University Press, Cambridge. World Wide Web Internet and Web Information Systems. doi:http://dx.doi.org/10.1017/CBO9780511626753.

Maxwell, R. M., W. E. Kastenberg, and Y. Rubin (1999), A methodology to integrate site characterization information into groundwater-driven health risk assessment, *Water Resour. Res.*, *35*(9), 2841–2855, doi:10.1029/1999WR900103.

Mohaghegh, S., O. Grujic, and S. Zargari, A. Kalantari-Dahaghi (2011), Modeling, history matching, forecasting and analysis of shale reservoirs performance using artificial intelligence, *Presented at the SPE Digital Energy Con-ference and Exhibition*, The Woodlands, Texas, 19–21 April, doi:10.2118/143875-MS.

Møller, I., V. H. Søndergaard, F. Jørgensen, E. Auken, and A. V. Christiansen (2009), Integrated management and utilization of hydrogeophysical data on a national scale, *Near Surf. Geophys.*, *7*(5–6), 647–659, doi:10.3997/1873-0604.2009031.

Morgan, D. R., J. W. Eheart, and A. J. Valocchi (1993), Aquifer remediation design under uncertainty using a new chance constrained programming technique, *Water Resour. Res.*, *29*(3), 551–561, doi:10.1029/92WR02130.

Muller, K., J. Vanderborght, A. Englert, A. Kemna, J. A. Huisman, J. Rings, and H. Vereecken (2010), Imaging and characterization of solute transport during two tracer tests in a shallow aquifer using electrical resistivity tomography and multilevel groundwater samplers, *Water Resour. Res.*, *46*(3), W03502, doi:10.1029/2008WR007595.

Oelkers, E. H., J. G. Hering, and C. Zhu (2011), Water: Is there a global crisis? *Elements*, *7*(3), 157–162, doi:10.2113/gselements.7.3.157.

Orozco, A. F., K. H. Williams, P. E. Long, S. S. Hubbard, and A. Kemna (2011), Using complex resistivity imaging to infer biogeochemical processes associated with bioremediation of an uranium-contaminated aquifer, *J. Geophys. Res. Biogeosci.*, *116*(3), G03001, doi:10.1029/2010JG001591.

Osborn, S. G., A. Vengosh, N. R. Warner, and R. B. Jackson (2011), Methane contamination of drinking water accompanying gas-well drilling and hydraulic fracturing, *Proc. Natl. Acad. Sci. USA*, *108*(20), 8172–8176, doi:10.1073/pnas.1100682108.

Palmisano, A., and T. Hazen (2003), *Bioremediation of Metals and Radionuclides: What It Is and How It Works*, Lawrence Berkeley National Laboratory, Berkeley, Calif.

Reed, P. M., D. Hadka, J. D. Herman, J. R. Kasprzyk, and J. B. Kollat (2013), Evolutionary multiobjective optimization in water resources: The past, present, and future, *Adv. Water Resour.*, *51*, 438–456, doi:10.1016/j.advwatres.2012.01.005.

Refsgaard, J. C., A. L. Højberg, I. Møller, M. Hansen, and V. Søndergaard (2010), Groundwater modeling in integrated water resources management: Visions for 2020, *Ground Water*, *48*(5), 633–648, doi:10.1111/j.1745-6584.2009.00634.x.

Sandersen, P. B. E., and F. Jørgensen (2003), Buried quaternary valleys in western denmark-occurrence and inferred implications for groundwater resources and vulnerability, *J. Appl. Geophys.*, *53*(4), 229–248, doi:10.1016/j.jappgeo.2003.08.006.

Sandersen, P. B. E., F. Jørgensen, N. K. Larsen, J. H. Westergaard, and E. Auken (2009), Rapid tunnel-valley formation beneath the receding Late Weichselian ice sheet in Vendsyssel,

Denmark, *Boreas*, *38*(4), 834–851, doi:10.1111/j.1502-3885.2009.00105.x.

Sanderson, E. W., M. Jaiteh, M. A. Levy, K. H. Redford, A. V. Wannebo, and G. Woolmer (2002), The human footprint and the last of the wild, *BioScience*, *52*(10), 891, doi:10.1641/0006-3568(2002)052[0891:THFATL]2.0.CO;2.

Sassen, D. S., S. S. Hubbard, S. A. Bea, J. Chen, N. Spycher, and M. E. Denham (2012), Reactive facies: An approach for parameterizing field-scale reactive transport models using geophysical methods, *Water Resour. Res.*, *48*(10), W10526, doi:10.1029/2011WR011047.

Siebert, S., J. Burke, J. M. Faures, K. Frenken, J. Hoogeveen, P. Döll, and F. T. Portmann (2010), Groundwater use for irrigation: A global inventory, *Hydrol. Earth Syst. Sci.*, *14*(10), 1863–1880, doi:10.5194/hess-14-1863-2010.

Singh, A. (2014a), Simulation-optimization modeling for conjunctive water use management, *Agric. Water Manag.*, *141*, 23–29, doi:10.1016/j.agwat.2014.04.003.

Singh, A. (2014b), Simulation and optimization modeling for the management of groundwater resources. II: Combined applications, *J. Irrig. Drain. Eng.*, *140*(4), 1–9, doi:10.1061/(ASCE)IR.1943-4774.0000689.

Singha, K., and S. M. Gorelick (2005), Saline tracer visualized with three-dimensional electrical resistivity tomography: Field-scale spatial moment analysis, *Water Resour. Res.*, *41* (5), 1–17, doi:10.1029/2004WR003460.

Slater, L., A. M. Binley, W. Daily, and R. Johnson (2000), Cross-hole electrical imaging of a controlled saline tracer injection, *J. Appl. Geophys.*, *44*(2–3), 85–102, doi:10.1016/S0926-9851(00)00002-1.

Sonnenborg, T. O., D. Seifert, and J. C. Refsgaard (2015), Climate model uncertainty versus conceptual geological uncertainty in hydrological modeling, *Hydrol. Earth Syst. Sci.*, *19* (9), 3891–3901, doi:10.5194/hess-19-3891-2015.

Sørensen, K. I., and E. Auken (2004), SkyTEM: A new high-resolution transient electromagnetic system, *Explor. Geophys.*, *35*(3), 191–199, doi:10.1071/EG04194.

Srinivasan, V., E. F. Lambin, S. M. Gorelick, B. H. Thompson, and S. Rozelle (2012), The nature and causes of the global water crisis: Syndromes from a meta-analysis of coupled human-water studies, *Water Resour. Res.*, *48*(10), W10516, doi:10.1029/2011WR011087.

Stauffer, F., P. Bayer, P. Blum, N. Giraldo, and W. Kinzelbach (2013), *Thermal Use of Shallow Groundwater*, CRC Press, Boca Raton, doi:10.1201/b16239.

Steefel, C. I., C. A. J. Appelo, B. Arora, D. Jacques, T. Kalbacher, O. Kolditz, V. Lagneau, P. C. Lichtner, K. U. Mayer, J. C. L. Meeussen, S. Molins, D. Moulton, H. Shao, J. Šimůnek, N. Spycher, S. B. Yabusaki, and G. T. Yeh (2015), Reactive transport codes for subsurface environmental simulation, *Comput. Geosci.*, *19*(3), 445–478, doi:10.1007/s10596-014-9443-x.

Thomas, D. (1995), Exploration limited since '70s in Libya's Sirte Basin, *Oil Gas J.*, *93*, 99–104.

Thomsen, R., V. H. Søndergaard, and K. I. Sørensen (2004), Hydrogeological mapping as a basis for establishing site-specific groundwater protection zones in Denmark, *Hydrogeol. J.*, *12*(5), 550–562.

Tujchneider, O., G. Christelis, and J. V. der Gun, (2013), Towards scientific and methodological innovation in transboundary aquifer resource management, *Environ. Dev.*, *7* (1), 6–16, doi:10.1016/j.envdev.2013.03.008.

United States. Environmental Protection Agency. Office of Solid Waste and Emergency Response (2005), A decision-making framework for cleanup of sites impacted with light non-aqueous phase liquids (LNAPL), Office of Solid Waste and Emergency Response.

Vanderborght, J., A. Kemna, H. Hardelauf, and H. Vereecken (2005), Potential of electrical resistivity tomography to infer aquifer transport characteristics from tracer studies: A synthetic case study, *Water Resour. Res.*, *41*(6), 1–23, doi:10.1029/2004WR003774.

Wada, Y., L. P. H. van Beek, C. M. van Kempen, J. W. T. M. Reckman, S. Vasak, and M. F. P. Bierkens (2010), Global depletion of groundwater resources, *Geophys. Res. Lett.*, *37* (20), doi:10.1029/2010GL044571.

Wagner, B. J., and S. M. Gorelick (1987), Optimal groundwater quality management under parameter uncertainty, *Water Res. Res.*, *23*(7), 1162–1174, doi:10.1029/WR023i007p01162.

Wang, Q., D. Kim, D. D. Dionysiou, G. A. Sorial, and D. Timberlake (2004), Sources and remediation for mercury contamination in aquatic systems: A literature review, *Environ. Pollut.*, *131*(2), 323–336, doi:10.1016/j.envpol.2004.01.010.

Whitaker, B. (2016), "Oklahoma's rise in quakes linked to man-made causes," 60 minutes, *CBS News*, 8 May.

Williams, K. H., A. Kemna, M. J. Wilkins, J. Druhan, E. Arntzen, A. L. N'Guessan, P. E. Long, S. S. Hubbard, and J. F. Banfield (2009), Geophysical monitoring of coupled microbial and geochemical processes during stimulated subsurface bioremediation, *Environ. Sci. Technol.*, *43*(17), 6717–6723, doi:10.1021/es900855j.

Williams, K. H., P. E. Long, J. A. Davis, M. J. Wilkins, A. L. N'Guessan, C. I. Steefel, L. Yang, D. Newcomer, F. Spane, and L. Kerkhof (2011), Acetate availability and its influence on sustainable bioremediation of uranium-contaminated groundwater, *Geomicrobiol. J.*, *28*(5–6), 519–539, doi:10.1080/01490451.2010.520074.

Williams, K., S. Hubbard, and D. Hawkes. (2013), Rifle–a community site of discovery and accomplishment source: Earth Science Division, Berkeley Lab, *News and Events*, 26 September.

Yabusaki, S. B., Y. Fang, P. E. Long, C. T. Resch, A. D. Peacock, J. Komlos, P. R. Jaffe, S. J. Morrison, R. D. Dayvault, D. C. White, and R. T. Anderson (2007), Uranium removal from groundwater via in situ biostimulation: Field-scale modeling of transport and biological processes, *J. Contam. Hydrol.* *93*(1–4), 216–235, doi:10.1016/j.jconhyd.2007.02.005.

Zachara, J. M., P. E. Long, J. Bargar, J. A. Davis, P. Fox, J. K. Fredrickson, M. D. Freshley, A. E. Konopka, C. Liu, and J. P. McKinley (2013), Persistence of uranium groundwater plumes: Contrasting mechanisms at two DOE sites in the groundwater-river interaction zone, *J. Contam. Hydrol.*, *147*, 45–72, doi:10.1016/j.jconhyd.2013.02.001.

2

Decision Making Under Uncertainty

2.1. INTRODUCTION

Making good decisions is important in many aspects of life. Decisions in the personal realm are made by individuals and usually consider the consequences of those decisions on others (e.g., family members). In organizations (e.g., corporations, governments, universities, etc.), individuals also play a critical role in decision making, but they are usually part of a group-based decision-making process. How does an individual or an organization know whether they are making a good decision at the time they are making that decision (without the benefit of hindsight)? Would you know a good decision if you saw one? Without any field-specific knowledge one could be inclined to define decision making as "choosing between many alternatives that best fits your goals." However, the evident questions then are (i) how to define what is best or optimal, requiring the definition of some criterion, which may change the decision if this criterion changes and (ii) what are the stated goals? Decision analysis theory provides axiomatic scientific tools for addressing these questions in a structured, repeatable way.

Uncertainty plays a very important role in making sound decisions. The existence of uncertainty does not preclude one from making a decision. Decisions can be made without perfect information. A poor way of proceeding is to make a decision first and then question whether particular events were uncertain. Decision making and uncertainty modeling is an integral and synergetic process, not a sequential set of steps.

In most meaningful circumstances, a decision can be defined as a conscious, irrevocable allocation of resources to achieve desired objectives [*Howard*, 1966]. This definition very much applies to any type of geo-engineering situation. The decision to drill a well, cleanup a site, construct aquifer storage and recovery facilities, or re-allocating water abstraction requires a clear commitment of resources. One may go even to a higher level and consider policy making by government or organizations as designed to affect decisions to achieve a certain objective.

Ron Howard who was at the forefront of decision making as a science describes this field as a "systematic procedure for transforming opaque decision problems into transparent decision problems by a sequence of transparent steps." Applying the field of decision analysis to subsurface systems is not trivial because it involves the following:

1. *Uncertainty*. While most of this book addresses the geoscience aspect of uncertainty as it pertains to the measurements and models we establish to make prediction and optimize profit or use of resources, there may be many other sources of uncertainty, more related to the economic portion of uncertainty (costs, prices, human resources) or human behavior that are not discussed in this book.
2. *Complexity*. Rarely does one make a single decision on a single decision question. Often a complex sequence of decisions needs to be made. This is certainly the case in oil field production where engineers need to make decisions on facility or well location and well-types as the field is being produced.
3. *Multiple objectives*. Often, competing objectives exist in decision making, for example as related to safety and environmental concern compared to the need for energy resources.
4. *Time component*. If it takes too much time to quantify uncertainty that tries to include all sorts of complexity, and the decision must be made in a much shorter time frame, then a complex model ends up having little input into the decision. This is often the case in a time-sensitive business or industries (competitive oil-field reserve calculations, for example). In such cases, one may want to employ simpler models of uncertainty over complex ones.

This chapter provides a basic overview of those elements of decision analysis that are important in the

Quantifying Uncertainty in Subsurface Systems, Geophysical Monograph 236,
First Edition. Céline Scheidt, Lewis Li, and Jef Caers.

context of the subsurface. The purpose is not to be exhaustive by any means, instead to be more illustrative of concepts that may be new to some readers. The following publications are recommended as introductory material:

Foundation of Decision Analysis [*Howard and Abbas*, 2015]. This book is based on course notes that Ronald Howard used for teaching decision analysis at Stanford. This is one of the gold standard texts in decision analysis.

Handbook of Decision Analysis [*Parnell et al.*, 2013]. As an excellent introduction to the topic, it covers next to the more axiomatic component of decision analysis, the soft skills that are needed to make good decisions.

Value of Information in the Earth Sciences [*Eidsvik et al.*, 2015]. This publication looks at decision analysis with a spatial context as well as values information analysis with many applications in the Earth sciences.

Making Good Decisions [*Bratvold and Begg*, 2010]. This book focuses on the petroleum industry, but it is an excellent easy read for those looking to be exposed to the subject matter.

2.2. INTRODUCTORY EXAMPLE: THE THUMBTACK GAME

To illustrate some basic concepts in decision making, let us play a game. Imagine you are offered an opportunity to win \$100. The game is simple. A thumbtack will be tossed with two possible outcomes, "pin up" and "pin down"; if you guess correctly, you win \$100, otherwise you get nothing. However, there is no free lunch; you need to go into competition with other players to buy your way into this opportunity to bet. In other words, the opportunity will be auctioned off. This auction can be done under varying rules: closed first price, closed second price (Vickrey auction, e.g., E-bay), open descending (Dutch auction), or open ascending (English auction).

Regardless of the auction, someone will get the opportunity and pay an amount for it. Imagine you won the auction by offering \$20. This \$20 is now gone, you will never see it again. In decision analysis, this is termed a "sunk cost." In rational decision making, sunk costs should be ignored; in other words, one should not have a sentimental attachment such as "I already invested so much in the project; that means I need to keep investing, because I feel committed to it." Future decisions will not and should not be affected by sunk costs; they will only affect net-profit. Sunk costs are about the past, decisions are about the future. Figure 2.1 describes this situation with a decision tree (a logical time-dependent arrangement of decisions, uncertainties, and payoffs, see Section 2.5).

The decision tree allows introducing a few more concepts:

1. *A scenario*. An instantiation of every decision situation, here it is the combination of your call with the outcome (four possibilities).
2. *A prospect*. How the decision maker views the future for each scenario. It is the equivalent of "outcome" in probability theory.
3. *A lottery (gambles or deals)*. A situation with uncertain prospects without a decision being made. For example, you could be told to call pin down, without having a say in this. Then you face a lottery.

After paying \$20, you get a certificate that gives you the right to bet on the game. Let us now consider the following question: What is the least amount of dollars you are willing to sell this certificate for? There is no objective answer, it depends on your willingness to sell it at a high or low price, and hence we need to introduce a second

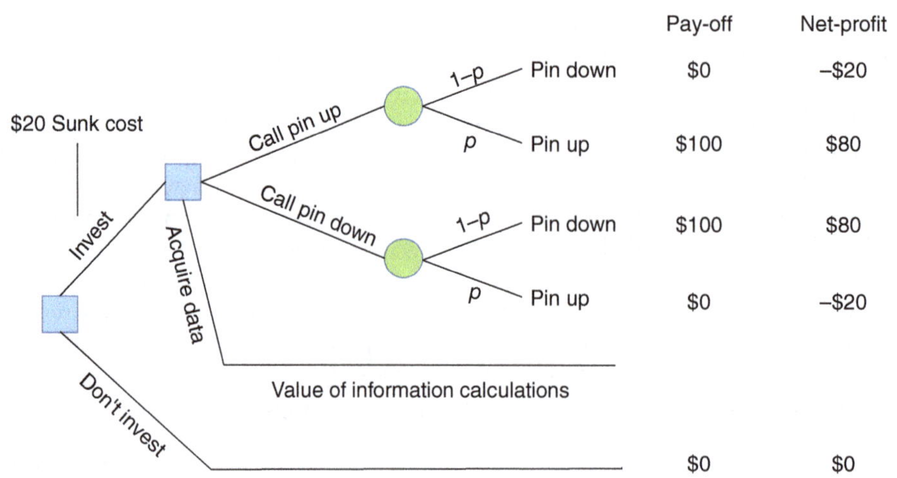

Figure 2.1 Decision tree for a simple game of investing and betting. Squares represent decisions nodes and circles represent uncertainty nodes.

important component: utility. All situations that face decisions and uncertainty require only two basics concepts: probability and utility. A utility represents the decision maker's preference for lotteries with uncertain value prospects. A risk neutral decision maker takes values as they are and takes the alternative that maximizes expected value. To account for risk averse or risk-seeking decision makers, a utility function is introduced to map the value into a new value termed "utility" [*Von Neumann and Morgenstern*, 1944], which is between 0 and 1 or 0 and 100 (see Figure 2.2).

Let us now return to the question of selling your certificate. We now introduce the concept of certainty equivalent (CE) which is the certain amount in your mind that you are willing to sell the certificate for. In more formal language, it is the amount where the decision maker is indifferent between selling it and retaining it. Logically then, the difference between the expected prospect and the CE reflects your attitude toward risk.

$$\text{Risk premium} = \text{expected value} - \text{certainty equivalent}$$

The expected value in this binary game is simply

$$E[\text{payoff}] = P(\text{correct call}) \times \$100$$

which is not known in this case because we do not know the probabilities related to tossing a thumbtack. In a Bayesian sense, we could assume some prior distribution on this (see Chapter 5 and the billiard table example). Therefore, this probability reflects our state of knowledge, it is not a property of the thumbtack. Indeed, if we know the outcome of the toss, then this probability is simply one. Risk neutral investors will have a CE equal to the expected value (they sell the certificate for a price equal to what they consider the expected payoff). Risk averse investors will be conservative and sell it for a low price to make sure they get paid at least some amount for certain (note that the sunk cost does not come into play here). Risk-seeking investors are willing to set high prices, with the risk of getting paid nothing by entering risky investments. For the same utility, risk seekers have higher CE (see Figure 2.2).

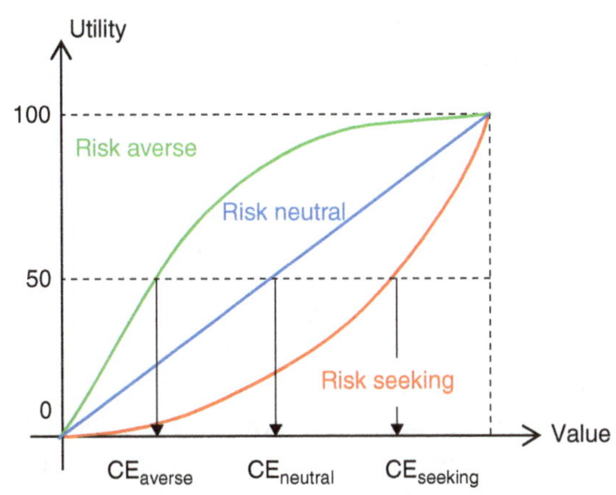

Figure 2.2 Illustration of the concept of value, utility, and certainty equivalent.

Decision makers who follow the rules of decision analysis (Section 2.4.2) take the alternative that maximizes expected utility.

You or an investor interested in buying your certificate may want to gather some information about the uncertain event, the outcome of tossing the thumbtack. What information would you get and how much would you pay for it? This is a "value of information" question. You may want to buy some tosses, say 50 cents per toss, or, you may want to buy a computer program that simulates tosses, and so on. All this information is, however, imperfect. It does not reveal the "truth," the ultimate toss outcome. Perfect information is tantamount to revealing the truth, here knowing the toss outcome. It makes logical sense that you would not pay more for imperfect information than for perfect information. Hence, the value of perfect information (VOPI) is an upper limit that you will never exceed when buying information. The VOPI is therefore

$$\begin{aligned}\text{VOPI} = {} & \text{value with perfect information} \\ & - \text{value without information}\end{aligned}$$

Clearly knowing the toss result will get you \$100 (=value with perfect information) and without any information you will get your CE, because that is the certain amount in your mind (knowing nothing else), hence

$$\text{VOPI} = \$100 - \text{certainty equivalent}$$

Let us now consider imperfect information or simply VOI:

$$\text{VOI} = \text{value with information} - \text{value without information}$$

Without yet getting into any calculations (these are discussed in Section 2.6.2), three main elements influence this value as per the following definitions:

1. *Prior*. What we know before, a base-level uncertainty. If we already know a lot, then additional information, data, or experiments will not add much value.
2. *Reliability*. How reliable is the information, meaning, how well it predicts what I need to know, the unknown truth.
3. *Decision*. There is no value in knowledge that does not affect the decision.

Deciding to gather information is therefore part of the decision model, in the sense that it adds one more alternative from which one can choose (see Figure 2.1). If the information gathering branch in Figure 2.1 has higher payoff, then one should decide to gather information, and then only make a decision. Note that the VOI does not depend on the cost of the information gathering. This is a common confusion. Information gathering costs are sunk costs because they precede the decision. Instead, the cost should be compared with the value of information; hence, a decision to acquire data should be made on that basis.

Imagine now that you are asked to call the toss and you made the right call. Did you make a good decision? Unfortunately, the skill of a person in terms of making decisions is often evaluated based on one outcome. Somebody got rich in life, he/she must really have made good decisions (?). But what if this outcome was purely based on chance, and whatever decision he/she made had in fact very little impact? Good decisions can be evaluated only in the long run. If one wants to compare decision making based on a gut-feeling versus decision making based on decision theory, then this can only be evaluated for a large set of decision-making processes and chance outcomes. Unfortunately, very few corporations or decision entities keep track of such success/failure rates.

2.3. CHALLENGES IN THE DECISION-MAKING PROCESS

2.3.1. The Decision Analyst

To make this section educational, the reader should now consider themselves as decision analyst/consultant observing and analyzing the decision-making process in the situations described in the following. This third-person view is important because readers may have different background and hence interpret the subjective decision process very differently, or may have been decision makers, or more likely subject matter experts. We will also illustrate the challenges with two very different situations that would be rather typical for the decision-making background in the context of this book. The first situation is a decision analyst visiting ExxonMobil in Houston (or any large oil/gas company) and the second situation concerns the Danish government (see Chapter 1 for general background information). We will abbreviate ExxonMobil as EM and the Danish government as DG. Clearly, as decision analyst and consultant, you will encounter two very different situations; hence, no single solution will fit all problems. However, in both cases you will likely interact with decision makers, stakeholders, and subject matter experts (SMEs), all of which are humans and therefore not necessarily rational. Important skills for any analyst are therefore not just technical (as most readers have) but also require certain soft skills such as understanding how people think (rational and irrational), how experts should be approached, how decision makers should or should not be aware of the technical context, how the group dynamic works, and so on. The analyst will also need to face some push-back against rational decision making or any decision-theoretic framework. Even today, with the advances in decision analysis as a science, many still rely on "my intuition" or "my gut-feeling," or "my rules of thumb." It is well documented that the rational decision-making process outperforms these one-at-a-time, anatomical decisions [*Parnell et al.*, 2013]. Few see decision analysis as something that will be beneficial in the long term, and many will judge the decision-making process on single outcomes ("see, I knew this all along, so we didn't need any technical, or advanced approaches, I could just have told you so"). That would be the same as judging one's black-jack playing skills from one single win (which is what makes casinos rich).

As decision analyst, you will not just focus on the technicalities involved in running decision models or structuring the decision axiomatically (e.g., probability theory, decision trees) but also be the facilitator integrating the complex and multiple objectives of stakeholders or the conflicting information provided by the various domain experts involved. As decision analyst, you will need to integrate technical knowledge with business knowledge. In the case of EM, you may need to have technical knowledge (such as about the subsurface). In other cases, in particularly when working with EM management, you may need to work with higher level technical managers or executives focusing on the entire decision-making process rather than on specific technical challenges. In the Danish case, you will need to be aware of the Danish democratic process, the sensitivities concerning sustainable agriculture within a changing environment, the dynamic between industry and local farmers, and the communities they live in.

2.3.2. Organizational Context

The decision-making process for EM and the DG are very different. In both the cases, however, the decision process is complex. In the case of EM, many stakeholders exist (board, stock-holders, government regulators, domain experts, executives, etc.). This is less the case for the DG which obtains some input from stakeholders, such as cities, farmers, agriculture, but because of the Danish style of democracy, the government is the central decision maker. "Stakeholders" refer to all parties with vested interest, not just decision makers or experts. A proper analysis of the stakeholders is required too since the nature of the fundamental and means objective (see Section 2.4.4) may depend on how they are defined.

Complexity is present for various reasons. First, there is the technical complexity, which the subject matter of this book. EM and also increasingly DG are using complex, multidisciplinary data acquisition and modeling approaches to inform their decisions. This by itself involves many domain experts in geology, geophysics, subsurface engineering, surface facilities, economist, and so on. With 73,500 employees (2015), EM organizational complexity is substantial. Decisions are made from the very high level to the daily operations, with a complex hierarchy, many stakeholders, contractors, and experts. Typically, various decision levels are present. Strategic decisions focus on the long-term goals of the organization: the "mission." Tactical decisions turn these strategic goals into measurable objectives. This book covers some of these types of decisions, such as how to allocate resources (cleaning, drilling, data acquisition) to obtain certain objectives. Day-to-day operational decisions are short term and usually reactive. The latter usually does not involve complex modeling or technical analysis.

Therefore, it is important for the decision analyst to understand the social aspect of such organizations, in particular the cultural differences that exist in these various situations. For DG, EM, and others, no single solution fits all. In fact, it may be dangerous to think of solutions, because it points to an overly technical and analytical approach to the problem, ignoring the human aspect. Public sectors often require transparency and clear oversight, while private companies, certainly those not publicly traded, may have a very closed decision-making process.

One of the common issues, certainly in large organizations is the lack of transparency around objectives; this leads to technical experts to perform their technical work without proper context. They have no stake in the decision but simply execute some task. This may lead to gathering data without real goals, or ambiguous goals, or just collect data because "that's what we always do." As such, technical experts may focus on a wrong problem or an unimportant problem or task.

In executing such tasks, there may be overconfidence in one's judgment, or overreliance on the limited domain of expertise. This is common in oil/gas companies. The domain expert will emphasize that their domain is important and relevant, "because it is." Some geophysicist may state that everything can be explained with geophysical data, or well-test engineers with his/her well-test data, as long as enough time is spent on analyzing the data or gathering more "perfect data." From a human point of view, this is understandable, since the domain expert may have anxiety about one's irrelevance with the larger context of the problem; hence, the focus is on the narrow problem only. This way of working often leads to some form of "decision paralysis," meaning postponing decisions until everything is fully understood (determinism). The problem is that in any sciences and in particular in the subsurface geological sciences, we will rarely fully understand everything; hence, geologist may find it difficult to move forward. This also makes their domain increasingly irrelevant, since usually some form of quantification is needed to make decisions meaningful.

Another issue, in particular in large organization, is that both the decision analysis and the domain experts are shielded from the decision maker. In fact, there is often a geographical problem as decision makers do not work in the same location (or even building) as the technical experts. As such, experts rarely understand the decision maker's preferences and therefore lack value-focused thinking (addressing their own small problems, instead of the organizations').

Cognitive bias is a problem when dealing with complex situations. "Cognitive biases are mental errors caused by our simplified information processing strategies. A cognitive bias is a mental error that is consistent and predictable" [*Heuer*, 1999]. A typical problem in both academia and industrial setting-related decision problems is the bandwagon effect, meaning doing things a certain way because that is what other people do, without asking questions as to whether this is appropriate. This bandwagon effect may be present on a small scale, such as experts always using a software in the same way, without question, because that is what the organization does or that is what the software provides, even if it makes no sense. At a larger scale, a bandwagon effect may affect entire industries or academic fields. In Chapter 5, we will discuss this extensively as "blindly following the paradigm." In this type of bandwagon effect, there is an undocumented consensus that things should be done in a certain way, and that any other way that questions on the very nature of the paradigm is simply cast aside (and hence never funded!).

Information and confirmation biases occur when information is gathered without knowing if it adds value or worsens it, to confirm a hypothesis rather than attempting to reject one (see Chapter 5 on inductionism vs. falsificationism). Another common trait is to anchor, meaning creating a best guess and anchoring uncertainty on that best guess, never questioning the anchor. In the subsurface, this is quite common. For example, a few wells are drilled, the average of some petrophysical property is estimated from logging or coring, and the uncertainty on that property is specified as the mean of the data plus or minus some standard deviation. Clearly, the mean may be completely incorrect, due to under-sampling, biases, measurement issues, and so on. Another common form of anchoring is to build a base case and design the entire future on it even in the presence of evidence that refutes the base case, or to make ad hoc modification to the base case. The issue of ignoring Bayes' rule and making ad hoc model choices, without assessing them against evidence, or evaluating the probability of such ad hoc modification, will be treated in extenso in Chapter 5.

2.4. DECISION ANALYSIS AS A SCIENCE

2.4.1. Why Decision Analysis Is a Science

What is science? This common question is also known as the "demarcation problem," first introduced by *Popper* [1959]. Science operates in certain ways; a complex interaction of axioms, hypothesis, conjectures, evidence gathering, experimentation, ways of reasoning, such as induction and deduction, and others such a Bayesianism. We will dedicate Chapter 5 to this topic.

Decision analysis uses axioms of probability theory and utility theory [*Howard*, 1968; *Raiffa*, 1968; *Keeney and Raiffa*, 1993]. Most decision analysis are Bayesian (see Chapter 5) in the sense that they work with subjective beliefs and Bayes' rule to update such beliefs with evidence. They accept the notion of conditional probability in doing so. In addition to these axioms, decision analysis relies on behavioral decision theory, an area of psychology [*Edwards*, 1954; *Kahneman et al.*, 1974]. For example, prospect theory (winning the Noble Prize in 2002) uses behavioral decision theory as an alternative to the well-known utility theory. Game theory is another important element in decision science [*Von Neumann and Morgenstern*, 1944].

2.4.2. Basic Rules

Decision analysis guides the decision maker to turn opaque situations into a clear set of actions based on beliefs and preferences. Therefore, it is both prescriptive and normative. The latter require invoking a set of rules/axioms. We already encountered one rule of decision analysis in the thumbtack example: maximize expected utility.

Parnell et al. [2013] state five basic rules under which any decision analysis should operate. They are as follows:

1. *Probability rule*. A formulation of subjective degrees of belief. Decision analysis is Bayesian and requires exhaustive and mutually exclusive events.

2. *Order rule*. It refers to the order of preferences for all prospects, such ordering also needs to be transitive: if you prefer X over Y and you prefer Y over Z, then you must prefer X over Z.

3. *Equivalence rule*. It refers to the hypothetical creation of a lottery involving the best and the worst prospects. Suppose there are three prospects X, Y, and Z. X is the worst, and Z is the best. Some probability p exists such that a deal gives you X with probability p, Z with probability $(1 - p)$, and you are receiving Y for sure (CE). This probability p is termed "the preference probability" because it depends on preferences rather than referring to real events.

4. *Substitution rule*. The decision maker should be willing to substitute any prospect with a lottery. In other words, your preference for a prospect will not change if an uncertain deal contained in the prospect is replaced by the CE.

5. *Choice rule*. The decision maker should choose the lottery with the highest probability of winning (i.e., a Bayesian classification). Simply if you prefer prospect X over Y, and if in deal A, $P(X) = 35\%$ and in deal B $P(X) = 20\%$, then you prefer deal A.

2.4.3. Definitions

As with any science, decision analysis operates with definitions and nomenclature. This will help with a clear structuring of the decision problem and with identifying the main "elements" and avoid any ambiguity.

Important to making a decision is to define the *decision context*, that is, the setting in which the decision occurs. Note that the same decision problem may occur in different contexts. The context will identify relevant alternatives and set the objectives. The context will also identify the decision maker, that is, that person whose objectives and preferences are required. In the context, the necessary assumptions and constraints need to be identified as well.

Decision: A conscious, irrevocable allocation of resources to achieve desired objectives. A good decision is, therefore, an action we take that is logically consistent with the objectives stated, the *alternatives* we believe there to be, the knowledge/information, datasets, and the preferences we have. Decision making is not possible if there are no (mutually exclusive) alternatives or choices to be decided on. Alternatives can range from the simple yes/no (e.g., cleanup or not), through the complex and sequential (e.g., oil and gas exploration, field development), to those with extremely large numbers of alternatives. Leaving out realistic alternatives has been identified as a fatal flaw, in hindsight, in important decisions. A decision is only as good as the alternative listed.

Rational decision making requires clear *objectives* that will be used to compare each alternative. An *objective* is defined as a specific goal whose achievement is desired.

A quantitative measure to determine how well each alternative achieves the stated objective is needed. This measure is often termed an *attribute*. A *payoff* or *performance score* is what finally happens with respect to an objective, as measured on its value scale, after all decisions have been made and all outcomes of uncertain events have been resolved. Payoffs may not be known exactly because of uncertainty and need to be predicted.

A value metric is then a quantitative scale that measures the value to the decision makers of the degree to which objectives are achieved. *Value functions* map performance scores to a value metric.

The following are the other common definitions:

1. *Risk preference*. A description of a decision maker's attitude toward risk, whether averse (EM), neutral (DG), or seeking.

2. *Utility metric*. It is a quantitative scale that expresses the decision maker's attitudes toward risk-taking for the value metric.

3. *Utility function*. It maps the utility metric to a value metric in the case of a single-dimensional utility function. Section 2.4.5 will provide an example to illustrate these various definitions.

2.4.4. Objectives

Decision problems may have a single objective (e.g., maximize share-holder value) or multiple, often competing objectives (maximize income, minimize environmental impact). In single objective cases, a common performance score is net present value (NPV). Monetary scales are often preferred as dollar values allow for easier comparison. Another common and easily interpretable scale is "time." Value functions are used to translate any nonmonetary performance score into a monetary value metric (see Section 2.4.5.2 for example). Single objective decisions also allow for risk-attitude in terms of single-dimensional utility functions. These functions need to be assessed by interviewing the decision maker. In risk-neutral cases, the expected value is maximized.

Many problems involve multiple objectives (see, e.g., the Danish groundwater case of Chapter 1). These objectives are organized using a value tree. This tree is generally developed by working from high-level to specific objectives. "Values" are general in nature: for example, values could be "be popular," "support UNICEF," "be healthy," "make money," while objectives are specific and could be of the form "maximize this" or "minimize that." One should distinguish between fundamental objectives that identify the basic reasons why a decision is important and means objectives that are ways of achieving a fundamental objective. Fundamental objectives should be independent and can be organized in a hierarchy. For example, "maximize profit" can be divided into "minimize cost" and "maximize revenue." Means objectives are not the fundamental reason for making a decision; a means objective could be, for example, to "create a clean environment" or to "have welfare programs." Indeed, welfare programs and a clean environment are only a means to population happiness. Figure 2.3 shows such a tree that could be relevant to a local government. Some objectives are fundamental (improve welfare), others are means (improve safety).

The next step is to measure the achievement of an objective. For certain objectives, there will be a natural scale, in either dollars or ppm or rates. For other, more descriptive objectives, a scale needs to be constructed, usually through numerical or other "levels" (high, medium, low). An objective such as "minimize tax" has a natural scale in dollars, while others such as "maximize ecosystem protection" can be measured using the constructed scale:

1 = no protection
2 = minimal monitoring
3 = monitoring/reactive
4 = monitoring/proactive
5 = special status as protected zone

2.4.5. Illustrative Example

2.4.5.1. Overview. Chapter 8 presents a real-life example of the above-mentioned ideas. Here we discuss a simple hypothetical case that will help clarify some of the concepts and definitions.

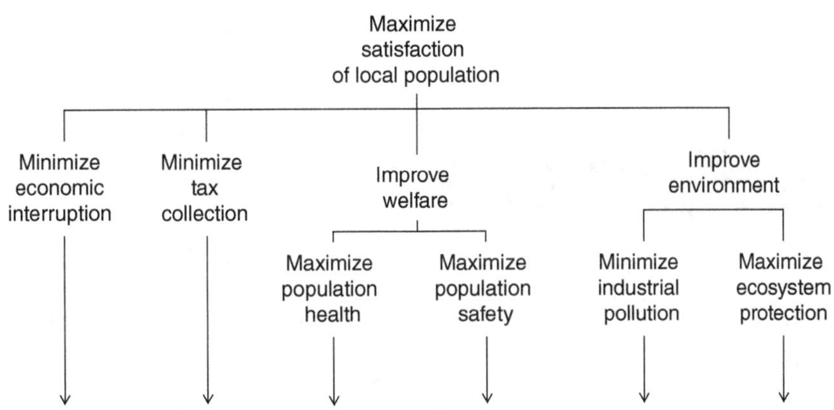

Figure 2.3 Example of a hierarchical tree with objectives.

Consider that in Denmark a leakage of chemicals was discovered in the subsurface close to an aquifer. Government analysts in collaboration with consultants are speculating that due to the geological nature of the subsurface, this pollution may travel to the aquifer and potentially be widely distributed. Models built on gathered data (e.g., SkyTEM) can be used to make any probabilistic forecasts of a property of interest.

The local government has to make a decision, which in this case is whether to act, and hence start a cleanup operation (which is costly for tax-payers) or do nothing, thereby avoiding the cleanup cost but potentially be asked to pay damages to local residents in case widespread contamination occurs. What decision would the local government make? Cleanup or not? How would they reach such a decision? Are there any other alternatives? For example, monitoring at certain locations, cleaning if contamination is detected, or importing "clean" water from another source? Is investing in such monitoring actually useful?

*2.4.5.2. **Performance Score Matrix.*** Recall that a performance score is a metric that quantifies how an objective is met after the decision is made and the outcomes of any uncertain events have been resolved. Therefore, performance scores are not known in advance and must be predicted. This is what most of this book is about. The objectives are listed in the tree of Figure 2.3 and the alternatives are whether to "cleanup" or "not cleanup." Assuming neither alternative will impact safety (cleaning up or not will not affect crime), then Table 2.1 could be an example of hypothetical performance scores matrix (also called payoff matrix) for this case. These are averages (expected values) which would be fine if the decision maker is risk neutral (a Scandinavian government may be more risk averse). In reality, probability density functions are obtained from the modeling study.

Tax collection will be impacted by such cleanup because of its cost (say $10 millions), which will affect the local budget. Ecosystem protection will increase (a constructed scale), while industrial pollution (in ppm) will be small (some pollutants may be left in the ground). In the case of not cleaning up, the tax collection also increases because of the requirement to import "clean" water to meet the needs of the population, assuming the government would have to pay for this (e.g., suppose the contamination was made by a government research lab). This number is more difficult to establish. Indeed, damage payoffs will occur when the geology is unfavorable or if the pollutant is very mobile in the specific environment, causing the pollution to leak into the aquifer. In a payoff matrix, it makes sense to only include objectives that distinguish among alternatives. Any other objectives should be removed, such as population safety in this case. Also, in a payoff matrix, one works across the rows of the payoff matrix rather than down its columns.

Table 2.1 Hypothetical performance score matrix in a binary decision problem.

Objectives	Alternatives: Cleanup	Alternatives: Do not cleanup
Tax collection (million $)	10	18
Industrial pollution (ppm/area)	30	500
Ecosystem protection (1–5)	4	1
Population health (1–5)	5	2
Economic interruption (days)	365	0

The next evident question is how to incorporate preferences into a single attribute scale and combine performance scores measured on different scales. This is addressed by the above-mentioned value functions. Value functions transform attributes to a common scale, say, from 0 to 100. The value function expresses how an increase in the score translates into an increase in value. Therefore, a linear value function (Figure 2.4) states that such an increase is proportional, such as for health, or inversely proportional, such as for pollution. A nonlinear function such as for taxes in Figure 2.4 states that an increase in dollars collected results in a smaller decrease in actual value (high value if less taxes are collected). This means that if tax becomes larger, then any increase in tax will leave the population not necessarily equally more displeased (low value); they are already displeased with such high taxes! For the ecosystem, one could argue for an opposite attitude: more pollution will eventually completely ruin the ecosystem, while a small increase can possibly be tolerable. Such nonlinearity in the function can therefore be interpreted as the attitude toward "risk" one may have about certain outcomes. For example, the attitude toward safety may be different than the attitude toward income. One's preference may be to risk more when money is involved (tax) than with the environment because such effects are often irrevocable (although governmental attitudes around the world may substantially vary in this aspect).

*2.4.5.3. **Swing Weighting.*** Different objectives may carry different weights. This allows the decision maker to inject his/her preference of one objective into another. For example, preference in environmental protection may supersede preference in being displeased with increased taxes. Note that preference is used here to compare

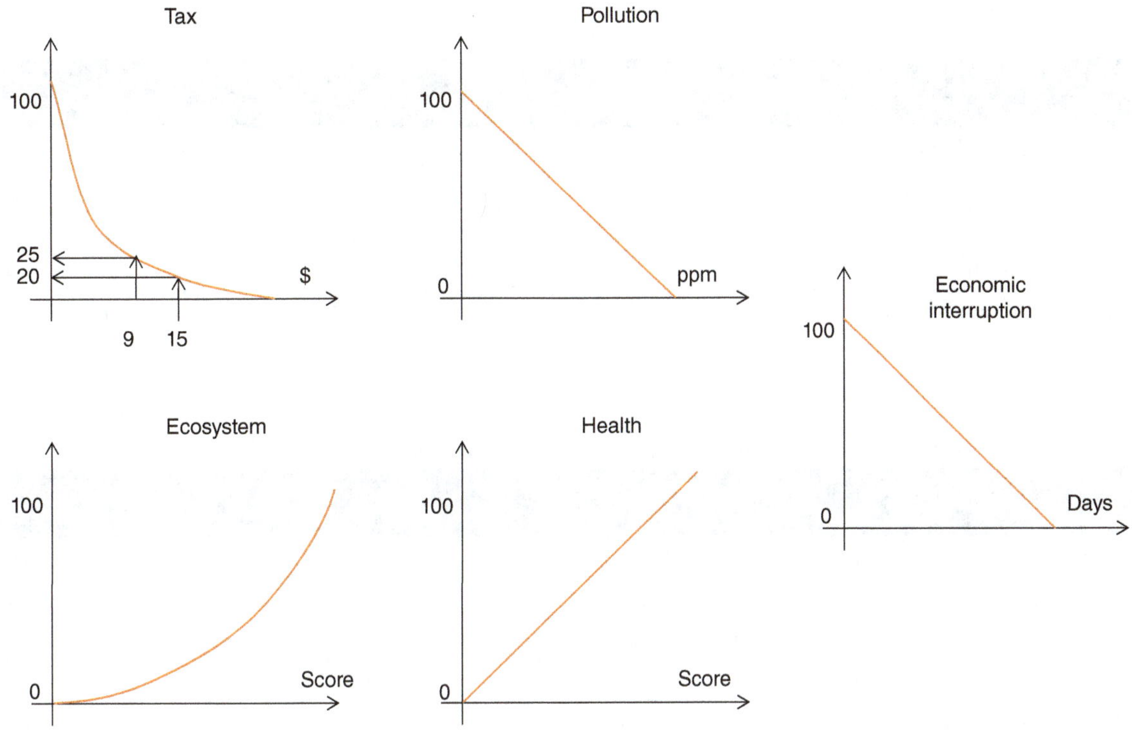

Figure 2.4 Hypothetical value functions, turning a score into a common scale.

various objectives, which is different from the previous sections where preference was used to describe "risk" toward various outcomes within a single objective. One may be tempted to use a simple relative weighting using the following: (i) Rank the various objectives, (ii) assign a number on scale 0–100, and (iii) normalize and standardize the score to unity.

Such an approach does not account for the performance scores on the alternatives. For example, a specific objective may be ranked high but may not have much effect on the various alternatives formulated. A direct weighting method, therefore, does not account for the ultimate purpose, that is, to decide among various alternatives. In practice, the problem can be overcome by using swing weighting, which considers the relative magnitudes of the performance scores. The objectives are first ranked by considering two hypothetical alternatives: one consisting of the worst possible payoff on all objectives (in terms of score, not value) and one consisting of the best possible payoff. The objective whose best score represents the greatest percentage gain over its worst score is given the best rank (i), and the methodology is repeated for the remaining objectives until all are ranked.

Since we are dealing with a binary decision problem, the weighting problem does not present itself (there is always a best and a worst). To illustrate the swing weighting, therefore, consider a slightly modified example where one adds two more alternatives: (i) a detailed cleanup that is costlier but removes more contaminant, therefore, protecting health and environment and (ii) a partial cleanup that leaves some pollutant behind with a decreased risk of drinking water contamination. Table 2.2 shows how swing weighting works. First, the best and worst scores for each objective are taken, then the relative differences are ranked, with 1 being the largest relative difference. Clearly, the tax impact is least discriminating amongst the alternative and therefore gets the smallest weight. Weights are then attributed to each objective following the rank order and then normalized. After weights and attributes are defined, we can combine scores on each objective to determine an overall value for each alternative. This is achieved by calculating the weighted sum of each column in the matrix:

$$v_j = \sum_{i=1}^{N_j} w_i v_{ij} \tag{2.1}$$

where w_i is the weight calculated for each objective and v_{ij} is the score of the j-th alternative for the i-th objective. This is done in Table 2.3 where attributes are now turned into values using some hypothetical value functions (not shown). Therefore, in summary, the cleanup alternative

Table 2.2 Calculating swing ranks.

Objectives	Detailed cleanup	Cleanup	Partial cleanup	Do not cleanup	Best	Worst	Swing rank
Tax collection (million $)	12	10	8	18	8	18	5
Industrial pollution (ppm/area)	25	30	200	500	25	500	2
Ecosystem protection (1–5)	5	4	2	1	5	1	3
Population health (1–5)	5	5	2	2	5	2	4
Economic interruption (days)	500	365	50	0	500	0	1

(Columns: Alternatives)

Table 2.3 Outcome of the hypothetical score matrix with cleanup being the winning alternative.

Objectives	Rank	Weight	Detailed cleanup	Cleanup	Partial cleanup	Do not cleanup	Type
Tax collection (million $)	5	0.07	30	20	100	0	Return/$ benefit
Industrial pollution (ppm/area)	2	0.27	100	99	40	0	Risk/$ cost
Ecosystem protection (1–5)	3	0.20	100	75	25	0	Risk/$ cost
Population health (1–5)	4	0.13	100	100	0	0	Risk/$ cost
Economic interruption (days)	1	0.33	0	33	90	100	Return/$ benefit
		Total	62.1	**67.0**	52.5	33.0	
	Return/$ benefit		2.1	12.3	36.7	33	
	Risk/$ cost		60	54.7	15.8	0	

Note: Calculation of risk versus return.

is the one that is logically consistent with maximizing the value of the decision, for given alternatives, objectives, weights, score predictions, and preferences expressed in value functions.

2.4.5.4. The Efficient Frontier. Conflicting objectives can make decision making hard. In this case the minimization of tax burden is opposite to the cost of maintaining a clean environment. Increasing returns (money) may come at the expense of increasing risks (health, safety, and environment). A term called "the efficient frontier" may help investigate what kind of trade-offs are made and possibly change a decision based on this insight. This is very common in portfolio management (choice of equities, i.e., shares in stocks of companies, and bonds). Portfolio management utilizes historical data on return of equities to form the basis for assessment or risk and return and use the past performance as a proxy for future performance.

To study trade-offs, two categories are created: one for the risks and one for the returns (or cost/benefit). Overall weighted scores are then calculated for each subset, in a similar fashion as described earlier, as shown in Table 2.3. Risk/return is plotted versus cost/benefit in Figure 2.5. From this plot, we can eliminate some obvious alternatives as follows. The alternative "do not cleanup" is clearly dominated by the alternative "partial cleanup." Indeed, "partial cleanup" has both more return and less risk. Therefore, the alternative "do not cleanup" can be eliminated because it results in taking on more risk relative to the return. "Do not cleanup" is the only alternative that can be eliminated as such; other alternatives involve a trade-off between risk and return. The curve connecting these points is the efficient frontier. The efficient frontier can be seen as the best set of trade-offs between risk and return for the current alternatives. Recall that a decision can only be as good as the alternatives formulated. Therefore, pushing the efficient frontier upward (i.e., up and toward the right in Figure 2.5) would require different alternatives, leading to a better set of trade-offs. Such alternatives are only as good as the imagination of those creating them.

An efficient frontier allows asking question such as "Am I willing to trade-off more risk for more return between any two alternatives?" For example, is the decrease of about five units of risk, worth the decrease

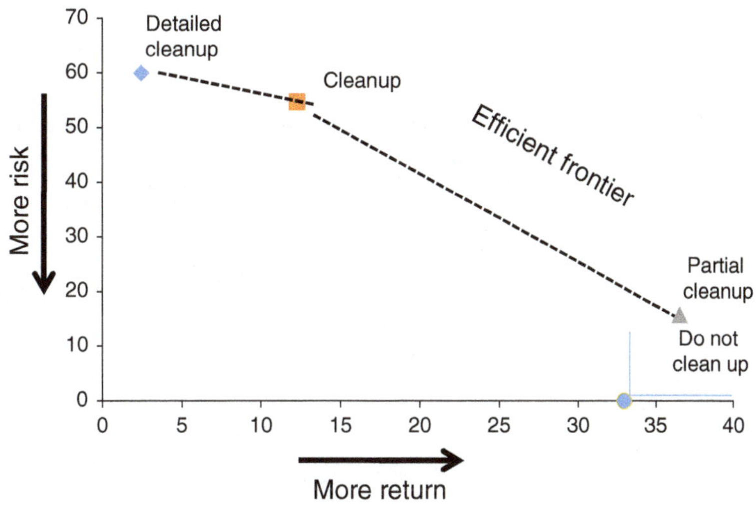

Figure 2.5 Efficient frontier.

in about ten units of return when going from detailed cleanup to "cleanup"? If all attributes were in dollar values than these would be actual dollar trade-offs, in our case these are only indicative trade-offs, basically forming a scale from "less preferred" to "more preferred" in terms of trade-off.

2.5. GRAPHICAL TOOLS

2.5.1. Decision Trees

Decision trees (Figure 2.6) are graphical models that organize logically, in time, the various decisions, alternatives, uncertainties, and payoffs (value for each prospect). The time component is critical here. First, we state the alternatives, then we face uncertainties (not the other way around); these uncertainties are then resolved sometime in the future, resulting in a payoff. This also means that the root is a decision. Any costs or uncertainties prior to the decision are irrelevant. The leaves in the decision tree represent the various scenarios that can occur.

To solve a decision tree, meaning find the best alternatives, we go the opposite way: we start at the leaves and resolve uncertainty nodes by taking consecutively expected values. If the decision maker is not risk neutral, then the solution involves utilities. At any decision nodes, we then take the alternative that is maximal (maximum expected utility). Figure 2.6 shows a hypothetical example. Some of the probabilities in this tree are prior probabilities, other may be the result of modeling (conditional probabilities).

A limitation of decision trees is that they become intractable for decisions with either a large set of alternatives or a more continuous type of uncertainty, rather than discrete outcomes such as in Figure 2.6.

2.5.2. Influence Diagrams

An influence diagram captures a decision situation by depicting relationships between decisions, uncertainties, and preferences [*Shachter*, 1986; *Eidsvik et al.*, 2015]. Consider as illustration the hypothetical example in Figure 2.7. A site is contaminated, which potentially poses a health risk. The decision is to clean (or not, or how to clean) and also the decision is to hire a consultant (or not). Depending on a report (negative/positive) certain actions will be taken. The outcome of the report depends on the unknown distribution of the subsurface plume. The costs will depend on how uncertainties are resolved. One distinguishes three kinds of nodes (uncertain nodes, decision nodes, and value nodes) and three kinds of arcs (conditional, information, and functional). Notice how there is no arc between "clean" and "contamination." The decision to clean does not affect the amount of contamination present before cleaning. Each uncertain node is associated with a probability table. For example, the contamination node can be associated with a table of "low," "medium," and "high" contamination and associated probabilities (obtained through measurements and models). A value node is associated with a value table indicating the payoffs of various scenarios. In that sense, calculations can be done with influence diagrams in the same way as with decision trees. Because the time component is not explicitly represented, it is easier to make mistakes with such diagrams, in particular when they become complex (e.g., Figure 2.8). For example, information should be

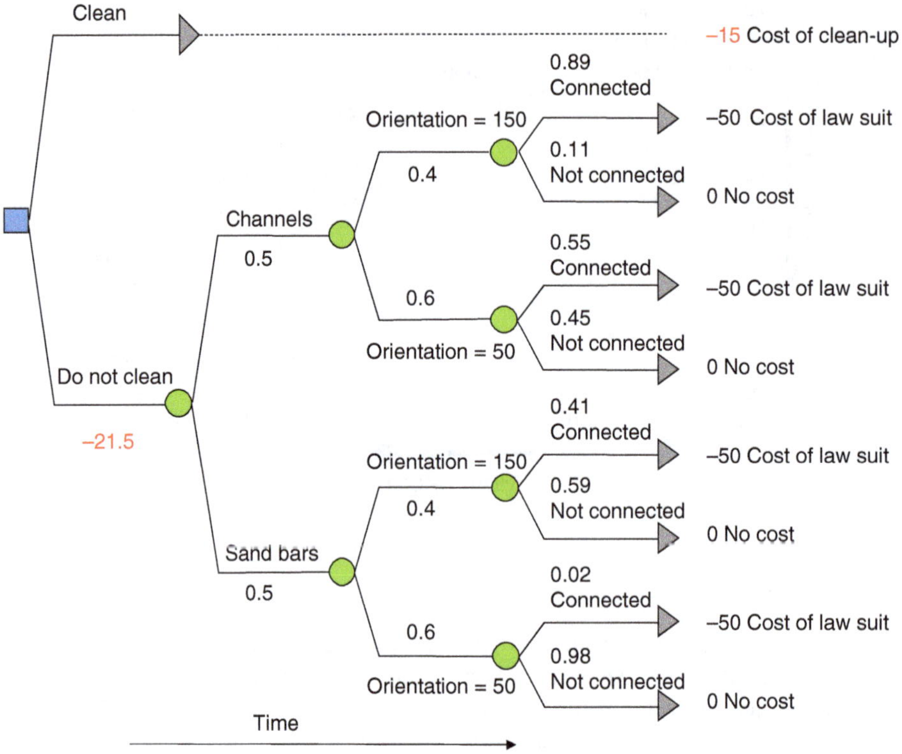

Figure 2.6 A hypothetical decision tree. The first node has to be a decision node. Uncertainties here are the type of geological system (channel vs. bar), the orientation of geological bodies and the degree of connectivity between them as measured by a probability. The latter is calculated from actual models. The best alternative is "clean." Adapted from *Caers* [2011].

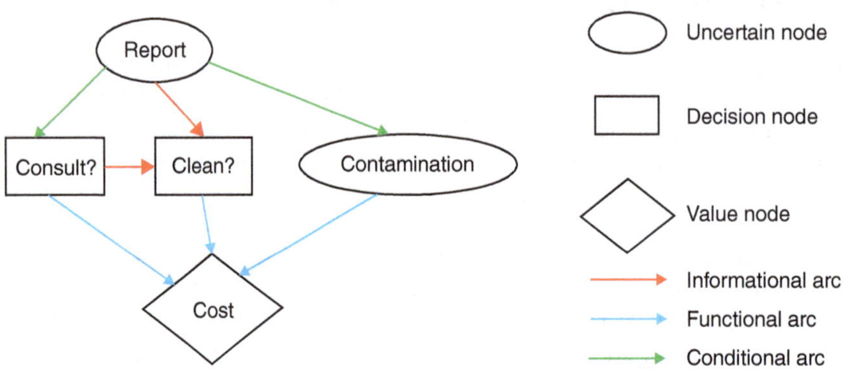

Figure 2.7 Example of a hypothetical influence diagram.

known to all decisions that are posterior to the information acquisition. It is not rational to forget information.

2.6. VALUE OF INFORMATION

2.6.1. Introduction

In many practical situations, we are faced with the following question: Given a decision situation between several alternatives, do we choose or do we decide first to gather information to help improve the decision? In the subsurface, this may mean many things: conducting a geophysical survey, drilling wells, doing more core experiments, doing a more detailed modeling study, hiring experts or consultant, and so on. The main driver is to reduce uncertainty on key decision variables. However, data acquisition may be costly. Questions that arise include the following:

1. Is the expected uncertainty reduction worth its cost?
2. If there are several potential sources of information, which one is the most valuable?

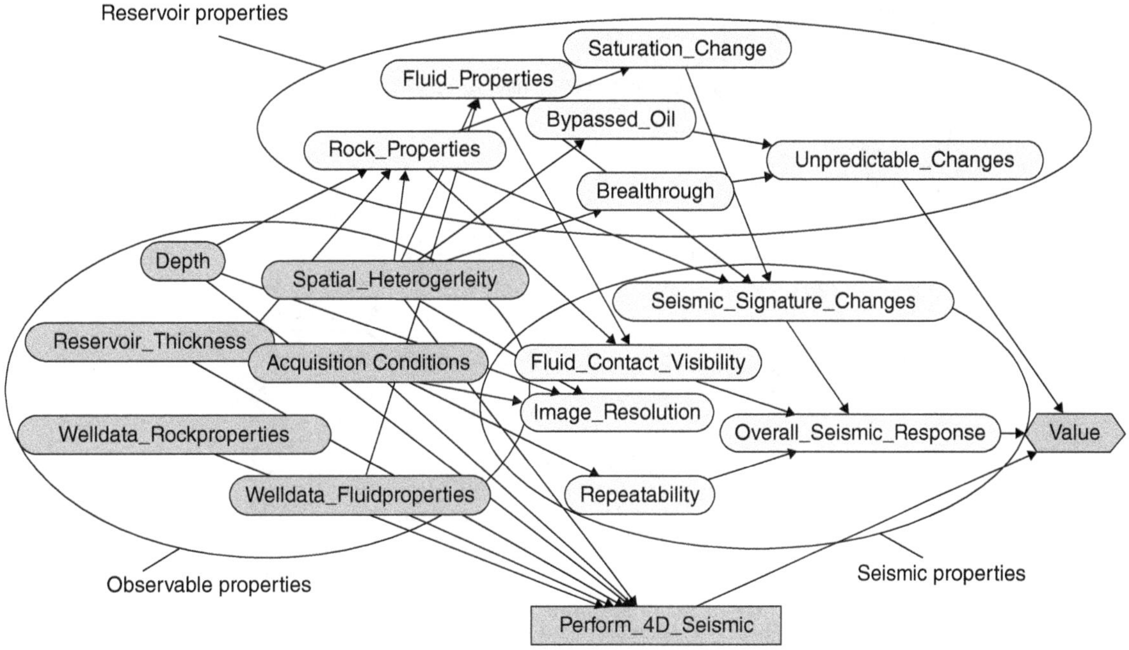

Figure 2.8 Example of a complex influence diagram for assessing the value of 4D seismic data. From *Eidsvik et al.* [2015].

3. Which sequence of information sources is optimal?

These types of questions are framed under "the value of information problem." These questions are not trivial to answer because we need to assess the value *before* any measurement is taken. You cannot first take the measurement and then decide whether it is valuable, because then you have already decided to invest in some sunk cost.

Decision analysis and VOI has been widely applied to decisions involving engineering designs and tests, such as assessing the risk of failure for buildings in earthquakes, components of the space shuttle, and offshore oil platforms. In those fields, gathering information consists in doing more "tests" and if those tests are useful, that is they reveal design flaws (or lack thereof), then such information may be valuable depending on the decision goal. This invokes some measure of "usefulness" of the test. Indeed, if the test conducted does not inform the decision variable of interest, then there is no point in conducting it. The "degree of usefulness" is termed the "reliability" of the test in the traditional value of information literature. In engineering sciences, the statistics on the accuracy of the tests or information sources that attempt to predict the performance of these designs or components are available, as they are typically made repeatedly in controlled environments such as a laboratory or testing facility. These statistics are required to complete a VOI calculation as they provide a probabilistic relationship between the information message (the data) and the state variables of the decision (the specifications of the engineering design or component).

Many challenges exist in applying this framework to spatial decisions pertaining to an unknown subsurface. *Eidsvik et al.* [2015] provide a thorough treatment on the topic including several case studies. For application to petroleum system in particular, see *Bratvold et al.* [2013]. Here we provide a short overview of the main elements.

2.6.2. Calculations

The aim of collecting more data is to reduce uncertainty on those parameters that are influential to the decision-making process. In the thumbtack example, we discussed that VOI should depend on three components:

1. The prior uncertainty of what one is trying to model. The more uncertain one is about some subsurface component the more the data can possibly contribute to resolving that uncertainty.

2. The information content of the data (this will be translated into data reliability or vice versa). If the data is uninformative, it will have no value. But even perfect data (data that resolves all uncertainty) may not help if that does not influence the decision question.

3. The decision problem. This drives the value assessment on which any VOI calculation is based.

The simplest way to illustrate VOI calculation is by means of a decision tree. Consider again, our simple illustrative "contamination" case. The top part of the tree in Figure 2.9 is the basic decision problem. It is binary and has one binary uncertainty: contamination is low (a_1)

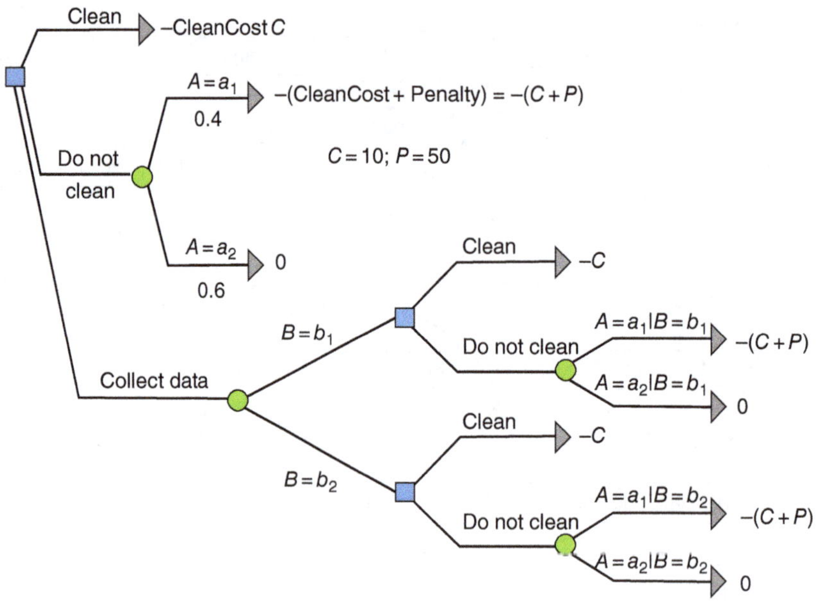

Figure 2.9 Example of a decision tree that involves collecting data as an alternative.

versus contamination is high (a_2). In VOI assessment, we consider acquiring data as an additional alternative. However, now we also face an additional uncertainty: What is the outcome of this data acquisition? Here we consider a binary outcome: positive versus negative. Positive means *indicating* "high contamination" (b_2) and negative means indicating "low contamination" (b_1). The term "indicating" is important here: we do not know for sure; the data may not necessarily be clairvoyant and reveal the truth. Next we notice in the tree that the basic decision problem is literally repeated after each possible data outcome. This is important. If this is not done properly then VOI may be negative, which makes no sense because you can always decide not to gather information. Instead, what has changed is the probability associated to the branches involving our uncertainty. We now have a conditional probability instead of a prior probability. The conditional probability $P(A_i = a_i| B_j = b_j)$ is termed the information content and is of the general form P(real world is| data says). In traditional VOI calculations, and we refer here to the original engineering test, the following probability is usually specified P(data says| real world is). This originated from the idea of doing tests under various "real world" conditions. The relationship between information content and reliability is simply Bayes' rule.

Let us assume perfect information, meaning that

$$P(A_1 = a_1|B_1 = b_1) = 1; P(A_1 = a_1|B_2 = b_2) = 0;$$
$$P(A_2 = a_2|B_1 = b_1) = 0; P(A_2 = a_2|B_2 = b_2) = 1 \quad (2.2)$$

The data is clairvoyant, that is, it will tell if we have low or high amount of contaminant. Considering the numbers in Figure 2.9, we find that for the basic decision problem "clean" has value −10 and "not clean" has value −7. If we have perfect information, then the "collect data" branch has value −4. Hence, the VOPI is −4 − (−10) = 6 in other works (assuming number in $K), we would never pay more than $6000 for any information. Plugging in any other values will result in the VOI. These reliabilities require modeling studies. It would require forward modeling of the data on some reference cases (or using Monte Carlo) and observing how well the data resolves the truth.

REFERENCES

Bratvold, R. B., and S. H. Begg (2010), *Making Good Decisions*, Society of Petroleum Engineers, Richardson, TX.

Bratvold, R. B., J. E. Bickel, and H. P. Lohne (2013), Value of information in the oil and gas industry: Past, present, and future, *SPE Reserv. Eval. Eng.*, *12*(4), 630–638, doi:10.2118/110378-PA.

Caers, J. (2011), *Modeling Uncertainty in the Earth Sciences*. Wiley, Hoboken, NJ.

Edwards, W. (1954), The theory of decision making, *Psychol. Bull.*, *51*(4), 380–417, doi:10.1037/h0053870.

Eidsvik, J., T. Mukerji, and D. Bhattacharyya (2015), *Value of Information in the Earth Sciences: Integrating Spatial Modeling and Decision Analysis*, Cambridge University Press, Cambridge.

Heuer, R. J. (1999), Improving intelligence analysis, in *Psychology of Intelligence*, edited by J. Davis, CIA, Washington, D.C., pp. 173.

Howard, R. A. (1966), Information value theory, *IEEE Trans. Syst. Sci. Cybern.*, *2*(1), 22–26, doi:10.1109/TSSC.1966.300074.

Howard, R. A. (1968), The foundations of decision analysis revisited, *IEEE Trans. Syst. Sci. Cybern.*, *4*(3), 211–219, doi:10.1017/CBO9780511611308.004.

Howard, R. A., and A. E. Abbas (2015). *Foundation of Decision Analysis*, Pearson, London.

Kahneman, D., P. Slovic, and A. Tversky (1974), Judgment under uncertainty: Heuristics and biases, *Science*, *185*(4157), 1124–1131, doi:10.1126/science.185.4157.1124.

Keeney, R., and H. Raiffa (1993), Decisions with multiple objectives–preferences and value tradeoffs, *Behav. Sci.*, *39*, doi:10.1002/bs.3830390206.

Parnell, G. S., T. A. Bresnick, S. N. Tani, and E. R. Johnson (2013). *Handbook of Decision Analysis*, Wiley, Hoboken, NJ, doi:10.1002/9781118515853.

Popper, K. R. (1959), The logic of scientific discovery. *Phys. Today*, *12*(11), 53, doi:10.1063/1.3060577.

Raiffa, H. (1968), Decision analysis: Introductory lectures on choices under uncertainty, *MD Comput.*, *10*(5), 312–328.

Shachter, R. D. (1986), Evaluating influence diagrams, *Oper. Res.*, *34*(6), 871–882, doi:10.1287/opre.34.6.871.

Von Neumann, J., and O. Morgenstern (1944), *Theory of Games and Economic Behavior*, Princeton University Press, pp. 625, doi:10.1177/1468795X06065810.

3

Data Science for Uncertainty Quantification

3.1. INTRODUCTORY EXAMPLE

3.1.1. Description

This chapter provides an overview of the relevant mathematical, statistical, and computer science components needed to develop and understand the various approaches to uncertainty quantification in subsequent chapters. The aim here is to place these various components within the unified context of uncertainty quantification, and as application to geoscientific fields and not just to provide a mere overview, which can be found in many excellent existing books. Instead, we provide some additional insight into how various, seemingly independent applied mathematical concepts share important common traits within the context of uncertainty quantification. The field of data science (statistics, machine learning, and computer vision) is growing fast. Most developments are driven by a need to quantify human behavior or interaction (e.g., Facebook). Here we deal with the physical world. Data scientific approaches will need to be tuned to the kind of challenges we face in this world, such as data sparsity, complex relationship between physical variables, high dimension, non-Gaussianity, nonlinearity, and so on.

As an aid to this overview, we develop a simple illustrative example. This example is not a real example by any stretch of the imagination but has elements common to uncertainty quantification as outlined by the real field studies in the previous chapter. This will also aid in developing notation in this book by providing common notation in the various applied mathematical and computer science disciplines involved. The reader should also refer to the notation in the next two sections.

The case here concerns the simple design of a water purification system by pumping and infiltrating river water into an aquifer by means of an infiltration basin (see Figure 3.1). Then, usable drinking water is retrieved from a pumping well. The aims of this artificial recharge is

1. to improve groundwater quality through filtration and bioactivity in the soil
2. to create a hydraulic barrier to divert any contaminated groundwater from an known industrial polluting area

However, the pumping wells needs to be shut off from time to time, either for saving energy or for maintenance. Too long of a shut-off period may result in pollutants infiltrating the filtration zone. Hence, knowing whether such contamination will take place and knowing when this will take place will help in designing shut-off periods. To aid in this design, head measurements from several wells in the area are available, as well as reports that the subsurface is substantially heterogeneous because of the depositional system induced by the nearby river (fluvial system).

3.1.2. Our Notation Convention

One of the challenges in scientific writing is to come up with a strategy for representing common objects, vectors, scalars, function, random functions, and so on. Many publications and books use different conventions and notations. Our material comes from different worlds with different notations, so we do not want to leave the reader guessing what "x," or "i," or "n" is; hence, we are quite explicit on notation in the hope this also unifies many concepts. Here is an overview what you need to know:

1. a, b, x: small italic is a scalar, an outcome, a sample
2. A, B, X: capital italic is either a matrix or a random variable. Those different contexts are usually clear.
3. $\mathbf{X}$: capital bold is a random vector (a vector of random variables)
4. $\mathbf{x}$: small bold is the outcome of a random vector or simply a vector
5. L: the number of samples in a statistical study
6. N: the dimension of the problem in general, for example, the dimension $\mathbf{X}$

Quantifying Uncertainty in Subsurface Systems, Geophysical Monograph 236,
First Edition. Céline Scheidt, Lewis Li, and Jef Caers.

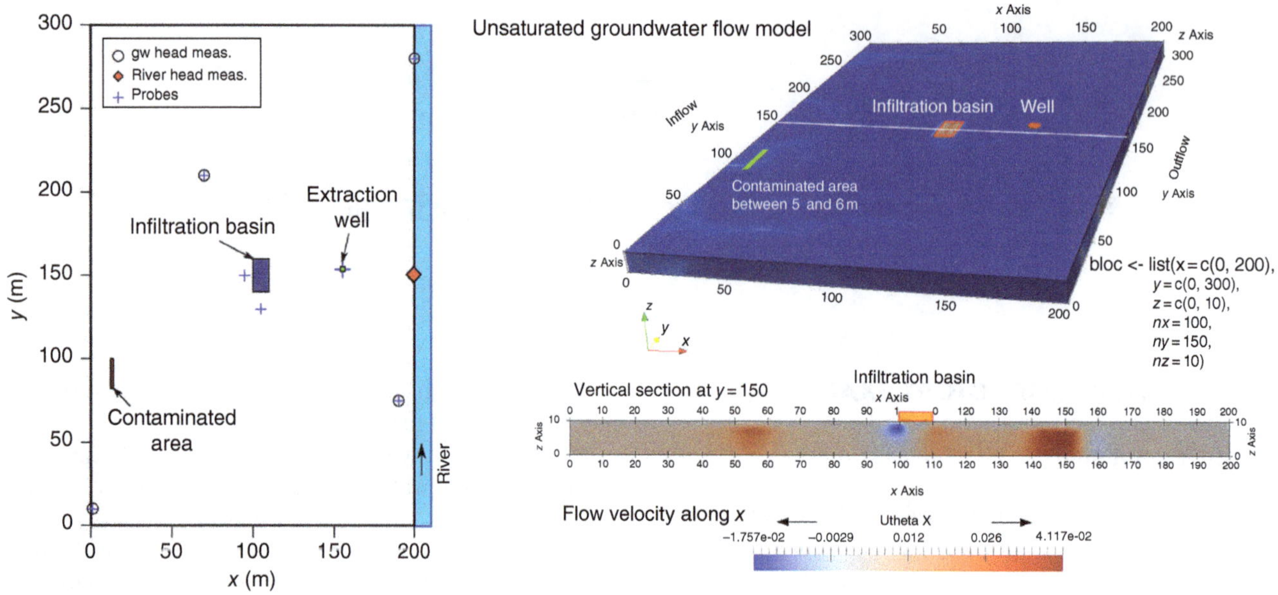

Figure 3.1 Setup of our simple illustration case.

7. *Counters*: we use small font and corresponding capital letter, for example, $n = 1, \ldots, N$ or $\ell = 1, \ldots, L$. We avoid using mixing letters such as $i = 1, \ldots, N$
8. f: probability density function (pdf) or just a function
9. F: cumulative distribution function (cdf)

3.1.3. Variables

Since we are interested in the future evolution of a system, we need to model and simulate the above-presented situation with computer models. Basically, any UQ exercise has a number of components as synthetized in Section 1.7. For any UQ problem, it is therefore critically important to clearly and rigorously define the data variables, the model variables, and the prediction variables. A common confusion is to think that the data variables are the observed data. They are not. In a typical probabilistic way of reasoning, the actual field data are seen as one particular realization, sample, or instantiation (whatever choice of nomenclature one is used to) of these data variables. Data variables are random variables whose outcomes are not known. How do we define such data variables? Data is caused by some action (a sampling, a measurement, a study, a drilling, a logging, etc.) on the subsurface system in question. Hence, in order to define data variables, we need to first define the variables that describe the (unknown) subsurface system. These are the model variables. These model variables are, therefore, the parameterization that the modeler believes to allow for a proper representation of all aspects of the subsurface system, whether these are fluxes, pressures, concentrations, chemical reactions kinetics, and so on.

In the hydro case mentioned earlier, we will represent the area in question with a number of grid cells. Many simulators, for example of multiphase flow, of reactive transport, and of geostatistical algorithms, require a grid and a definition of the size of these cells. These cells may be on a regular grid or on any grid, depending on how accurately models need to be simulated. Here we assume the grid is regular and has a certain cell size. To model this system, we will need to specify spatially distributed values such as porosity and hydraulic conductivity. A subset of the model variables are the relevant properties for each cell value. However, property values in different grid cells are not independent (statistically). The geological depositional system has induced spatial correlation and such spatial correlation is often modeled using geostatistical models or algorithms. As such, the gridded model variables depend on the definition of some statistical model, which may have its own parameter/model variables, for example the mean porosity or in the case of a correlated spatial porosity, the spatial covariance or variogram model parameters. In addition, the above model requires defining initial conditions and boundary conditions, both of which may be uncertain and also modeled using probability distributions with their own set of parameters.

In this book, because of the specific spatial (or spatio-temporal) nature of subsurface models, we split the model variables in two groups: (i) those model variables that comprise the spatial distribution of properties on a grid, for example concentration, porosity, and permeability at each grid location and (ii) the parameters that were used to generate these spatial model variables, or any other model variables or parameters that are not defined on the grid. For the model parameterization, as a whole, we use the notation **m** comprising of (i) gridded model variables and (ii) non-gridded model variables.

$$\mathbf{m} = (\mathbf{m}_{\text{grid}}, \mathbf{p}) \quad (3.1)$$

Another part of the model that will be dealt with separately is the physical/chemical/biological process that is modeled using ODEs or PDEs or whatever other mathematical representation one deems appropriate. Such equations can be seen as "theories" (in the mold of *Tarantola* [1987]) that provide information on the relationship between several aspects of the subsurface system, for example stating the theoretical link between data and model variables. In this book, we will limit ourselves to expressing these relationships using forward models. A forward model is represented by an explicit function (derived from an implicit physical relationship). A first forward model is between the data variables and the model variables

$$f_d(\mathbf{m}, \mathbf{d}) = 0 \Rightarrow \mathbf{d} = g_d(\mathbf{m}) = g_d(\mathbf{m}_{\text{grid}}, \mathbf{p}) \quad (3.2)$$

For example, f_d could be a set of partial differential equations and g_d the numerical implementation that outputs the desired data variables. Therefore, g_d represents the data forward model, a simulation model that takes as input the model variables and outputs the data variables. For the hydro case in question, we can now specify both model variables and data variables. As model variables, we have the following (summarized in Table 3.1):

1. *Hydraulic conductivity* (a gridded property). The parameters used to generate a gridded hydraulic conductivity model are uncertain. Here we assume that hydraulic conductivity can be modeled using a Gaussian process (see Section 3.7.5). Such process requires specifying the mean, standard deviation, and a set of spatial covariance parameters such as range, nugget, and anisotropy, as well as type of covariance model.

2. *Boundary conditions.* Boundary conditions are simulated by means of Gaussian process regression. The Gaussian process regression is here defined by a prior mean and Matérn covariance function. This covariance is not known, and its uncertainty is parametrized as uncertainty in the variance, range, and smoothness. The prior mean function is specified by three monomial basis functions with unknown parameters with a vague prior. The Gaussian process is conditioned to four groundwater

Table 3.1 Overview of the various parameters and their uncertainty for hydro case.

	Parameter code	Description	Variable type	Distribution
Hydraulic conductivity representation	K_{mean}	Mean value of hydraulic conductivity *K* (m/s)	Continuous	U(7e−4, 10^{-3})
	K_{sd}	Standard deviation of log(*K*) (m/s)	Continuous	U(0.05, 0.3)
	K_{Cov}	Type of covariance model for simulation of *K*	Discrete	Gaussian or spherical
	K_{angle}	Horizontal anisotropy angle for *K* (degree)	Continuous	U(110, 150)
	K_{range}	Correlation length along the principle direction for *K* (m)	Continuous	U(10, 100)
	K_{anixy_ratio}	Anisotropy, horizontal stretching ratio	Continuous	U(l/20, 1/2)
	K_{aniz_ratio}	Anisotropy, vertical stretching ratio	Continuous	U(15, 30)
	K_{nugget}	Nugget for *K* (m/s)	Continuous	U(0, 0.1)
Boundary conditions representation	H_{sd}	STD of the Matern covariance model for simulation of boundary conditions	Continuous	U(0.01, 0.1)
	H_{range}	Correlation length of the Matern covariance model for simulation of boundary condition	Continuous	U(20, 40)
	H_{nu}	Smoothness of the Matern covariance model for simulation of boundary conditions	Continuous	U(1.5, 3.5)
	$H_{rivGrad}$	Gradient of the river	Continuous	N(−0.0015, 0.0001)
Measurement error	H_{rivRef}	River hydraulic head (meters)	Continuous	N(7, 0.05)
	H_{nugget}	Measurement error groundwater hydraulic heads (meters)	Continuous	U(0.02, 0.1)

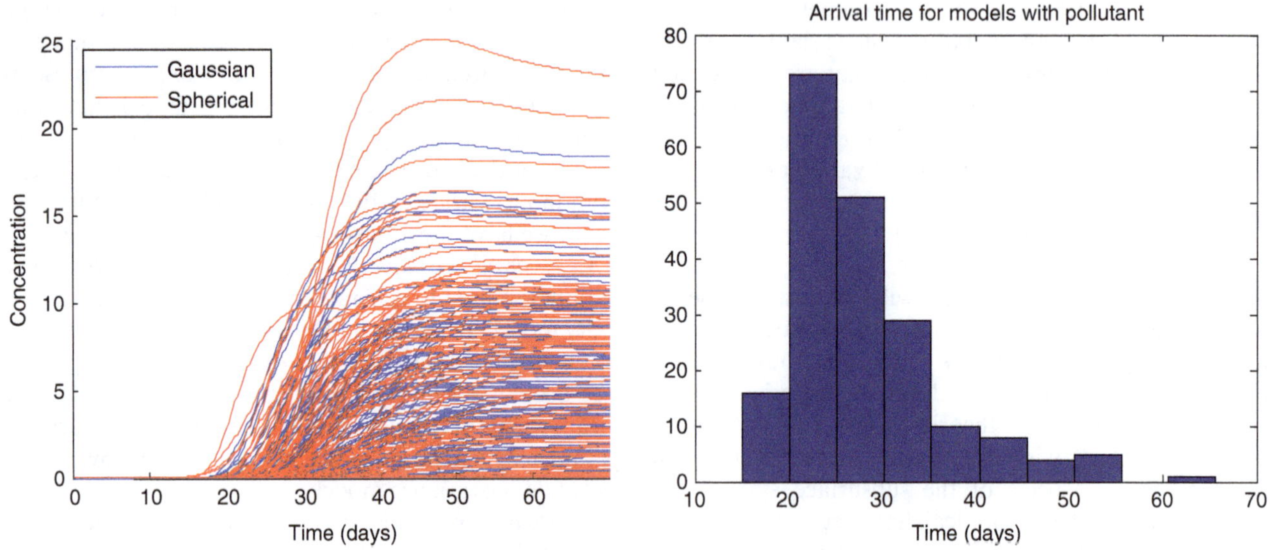

Figure 3.2 Concentration of DNAPL at the drinking well location with histogram of arrival times.

hydraulic head measurements as well as to the river heads (with measurement error, see the following text). The river head gradient is uncertain as well.

3. *Measurement error*. The hydraulic head measurements are subject to noise, modeled as a noise variance (entering as a nugget effect when estimating the boundary conditions). The hydraulic head of the river is uncertain as well.

Essentially, the model is infinite dimensional, unless one discretizes space and time. Because both space (grid cells) and time (time-steps) are usually discretized, the model is very high dimensional, namely assuming all model variables are time-varying

$$\dim(\mathbf{m}) = N_{\text{timesteps}}\left(N_{\text{gridcells}} \times N_{\text{properties}} + \dim(\mathbf{p})\right) \quad (3.3)$$

Not all model variables may be time-varying, but regardless of this fact, dim(**m**) can be extremely large for real-world applications (10^6–10^9).

We now return to the data variables. The data variables are obtained by generating one model realization (e.g., using Monte Carlo, see Section 3.10) and applying the forward model representing the physical relationship between the model and the data variable. Here the data consists of hydraulic head data at four locations. The observed data are denoted as $\mathbf{d}_{\text{obs}}$: the head measurements in four wells.

The design of the system will be based on the evolution of the contaminant in the future as the recharge is terminated. Therefore, key prediction variables, generically denoted as **h**, are as follows:

1. The future concentration (over time) of DNAPL in the drinking well
2. The DNAPL arrival time in the critical zone

These can be forward modeled as well (see Figure 3.2) based on models:

$$f_h(\mathbf{m},\mathbf{h}) = 0 \Rightarrow \mathbf{h} = g_h(\mathbf{m}) = g_h\left(\mathbf{m}_{\text{grid}},\mathbf{p}\right) \quad (3.4)$$

3.2. BASIC ALGEBRA

3.2.1. Matrix Algebra Notation

Some basic algebra notations are reviewed in this section, as well as eigenvalues and eigenvectors, which lie at the foundation of multivariate (high dimension) problems in both mathematics and statistics. Ultimately, regardless of the technique or method developed, all operations can be summarized as matrix operations and most of them rely on some form of orthogonalization. Specific to multivariate modeling, an important matrix is the "data matrix," which consists of multiple observations of a random vector of a certain size. "Data" should not be limited to any field data or actual observation or to data variables; such matrix may contain model realizations. For example, one may generate a set of L model realizations with a total amount of model variables N. Throughout the book, unless otherwise stated, we will use N to represent the dimension of "data," whether models, samples, realizations, and observations, while L represents the amount of "data." For example, consider a model realization,

$$\mathbf{m} = (m_1, m_2, \ldots, m_N) \quad (3.5)$$

Notice the notation as follows: bold for vectors and italic for scalar variables. Now also consider that L model realizations have been generated, in our notation:

$$\mathbf{m}^{(\ell)} = \left(m_1^{(\ell)}, m_2^{(\ell)}, \ldots, m_N^{(\ell)}\right),\ \ell = 1, \ldots, L \quad (3.6)$$

Then, a matrix of model realizations becomes

$$\begin{pmatrix} m_1^{(1)} & m_2^{(1)} & \dots & \dots & \dots & m_N^{(1)} \\ \vdots & m_2^{(2)} & & & & \vdots \\ \vdots & & \ddots & & & \vdots \\ \vdots & & & \ddots & & \vdots \\ \vdots & & & & \ddots & \vdots \\ m_1^{(L)} & m_2^{(L)} & \dots & \dots & \dots & m_N^{(L)} \end{pmatrix} \tag{3.7}$$

Consider now applying a forward model, for example the data forward model on each model, to generate the data variable outcomes (such as hydraulic heads in our simple case):

$$\mathbf{d}^{(\ell)} = g_d\left(\mathbf{m}^{(\ell)}\right) = g_d\left(m_1^{(\ell)}, m_2^{(\ell)}, \dots, m_N^{(\ell)}\right),\ \ell = 1, \dots, L$$

Then, another matrix can be generated as

$$\begin{pmatrix} d_1^{(1)} & d_2^{(1)} & \dots & \dots & \dots & d_{N_d}^{(1)} \\ \vdots & d_2^{(2)} & & & & \vdots \\ \vdots & & \ddots & & & \vdots \\ \vdots & & & \ddots & & \vdots \\ \vdots & & & & \ddots & \vdots \\ d_1^{(L)} & d_2^{(L)} & \dots & \dots & \dots & d_{N_d}^{(L)} \end{pmatrix} \tag{3.8}$$

with N_d the dimension of the data variables, to distinguish it from the dimension of the model variables.

The following notation and conventions are adapted for matrix algebra, including some specific type of matrices:

1. $X = (x_{ij})$ an $N \times L$ matrix consisting of scalars x_{ij}
2. A row vector $\mathbf{x}^T$ with $\mathbf{x}$ a column vector
3. $\mathbf{1}_N = (1, \dots, 1)^T$ a vector of ones with length N
4. $\mathbf{0}_N = (0, \dots, 0)^T$ a vector of zeros with length N
5. a diagonal matrix diag(x_{ii}), $x_{ij} \forall i \neq j$
6. an identity matrix diag(1, …, 1) = I_N
7. a unity matrix $\mathbf{1}_N \mathbf{1}_N^T$
8. Trace: tr(X)
9. Determinant det(X)

3.2.2. Eigenvalues and Eigenvectors

When multiplying a matrix of size $N \times L$ with a vector of size $L \times 1$ (or simply L), a vector of size $N \times 1$ is obtained. In one specific interpretation, when dealing with Cartesian axis systems (more about spaces and geometries later), such operation can be seen as mapping or projection of a vector in a Cartesian axis system of dimension N into a new Cartesian axis system with dimension L. When $L < N$ then this amounts to a dimension reduction, otherwise a dimension increase.

Consider now the special case where such matrix is a square matrix. For example, a linear relationship between model and data exists (or be assumed):

$$G\mathbf{m} = \mathbf{d} \tag{3.9}$$

Hence, the forward model operator, as a linear model, is expressed in matrix G. We consider for now that the dimension of data and model are the same:

$$N = N_m = N_d \tag{3.10}$$

Now we wish to analyze the properties of G. This is relevant, since the specifics of G will determine how L model realizations are mapped into L data realizations (see Figure 3.3). Since G is a linear operator we do not expect such mapping to change the topology of the space. Indeed, we expect that the model point cloud in Figure 3.3 is stretched/squeezed and rotated, perhaps

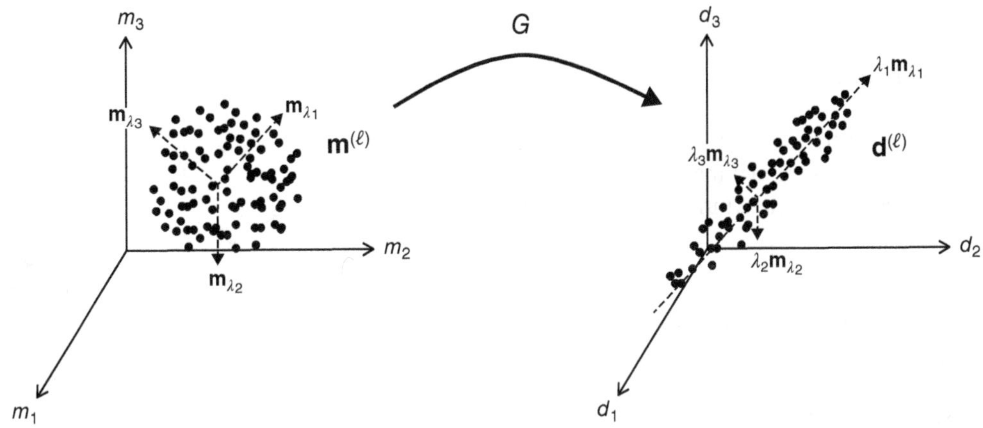

Figure 3.3 A matrix multiplication representing a mapping from one Cartesian axis space to another Cartesian axis space. Eigenvalues and eigenvectors characterize the nature of such transformation.

translated into a new point cloud (the data point cloud). It is, therefore, interesting to study which model realizations (model vectors) are modified/transformed only up to a scalar. These special model realizations are termed eigenvectors of the transformation G.

Formally, consider a square matrix G, if a scalar λ and vector $\mathbf{m}_\lambda$ exists such that

$$G\mathbf{m}_\lambda = \lambda \mathbf{m}_\lambda \tag{3.11}$$

then λ is termed an eigenvalue and $\mathbf{m}_\lambda$ an eigenvector. In this case, eigenvalues can be calculated as roots of and Nth order polynomial $\det(G - \lambda I_N)$; hence, up to N eigenvalues exist $\lambda_1, \ldots, \lambda_N$, for each eigenvalue a corresponding eigenvector exists $\mathbf{m}_{\lambda_1}, \ldots, \mathbf{m}_{\lambda_N}$. The determinant and trace can be written as function of eigenvalues:

$$\det(G) = \prod_n^N \lambda_n \tag{3.12}$$

$$\mathrm{tr}(G) = \sum_n^N \lambda_n \tag{3.13}$$

An intuitive explanation by means of geometric description for trace and determinant is as follows. The determinant of a matrix represents the signed volume change of a unit cube into a parallelepiped after projection with G. The sign indicates how the volume is rotated (clockwise or counterclockwise). The sign and volume change depends on the sign of the eigenvalues as well as their magnitudes, where increases in volume occur when λ are larger than unity in absolute value and decreases for absolute values smaller than unity.

The trace has a geometric interpretation of a change of that volume after projection under an infinitesimal change before projection; hence, the trace determines "how big" the projection is. Both determinant and trace have important application in UQ such as in the analysis of covariance matrices, least-squares methods, or in dimension reduction methods.

3.2.3. Spectral Decomposition

3.2.3.1. Theory. The spectral decomposition or Jordan decomposition links the structure of a matrix to the eigenvalues and the eigenvectors. Each symmetric matrix can be written as

$$A = V\Lambda V^T \tag{3.14}$$

For example, when considering the forward model G of the previous section,

$$G = V\Lambda V^T = \sum_{n=1}^N \lambda_n \mathbf{m}_{\lambda_n}\mathbf{m}_{\lambda_n}^T \tag{3.15}$$

$$\Lambda = \mathrm{diag}(\lambda_1, \ldots, \lambda_N) \tag{3.16}$$

$$V = (\mathbf{m}_{\lambda_1}, \ldots, \mathbf{m}_{\lambda_N}) \quad VV^T = I \tag{3.17}$$

If all eigenvalues are positive then

$$G^{-1} = V\Lambda^{-1}V^T \tag{3.18}$$

In the more general case, an operator may not be a square matrix, for example when the data variable dimension is different from the model dimension variable ($N_m \neq N_d$), then each (non-square) matrix with rank r can be decomposed as

$$G = V\Sigma U^T \quad VV^T = UU^T = I_r \tag{3.19}$$

with V a matrix of size $(N_m \times r)$ and U a matrix of size $(N_d \times r)$.

$$\Sigma = \mathrm{diag}\left(\sqrt{\lambda_1}, \ldots, \sqrt{\lambda_r}\right), \quad \lambda_i > 0 \tag{3.20}$$

The values $\lambda_1, \ldots, \lambda_r$ are the nonzero eigenvalues of GG^T and G^TG with V and U containing the corresponding eigenvalues of these matrices.

3.2.3.2. Geometric interpretation. Singular value decomposition (SVD) lies at the heart of many useful methods of UQ covered later in this book. SVD decomposes a matrix (whether a data matrix, a covariance matrix) into simpler, more easily interpretable and meaningful parts. To understand this, again geometrically, we consider that a matrix A is a linear mapping from one Cartesian space to another (see Figure 3.4). Consider the case of a simple matrix

$$A = \begin{pmatrix} 1 & 1 \\ 0 & 1 \end{pmatrix} \tag{3.21}$$

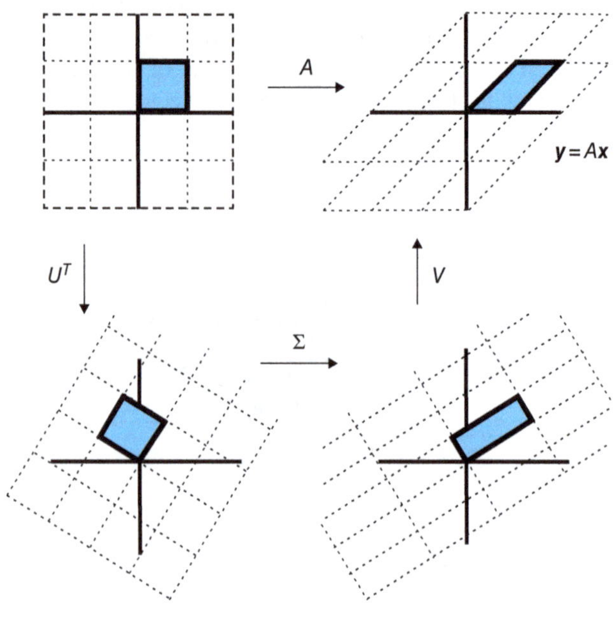

Figure 3.4 Geometric explanation of SVD as a transformation from one orthogonal system to another orthogonal system.

which results in shearing a unit cube when mapping the vectors of that cube using A. SVD allows writing this transformation as a series of affine corrections between one orthogonal space and another orthogonal space. The fact that we map into a space that is also orthogonal means that we can use the usual tools such as Euclidean distance, norms, and so on. To map a vector $\mathbf{x}$ with A is now equivalent to

$$\mathbf{y} = A\mathbf{x} = V\Sigma U^T\mathbf{x} \tag{3.22}$$

which is equivalent to applying a rotation U^T (here by 58.28°), a stretching (Σ), and another rotations V (again by 58.28°). A here is a full rank matrix. If the matrix is not full rank then the mapping (rank = 1) is onto a single vector. However, the formulation is still the same, meaning one can achieve such mapping, always, by means of rotations and stretching (even if stretched to infinity).

3.2.4. Quadratic Forms

Quadratic forms in higher dimensions are a useful way to study properties of data. These forms are often associated with least squares, inverse modeling, or the multi-Gaussian distribution. Calculating the derivative of a quadratic form in the context of optimization (e.g., maximum likelihood method) leads to a linear system of equations that can then be solved with standard techniques. Since the quadratic form is associated with a symmetric matrix, any application that involves such matrix, such as a covariance matrix, relies on these forms.

Consider $\mathbf{x} \in \mathbb{R}^L$, then a quadratic form is built from a symmetric matrix A as follows:

$$Q(\mathbf{x}) = \mathbf{x}^T A\mathbf{x} = \sum_{\ell=1}^{L}\sum_{\ell'=1}^{L} a_{\ell\ell'}x_\ell x_{\ell'} \tag{3.23}$$

If $Q(\mathbf{x}) > 0, \quad \mathbf{x} \neq 0$ then the quadratic form is positive definite. A is then positive definite if the corresponding quadratic form is positive definite. Applications occur when A is a covariance matrix of $\mathbf{x}$ (see later) or when A is the matrix of second derivatives on $\mathbf{x}$. In the latter case, the quadratic form measures the curvature in $\mathbf{x}$. The eigenvalues of A determine the shape of the quadratic form. It is easy to show that with λ_ℓ as eigenvalues and V as eigenvectors that

$$\mathbf{x}^T A\mathbf{x} = \sum_{\ell=1}^{L} \lambda_\ell y_\ell^2 \text{ with } \mathbf{y} = V^T\mathbf{x} \tag{3.24}$$

An interesting property of quadratic forms relates to the extrema of these forms. Consider two symmetric matrices A and B, then

$$\max_{\mathbf{x}} \frac{\mathbf{x}^T A\mathbf{x}}{\mathbf{x}^T B\mathbf{x}} = \lambda_1 \geq \lambda_2 \geq \cdots \geq \lambda_L = \min_{\mathbf{x}} \frac{\mathbf{x}^T A\mathbf{x}}{\mathbf{x}^T B\mathbf{x}} \tag{3.25}$$

with λ_1 the eigenvalues of $B^{-1}A$. In the specific case when $\mathbf{x}^T B\mathbf{x}$=1, $\forall\, \mathbf{x}$ we get

$$\max_{\mathbf{x}} \mathbf{x}^T A\mathbf{x} = \lambda_1 \geq \lambda_2 \geq \cdots \geq \lambda_L = \min_{\mathbf{x}} \mathbf{x}^T A\mathbf{x} \tag{3.26}$$

This property will be useful later when searching for linear combinations in the data $A\mathbf{x}$ that maximally explain variance (variance = a square form) of that data.

3.2.5. Distances

3.2.5.1. Basics. Distances form an important component of many of the UQ tools presented in this book. Models of the subsurface are usually complex and high dimensional. Hence, representing them mathematically in some extremely high-dimensional Cartesian space is not feasible. In addition, Cartesian spaces may not be the best choice for physical properties (see Chapter 6). We may not at all be interested in the model itself, but we may be interested in the difference between one model and another model. After all, UQ is only meaningful if models "differ" in some sense or another. Of relevance is that they differ in the prediction calculated from them or in some other summary statistics. The mathematical foundation of difference is "distance" and the space created is metric space.

A metric space is a set for which the distances between the members of the set are defined. For example, a set of L model realizations is such a set or the L data variables or response variables calculated from the models is also a set. If we call this set $X = (\mathbf{x}^{(1)}, \ldots, \mathbf{x}^{(L)})$, then the following axioms of distance are formulated for a metric space:

$$\begin{aligned}
&d\left(\mathbf{x}^{(\ell)},\mathbf{x}^{(\ell')}\right) \geq 0 && \text{non-negativity}\\
&d\left(\mathbf{x}^{(\ell)},\mathbf{x}^{(\ell')}\right) = 0 \Leftrightarrow \mathbf{x}^{(\ell)} = \mathbf{x}^{(\ell')} && \text{identity}\\
&d\left(\mathbf{x}^{(\ell)},\mathbf{x}^{(\ell')}\right) = d\left(\mathbf{x}^{(\ell')},\mathbf{x}^{(\ell)}\right) && \text{symmetry}\\
&d\left(\mathbf{x}^{(\ell)},\mathbf{x}^{(\ell'')}\right) \leq d\left(\mathbf{x}^{(\ell)},\mathbf{x}^{(\ell')}\right) + d\left(\mathbf{x}^{(\ell')},\mathbf{x}^{(\ell'')}\right) && \text{triangular inequality}
\end{aligned} \tag{3.27}$$

Here d is called the distance function. Well-known metric spaces are the real numbers with absolute difference or any Euclidean space with a Euclidean distance defined as

$$d\left(\mathbf{x}^{(\ell)},\mathbf{x}^{(\ell')}\right) = \sqrt{\left(\mathbf{x}^{(\ell)} - \mathbf{x}^{(\ell')}\right)^T\left(\mathbf{x}^{(\ell)} - \mathbf{x}^{(\ell')}\right)} \tag{3.28}$$

3.2.5.2. Useful Distances for UQ

3.2.5.2.1. Univariate Variables. A wide class of differences are generated based on norms:

$$d\left(\mathbf{x}^{(1)},\mathbf{x}^{(2)}\right) = \|\mathbf{x}^{(1)}-\mathbf{x}^{(2)}\|_r = \left(\sum_{n=1}^{N}|x_n^{(1)}-x_n^{(2)}|^r\right)^{1/r} \tag{3.29}$$

Hence, any normed vector space is a metric space. An assumption here is that the components of $\mathbf{x}$ are on the same scale. The Manhattan norm gives rise to the Manhattan distance (also termed $L1$ distance), where the distance between any two vectors is the sum of the differences between corresponding components. The maximum norm gives rise to the Chebyshev distance or chessboard distance.

The norms above are appropriate when dealing with continuous variables; however, they become problematic for categorical variables. Categorical variables may not have ordinality. Consider the following examples of geological sequences:

$\mathbf{S}_1 : D\,F\,F\,E\,D\,E \quad D=\text{delta}, \quad F=\text{fluvial}, \quad E=\text{estuarine}$

$\mathbf{S}_2 : F\,D\,F\,E\,F\,E$

What is a measure of their difference? In the absence of order, there should be no difference between D E and D F. To alleviate this, we need to create indicator variables

$$I_D = \begin{cases} 1 & \text{if } S=D \\ 0 & \text{else} \end{cases} \quad I_E = \begin{cases} 1 & \text{if } S=E \\ 0 & \text{else} \end{cases} \quad I_F = \begin{cases} 1 & \text{if } S=F \\ 0 & \text{else} \end{cases} \tag{3.30}$$

and hence an appropriate distance is

$$d(\mathbf{S}_1,\mathbf{S}_2) = \frac{1}{3}\sum\nolimits_{s\in\{D,E,F\}} I_s \tag{3.31}$$

3.2.5.2.2. Hausdorff Distance. The Hausdorff distance is popular in image analysis for measuring the dissimilarity between two sets of data $\mathbf{S}^{(1)}$ and $\mathbf{S}^{(2)}$ [*Huttenlocher et al.*, 1993]. Individual elements of $\mathbf{S}^{(1)}$ are denoted as $s_i^{(1)}$ and individual elements of $\mathbf{S}^{(2)}$ are denoted as $s_i^{(2)}$. $\mathbf{S}^{(1)}$and $\mathbf{S}^{(2)}$ are deemed similar if each point in one set is close to all other points in the other set. Consider two objects in Figure 3.5. What is the distance between these objects? First, each has been rasterized into a set of points. Consider first calculating $d\left(s_i^{(1)},\mathbf{S}^{(2)}\right)$ between one point and the set. This distance is defined as the minimum:

$$d\left(s_i^{(1)},\mathbf{S}^{(2)}\right) = \min_j d\left(s_i^{(1)},s_j^{(2)}\right) \tag{3.32}$$

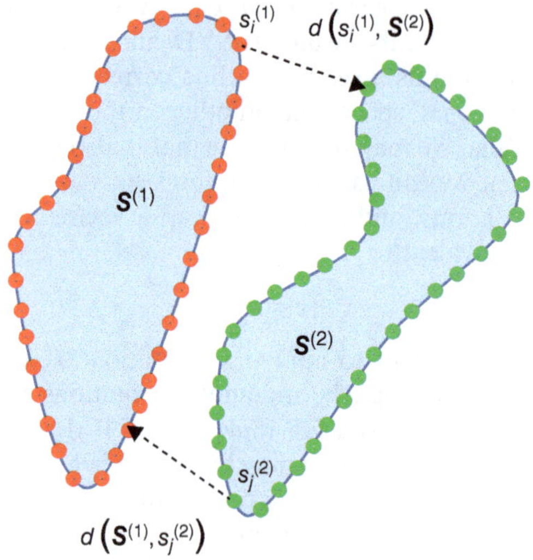

Figure 3.5 Creating a Hausdorff distance between two objects that are discretized with points.

Then, to calculate $d(\mathbf{S}^{(1)}, \mathbf{S}^{(2)})$, *Dubuisson and Jain* [1994] propose to average over the points in set $\mathbf{S}^{(1)}$:

$$d\left(\mathbf{S}^{(1)},\mathbf{S}^{(2)}\right) = \frac{1}{N_1}\sum_{i=1}^{N_1}\min_j d\left(s_i^{(1)},s_j^{(2)}\right) \tag{3.33}$$

The resulting measure is not symmetric $d(\mathbf{S}^{(1)}, \mathbf{S}^{(2)}) \neq d(\mathbf{S}^{(2)}, \mathbf{S}^{(1)})$, hence not a distance, so technically we cannot use the notation d. *Dubuisson and Jain* [1994] propose to use the maximum to symmetrize the distance of Eq. (3.34)

$$\begin{aligned} d\left(\mathbf{S}^{(1)},\mathbf{S}^{(2)}\right) &= d\left(\mathbf{S}^{(2)},\mathbf{S}^{(1)}\right) \\ &= \max\left[\frac{1}{N_1}\sum_{i=1}^{N_1}\min_j d\left(s_i^{(1)},s_j^{(2)}\right), \frac{1}{N_2}\sum_{j=1}^{N_2}\min_i d\left(s_i^{(1)},s_j^{(2)}\right)\right] \end{aligned} \tag{3.34}$$

Figure 3.6 shows an example of how the Hausdorff distance outperforms the Euclidean distance in discriminating between objects.

3.2.5.2.3. Distances Based on Transformations. Simple Euclidean distances between images, or spatial models, often are not very informative about their actual difference. Consider a simple case of three simple images with one line (e.g., a fracture or fault). In the second image, the line is slightly offset, hence there are no overlapping pixels, while the third image is orthogonal to the first image, and therefore it has at least one location in common. The Euclidean distance cannot capture the significant similarity between images 1 and 2.

A solution is to transform the categorical variable into a continuous variable that informs the distance to the edge

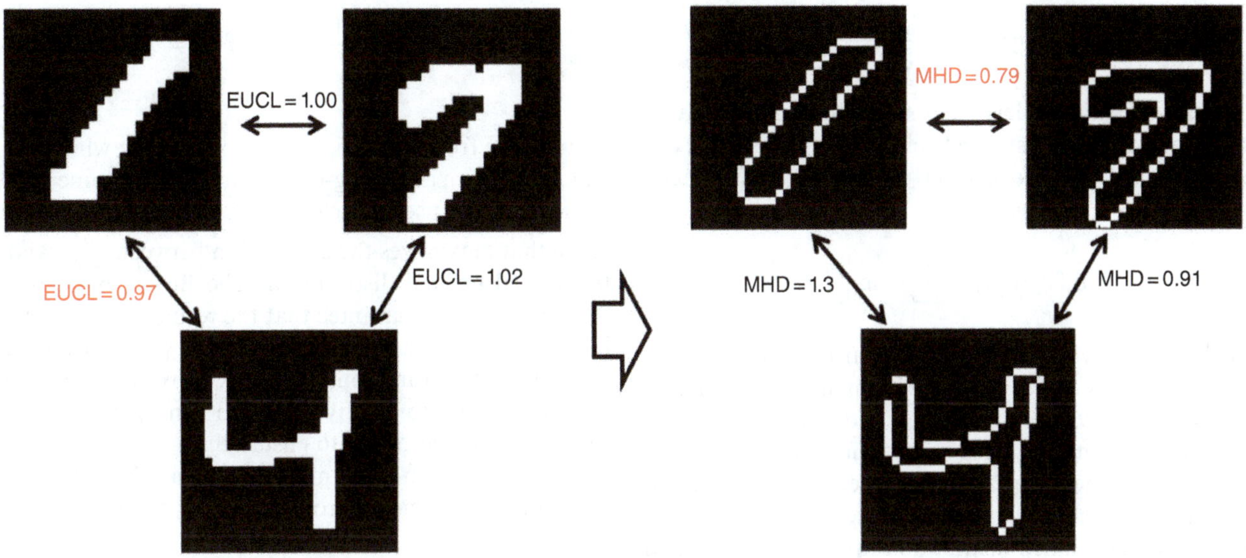

Figure 3.6 Evaluation of the modified Hausdorff distance for images of numbers 1, 4, and 7. Modified Hassdorf distance (MHD) requires the rasterization of the images into points sets (shown on the right). The smallest MHD is observed between images of 1 and 7, which is consistent with visual inspection. This is not the case for the pixel by pixel Euclidean distance (left).

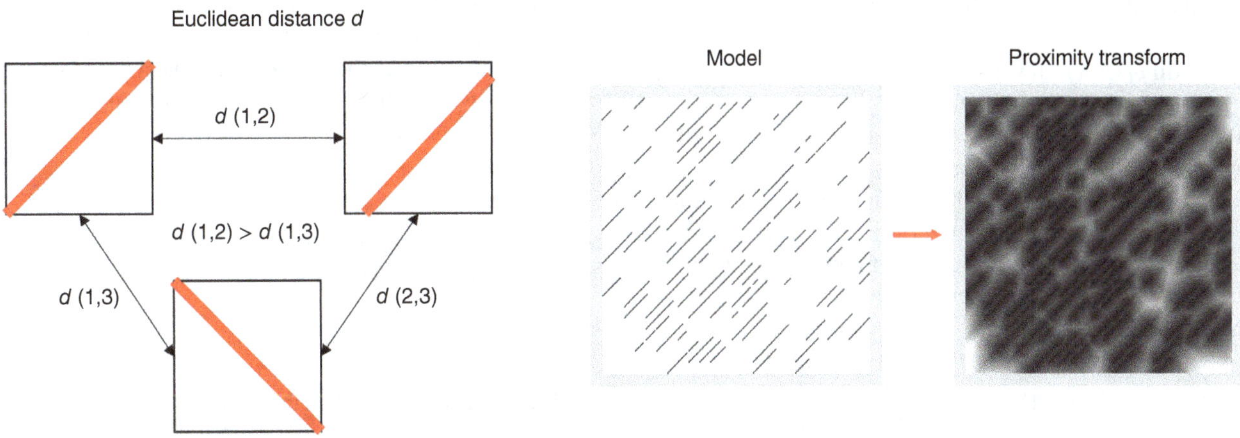

Figure 3.7 The proximity transforms to measure differences in images constituted by discrete objects.

of the feature of interest. This proximity transform results in a "distance map" as shown Figure 3.7. It results in a transformed variable where the grayscale levels indicate the distance to the diagonal black object. The computation of a distance between both distance maps, using for example a Euclidean norm, is then more meaningful than the distance between categorical patterns.

3.2.5.2.4. Distances Between Distributions. Is the prior distribution different from the posterior? Is the posterior distribution generated by a Markov chain Monte Carlo method similar to the theoretical posterior, or the posterior of another method? Is the histogram of one model realization (e.g., porosity) significantly different from another realization? All these questions call for a measure of difference between distributions. The statistical literature offers various methods to test whether two distributions are statistically significantly different. In this section, we will limit ourselves to comparing univariate distributions. In the application chapters, we will see how comparisons between multivariate distributions are achieved by comparing distributions of orthogonal components of the random vector in question. Consider to that end two discrete probability distributions represented by the discrete probabilities p_k, q_k, $k = 1, \ldots, K$. A well-known distance is the chi-squared distance

$$d_{\chi^2}(\mathbf{p},\mathbf{q}) = \frac{1}{2}\sum_{k=1}^{K}\frac{(p_k - q_k)^2}{(p_k + q_k)} \tag{3.35}$$

This distance may underweight small differences because of the square, but it is symmetric. Another measure of difference related to information theory is the Kullback–Liebler (KL) divergence

$$\text{dif}_{KL}(\mathbf{p},\mathbf{q}) = \sum_{k=1}^{K} p_k \log\frac{p_k}{q_k} \tag{3.36}$$

which is the expected value of the logarithmic differences. This measure is not symmetric; it emanates from information theory where information is optimally coded by assigning the smallest code to the most frequent letter/message. This measure can be interpreted as the expected extra message-length that is communicated if a code optimal for some assumed distribution (**q**) is used, compared to a code that is based on the underlying, unknown true distribution (**p**). The symmetric form of the *KL* difference is the Jensen–Shannon divergence:

$$d_{JS}(\mathbf{p},\mathbf{q}) = \frac{1}{2}\text{dif}_{KL}(\mathbf{p},\mathbf{q}) + \frac{1}{2}\text{dif}_{KL}(\mathbf{q},\mathbf{p}) \tag{3.37}$$

Note that the continuous form of the *KL* distance is

$$\text{dif}_{KL}(p(x),\ q(x)) = \int p(x)\log\frac{p(x)}{q(x)}dx \tag{3.38}$$

with $p(x)$ and $q(x)$ densities. Another distance is the earth movers distance (EMD) where pdfs are seen as two piles of material. The EMD is then defined as the minimum cost of turning one pile into the other. The cost is defined as the amount of material moved times the distance by which it is moved. A last example based on pdfs is the Bhattacharyya distance:

$$d_{BC}(\mathbf{p},\mathbf{q}) = -\log\left(\sum_{k=1}^{K}\sqrt{p_k q_k}\right) \tag{3.39}$$

In terms of cdfs one can use the $L1$ norm, which is basically the area between the two cdfs.

3.3. BASICS OF UNIVARIATE AND MULTIVARIATE PROBABILITY THEORY AND STATISTICS

In this section, we present some basic elements of multivariate probability theory, mostly to cover notation and conventions and for those who need a brief refresher.

3.3.1. Univariate Transformations

The Box–Cox transform is a procedure for transforming data into a normal shape. It uses a single parameter λ, such that for each sample $x^{(\ell)}$

$$x_{\text{trans}}^{(\ell)} = \begin{cases} \dfrac{\left(x^{(\ell)}\right)^{\lambda} - 1}{\lambda}, & \lambda \neq 0 \\ \ln x^{(\ell)}, & \lambda = 0 \end{cases} \tag{3.40}$$

λ can range from -5 to 5, with a special case when $\lambda = 0$, which is known as the log-transform. To determine which value of λ to be used, we can search over the range for the value that maximizes the correlation between $x_{\text{trans}}^{(\ell)}$ and a theoretical normal distribution. The Box–Cox transformation does not guarantee that the resulting distribution is actually normal, so it is essential to perform a check after the transformation. The Box–Cox transform can be applied only to positive data, so it may be necessary to add a constant to $x^{(\ell)}$ to ensure this.

Another useful transform that is useful when $x^{(\ell)}$ varies from 0 to 1 such as for proportions or percentages.

$$x_{\text{trans}}^{(\ell)} = \sin^{-1}\left(\sqrt{x^{(\ell)}}\right) \tag{3.41}$$

The result of the transform is given in radians ranging from $-\pi/2$ to $\pi/2$. The arcsin transform is helpful when the variance of the variable is uneven (smaller near 0 and 1) by spreading the variance over the entire range and can make the variable more normal.

The rank transform is a common way to transform into any distribution type, such as for example a normal score transform. It also allows any easy back-transformation. We first rank the data

$$x_{(1)} \leq x_{(2)} \leq \cdots \leq x_{(L)} \tag{3.42}$$

the subscript indicating the rank. The cumulative frequency for each ranked sample:

$$p_\ell = \frac{\ell}{L} - \frac{1}{2L} \tag{3.43}$$

then for any cumulative distribution function G (such as a normal distribution)

$$x_{\text{trans}}^{(\ell)} = G^{-1}(p_\ell) \tag{3.44}$$

3.3.2. Kernel Density Estimation

The goal of density estimation is to estimate a probability density function using only samples drawn from it. The simplest form of density estimation is the histogram. By dividing sample spaces into a fixed number of equally sized bins and counting the fraction of samples that fall within each bin, an estimate of the density over the bin interval is obtained. This histogram is straightforward to compute, but it results in discontinuities in the estimated density because of the discrete nature of the bins. Furthermore, as the dimension of the space increases, the number of bins increases exponentially. Another way offers itself by density estimation based on the kernel method and is widely used in the later chapters.

Based on Parzen windows, which places a bin centered at each sample, the kernel density estimate at x is then the sum of the number of bins that encompass it. Therefore, for a given value x, the density is expressed as

$$\hat{f}(x) = \frac{1}{L}\sum_{\ell=1}^{L} K\left(\frac{x^{(\ell)} - x}{w}\right) = \frac{1}{L}\sum_{\ell=1}^{L} K\left(\frac{z_\ell}{w}\right) \tag{3.45}$$

where z_ℓ is the distance between the location at which the density is being evaluated and sample point $x^{(\ell)}$, and K is the kernel or the shape of the bin. For Parzen windows, the kernel is a rectangular function, and the width is set by the w parameter. This approach fixes the number of bins to the number of samples, which alleviates the problem of exponentially increasing number of bins in high dimensions. However, Parzen windows do not address the issue of discontinuities because of the shape of the kernel. A smoothly varying function such as the Gaussian kernel is more frequently used:

$$K(z) = \frac{1}{\sqrt{2\pi}} \exp\left(-\frac{z^2}{2}\right) \tag{3.46}$$

Other popular kernels are the uniform, triangular, and Epanechnikov kernels. The method is named *kernel density estimation*. w is the smoothing parameter or *bandwidth* of the estimator. In higher dimensions, the bandwidth is the covariance matrix of the Gaussian kernel. The bandwidth controls the extent of the influence each sample has on estimating the density. The choice of bandwidth has a very strong influence on the resulting density estimate. A small bandwidth will result in "spiky" density estimates, while a large choice will over smooth the estimate and obscure its structure.

A popular choice for w is Silverman's rule of thumb, which assumes that the underlying distribution is Gaussian. The diagonal components of the matrix are

$$w_n = \left(\frac{4}{N+2}\right)^{\frac{1}{N+2}} L^{-\frac{1}{N+4}} \hat{\sigma}_n \tag{3.47}$$

where $\hat{\sigma}_n$ is an estimate of the standard deviation of the n-th variate, and N is the dimension of the problem.

In theory, kernel density estimation can be extended to any number of dimensions. The kernel function in Eq. (3.47) is simply extended to higher-dimensional functions. However, in practice because of the curse of dimensionality, the number of samples required for accurate estimation grows exponentially with dimension. *Silverman* [1986] provides an optimistic upper bound on the number of dimensions as five, before the number of required samples for accurate joint density estimation often becomes impractically large. Nonetheless, smoothing techniques provide a powerful way of gaining insight into complex distributions. This additional flexibility does, however, come with the challenging task of specifying the bandwidth, as well as the limitation of only working effectively in low dimensions.

3.3.3. Properties of Multivariate Distributions

In multivariate statistics, we study random vectors, generically written as $\mathbf{X} = (X_1, \ldots, X_N)$. The stochastic variation of these random vectors is fully described by the joint cumulative distribution

$$P(X_1 \le x_1, X_2 \le x_2, \ldots, X_N \le x_N) = F(x_1, x_2, \ldots, x_N) \tag{3.48}$$

or the corresponding joint (multivariate) density function

$$f(x_1, x_2, \ldots, x_N) = \frac{\partial^N F(x_1, x_2, \ldots, x_N)}{\partial x_1 \partial x_2 \ldots \partial x_N} \tag{3.49}$$

From the full multivariate distribution, one can deduce any marginal distribution, such as a univariate or bivariate distribution

$$\begin{aligned} F(x_n) &= \int dx_1 \cdots \int dx_{n-1} \int dx_{n+1} \cdots \int f(x_1, \ldots, x_N) dx_N \\ F(x_1, x_2) &= \int dx_3 \int dx_4 \cdots \int f(x_1, \ldots, x_N) dx_N \end{aligned} \tag{3.50}$$

or any conditional distribution

$$\begin{aligned} &F(x_n | X_{n'} = x_{n'} \forall n' \neq n) = P(X_n \le x_n | X_{n'} = x_{n'} \forall n' \neq n) \\ &= \frac{\int_{-\infty}^{x_{n'}} f\left(x_1, \ldots, x_{n'-1}, x'_{n'}, x_{n'+1}, \ldots, x_N\right) dx'_{n'}}{f\left(x_1, \ldots, x_{n'-1}, x_{n'+1}, \ldots, x_N\right)} \end{aligned} \tag{3.51}$$

The problem in reality is that very few analytical expressions exist for the joint multivariate distribution (except for the Gaussian, see later). The focus instead lies on lower-order statistics, such as the bivariate distributions from which then moments such as the variogram γ and covariance, cov, can be derived (and estimated with data)

$$\begin{aligned} \gamma(X_n, X_{n'}) &= \frac{1}{2} E(X_n - X_{n'})^2 \\ &= \frac{1}{2} \int\int (x_n - x_{n'})^2 f(x_n, x_{n'}) dx_n dx_{n'} \end{aligned} \tag{3.52}$$

$$\begin{aligned} \mathrm{cov}(X_n, X_{n'}) &= E[(X_n - E[X_n])(X_{n'} - E[X_{n'}])]^2 \\ &= \int\int (x_n - E[X_n])(x_{n'} - E[X_{n'}]) f(x_n, x_{n'}) dx_n dx_{n'} \end{aligned} \tag{3.53}$$

From the latter we define correlation as

$$\rho(X_n, X_{n'}) = \frac{\mathrm{cov}(X_n, X_{n'})}{\sqrt{\mathrm{var}(X_n)\,\mathrm{var}(X_{n'})}} \tag{3.54}$$

When the $(X_1, \ldots, X_N)$ are (globally) independent then

$$f(x_1, x_2, \ldots, x_N) = \prod_{n=1}^{N} f(x_n) \tag{3.55}$$

Global independency entails also pair-wise independence but not vice versa. Pair-wise independence means that any bivariate distribution becomes a product of marginal, moreover then:

$$\rho(X_n, X_{n'}) = 0 \ \forall n, n' \tag{3.56}$$

Equation (3.56) entails linear independence only. When the relationship is nonlinear, then Eq. (3.56) may still hold, even if pair-wise dependency exists. Another form of independence is conditional independence, which in the univariate case becomes

$$f(x_n, x_{n'} | x_{n''}) = f(x_n | x_{n''}) f(x_{n'} | x_{n''}) \tag{3.57}$$

or equivalently

$$f(x_n | x_{n'}, x_{n''}) = f(x_n | x_{n'}) = f(x_n | x_{n''}) \tag{3.58}$$

This concept can be extended to any mutually exclusive subset of random variables that comprise the random vector $\mathbf{X}$. The concept of conditional independence is used throughout UQ. For example, one may have various types of data that inform the subsurface $\mathbf{d}_1$, $\mathbf{d}_2$, …, $\mathbf{d}_n$. These data may be from different origin, such as one data source from geophysics, another from drilling, or yet another from testing the subsurface formation. The conditional independence assumption that is used is then as follows:

$$f(\mathbf{d}_1, \mathbf{d}_2, \ldots | \mathbf{m}) = \prod_i f(\mathbf{d}_i | \mathbf{m}) \tag{3.59}$$

Basically this means that if we knew the real earth, then these data sets can be treated as independent. In other words, knowing the real earth is enough to model each data set separately through individual likelihoods $f(\mathbf{d}_i|\mathbf{m})$. This appears quite reasonable for data sets that are very different, for example whose forward models capture very different physics.

3.3.4. Characteristic Property

Recall that among all outcomes of a random variable, the expected value is known to minimize the average square error (essentially a variance):

$$E[X] = \min_{x_0} E\left[(X - x_0)^2\right] \tag{3.60}$$

This property lies at the heart of all least-square method: the expectation is the best least-square estimate of an unknown RV. Similarly, consider the conditional mean as a $N-1$ dimensional function

$$E[X_n | X_{n'} = x_{n'}, \forall n' \neq n] = \psi(x_{n'}, n' \neq n) \tag{3.61}$$

Among all $N-1$ variate functions, the conditional expectation in Eq. (3.61) minimizes the estimation variance

$$\psi(x_{n'} \, n' \neq n) = \min_{\psi_0(x_{n'}, \ n' \neq n)} E\left[(X_n - \psi_0(x_{n'}, n' \neq n) | X_{n'} = x_{n'} \forall n' \neq n)^2\right] \tag{3.62}$$

or any function that minimizes a variance (least square) is an expectation, whether conditional or unconditional. This property will be used throughout the book when dealing with least squares, Gaussian processes (see Section 3.7), linear inverse problems, and so on.

3.3.5. The Multivariate Normal Distribution

The multivariate normal is a very popular model in multivariate statistics as well as in UQ. The reason lies in the mathematical convenience of this model, its arranged marriage will least-square and maximum likelihood estimation methods as well as linear modeling. When $\mathbf{X} = (X_1, \ldots, X_N)$ is multivariate normal then

$$f(x_1, \ldots, x_N) = \frac{1}{\sqrt{(2\pi)^N \det(C)}} \exp\left(-\frac{1}{2}(\mathbf{x} - E[\mathbf{X}])^T C^{-1} (\mathbf{x} - E[\mathbf{X}])\right) \tag{3.63}$$

One notices the quadratic form in the exponent (basically a second-order polynomial in $\mathbf{x}$). This form is relevant because the log of the density is quadratic, and hence the derivative is linear. C is the covariance matrix with elements

$$[C]_{nn'} = c_{nn'} = \text{cov}(X_n, X_{n'}), \quad n, n' = 1, \ldots, N \tag{3.64}$$

$(\mathbf{x} - E[\mathbf{X}])^T C^{-1} (\mathbf{x} - E[\mathbf{X}])$ is also termed the Mahalanobis distance: a distance of a point $\mathbf{x}$ from some center $E[\mathbf{X}]$. It can be generalized to a distance as

$$d(\mathbf{x}, \mathbf{x}') = (\mathbf{x} - \mathbf{x}')^T C^{-1} (\mathbf{x} - \mathbf{x}') \tag{3.65}$$

This distance accounts for the correlation that may exist. If no correlation exists then $C = I$, and we get the Euclidean distance. The more $\mathbf{x}$ is correlated with $\mathbf{x}'$, the less that component will contribute to the distance (they look closer, because they are more similar). The Mahalanobis transformation renders the elements (linearly) independent:

$$\mathbf{Y} = C^{-1/2}(\mathbf{X} - E[\mathbf{X}]) \tag{3.66}$$

The elements of $\mathbf{Y}$ have no (linear) correlation.

Some useful properties of the multivariate normal are as follows:

1. All $N-n$ variate distribution are multivariate normal.
2. All conditional distribution are multivariate normal.

3. In case of a univariate conditional distribution, we find that the corresponding conditional expectation are linear functions of the conditioning values, for example

$$E[X_n|X_{n'} = x_{n'}, \forall n' \neq n] = \sum_{n' \neq n} w_{n'} x_{n'} \tag{3.67}$$

or for any subset of conditioning values.

4. The joint distribution of any subset of $\mathbf{X}$ and linear combinations of another mutually exclusive subset is also multivariate normal, for example

$$\left(\sum_{n' \neq n} w_{n'} X_{n'}, X_n\right) = \left(\hat{X}_n, X_n\right) \sim \text{bivariate normal} \tag{3.68}$$

This results in the so-called conditional unbiasedness property

$$E\left[X_n|\hat{X}_n = x\right] = x \tag{3.69}$$

These properties make the multivariate Gaussian useful in many applications. However, assuming multivariate Gaussian models is not without risk. First, the assumption can rarely be verified with actual data, simply because of the lack of data in higher dimensions. More importantly, assuming a multivariate Gaussian distribution imposes a higher-order dependency structure that goes beyond the covariance defining that distribution. In spatial modeling (where the X_n are properties in a grid), limitations of this distribution are well known [*Gómez-Hernández and Wen*, 1998; *Zinn and Harvey*, 2003; *Feyen and Caers*, 2006]. More specifically, the maximum entropy property [*Journel and Deutsch*, 1993] entails that extremes become rapidly uncorrelated. Additionally, the multivariate distribution is clearly tied to linear modeling; hence, any relationships between the X_n that is not linear becomes problematic.

3.4. DECOMPOSITION OF DATA

3.4.1. Data Spaces

Subsurface models are complex, and data can be extensive, such as for example seismic data; hence, forms of dimension reduction are critical to render such complex problems manageable. Dimension reduction lies at the foundation of many of the methods covered in subsequent chapters. Here we provide the very foundation of most dimension reduction techniques.

Consider again the "data matrix" of size $N \times L$, generically written as

$$X = \begin{pmatrix} x_1^{(1)} & x_2^{(1)} & \cdots & \cdots & \cdots & x_N^{(1)} \\ \vdots & x_2^{(2)} & & & & \vdots \\ \vdots & & \ddots & & & \vdots \\ \vdots & & & \ddots & & \vdots \\ \vdots & & & & \ddots & \vdots \\ x_1^{(L)} & x_2^{(L)} & \cdots & \cdots & \cdots & x_N^{(L)} \end{pmatrix} \tag{3.70}$$

This matrix can be viewed in two ways: row by row (per "data sample") or column by column (per each dimension of the "data sample"). For example, if the data matrix stores model realizations, then we can look either at the collection of model (rows) or the samples of each individual variable of the model (columns). Hence, from a geometric point of view, one can take two views and create two alternative Cartesian axis systems (see Figure 3.8).

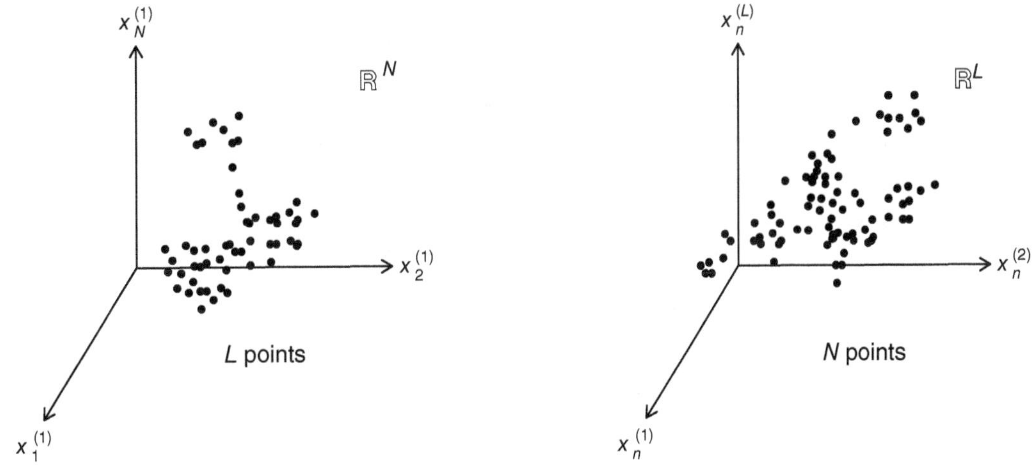

Figure 3.8 Two views of the same data matrix.

1. (Sample dimension) Space $\mathbb{R}^N$: The space containing vectors $\mathbf{x}_N^{(\ell)} = \left(x_1^{(\ell)}, x_2^{(\ell)}, \ldots, x_N^{(\ell)}\right)$, a cloud of $\ell = 1, \ldots, L$ points.
2. (Sample size) Space $\mathbb{R}^L$: The space containing vectors $\mathbf{x}_n^{(L)} = \left(x_n^{(1)}, x_n^{(2)}, \ldots, x_n^{(L)}\right)$, a cloud of $n = 1, \ldots, N$ points.

While both spaces convey the same information, choosing which axis system to work with (model, sample, fit, regress) is of critical importance to many of the methods in this book. In fact, we will illustrate several dualities between these two spaces and the techniques that are applied in such spaces. The choice of the axis system largely depends on the dimension of the space. Clearly, a low-dimensional space with a lot of points in the point cloud is easier to manage than a high-dimensional space with very few. High dimensions are challenging because they become very rapidly "empty" as dimension increases. Consider a simple example of a circle within a square and calculate the ratio between the areas occupied by the circle over the area occupied by the square; consider increasing the dimension and calculating the same ratio, now between hypercube and hypersphere. In Figure 3.9 one notices that the ratio is basically zero after dimension 10. In other words, an exponential increase in volume exists with each dimension added to a Cartesian space. This also means that an exponentially increasing amount of samples is needed to cover high-dimensional spaces as one would cover/sample low-dimensional spaces. One characteristic of uncertainty quantification is that subsurface models are complex, spatial, or spatiotemporal; hence it is of very high dimension N. As shown in Chapter 1, the key target variables, on which decisions are made and whose uncertainty is most relevant, are often of much lower dimension than the models. For quantifying their uncertainty (as simple as a volume for example), much less samples L are needed. For most UQ problems, certainly, those treated in Chapter 1, we can safely state that

$$L \ll N \tag{3.71}$$

This is not an observation without considerable consequence. This property is different from data problems that involve people (e.g., Facebook, Google) where the sample size is very large (millions/billions) and the model dimensions (e.g., people's preference) are much smaller. Hence, blindly applying data science in these areas to the problems in this book will be ineffective and inefficient.

3.4.2. Cartesian Space of Size L: The Sample Size

In $\mathbb{R}^L$ we need to embed N data points, when considering the data matrix of Eq. (3.70). To reduce dimension, we need to project points into a lower-dimensional space. A straightforward way would be just to ignore a model realization, but this would lead to too much of a loss of information, and potential removal of important realizations from the analysis. Instead, we attempt to project into a lower-dimensional space by minimizing the loss of information caused by such projection. Consider, therefore, first projecting the cloud into one dimension, in other words, a line. Since a line can be defined through a unit vector, $\mathbf{u}_1$, $\|\mathbf{u}_1\| = 1$, we need to find the "optimal" $\mathbf{u}_1$ that best represents the L-dimensional cloud. This is illustrated in Figure 3.10. The coordinate of a projection of a point is

$$\left(\mathbf{x}_N^{(\ell)}\right)^* = \left(\mathbf{x}_N^{(\ell)}\right)^T \mathbf{u}_1 \tag{3.72}$$

As measure of "loss of information" due to such projection, we use a least-square formulation and minimization:

$$\sum_{\ell=1}^{L} \left\| \left(\mathbf{x}_N^{(\ell)}\right)^* - \mathbf{x}_N^{(\ell)} \right\|^2 \tag{3.73}$$

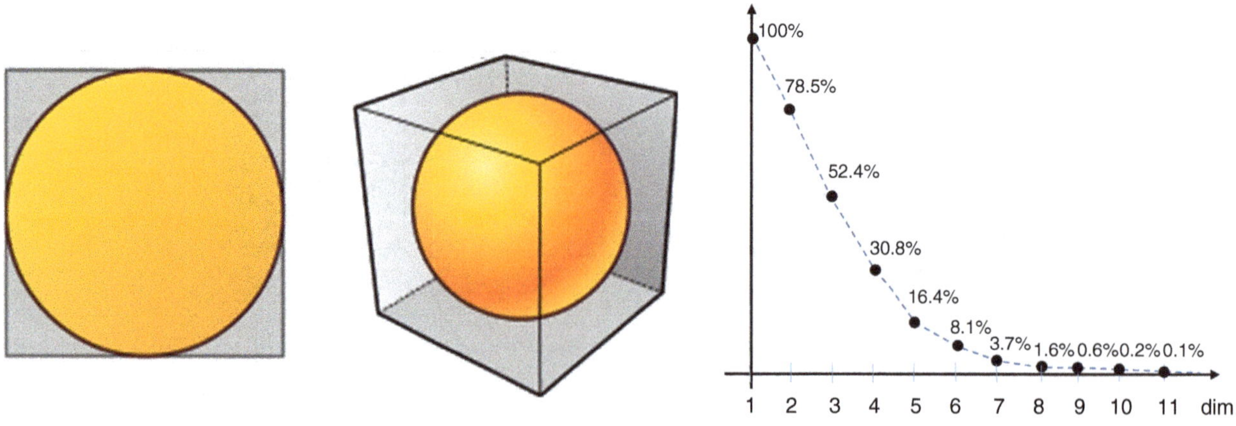

Figure 3.9 Curse of dimensionality, the ratio of hypersphere with hypercube. Space becomes virtually empty after dimension 10. Modified from Wikipedia.

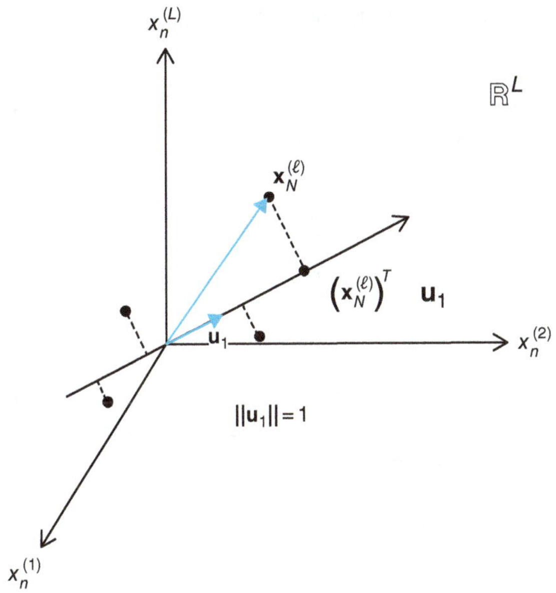

Figure 3.10 Projection of data onto an eigenvector.

The problem of minimizing Eq. (3.73) is equivalent to maximizing

$$\sum_{\ell=1}^{L} \left\| \left(\mathbf{x}_N^{(\ell)}\right)^* \right\|^2 \tag{3.74}$$

which is a convenience resulting from using a square loss function. Because

$$\begin{pmatrix} \left(\mathbf{x}_N^{(1)}\right)^* \\ \left(\mathbf{x}_N^{(2)}\right)^* \\ \vdots \\ \left(\mathbf{x}_N^{(L)}\right)^* \end{pmatrix} = \begin{pmatrix} \left(\mathbf{x}_N^{(1)}\right)^T \mathbf{u}_1 \\ \left(\mathbf{x}_N^{(2)}\right)^T \mathbf{u}_1 \\ \vdots \\ \left(\mathbf{x}_N^{(L)}\right)^T \mathbf{u}_1 \end{pmatrix} = X\mathbf{u}_1 \tag{3.75}$$

maximizing Eq. (3.74) is equivalent to solving

$$\max_{\mathbf{u}_1} (X\mathbf{u}_1)^T (X\mathbf{u}_1) = \max_{\mathbf{u}_1} \mathbf{u}_1^T X^T X \mathbf{u}_1 \text{ subject to } \|\mathbf{u}_1\| = 1 \tag{3.76}$$

According to our previous analysis about quadratic forms such as this, we know that the vector that maximizes Eq. (3.76) is the first eigenvector of $X^T X$ associated with the first eigenvalue.

An important observation is made when the data is centered, meaning that $\sum_{\ell=1}^{L} \frac{\mathbf{x}_N^{(\ell)}}{L} = 0$. Then $1/L\ X^T X$ is the empirical covariance between the L realizations consisting of N variables. Hence, in making such projection, we are seeking to make use of correlations among the variables in X. Obviously, if that covariance is zero (and variance some constant), then we cannot reduce the problem to a lower-dimensional problem, all realizations (L) will need to be used.

The above analysis extends to projections in more than one dimension. The solution is simply found by calculating the eigenvalues λ_n, $n = 1, \ldots, N$ and corresponding eigenvectors $\mathbf{u}_n$ of the matrix $X^T X$, with its eigenvalues ranked from large to small. Note that in our type of problem $X^T X$ of size $N \times N$ is very large. Directly calculating eigenvalues on $X^T X$ may be impossible for large problems. We will see later that ways around this problem exist.

3.4.3. Cartesian Space of Size N: Sample Dimension

Now we turn to the space $\mathbb{R}^N$. We need to embed L data realizations. Hence, any reduction in this space means reducing the number of variables. Clearly, just ignoring individual variables will usually not be efficient, instead we will, as before, rely on projections; basically, linear combinations of the variables in question. The solution to this projection problem is exactly the same as in the previous section, but now replace X by X^T, namely

$$\begin{aligned} \sum_{n=1}^{N} \left\| \left(\mathbf{x}_n^{(L)}\right)^* \right\|^2 &= \max_{\mathbf{v}_1} (X\mathbf{v}_1)(X\mathbf{v}_1)^T \\ &= \max_{\mathbf{v}_1} \mathbf{v}_1^T X X^T \mathbf{v}_1 \text{ subject to } \|\mathbf{v}_1\| = 1 \end{aligned} \tag{3.77}$$

Hence, we seek the eigenvalues μ_ℓ, $\ell = 1, \ldots, L$ and corresponding eigenvectors v_ℓ of XX^T from largest to smallest to create projections in lower dimensions. Note that the rank

$$r = \text{rank}(XX^T) = \text{rank}(X^T X) = \text{rank}(X) \tag{3.78}$$

In other words, we cannot obtain a dimension larger than r. While $X^T X$ is related to the covariance of data variables, calculated from data realizations, XX^T is related to the dot-product of data realizations calculated from data variables. Recall that the dot-product is the projection (a length is a scalar) of one vector onto another vector or, in this case, a projection of one (data) sample onto another (data) sample. Since we have L samples one can calculate $L \times L$ of such dot-products; hence, the dot-product matrix is, in our context, much smaller in size than the covariance matrix.

3.4.4. Relationship Between Two Spaces

Because both spaces represent the exact same information, but each with different projections into lower dimensions, a relationship must exist between these two projections. Consider the eigenvectors and eigenvalues in $\mathbb{R}^N$

$$XX^T \mathbf{v}_\ell = \mu_\ell \mathbf{v}_\ell \quad \ell \le r \tag{3.79}$$

By multiplying each side with X^T, we find

$$(X^T X)(X^T \mathbf{v}_\ell) = \mu_\ell (X^T \mathbf{v}_\ell) \quad \ell \le r \tag{3.80}$$

Hence, each eigenvector $\mathbf{v}_\ell$ of $(X^T X)$ in $\mathbb{R}^N$ corresponds to the eigenvector $X^T \mathbf{v}_\ell$ in $\mathbb{R}^L$, or

$$\mathbf{u}_\ell \sim X^T \mathbf{v}_\ell \tag{3.81}$$

and associated with the same eigenvalue μ_ℓ. Every non-zero eigenvalue of XX^T is also an eigenvalue of $X^T X$. Similarly, when multiplying

$$X^T X \mathbf{u}_n = \lambda_n \mathbf{u}_n \quad n \le r \tag{3.82}$$

Therefore, each eigenvector $\mathbf{u}_n$ of XX^T in $\mathbb{R}^L$ corresponds to the eigenvector $X\mathbf{u}_n$ in $\mathbb{R}^N$, or

$$\mathbf{v}_n \sim X \mathbf{u}_n \tag{3.83}$$

Since the eigenvectors $\mathbf{u}$ and $\mathbf{v}$ need to be unit vectors, the proportionality constant in both cases is $1/\sqrt{\lambda_k} = 1/\sqrt{\mu_k}$.

This leads to stating the following duality: for a data matrix X of size $N \times L$ with rank r, the eigenvalues (number of eigenvalues $\le r$) of $X^T X$ and XX^T are the same and the eigenvectors are related as follows:

$$\begin{aligned} \mathbf{u}_k &= \frac{1}{\sqrt{\lambda_k}} X^T \mathbf{v}_k \quad k \le r \\ \mathbf{v}_k &= \frac{1}{\sqrt{\lambda_k}} X \mathbf{u}_k \end{aligned} \tag{3.84}$$

Additionally, it can be shown that with $U = (\mathbf{u}_1 \; \mathbf{u}_2 \cdots \mathbf{u}_r)$, $V = (\mathbf{v}_1 \; \mathbf{v}_2 \cdots \mathbf{v}_r)$, and $\Sigma = \text{diag}(\sqrt{\lambda_1}, \ldots, \sqrt{\lambda_r})$, we obtain the SVD of X:

$$X = V \Sigma U^T \tag{3.85}$$

3.5. ORTHOGONAL COMPONENT ANALYSIS

As discussed in the previous section, data matrices can be represented in two spaces. Analysis of the characteristics of the data cloud can be done by means of eigenvalues or eigenvectors, if one adapts a least-square-based projection method. The treatment was done from a pure algebraic point of view. In this section, we will include a statistical interpretation of the data matrix and collection of samples of a certain dimension. The least-square minimization then becomes a variance maximization. The "data" are no longer simple algebraic values but are now considered outcomes of some multivariate distribution. The workhorse of multivariate statistical analysis is principal component analysis (PCA), which follows immediately from the previous section. It lies at the foundation of other types of data analysis, certainly those that deal with orthogonal axis systems, least squares, covariances, linear regressions, and anything within that realm.

3.5.1. Principal Component Analysis

3.5.1.1. Theory. We now consider the realizations in the data matrix to be realizations of a random vector of dimension N. Each entry in this random vector $\mathbf{X}_N$ is a random variable X_n, or

$$\mathbf{X}_N = (X_1, X_2, \ldots, X_N)^T \tag{3.86}$$

of which we have L samples/realizations

$$\mathbf{x}_N^{(\ell)} = \left(x_1^{(\ell)}, x_2^{(\ell)}, \ldots, x_N^{(\ell)}\right)^T \quad \ell = 1, \ldots, L \tag{3.87}$$

From a statistical point of view, dimension reduction can be achieved by creating standardized linear combinations (SLC of the elements in random vectors):

$$\text{SLC} = \mathbf{u}^T \mathbf{X}_N = \sum_{n=1}^{N} u_n X_n \quad \text{with} \quad \sum_{n=1}^{N} u_n^2 = 1 \tag{3.88}$$

The goal is to simplify the relationships between the various X_n, such that the newly obtained random vector contains random variables with simpler and easier representation of statistical properties; for example, a transformation that minimizes the correlation between the components in Eq. (3.86). In Section 3.4.2 we used a least-square criterion to minimize the loss of information. Here we treat the same problem differently, namely we would like this new random vector to be as close as possible in variance to the original random variable. Information is now as expressed as a variance (note that a variance is least square (!) deviation from mean, so a comparison with two-norms in Eq. (3.73) is not coincidental as it looks).

Practically, we maximize

$$\text{Var}(\mathbf{u}^T \mathbf{X}_N) = \mathbf{u}^T \text{Var}(\mathbf{X}_N) \mathbf{u} = \mathbf{u}^T C_N \mathbf{u} \tag{3.89}$$

Here C_N is the covariance matrix of the N random variables and has size $N \times N$. Hence, we need to solve

$$\max_{\mathbf{u}} \mathbf{u}^T C_N \mathbf{u} \text{ subject to } \|\mathbf{u}\| = 1 \tag{3.90}$$

This problem is similar to the algebraic projection problem in $\mathbb{R}^N$ as outlined in Section 3.4.3, where the matrix to be decomposed was $X^T X$, which basically serves as the basis for calculating the (empirical) covariance matrix (although in algebra such interpretation is not given). Therefore, the solution to Eq. (3.90) is found by calculating the eigenvalues and eigenvectors of the covariance matrix, which can be estimated from the data (see next section). Note that this covariance matrix may be extremely large.

Without yet considering any practical calculation, we write the probabilistic result of this "principal component" transformation as

$$\mathbf{Y}_N = (Y_1, Y_2, \ldots, Y_N)^T = U^T (\mathbf{X}_N - E[\mathbf{X}_N]) \tag{3.91}$$

where we centered the original random variable. By noting that V contains the eigenvectors of C_N with eigenvalues λ_n, $n = 1, \ldots, N$, we have the following properties for $\mathbf{Y}_N$:

$$\begin{aligned} &E[Y_n] = 0 \qquad n = 1,\ldots,N \\ &\mathrm{Cov}[Y_n, Y_{n'}] = 0 \quad n, n' = 1,\ldots,N \;\; n \neq n' \\ &\mathrm{Var}[Y_n] = \lambda_n \qquad n = 1,\ldots,N \\ &\mathrm{Var}[Y_1] \geq \mathrm{Var}[Y_2] \geq \cdots \geq \mathrm{Var}[Y_N] \geq 0 \end{aligned} \tag{3.92}$$

In other words, we obtain a new random vector, whose mean is zero, whose components are not correlated, and whose variances are ranked from high to low, and given by the eigenvalues of the covariance of the original random vector.

3.5.1.2. Practice. The above formulation relies on the theoretical covariance matrix, as defined by expectations. In practice, all the expectations need to be substituted with empirical estimates, subject to their own variances/errors. Consider the estimates of mean and covariance as

$$\begin{aligned} E[\mathbf{X}_N] &\rightarrow \bar{\mathbf{x}}_N \\ C_N &\rightarrow S_N \end{aligned} \tag{3.93}$$

Then using the eigenvalue decomposition on S_N: $S_N = U^T \Lambda_S U$, we obtain the principal components as

$$\mathbf{y}_N^{(\ell)} = U^T \left(\mathbf{x}_N^{(\ell)} - \mathbf{1}_N \bar{\mathbf{x}}_N^T \right) \tag{3.94}$$

With $\Lambda_S = \mathrm{diag}(\lambda_{S,1}, \ldots, \lambda_{S,N})$, the following property follows

$$s_{y_n}^2 = \lambda_{S,n} \; n = 1,\ldots,N \tag{3.95}$$

meaning that the empirical variances $s_{y_n}^2$ depend on the eigenvalue of S_N, also the empirical covariances are zero. Because covariances and variances depend on the scale of a variable, PCA is sensitive to scale changes, such as simply multiplying a variable with a constant. To mitigate this effect, one will need to scale the variables to the same or similar scale. In summary,

$$\begin{aligned} &\text{Calculate } \bar{\mathbf{x}}_N = \frac{1}{L}\sum_{\ell=1}^{L} \mathbf{x}_N^{(\ell)} \\ &\text{Center } S_N = \frac{1}{L}\sum_{\ell=1}^{L} \left(\mathbf{x}_N^{(\ell)} - \bar{\mathbf{x}}_N\right)\left(\mathbf{x}_N^{(\ell)} - \bar{\mathbf{x}}_N\right)^T \\ &\text{Decompose } S_N = U^T \Lambda_S U \\ &\text{Project } \mathbf{y}_N^{(\ell)} = U^T\left(\mathbf{x}_N^{(\ell)} - \mathbf{1}_N \bar{\mathbf{x}}_N^T\right), \; \ell = 1,\ldots,L \end{aligned} \tag{3.96}$$

PCA is not just a "trick" to orthogonalize data and look for combinations of maximum variance. The resulting linear combination and variances contain interesting information that need to be interpreted. Since PCA relies on a linear combination of components of a random vector, it is imperative to look at the resulting weights. The weighting informs which directions best explain variance; hence, it is useful to plot a measure of how well the first n' components explain that variance. This proportion is given as the ratio:

$$\frac{\sum_{n=1}^{n'} \lambda_{S,n}}{\sum_{n=1}^{N} \lambda_{S,n}} = \frac{\sum_{n=1}^{n'} s_{y_n}^2}{\sum_{n=1}^{N} s_{y_n}^2} \tag{3.97}$$

To summarize this information for all $n' = 1, \ldots, N$, a so-called scree plot is created (either as individual contribution or as cumulative contribution). It allows for a direct visual inspection of how much variance the first n' components explain. Then for some given desired variance, one can read off the corresponding n'.

3.5.1.3. Application of PCA to Spatial Models. To illustrate PCA and its relevance to UQ, consider a case where the matrix X contains $L = 1000$ models of some spatial variable (e.g., porosity, hydraulic conductive, saturation). The dimension of a model is 125 × 100, hence $N = 12{,}500$. Figure 3.11 shows a few models; the models appear to exhibit smooth spatial variability. Figure 3.12 shows the

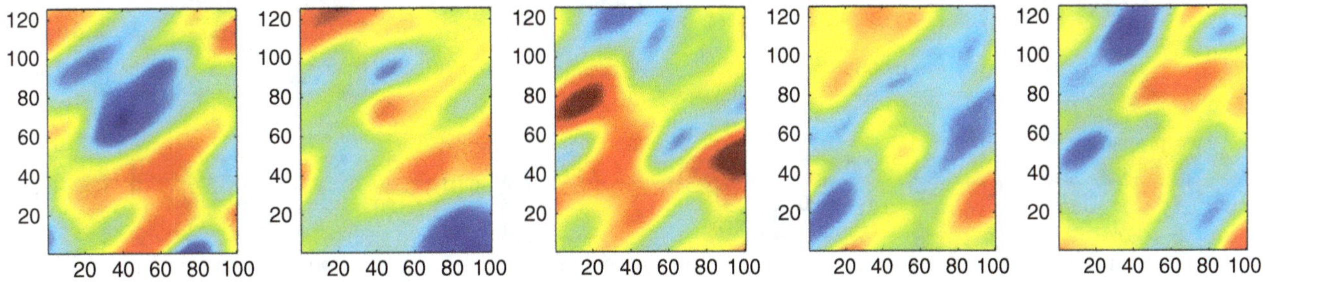

Figure 3.11 Five spatial models out of a set of 1000.

resulting PCA. First, we calculate the cumulative contribution of each principal component (PC) to the variance and create the scree plot. From this plot, we can deduce that 34 PCs are required to explain 95% of variance. Although the size of the covariance matrix is $N \times N$, only a maximum of $L-1$ eigenvalues exist, because of the limited amount of models used.

We can also create a score plot, here the first versus second score (first two entries in $\mathbf{y}_N$), as axes label (and this a convention throughout the book), we list the percent contribution to the variance, 10 and 8% respectively in Figure 3.12. In the score plot, each dot represents an image. The scatter appears to be a bivariate Gaussian distribution with zero correlation. The mean of the realizations is also shown and is close to zero almost everywhere. Next we plot the PCs, the vectors $\mathbf{u}$ of U. Since X contains "images," the resulting PCs are images as well. They also appear to have a meaning: the first PC seems to contain a spatial basis function that is elliptical, somewhat centered in the middle. The second PC seems to contain two ellipses (red and blue areas), one positive and one negative; hence, this represents some gradient in the image. The next PCs contain more ellipses with increasing spatial frequency. Why is this? The mathematical reason for this will be studied in Section 3.7.5. The spatial models here are realizations of a Gaussian process (Section 3.7.5), which requires specifying a mean and spatial covariance. It seems that any realization of such process can be written as a combination (with standard Gaussian coefficients) of filters or basis functions (the

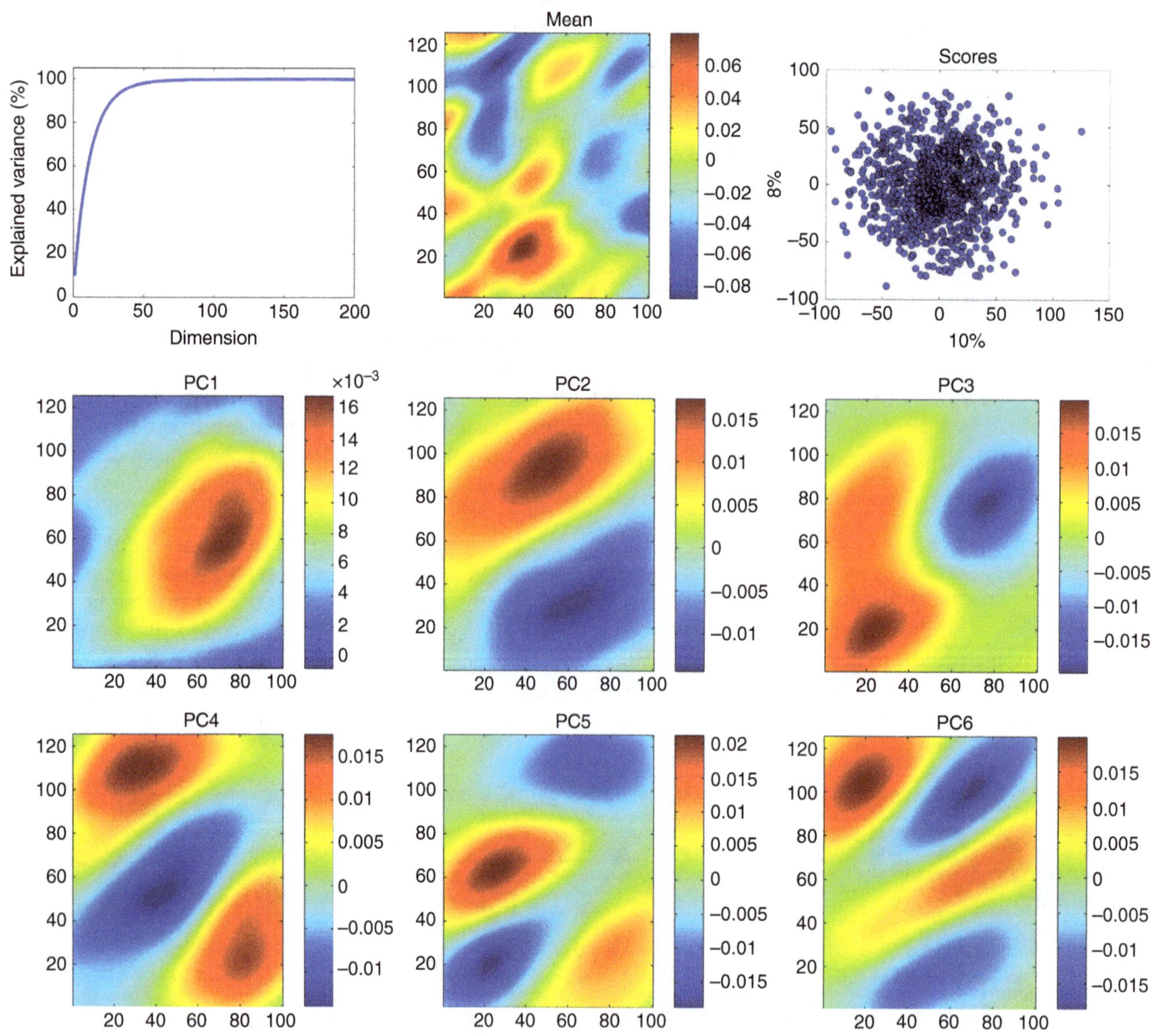

Figure 3.12 Scree plot, average of all models, score plot and 6 out of 1000 PCs, each of dimension 125 × 100.

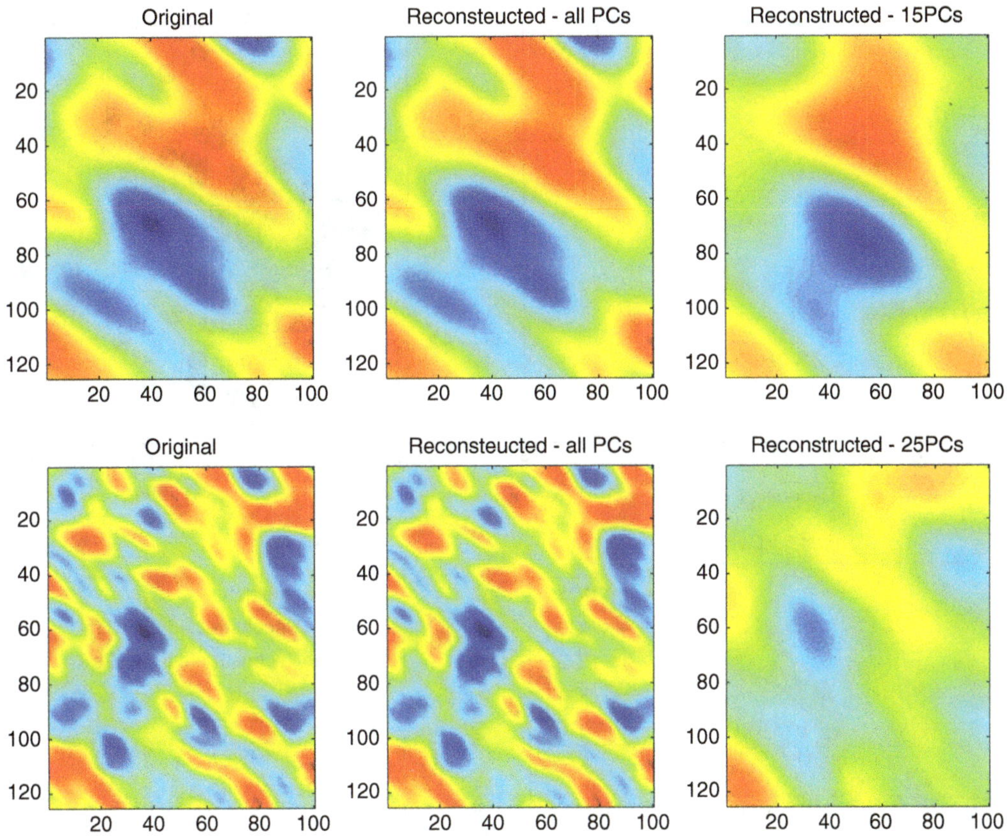

Figure 3.13 Examples of reconstruction of two spatial models with different spatial variability.

PCs) with different spatial frequencies. Hence, it provides a decomposition of the images into a basis and a set of random coefficients.

Figure 3.13 illustrates the bijective nature of PCA, meaning that an image can be reconstructed back, once all PCs and scores are used. By using an amount of PC less than L, we find only an approximation of the original images:

$$\mathbf{x}_{N,\,\text{reconstructed}}^{(\ell)} = \sum_{n=1}^{n'} \left[\mathbf{y}_N^{(\ell)}\right]_n \mathbf{u}_n + \mathbf{1}_N \bar{\mathbf{x}}_N^T \tag{3.98}$$

Even when using a limited amount of PCs ($n' = 15$) we find a reasonable approximation of the original image $\mathbf{x}_N^{(\ell)}$. This is less so when the spatial variability becomes less smooth, see the bottom row of Figure 3.13. This makes intuitive sense: higher-frequency models require more frequencies (PCs) to represent them.

Decomposition and reconstruction with PCA is appealing because it is bijective. However, the method does not work so well for variables that are not Gaussian, or variables that have physical bounds, such as the concentrations. Figure 3.14 illustrates this, where PCA is applied to a set of concentration curves. PCA reveals that only a few PCs are required to explain variance. This makes sense, the variation in these curves is not complex; one could easily fit a parametric function with a few parameters. However, the PCs, which are unbounded, have negative values as shown in Figure 3.14. As a result, any reconstruction with limited PCs will lead to unphysical values (negative concentrations).

3.5.2. Multidimensional Scaling

3.5.2.1. Basic Example. In Section 3.4.4, we presented a dual algebraic view on the data matrix, one based on X^TX and one based on XX^T. The statistical variation of the same type of orthogonal analysis, termed PCA is based on X^TX which is essentially the empirical covariance matrix (up to a factor of $1/L$). It therefore makes sense when $L \ll N$ to look at the dual of PCA, which will now be based on the dot-product XX^T. This method is termed multidimensional scaling (MDS).

In presenting MDS, we will not start from a Cartesian axis system and space, but we will start from a space in which only distances are defined: a metric space. As an illustrative example, consider L cities and a distance

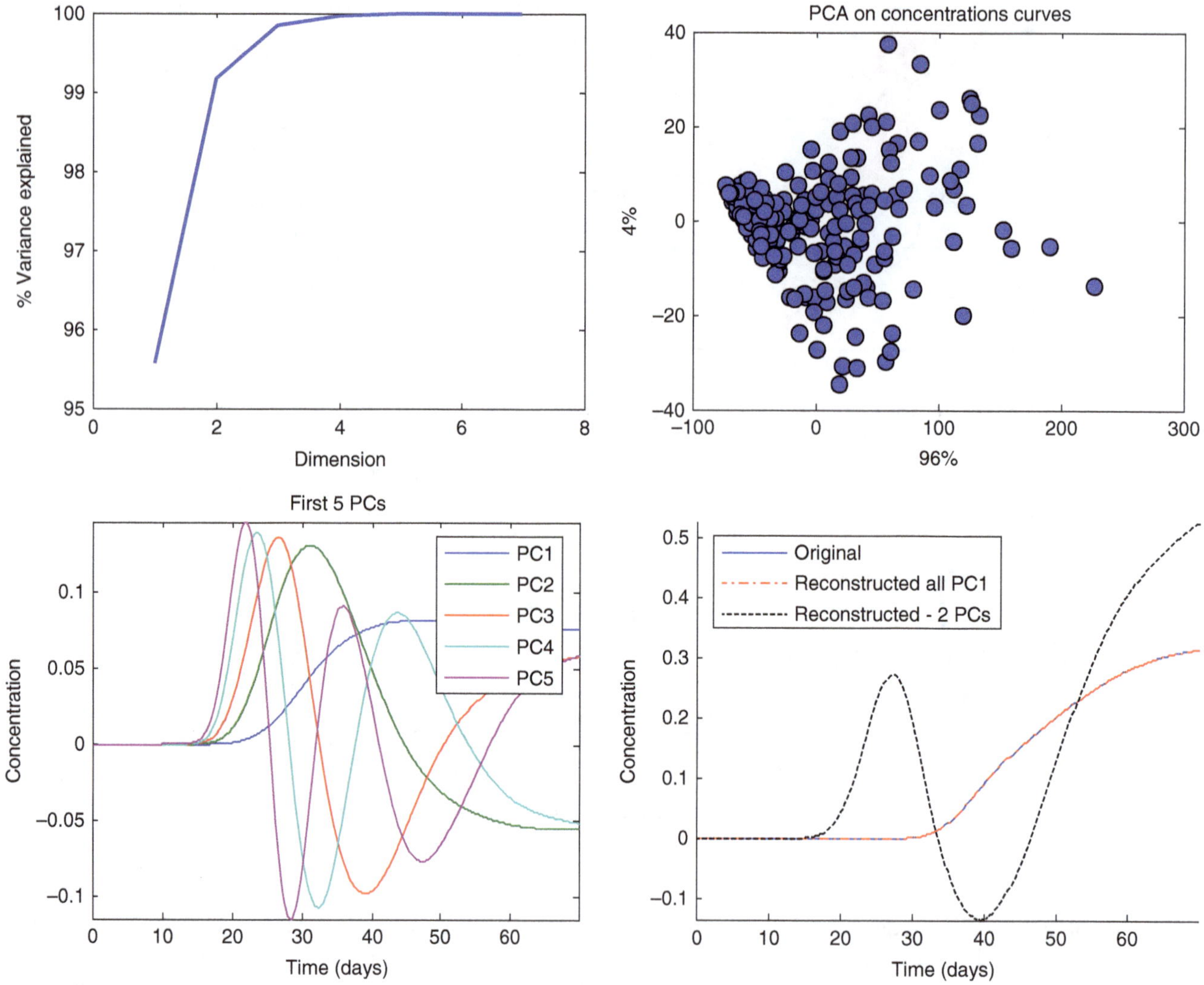

Figure 3.14 PCA applied to the set of concentration curve of Figure 3.2.

table between those cities (see Figure 3.15). Like PCA, MDS is a projection method, but now such projection occurs from a metric space into a Cartesian space. More specifically, MDS finds the projection that best matches the distance in the distance table. Like PCA, such projection can be done in $1D$, $2D$, ..., LD. Note that the maximal projected Cartesian space has dimension L (and not N as for PCA). In that sense, it is also a dimension reduction method.

3.5.2.2. MDS Projection. In MDS, we start from an $L \times L$ distance matrix. This could be the differences between any two models, any two data responses, or any two predictions. What is important is that from now on we will work with $\mathbf{x}_N^{(\ell)}, \ell = 1, \ldots, L$ (column-wise) and not $\mathbf{x}_n^{(L)}, n = 1, \ldots, N$ (row-wise) as in PCA. To lighten the notation, we will write $\mathbf{x}^{(\ell)}$ instead of $\mathbf{x}_N^{(\ell)}$. Consider first the Euclidean distance:

$$d_{\ell\ell'}^2 = \left(\mathbf{x}^{(\ell)} - \mathbf{x}^{(\ell')}\right)^T \left(\mathbf{x}^{(\ell)} - \mathbf{x}^{(\ell')}\right),\ 1 \le \ell, \ell' \le L\ \ \mathbf{x}^{(\ell)}, \mathbf{x}^{(\ell')} \in \mathbb{R}^N \tag{3.99}$$

The usual notation of $d_{ij}^2 = \left(\mathbf{x}_i - \mathbf{x}_j\right)^T \left(\mathbf{x}_i - \mathbf{x}_j\right)$ is now put in the specific context of UQ problems. The aim of MDS is to recover in lower dimensions, the original Cartesian coordinates of $\mathbf{x}^{(\ell)}$ from a distance matrix, in this case containing Euclidean distances. However, since our dual treatment of the data matrix only works on either the covariance or the dot-product matrices, we need to retrieve the dot-product matrix from the Euclidean

	Boston	NYC	DC	Miami	Chic	Seattle	SF	LA	Denver
Boston	0	206	429	1504	963	2976	3095	2979	1949
NYC	206	0	233	1308	802	2815	2934	2786	1771
DC	429	233	0	1075	671	2684	2799	2631	1616
Miami	1504	1308	1075	0	1329	3273	3053	2687	2037
Chic	963	802	671	1329	0	2013	2142	2054	996
Seattle	2976	2815	2684	3273	2013	0	808	1131	1307
SF	3095	2934	2799	3063	2142	808	0	379	1235
LA	2979	2786	2631	2687	2054	1131	379	0	1059
Denver	1949	1771	1616	2037	996	1307	1235	1059	0

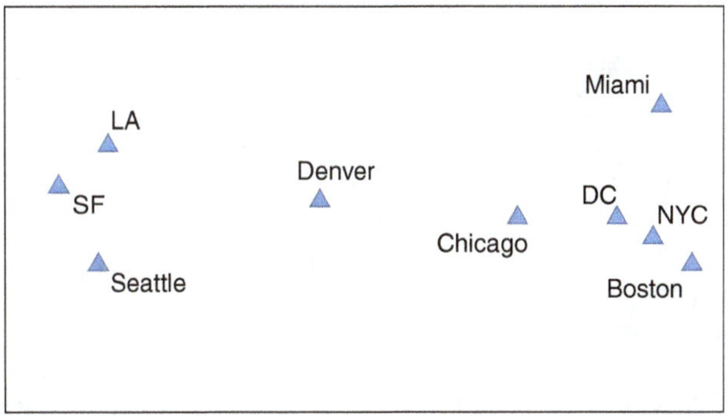

Figure 3.15 MDS: turning a distance table between cities into a 2D map.

distance matrix. To achieve this, we rely on the following relationship:

$$d^2_{\ell\ell'} = \mathbf{x}^{(\ell)^T}\mathbf{x}^{(\ell)} + \mathbf{x}^{(\ell')^T}\mathbf{x}^{(\ell')} - 2\mathbf{x}^{(\ell)^T}\mathbf{x}^{(\ell')} = b_{\ell\ell} + b_{\ell'\ell'} - 2b_{\ell\ell'} \tag{3.100}$$

with $b_{\ell\ell'}$ a dot-product between $\mathbf{x}^{(\ell)}$ and $\mathbf{x}^{(\ell')}$. Similar to centering operations on the covariance matrix in PCA, we need to center the dot-product matrix to obtain

$$\sum_{\ell=1}^{L} b_{\ell\ell'} = 0 \tag{3.101}$$

This is not as trivial because we can no longer calculate a mean directly from data. Instead, we impose this constraint as follows:

$$\begin{aligned} \frac{1}{L}\sum_{\ell=1}^{L} d^2_{\ell\ell'} &= \frac{1}{L}\sum_{\ell=1}^{L} b_{\ell\ell} + b_{\ell'\ell'} \\ \frac{1}{L}\sum_{\ell'=1}^{L} d^2_{\ell\ell'} &= b_{\ell\ell} + \frac{1}{L}\sum_{\ell=1}^{L} b_{\ell'\ell'} \\ \frac{1}{L^2}\sum_{\ell'=1}^{L}\sum_{\ell=1}^{L} d^2_{\ell\ell'} &= \frac{2}{L}\sum_{\ell=1}^{L} b_{\ell\ell} \end{aligned} \tag{3.102}$$

With this centering, we can now write the dot-product as a function of the Euclidean distance:

$$b_{\ell'\ell'} = -\frac{1}{2}d^2_{\ell\ell'} - \frac{1}{2}\left(d^2_{\ell\bullet} + d^2_{\bullet\ell'} - d^2_{\bullet\bullet}\right) \tag{3.103}$$

where the • refers to the summing over the indices Eq. (3.103). We can also put the same expression in matrix notation. Consider, then, A the matrix containing the entries $-\frac{1}{2}d^2_{\ell\ell'}$ and H the centering matrix, in this case, $= I_L - \frac{1}{L}\mathbf{1}_L\mathbf{1}_L^T$. Then the dot-product matrix B can be written as

$$\begin{aligned} B &= (HA)(HA)^T \\ &= XX^T \end{aligned} \tag{3.104}$$

Now consider the eigenvalue decomposition of B

$$\begin{aligned} B &= V_B\Lambda_B V_B^T \\ &= \left(V_B\Lambda_B^{1/2}\right)\left(V_B\Lambda_B^{1/2}\right)^T \end{aligned} \tag{3.105}$$

Hence, the coordinates of X can be reconstructed by means of

$$X = V_B\Lambda_B^{1/2} \tag{3.106}$$

One of the strengths of MDS is that any distance can be used, not just the Euclidean distance. Figure 3.16 illustrates overhead images of 136 from a flume experiment (see Chapter 5 for details on this data set). The modified Hausdorff distance is calculated for each pair of images. MDS is then used to create a 2D map of these images (just like mapping of cities). In this map, images that look alike (small distance) are plotted close by.

3.5.2.3. Kernel Density Estimation in Metric Space.

MDS generates a Euclidean space of dimension $n \ll L$. The higher the dimension the better that space approximates the variability of the physical variable as modeled by the distance defined. The Euclidean space approximates how close variables are with respect to each other; hence, if the original variable is a stochastic variable (of very high dimension possibly), then MDS creates an empirical density of these variables in lower-dimensional space. If the approximation can be done in a low dimension (e.g., <6), then one can estimate the pdf from the cloud of points. A useful method in that regard is kernel density estimation [*Silverman*, 1986, see Section 3.3.2]. When applied to a set of reconstructed coordinates $\mathbf{x}_n^{(\ell)}, \ell = 1, \ldots, L$ with some low dimension $n \ll L$, then a kernel density estimate is

$$f(\mathbf{x}_n) = \frac{1}{L}\sum_{\ell=1}^{L} K\left(\frac{\mathbf{x}_n - \mathbf{x}_n^{(\ell)}}{\sigma}\right) \tag{3.107}$$

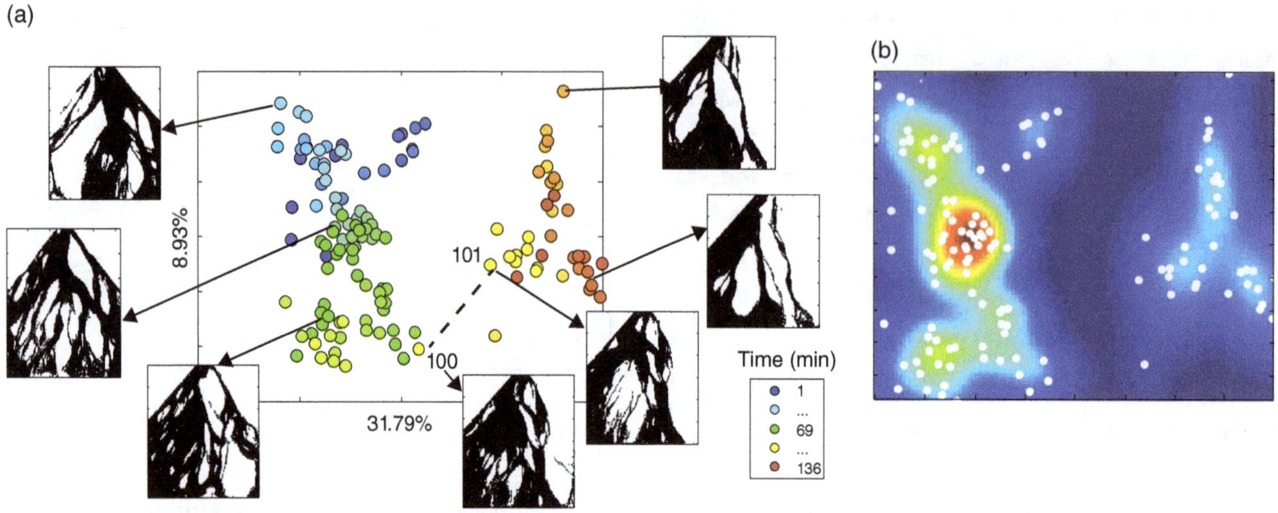

Figure 3.16 (a) MDS projection of 136 binary overhead snapshots of a flume experiment, using the modified Hausdorff distance. (b) Kernel density estimation in MDS projection.

Figure 3.16 shows an example of kernel density estimation applied to the flume experiment images.

As another example, consider again the simple hydro case. First, we calculate the Euclidean difference between any two model realizations in our hydro case:

$$d_{\ell\ell'}^2 = \left(\mathbf{m}^{(\ell)} - \mathbf{m}^{(\ell')}\right)^T \left(\mathbf{m}^{(\ell)} - \mathbf{m}^{(\ell')}\right), \quad 1 \le \ell, \ell' \le L \quad (3.108)$$

or simply the square difference between values in the grid. Note that these realizations have different variogram ranges, nuggets, and types. Figure 3.17 shows the score plot colored by the type of variogram (spherical vs. Gaussian).

Consider now calculating the distance between the concentration responses calculated from these hydraulic conductivity (Figure 3.17). The score plot of Figure 3.17 looks a lot like the score plot of Figure 3.14. This makes sense, because of the duality of MDS with Euclidean distance and PCA. An interesting observation in Figure 3.17 is the coloring of the dots with parameter values that were used to generate the models. In the first plot, we observe that the red and blue dots occur randomly, while in the second plot (mean k), we observe a clear trend. The interpretation is that mean k impacts the response while the variogram does not. This will be used in Chapter 4 to develop methods of sensitivity analysis.

3.5.3. Canonical Correlation Analysis

Continuing our study of "data" and the analysis of data matrices from an algebraic and statistical point of view, we now study two data matrices jointly and how they relate to each other. We will again rely on a least-squares framework with Cartesian axis, Euclidean distances, and orthogonal projections. A method of multivariate analysis for exploration and quantification of the relationships between two data matrices under these principles is termed canonical correlation analysis (CCA). Like PCA and MDS, CCA is an orthogonal method that relies on projections in lower dimensions.

3.5.3.1. Theory. Consider two random vectors $\mathbf{X}, \mathbf{Y} \in \mathbb{R}^N$ (or $\mathbb{R}^L$, or they can be of different dimension). We consider the linear combinations

$$\begin{array}{c}\mathbf{a}^T\mathbf{X} \\ \mathbf{b}^T\mathbf{Y}\end{array} \quad (3.109)$$

The correlation between these two random variables can be derived as

$$\rho\left(\mathbf{a}^T\mathbf{X}, \mathbf{b}^T\mathbf{Y}\right) = \frac{\mathbf{a}^T \operatorname{cov}(\mathbf{X}, \mathbf{Y})\mathbf{b}}{\sqrt{\mathbf{a}^T \operatorname{var}(\mathbf{X})\mathbf{a}}\sqrt{\mathbf{b}^T \operatorname{var}(\mathbf{Y})\mathbf{b}}} = \frac{\mathbf{a}^T C_{XY}\mathbf{b}}{\sqrt{\mathbf{a}^T C_{XX}\mathbf{a}}\sqrt{\mathbf{b}^T C_{YY}\mathbf{b}}} \quad (3.110)$$

In CCA we find the maximum of this correlation. This maximum depends on the singular value decomposition (eigenvalues in case dim($\mathbf{X}$) = dim($\mathbf{Y}$)) of the following matrix

$$C = C_{XX}^{1/2} C_{XY} C_{XX}^{1/2} = V \Lambda U^T \quad (3.111)$$

namely, the linear combinations that result in maximal correlation are the canonical correlation vectors

$$\begin{array}{ll}\mathbf{a}_k^c = C_{XX}^{-1/2}\mathbf{v}_k & k = 1, \ldots, \operatorname{rank}(C) \\ \mathbf{b}_k^c = C_{YY}^{-1/2}\mathbf{u}_k & \end{array} \quad (3.112)$$

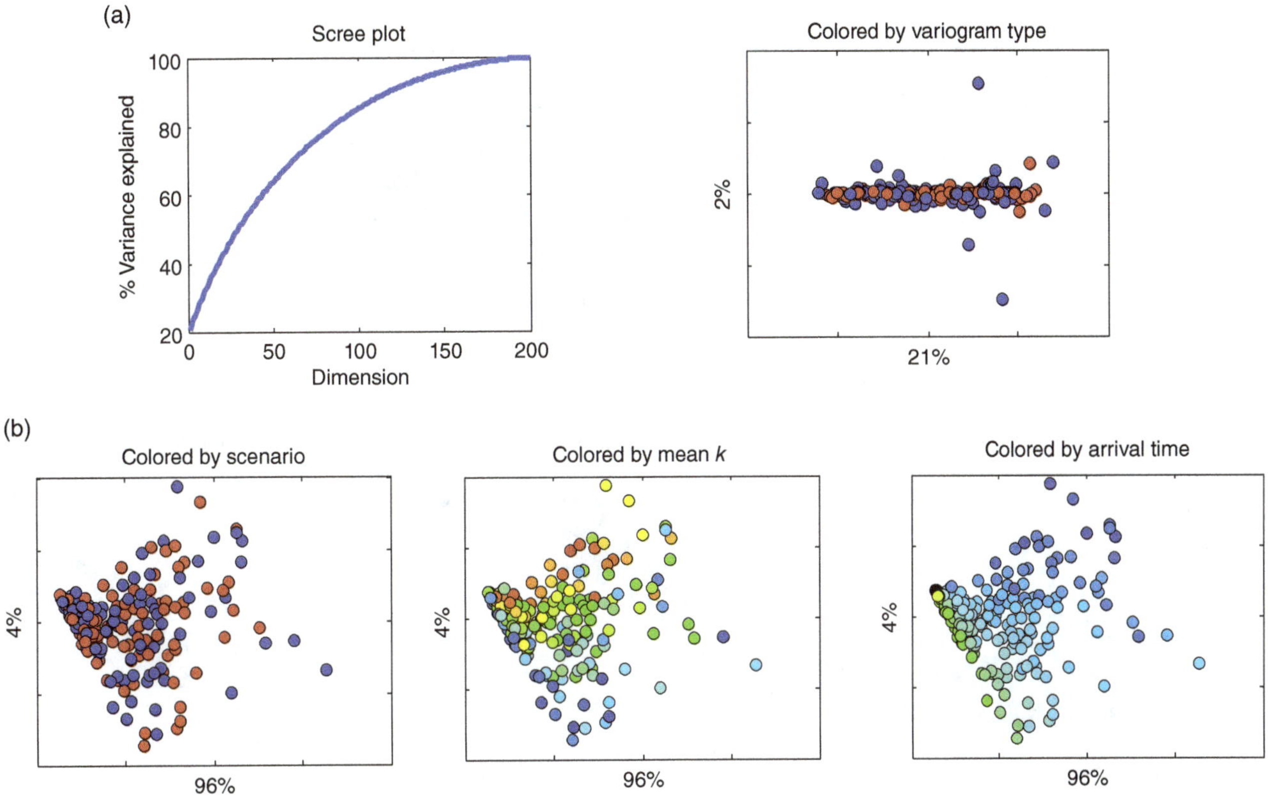

Figure 3.17 (a) Scree plot and MDS projection when the distance is the Euclidian distance between hydraulic conductivity realizations. (b) MDS projection when the distance is the Euclidian distance between realizations of concentration in time. The plots are colored by model parameters.

This results in a set of new vectors whose pair-wise components are maximally correlated (and hence cross-components are minimally correlated):

$$X_k^c = \left(\mathbf{a}_k^c\right)^T \mathbf{X} \qquad k = 1, \ldots, \mathrm{rank}(C)$$
$$Y_k^c = \left(\mathbf{b}_k^c\right)^T \mathbf{Y} \tag{3.113}$$

The maximal correlations are function of the eigenvalues

$$\rho_{\max,\mathrm{k}}^c = \sqrt{\lambda_k} \tag{3.114}$$

3.5.3.2. Example. In constraining predictions, we often use data. For example, we use pressure data to constrain permeability or head data to constrain hydraulic conductivity. The hope is that by building calibrated models, the prediction will have less uncertainty than not doing so. This kind of calibration (inversion, Chapter 6) becomes very difficult when models and data become very complex and high dimensional. In Chapter 7, we will develop an alternative approach that "learns" prediction directly from data. In such learning CCA will be used as one method to learn. A simple example is as follows. Consider the models generated in our hydro case and consider that both data variables and prediction variables have been evaluated. The data consists of seven measurements of hydraulic head, the prediction is the concentration. They should be related. The question is how this can be visualized, after all we are dealing with a seven-dimensional variable and a function. To do this, we apply first PCA on each variable (see Figure 3.18). The first PC of the data does not correlate with the first PC of the prediction. CCA finds the optimal linear combinations that find this correlation. Consider that for the data we retain two PCs (about 90%) and for the prediction we retain five PCs. This means that a maximum of two canonical components can be calculated, as shown in Figure 3.18. We note the high correlation of 0.86 in the first components, which is encouraging since it means that data are correlated with the prediction, so reduction in uncertainty on the prediction is possible. In Chapter 7, we will use CCA to build linear relationship between complex objects, whether vectors, functions, or 2D/3D maps.

3.6. FUNCTIONAL DATA ANALYSIS

3.6.1. Introduction

Multivariate data analysis comprises methodologies to address the statistical analysis of data and inference with

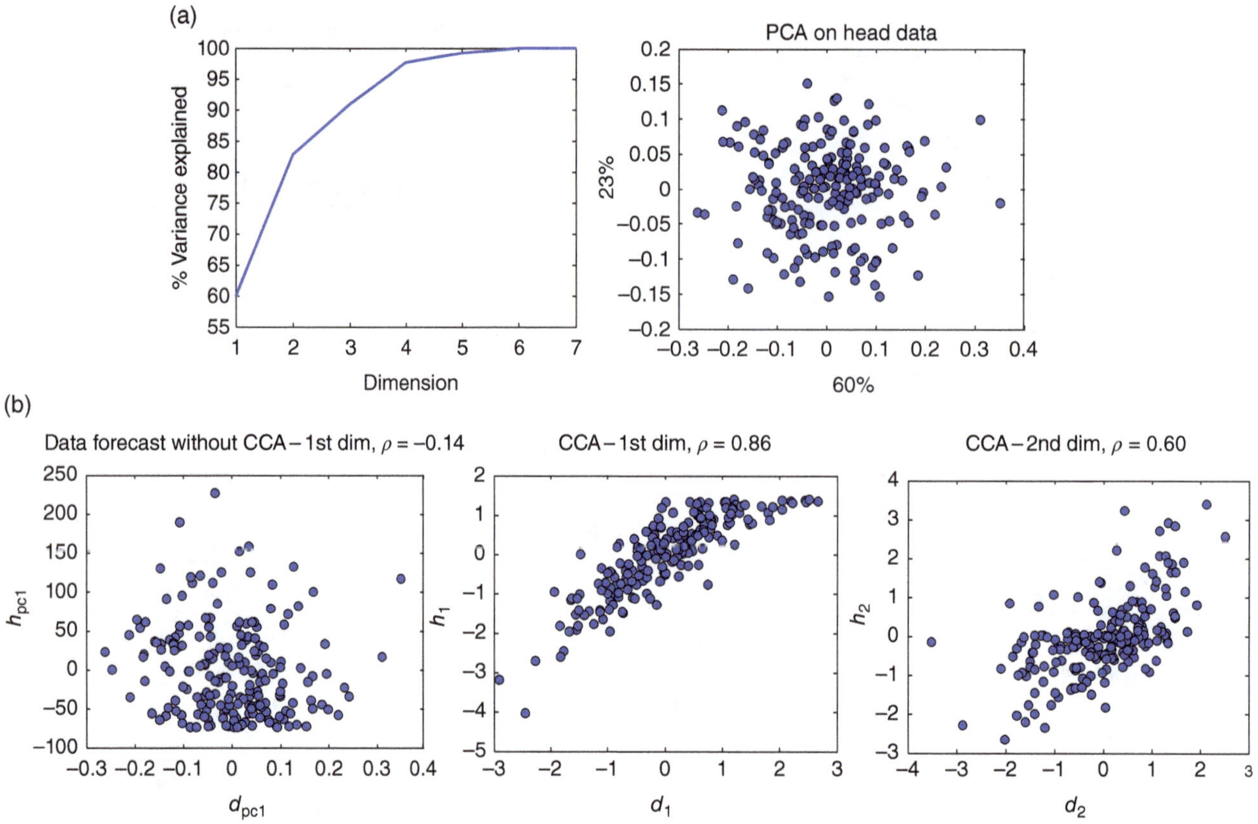

Figure 3.18 (a) PCA of seven head measurements and score plot. (b) Correlation between PCs is almost zero. CCA shows correlation of 0.86 and 0.6 between the canonical components of data and prediction.

high-dimensional data and models. In certain cases, it is however more effective to resort to functional data analysis (FDA) [*Ramsay and Silverman*, 2005]. The decision between the two is a subjective choice. Broadly speaking, FDA relies on the notion that the data has some systematic variation. In the context of this book, this can refer to cases where physical dynamics are driving the system, but because of uncertainty, the exact nature is not fully known; only what that systematic part looks like. For example, a concentration of some substance starts at zero and then gradually increases to level off at 100%, but we do not know when that starts or how fast or when it levels off. In the latter case, the variable is a function of time, and hence in theory, infinite-dimensional. In multivariate analysis, we would instead discretize time into small intervals and treat time instance variables as high-dimensional (and highly correlated). This is not required in FDA and is a major advantage. FDA is a non-parametric approach. It does not rely on fitting functions that represent physics. For example, in the production of shale gas (see Figure 3.19) the function is simply a peak (peak gas) followed by some decline. The parametric approach would be to fit these parametric functions to observed decline data to get parameter estimates. The problem is that the functional form has to be known and currently it is not clear, for shale gas production, which forms are appropriate. FDA avoids this by using standard basis functions that can be used in many applications.

3.6.2. A Functional Basis

More formally, FDA assumes that changes in any measurement of a physical variable over space or time is based on an underlying smooth physical process that in turn can be mathematically represented using a continuous and differentiable mathematical function and that this function not always needs to be known for analyzing measurements. This assumption allows for the decomposition of any time series measurement into a linear combination of underlying continuous functions called basis functions, forming a functional basis. Multiple functional bases such as a sinusoidal basis, a Fourier basis, a polynomial basis, an exponential basis, or a spline basis are available and the choice between them is application driven. The spline basis has an advantage over the others because of its versatility in terms of computational ease

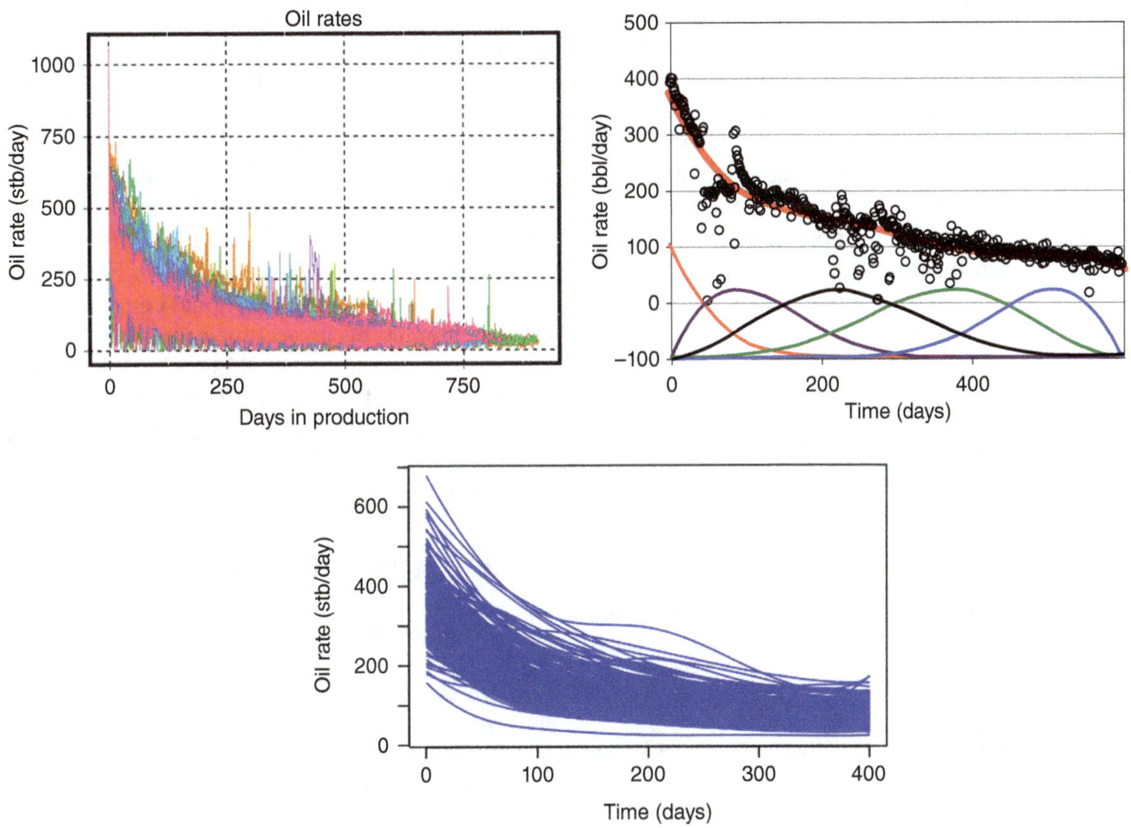

Figure 3.19 Raw data of oil rate decline in 200 wells; basis function and fitting with smoothing; all 200 smoothed data.

of evaluation, as well as their derivatives. This flexibility will be used throughout this book. Consider a time series $x(t)$ as an example. Using a spline basis of K spline functions $\{\psi_1(t), \psi_2(t), \ldots, \psi_K(t)\}$, the time series is approximated by a linear combination

$$x(t) \cong \sum_{k=1}^{K} c_{\psi,k}\psi_k(t) \tag{3.115}$$

where the FDA components $c_{\psi,k}$ are the scalar linear combination coefficients of the spline function $\psi_k(t)$. In terms of matrix notation, consider matrix X containing L samples of time series sampled at N time instances, the FDA composition can be written as

$$X \simeq C^{\Psi}\Psi \tag{3.116}$$

where the $K \times N$ matrix Ψ contains the values of the K spline basis functions at the N values of time $t = t_1, t_2, \ldots, t_N$ as row vectors, and the $L \times K$ matrix C^{Ψ} contains the K coefficients or FDA components of each of the L time series as row vectors.

Consider the example of shale oil production from 200 wells as shown in Figure 3.19. The coefficients in Eq. (3.115) are found by least-square fitting with a regularization term

$$\underset{c_{\psi,k}}{\operatorname{argmin}} \sum_{\ell=1}^{L}\left(x(t) - \sum_{k=1}^{K} c_{\psi,k}\psi_k(t)\right)^2 + \lambda \int \frac{d^2}{dt^2}\left(\sum_{k=1}^{K} c_{\psi,k}\psi_k(t)\right)^2 dt \tag{3.117}$$

The regularization term avoids overfitting noisy data by adding a roughness penalty in terms of the second derivative. FDA therefore requires specifying two tuning parameters: λ the amount of smoothing and K the number of basis functions. These are usually obtained by means of cross-validation [*Ramsay and Silverman*, 2005].

3.6.3. Functional PCA

As discussed in Section 3.5.1, PCA is a multivariate analysis procedure that uses an orthogonal transformation to convert a set of observations of possibly correlated variables into a set of values of linearly uncorrelated variables. PCA identifies the principal modes of variation from the eigen-vectors of the covariance matrix. Functional PCA (FPCA) simply consists of doing eigenvalue decomposition on the vectors of component data $c_{\psi,k}$ obtained after functional decomposition. Hence FDA turns a functional problem into a vector problem on which classical multivariate techniques apply. The end result is that the function can be approximated by a linear combination of eigen-functions $\phi_m(t)$:

$$x(t) \cong \sum_{m=1}^{M} c_m^f \phi_m(t) \tag{3.118}$$

Conventionally $M \ll K$, so the PCA step of FPCA achieves a dimension reduction. How do we get to (3.118)? Recall that PCA relies on an eigenvalue decomposition of the covariance matrix of a random vector X:

$$C_N \mathbf{u} = \lambda \mathbf{u} \text{ or } C_N = U^T \Lambda U \tag{3.119}$$

The covariance matrix is of size $N \times N$. In a functional space, however, we define the empirical covariance function of a set of sample functions $\{x^{(1)}(t), \ldots, x^{(L)}(t)\}$:

$$c(t_n, t_{n'}) = \frac{1}{L} \sum_{\ell=1}^{L} x^{(\ell)}(t_n) x^{(\ell)}(t_{n'}) \tag{3.120}$$

The eigen-problem is now defined as an integral

$$\int c(t_n, t_{n'}) u(t_{n'}) dt_{n'} = \lambda u(t_n) \tag{3.121}$$

If we define

$$C\mathbf{u} = \int c(\cdot, t_{n'}) u(t_{n'}) dt_{n'} \tag{3.122}$$

which is an integral transform C of the function u. C is now a covariance operator instead of a covariance matrix, hence

$$C\mathbf{u} = \lambda \mathbf{u} \tag{3.123}$$

which looks like PCA but now with functions. One major difference, however, lies in the rank of the covariance matrix versus the integral transform. In PCA, the rank of the empirical covariance matrix is maximally equal with $L-1$, because $L \ll N$ the dimension of the vector. In functional analysis the function can be sampled infinitely; hence, N can be very large. Practically, we need to retain only a few dimensions in most cases, here denoted as M. Expression (3.118) is obtained by equating the first M eigen-functions $u_m(t)$, $m = 1, \ldots, M$ with $\phi_m(t)$, $m = 1, \ldots, M$ and the weights (the component scores) as

$$c_m^f = \int \phi_m(t) x(t) dt \tag{3.124}$$

Evidently, all integrals are approximated by sums.

Figure 3.20 illustrates the use of FPCA to the shale oil decline curves. Similar to PCA, a score plot of the FPC can be produced. Additionally, one can plot the eigenfunctions, which are often termed "harmonics" referring to components of vibration of a string fixed at each end. Here these "vibrations" are the changes of the function around the mean. To visualize this better, one often plots this mean, added and subtracted with some multiple of the eigen-function. This is shown in Figure 3.20. Now one notices how the first functional component (75.3% of variance) represents a variation around the mean, the second functional component (12.6%) has a stationary point around 50 days. This can be attributed to a change in production from flow due to fracturation to flow through the geological medium. The third component has two stationary points, one early and one around 100 days.

3.7. REGRESSION AND CLASSIFICATION

3.7.1. Introduction

Regression and classification methods are important elements of statistical learning [*Hastie et al.*, 2009]. Numerous methods have been developed, from simple linear regression to nonlinear methods such a neural network or deep learning [see e.g., *Bishop*, 1995]. In general, these are denoted as methods for "predictive learning from data." In selecting a suitable method, one will need to account for a variety of criteria. Table 3.2 provides an overview of some popular methods, together with criteria against which each method can be evaluated. Artificial neural networks (ANN) constitute a large family of nonlinear regression models that create a nonlinear mapping between input and output variables. The initial idea of ANN was to solve problems in the same way the human brain would. A support vector machine (SVM) uses supervised learning to create a classifier that separated regions in space by means of hyperplanes. The k-nearest neighbor is a local classification or regression method that uses an amount k of nearest data near a location in input space that needs to be classified and regressed. The result is simply a majority vote (classification) or an average (regression). Kriging accounts for the correlation between predictors and allows to interpolate data. In this section, we will focus on several of these methods and how they apply in the context of UQ.

3.7.2. Multiple Linear Regression

The simplest form of linear regression is multiple linear regression. It is used to model the relationship between one or more data variables $\mathbf{X} = \{X_1, X_2, \ldots, X_N\}$ and a single continuous prediction variable Y. The fundamental assumption of multiple linear regression is that the

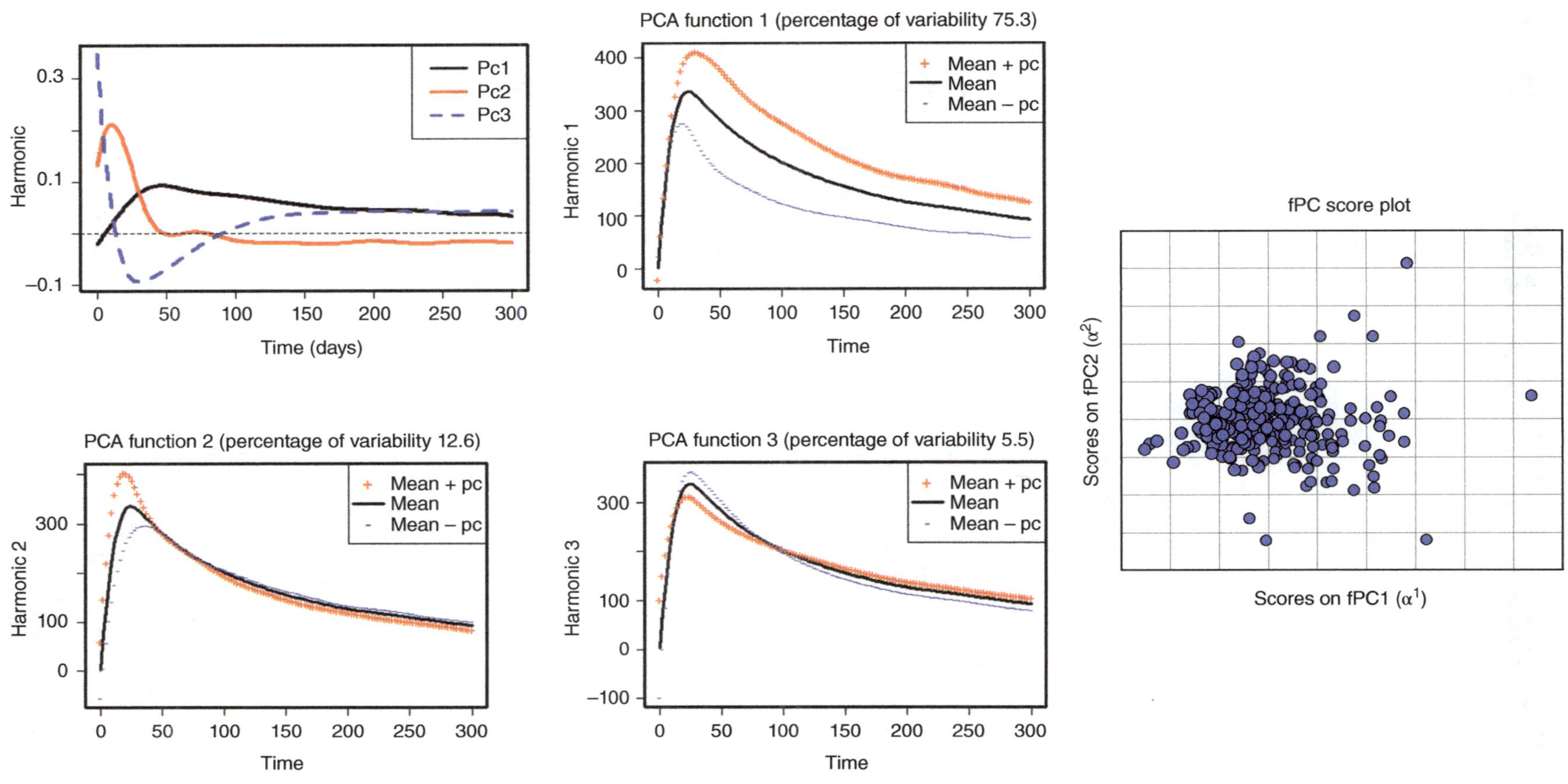

Figure 3.20 FPCA with harmonics, PC functions, and score plot.

Table 3.2 Overview of methods of regression and classification and how they score on various criteria.

Criteria	Neural net	Kriging	SVM	Kernel *k*-NN	Trees	Boosted trees
Mixed-type data	–	+	–	–	+	+
Missing data	–	+	–	+	+	+
Robust to outliers	–	0	–	+	+	+
Computational scalability	–	0	–	–	+	+
Deal with irrelevant input	–	0	–	–	+	+
Ease of interpretation	–	+	–	–	0	–
Predictive power	+	0	+	+	–	+

Source: Idea adapted and extended from *Hastie et al.* [2009].
– = poor; + = good; 0 = average.

conditional expectation function $E(Y|\mathbf{X})$ can be expressed using the linear relation:

$$E(Y|\mathbf{X}) = \beta_0 + \beta_1 X_1 + \beta_2 X_2 + \cdots + \beta_N X_N \tag{3.125}$$

β_0 is commonly referred to as the intercept while $\beta_1, \ldots, \beta_n$ are termed coefficients. We also assume the existence of an unobserved random variable ϵ that adds noise to the linear relationship. Given L measurements of $\mathbf{X}$ and Y that are independently and identically distributed, the modeled relationship can be expressed as

$$y^{(\ell)} = \beta_0 + \beta_1 x_1^{(\ell)} + \beta_2 x_2^{(\ell)} + \cdots + \beta_N x_N^{(\ell)} + \epsilon^{(\ell)} \text{ for } \ell = 1, \ldots, L \tag{3.126}$$

In matrix notation, this is written as

$$\mathbf{y} = X\boldsymbol{\beta} + \boldsymbol{\epsilon} \tag{3.127}$$

where each of the matrices is given as

$$X = \begin{pmatrix} 1 & x_1^{(1)} & \cdots & x_N^{(1)} \\ \vdots & \vdots & \ddots & \vdots \\ 1 & x_1^{(L)} & \cdots & x_N^{(L)} \end{pmatrix} \tag{3.128}$$

$$\boldsymbol{\beta} = \begin{pmatrix} \beta_0 \\ \vdots \\ \beta_N \end{pmatrix}, \mathbf{y} = \begin{pmatrix} y^{(1)} \\ \vdots \\ y^{(L)} \end{pmatrix}, \boldsymbol{\epsilon} = \begin{pmatrix} \epsilon^{(1)} \\ \vdots \\ \epsilon^{(L)} \end{pmatrix} \tag{3.129}$$

To solve for the unknown coefficients and intercept, an estimation method is required. A variety of different techniques can be used, for example ordinary least squares (OLS), generalized least squares, ridge regression, maximum likelihood estimation, and so on. Each method varies in terms of computational complexity and underlying assumptions. Refer to *Hastie et al.* [2009] for a discussion of various estimators and their properties. We will consider the simplest and most commonly used estimator: OLS. The solution from OLS in matrix notation is as follows:

$$\hat{\boldsymbol{\beta}} = \left(X^T X\right)^{-1} X^T \mathbf{y} \tag{3.130}$$

Using this estimated $\hat{\boldsymbol{\beta}}$, the predicted values of Y can be computed as $\hat{\mathbf{y}} = X\hat{\boldsymbol{\beta}}$. The difference between these predicted and observed values of Y is called the residual:

$$\hat{r} = \mathbf{y} - \hat{\mathbf{y}} = \left(I - X\left(X^T X\right)^{-1} X^T\right)\mathbf{y} \tag{3.131}$$

The OLS solution is the one that minimizes the sum of the squares of the residuals. It does however, require additional assumptions. First of all, the error random variable should have a zero mean, $E[\boldsymbol{\epsilon}] = 0$, and be uncorrelated with the data variables $E[X^T\boldsymbol{\epsilon}] = 0$. Furthermore, the data variables in X must be linearly independent, the failure of which is termed multicollinearity. Finally, the error term must be homoscedastic, that is, the variance of the errors should not change between different observations. This is expressed mathematically as $E[(\epsilon^{(\ell)})^2 | X] = \sigma^2$ for $\ell = 1, \ldots, L$, where σ^2 is finite.

When each of the previous conditions are met, OLS can be used to make estimates of the prediction variable for new data observations X_{new}.

$$\hat{\mathbf{y}}_{\text{new}} = X_{\text{new}}\hat{\boldsymbol{\beta}} \tag{3.132}$$

This is also known as the minimum-variance unbiased estimator. By performing this analysis, we can make estimates of a prediction variable using multiple data variables.

3.7.3. Support Vector Machines

SVMs are powerful and popular statistical models used for both regression and classification. It was originally invented as a linear binary classifier [*Vapnik and Lerner*, 1963] but has since seen numerous improvements to handle nonlinear boundaries, soft margins, regression, and multiple classes. Given a set of training samples, each labeled as being in one of the two classes, the basic SVM aims to train a linear hyperplane that separates the samples according to its class.

The basic SVM is applied when we have a set of real predictor variables X and binary predictor variable $Y \in \{-1, 1\}$. SVM attempts to fit a linear hyperplane $\mathbf{w}^T\mathbf{x} + \mathrm{b} = 0$ that separates the samples into two classes. This means finding the value of $\mathbf{w}$ can then be expressed as an optimization problem:

$$\min_{w} \|\mathbf{w}\|$$

subject to

$$y^{(\ell)}\left(\mathbf{w}^T\mathbf{x}^{(\ell)} + \mathrm{b}\right) \geq 1 \quad \forall \ell = 1, \ldots, L \tag{3.133}$$

This can be solved using quadratic programming. Once the hyperplane has been estimated, predictions for a new sample $\mathbf{x}^*$ can be made by evaluating

$$f(\mathbf{x}^*) = \mathrm{sgn}\left(\mathbf{w}^T\mathbf{x}^* + \mathrm{b}\right) \tag{3.134}$$

This optimization problem can be solved using Lagrangian multipliers, in which a multiplier $\lambda^{(\ell)}$ is assigned to each of the constraints. The resulting decision boundary can then be expressed as

$$f(\mathbf{x}^*) = \mathrm{sgn}\left(\sum_{\ell=1}^{L} \lambda^{(\ell)} y^{(\ell)} K\left(\mathbf{x}^*, \mathbf{x}^{(\ell)}\right) + \mathrm{b}\right) \tag{3.135}$$

The function $K(\mathbf{x}^*, \mathbf{x}^{(\ell)}) = \phi(\mathbf{x}^*)^T\phi(\mathbf{x}^{(\ell)})$ is known as the kernel function. ϕ is a function that projects $\mathbf{x}^{(\ell)}$ into a new space. This is particularly useful when the samples are not linearly separable in the original $\mathbf{x}^{(\ell)}$ space; ϕ can be used to map the samples into one in which they can be linearly separated. This newly transformed space can be of any arbitrary dimension and yield complex separating hyperplanes that may be computationally expensive to solve. However, by realizing that the decision boundary only requires the dot-product of the vectors in the transformed space, we can replace it with a chosen kernel function and forgo the transformation. This is known as the kernel trick and is what allows SVMs to work with nonlinearly separable data. Section 3.8 describes some possible choices of kernel functions.

Another use of SVMs is that of anomaly detection. This occurs when the training samples are only of a single class, but we need to determine if new samples fall within this class or represent an anomaly. Essentially, the idea is to find a minimal volume hypersphere around the training sample. Any new samples that fall outside the hypersphere are classified as anomalies. This hypersphere is fitted in a transformed space using the kernel trick and is parametrized by its center coordinate $\mathbf{a}$ and its radius R. In a one-class SVM, we cannot maximize the distance between the prior and the unknown non-prior class. Instead, we fit the hyperplane such that all the prior samples are on one side, and we maximize the distance between the hyperplane and the origin of the data space.

$$\min_{R,\mathbf{a}} \|R\|$$

subject to

$$\left\|\mathbf{x}^{(\ell)} - \mathbf{a}\right\|^2 \leq R^2 \quad \forall \ell = 1, \ldots, L \tag{3.136}$$

The decision boundary then becomes

$$f(\mathbf{x}^*) = \mathrm{sgn}\left(\sum_{\ell=1}^{L} \alpha_i K\left(\mathbf{x}^*, \mathbf{x}^{(\ell)}\right) - R^2\right) \tag{3.137}$$

This one-class SVM is useful for detecting when new samples are not part of the class in which the training samples are part of. Chapter 7 has application of one-class SVM for detecting if a prior distribution within a Bayesian context is consistent (or not) with some observed data.

3.7.4. CART: Classification and Regression Trees

3.7.4.1. Single Tree Methods. The basic idea of tree-based methods is simple, yet powerful. Figure 3.21 shows a simple tutorial example with two predictor variables X_1 and X_2. Trees rely on a binary partition of the input space and model Y by means of a constant in the partitioned regions. In other words, the model of Y is piecewise discontinuous. The hierarchy of this partitioning can be represented with a tree-like topology, such as Figure 3.21. The estimate for Y is simply the average of the $y^{(\ell)}$ in each region. The main question is, therefore, how to split the input space in these regions (what is the topology of the tree?).

For regression, the tree model is

$$y = \sum_{m=1}^{M} c_m I(\mathbf{x} \in R_m) \tag{3.138}$$

which is simply a linear combination of discrete indicator functions (I) over M regions R_m. Based on the observations $(\mathbf{x}^{(\ell)}, y^{(\ell)})$, $\ell = 1, \ldots, L$, a regression model is then estimated by calculating the coefficients as follows:

$$\begin{gathered} \hat{c}_m = \frac{1}{\#\left(y^{(\ell)} | \mathbf{x}^{(\ell)} \in R_m\right)} \sum_{y^{(\ell)} | \mathbf{x}^{(\ell)} \in R_m} y^{(\ell)} \\ \Rightarrow \hat{y} = \sum_{m=1}^{M} \hat{c}_m I(\mathbf{x} \in R_m) \end{gathered} \tag{3.139}$$

Basically, this is simply the average of the sample values of y in each region. For categorical variables that are

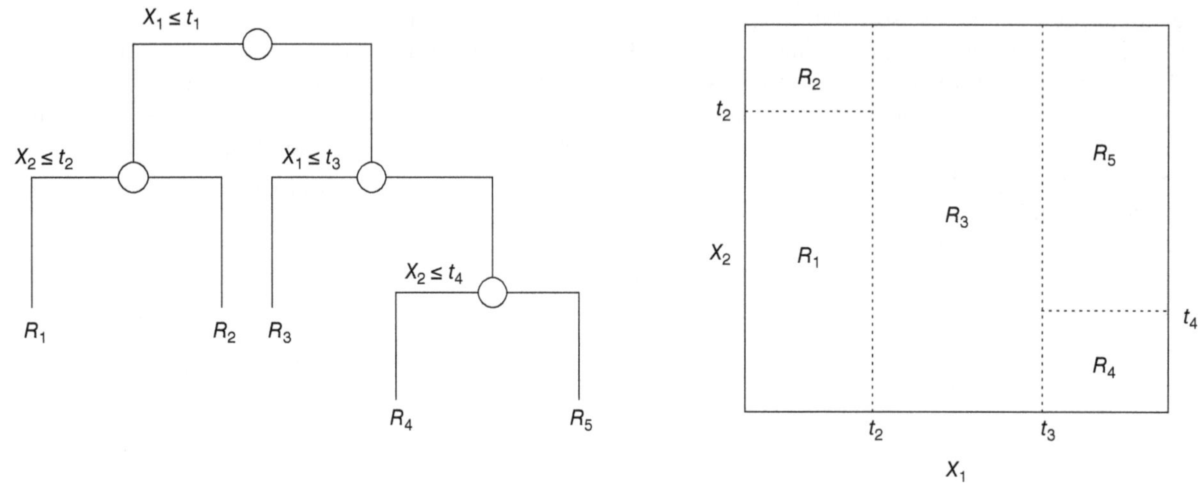

Figure 3.21 Example of a regression tree with two variables and four splits.

ordinal, the splitting occurs at discrete instances. However, for categorical variables that are not ordinal (e.g., a geological scenario), one needs to explore all combinatorials of splitting, which may become large.

To find an optimal partition, one could formulate a least-square criterion between the model and the data, then try out every possible partition. This is computationally infeasible. Instead, a greedy algorithm is used where the least squares is defined based on first the split variable then on the split-point for that choice of variable. The reason for this is that sorting through the universe of variables is easier than the possibilities of split-points. The question now is how large to grow the tree: too large of a tree will lead to overfitting the data. The tree size determines the complexity of the model of Eq. (3.138). The main idea here is to grow a large tree, then "prune" the tree. To achieve this, one formulates a new least-square criterion but now adding a regularization term that includes the number of terminal nodes (see *Breiman et al.* [1984] for details).

For classification the same idea is used except that the least-squares criterion is dropped in favor of a more suitable one for classification. For classification, one models in each region R_m the estimated proportions of each class $y \in \{k = 1, \ldots, K\}$, as $\hat{p}_{mk}$. In such regions, the class with the largest proportion is taken as the estimated class. An example of a suitable criterion (instead of least squares) for establishing an optimal tree is, for example, to minimize entropy

$$E = -\sum_{k=1}^{K} \hat{p}_{mk} \log(\hat{p}_{mk}) \tag{3.140}$$

Consider now the application of the regression tree to the hydro case of Section 3.1. The aim here is to emulate the forward model of predicting contaminant arrival time (y) using six input (predictor) variables. Four parameters related to the hydraulic conductivity, K_{mean}, K_{sd}, K_{range}, and K_{Cov}, and two parameters related to boundary conditions, H_{rivGrad}, H_{range}, are used. The tree model is shown in Figure 3.22, split first on K_{mean}, then on H_{rivGrad}, and so on.

3.7.4.2. Boosted Trees. As shown in Table 3.2, the prediction power of trees is lesser than most other methods, they seldom have more accuracy than what can be achieved with the data itself [*Hastie et al.*, 2009]. Figure 3.23 shows that in the hydro case the prediction performance of regression of arrival time in terms of correlation with a test set is only 0.43. This is partly due to the piecewise continuous nature of the model. It is, therefore, considered a weak classifier, meaning it would not do much better than random selection. A powerful method for dealing with such cases is "boosting." The idea of boosting is that it uses outcomes of weak classifiers and produces a powerful "committee" by combining the strengths of each weaker classifier (the whole is better than each of the individuals). The idea of boosting is very simple. It works for both regression and classification, but consider classification as an example here. A classifier, such as a tree, takes some input $\mathbf{x}$ and classifies it using $y(\mathbf{x})$, with possible class outcomes $k = 1, \ldots, K$. The idea of boosting is to generate a sequence of classifiers on repeatedly modified versions of the same data.

Intuitively it works as follows. Consider a data set $(\mathbf{x}^{(\ell)}, y^{(\ell)})$, $\ell = 1, \ldots, L$ of i.i.d. samples, meaning each has equal weight $w_\ell = 1/L$. A base classifier, such as the tree, will perform better on certain pairs $(\mathbf{x}^{(\ell)}, y^{(\ell)})$ than on other pairs. Boosting consist of generating a new tree classifier but now using an error model that has increased weights on

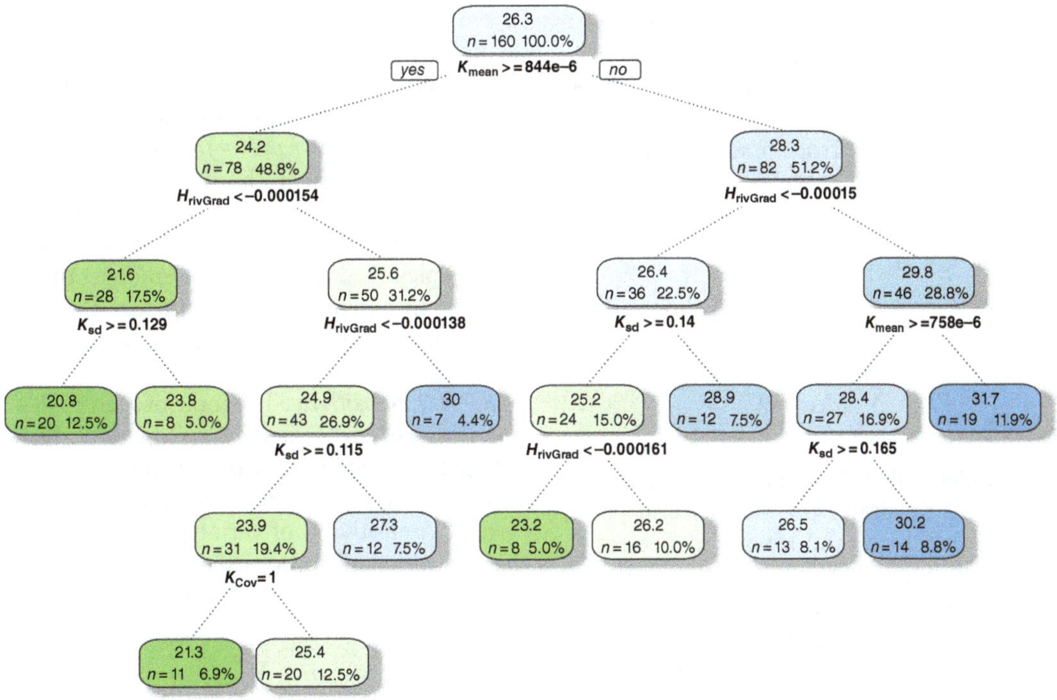

Figure 3.22 Regression tree for the case of Section 3.1, involving six variables and 200 runs of the model. The color is the average arrival time. The color indicates the deviation from the global average (26.3).

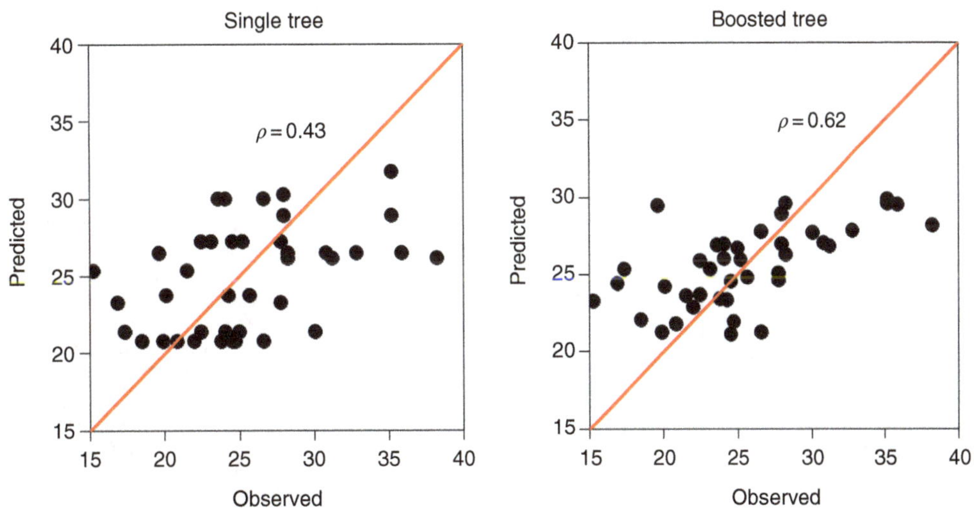

Figure 3.23 Application of a single tree and boosted tree to a test set, shown are observed and predicted arrival time.

samples with large errors. The classifier is, therefore, forced to concentrate on samples that are hard to classify. Each of the generated classifiers in this sequence is still weak classifiers (increasing weights may also decrease performance on others). Therefore, the ultimate classifier is taken as a weighted combination of the sequence of classifiers.

Mathematically, this means that a specification of error is needed for classifier y_m in the sequence $m = 1, \ldots, M$

$$\text{err}(m) = \frac{\sum_{\ell=1}^{L} w_\ell I\left(y^{(\ell)} \neq y_m\left(\mathbf{x}^{(\ell)}\right)\right)}{\sum_{\ell=1}^{L} w_\ell} \tag{3.141}$$

that this error is turned into a weight for that classifier y_m

$$\alpha_m = \log\left(\frac{1-\mathrm{err}(m)}{\mathrm{err}(m)}\right) \tag{3.142}$$

and that this weight and the error are used to determine the re-weighted samples

$$w_\ell \leftarrow w_\ell \exp\left(\alpha_m I\left(y^{(\ell)} \neq y_m\left(\mathbf{x}^{(\ell)}\right)\right)\right), \ell = 1,\ldots,L \tag{3.143}$$

for some initial set of weights $w_\ell = 1/L$. The final classifier is then

$$y(\mathbf{x}) = \sum_{m=1}^{M} \alpha_m y_m(\mathbf{x}) \tag{3.144}$$

A boosted tree model is, therefore, simply a weighted sum of trees. Figure 3.23 illustrates the difference in performance between a single tree and a boosted tree for the hydro regression problem. Here 80% of the data (160 runs) are used to establish the tree, while 20% (40 runs) are withheld to evaluate the tree. Even if improved, the rather low correlation also shows that other variables, not used in fitting the tree, affect the response, in particular the heterogeneity of the hydraulic conductivity. In Chapter 4, we will introduce methods that can include these variability tree methods and hence even further their performance.

Tree methods span a large variety of ideas on the same theme. Bagging or bootstrap aggregation with trees consists of generating bootstrap samples of the data set, then averaging the estimator generated with these bootstrap sample. We will discuss *The Bootstrap* in Section 3.13. Unlike bagging, boosting, a so-called committee of weak learners, varies over time. Another modification of boosting and bagging are *random forests*. Here we also generate bootstrap samples of the data, but in addition, when growing the tree, at each terminal node of the tree, we select randomly a subset of variables of the N variables x_n and pick the best split-point amongst that subset. This creates an ensemble of trees (a forest). In a regression problem, one simply averages the trees.

3.7.4.3. Sensitivity. The topic of "sensitivity" of predictor variables on predictants will be extensively treated in Chapter 4. Regression trees constitute one such method since it allows ranking the relative importance (or broadly, sensitivity) of variables $\mathbf{X}$ in predicting Y. *Breiman et al.* [1984] proposed the following quantification of the importance for each variable x_n:

$$I_n^2 = \sum_{j=1}^{J-1} i_j^2 I(v(j) = n) \tag{3.145}$$

To explain this intuitively, consider Figure 3.24. The summing here is over internal nodes, a total of $J-1$. The importance of a variable is quantified by how much the square error improves when splitting the tree on that variable. For example, in the first internal node $j = 1$, we split on the second variable $v(1) = 2$, and calculate how much

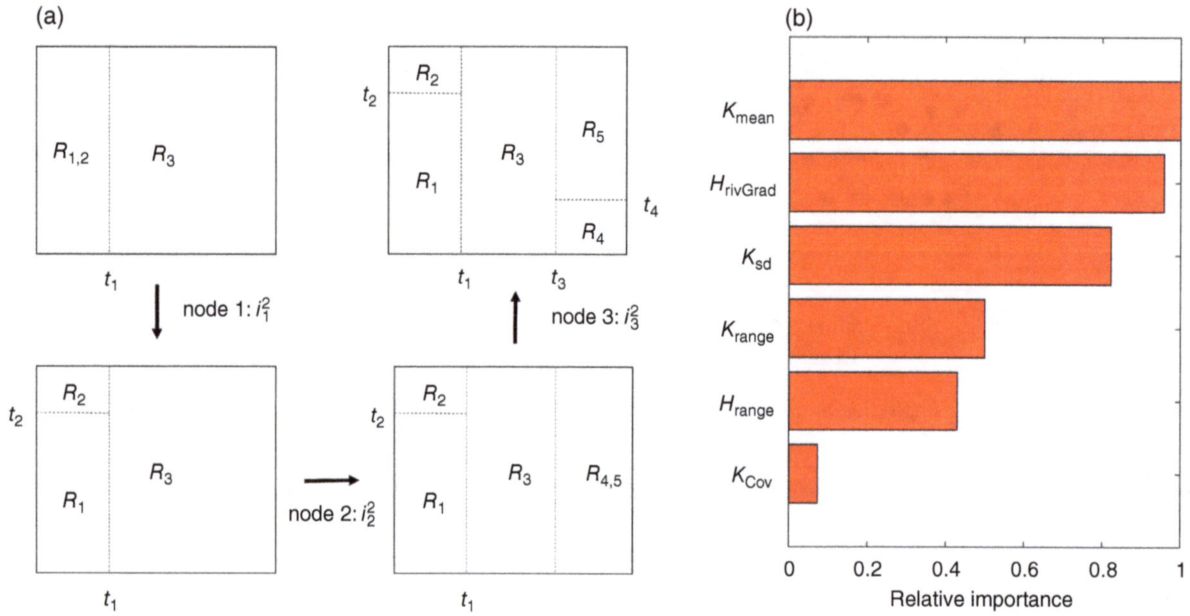

Figure 3.24 (a) Example calculation of variable importance (sensitivity) for the case of Figure 3.21 using regression trees. (b) Relative importance (sensitivity) using boosted tree model for the arrival time response. Relative importance is expressed in percent.

the square error improves as i_2^2. This is continued over all internal nodes, for Figure 3.24, we therefore get

$$I_1^2 = i_2^2 \quad I_2^2 = i_1^2 + i_3^2 \tag{3.146}$$

In Chapter 4, we will use this basic notion and extend it to perform sensitivity analysis on much more complex situations than a few simple scalars and develop methods of sensitivity for objects such as curves, maps, or volumes, given any type of input variable.

When using boosted trees, the relative importance is averaged over all trees in the sequence. Figure 3.24 shows the application of this to the arrival time response case. The mean hydraulic conductivity together with the river gradient have the largest relative importance.

3.7.5. Gaussian Process Regression: Kriging

3.7.5.1. Introduction. Least-square problems comprise a large family of statistical learning and inference problems. The "square" in the names is relevant as differentiation of squared functions lead to linear functions, which then allows for straightforward linear algebra. In addition, the Gaussian distribution contained a squared function (after taking the log), leading to a preponderant role of the Gaussian family of distributions in least-square problems. In presenting these methods, we provide a survey of two literature, which treat the same problem: Gaussian process regression (used in the statistical community) and simple kriging (developed somewhat independently in the geostatistical community).

In uncertainty quantification, least-square problems take an important role, in particular when relationships between models and data or data and prediction are linear. In all such cases, the inference problem is reduced to estimating first- and second-order statistics, in particular conditional means, variance, and covariance of, for example, the model parameters given the data observation. Closed form expressions for the posterior distribution modeling the uncertainty are available in such cases. Evidently, most real-world problems are not Gaussian, nor linear, and hence more advanced methods rely on extending these linear method to more general cases, which we present in Chapter 6.

3.7.5.2. What Is a Gaussian Process?. A Gaussian process is a stochastic process defined over a continuous domain of some finite dimension (e.g., 1D-time, 2D/3D = space). The stochastic nature means that we do not know their outcome at each point of the domain, but we assume it has a Gaussian distribution as a model of uncertainty of that unknown value. We also assume that any finite set of unknown values follows a multivariate Gaussian distribution and, as a consequence, any finite set of linear combinations of unknown values is also multivariate Gaussian. Note that Gaussian processes can be defined over any high-dimensional domain, not just 1D, 2D, or 3D. In 3D, a Gaussian process is also termed a Gaussian "random field." For example, we may have some unknown model variables that are spatially distributed over a grid:

$$\mathbf{m} = \left(m(\mathbf{s}_1), \ldots, m\left(\mathbf{s}_{N_{\text{gr}}}\right)\right) \tag{3.147}$$

with location in space $\mathbf{s}_n$. Consider now any pair of unknown model variables, then the covariance function is

$$C(\mathbf{s}_n, \mathbf{s}_{n'}) = \text{cov}(m(\mathbf{s}_n), m(\mathbf{s}_{n'})) \quad \forall n, n' \tag{3.148}$$

under stationarity assumptions over the domain this reduces

$$C(\mathbf{s}_n, \mathbf{s}_{n'}) = \text{cov}(\mathbf{s}_n - \mathbf{s}_{n'}) \quad \forall n, n' \tag{3.149}$$

with $\mathbf{s}_n - \mathbf{s}_{n'}$ the distance between these two locations in the domain. In geostatistics, one usually uses the variogram instead of the covariance:

$$\gamma(\mathbf{s}_n - \mathbf{s}_{n'}) = \text{var} - \text{cov}(\mathbf{s}_n - \mathbf{s}_{n'}); \quad \text{var} = \text{cov}(\mathbf{0})$$

The Gaussian process is completely defined by the second-order statistics, mean, and covariance function (or variogram). Common covariance functions used are exponential, Matern, or linear, each leading to Gaussian processes with different characteristics. Examples of some realizations of Gaussian processes are shown in Figure 3.25.

An important property of a Gaussian process relates to how Gaussian processes or data that follow such processes can be orthogonalized, meaning having their correlation removed (as was done using PCA). More specifically, consider the orthogonal decomposition of the covariance function as follows (Mercer's theorem):

$$C(\mathbf{s}_n, \mathbf{s}_{n'}) = \sum_{j=1}^{\infty} \lambda_j \psi_j(\mathbf{s}_n) \psi_j(\mathbf{s}_{n'}) \tag{3.150}$$

where $\psi_j(\mathbf{s})$ are termed eigen-functions and λ_j eigenvalues such that

$$\lambda_j \psi_j(\mathbf{s}_n) = \int_{-\infty}^{\infty} C(\mathbf{s}_n, \mathbf{s}_{n'}) \psi_j(\mathbf{s}_{n'}) d\mathbf{s}_{n'} \tag{3.151}$$

This looks similar to PCA (eigen-decomposition of the covariance matrix) but now written with integrals and functions instead of sums and vectors. Mercer's theorem can be used to turn a Gaussian process into a sum of uncorrelated random variables.

$$X(\mathbf{s}_n) = \sum_{j=1}^{\infty} \sqrt{\lambda_j} \psi_j(\mathbf{s}_n) Z_j \quad Z_j \sim N(0,1) \tag{3.152}$$

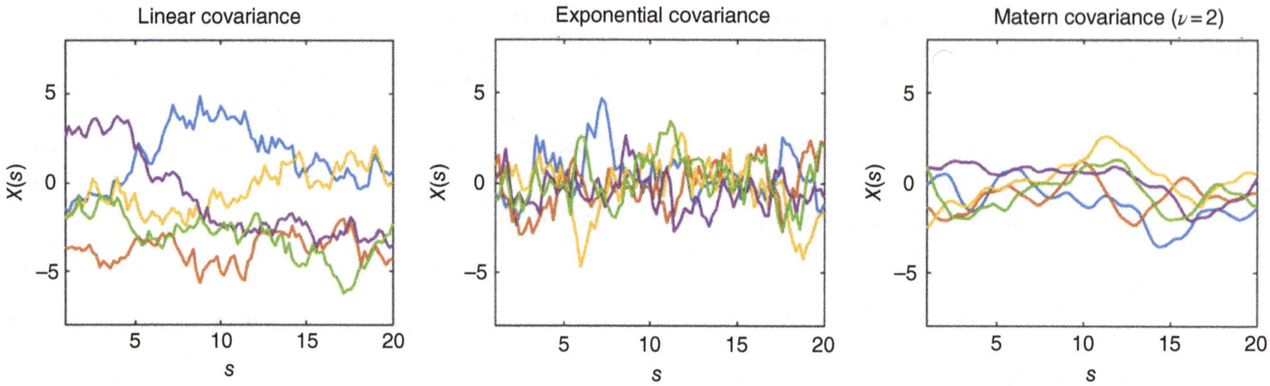

Figure 3.25 Five realizations of Gaussian processes for different covariance models.

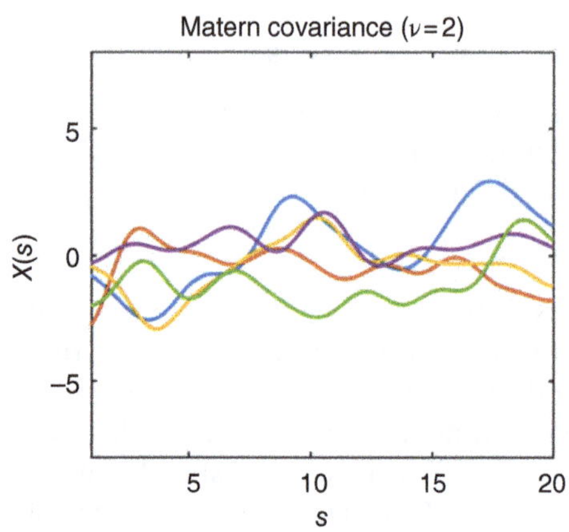

Figure 3.26 Example of approximate realizations of a Gaussian process generated using Karhunen–Loeve expansion (compare with the left figure in Figure 3.25).

Within the context of this book $X(\mathbf{s})$ can represent anything, whether model, data, or prediction variable, hence of any dimension. This decomposition is also termed the Karhunen–Loeve expansion, which is basically a generalization of SVD (PCA) to continuous spaces. A Gaussian process can be approximated by a finite sum by truncating the sum based on decreasing eigenvalues (see Figure 3.26):

$$X(\mathbf{s}_n) \cong \sum_{j=1}^{J} \sqrt{\lambda_j}\psi_j(\mathbf{s}_n)Z_j \quad Z_j \sim N(0,1) \tag{3.153}$$

Generating (approximate) realizations of the Gaussian process can now proceed simply by drawing Gaussian deviate z_j.

3.7.5.3. Prediction with Gaussian Processes. Consider first the general problem of predicting some unknown value, assuming that the unknown values as well as any observed value were somehow generated from a Gaussian process. Note that this cannot be verified [*Mariethoz and Caers*, 2014], since we do not have any replicates of the process. Still, it is interesting to study prediction (and hence uncertainty quantification) under such conditions.

Suppose we have observed the process at certain points $x(\mathbf{s}_1),\ldots,x(\mathbf{s}_N)$ and we want to predict X at some desired location $\mathbf{s}_0$ (with unobserved value of the process). Given the assumption of a Gaussian process, we assume the random vector $(X(\mathbf{s}_1),\ldots,X(\mathbf{s}_N),X(\mathbf{s}_0))$ is multivariate Gaussian. We also assume for convenience that the mean of the process is zero (and known). The covariance matrix of that multivariate Gaussian distribution is partitioned as follows:

$$K_+ = \begin{pmatrix} K & \mathbf{k} \\ \mathbf{k} & k_0 \end{pmatrix} \tag{3.154}$$

with

K : covariance between any two data locations
$\mathbf{k}$: vector of covariance between any data location and the unknown
k_0 : (prior) variance at the location to be predicted

The multivariate distribution of $(X(\mathbf{s}_1),\ldots,X(\mathbf{s}_N), X(\mathbf{s}_0))$ is completely known and needs to be conditioned on specific observations $(x(\mathbf{s}_1),\ldots,x(\mathbf{s}_N))$. The resulting conditional distribution of $X(s_0)$ is also Gaussian and has conditional mean and covariance [*von Mises*, 1964]

$$\begin{aligned} E[X(s_0)|x(\mathbf{s}_1),\ldots,x(\mathbf{s}_N)] &= \mathbf{k}^T K^{-1}\mathbf{x} \\ \mathbf{x} &= (x(\mathbf{s}_1),\ldots,x(\mathbf{s}_N)) \\ \operatorname{var}(X(\mathbf{s}_0)|x(\mathbf{s}_1),\ldots,x(\mathbf{s}_N)) &= k_0 - \mathbf{k}^T K^{-1}\mathbf{k} \end{aligned} \tag{3.155}$$

Important to note is that the expected value is a linear combination of the data and that the conditional variance is not a function of the observed values. Instead, it uses the same linear combination ($\mathbf{k}^T K^{-1}$), but it is now applied to the covariances $\mathbf{k}$ and compared (subtracted) from the

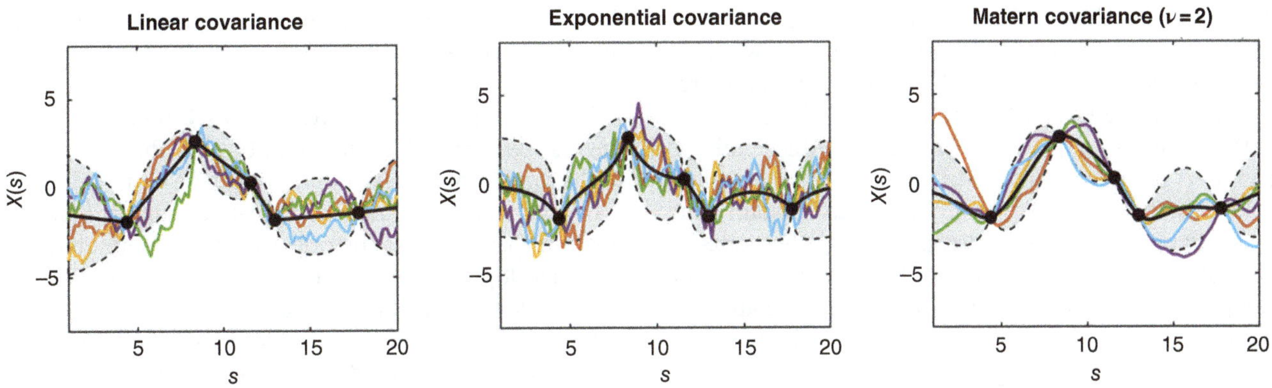

Figure 3.27 Five realizations of Gaussian processes for different covariance models that are conditioned to available data (represented as black dots). The black line represents the mean value and the shaded area represents the 95th confidence interval. This illustrates that Gaussian process regression (kriging) is an exact interpolator.

prior variance (not knowing anything, except that the process is Gaussian with some variance). A useful property of this result is that prediction is exact at the data locations. This means that the expected value, as function of **s**, interpolates exactly the data, as shown in Figure 3.27. This property makes Gaussian process regression an excellent candidate as a surrogate model to emulate a computer model based on a few sample runs. This form of regression will exactly replicate the output of the sample runs, and hence form an interpolator for runs that have not been evaluated, yet its linear formulation avoids overfitting, as compared to nonlinear models (such as ANN).

3.7.5.4. Generalized Linear Regression. In the previous section, we focused on the general problem of predicting with a Gaussian process. It turned out that the conditional expectation of some unknown given observations of the process elsewhere is a linear combination of the observations. The previous method is one of the most general forms of linear prediction in the sense that it uses knowledge about the Gaussian process in terms of a covariance function and uses the covariance between the predictors (the knowns) and the covariance between predictors and the predictant. The resultant weights are, therefore, functions of this covariance.

We now turn to a specific form of linear regression and the particular Gaussian process derived from it (hence we start from specifying the regression first, then make links with Gaussian processes). Consider fitting the data using a generalized linear model:

$$y(\mathbf{s},\mathbf{w}) = \sum_{j=1}^{J} w_j \phi(\mathbf{s}_j) \tag{3.156}$$

We use the classical regression notation of y for predictant. Essentially, we are trying to fit a surface using some observations but within a statistical framework. To do so, we introduce (again) a Gaussian assumption, now not on the unobserved process but on the weights. The weights are considered random variables **W** that need to be estimated. Additionally, we consider this regression within a Bayesian context, meaning we assume a prior distribution on the weights, namely a multivariate Gaussian with mean zero and covariance C_w:

$$\mathbf{W} \sim N(\mathbf{0}, C_w) \tag{3.157}$$

Observations may be subject to noise. To model this, we assume the following observation model with random uncorrelated and unbiased error

$$t(\mathbf{s}) = y(\mathbf{s}) + \varepsilon \quad \mathrm{var}(\varepsilon) = \sigma_\varepsilon^2 \tag{3.158}$$

and hence the observations $(t(\mathbf{s}_1), \ldots, t(\mathbf{s}_N))$. We can state a likelihood model for the data under the model of Eq. (3.158), namely

$$L(t(\mathbf{s}_1),\ldots,t(\mathbf{s}_N)|\mathbf{w}) = \frac{1}{\sqrt[N]{(2\pi\sigma_\varepsilon^2)}} \prod_{n=1}^{N} \exp\left(-\frac{(t(\mathbf{s}_n) - y(\mathbf{s}_n,\mathbf{w}))^2}{\sigma_\varepsilon^2}\right) \tag{3.159}$$

According to Bayes' rule, the posterior of the weights is also Gaussian (because both prior and likelihood are Gaussian, with mean

$$E[W|t(\mathbf{s}_1),\ldots,t(\mathbf{s}_N)] = \frac{1}{\sigma_\varepsilon^2}\left(C_w^{-1} + \frac{1}{\sigma_\varepsilon^2}\phi^T\phi\right)\phi^T\mathbf{t} \tag{3.160}$$

with

$$\mathbf{t} = (t(\mathbf{s}_1),\ldots,t(\mathbf{s}_N)) \tag{3.161}$$

and the design matrix

$$\phi = \begin{pmatrix} \phi_1(\mathbf{s}_1) & \cdots & \phi_J(\mathbf{s}_1) \\ \vdots & & \\ \phi_1(\mathbf{s}_N) & \cdots & \phi_J(\mathbf{s}_N) \end{pmatrix} \tag{3.162}$$

This classical solution also has an interpretation in terms of Gaussian processes. The relationship between the weight space view (weights as RV) and the Gaussian process view can be established by choosing C_w to approximate a Gaussian process. To establish that link, we consider unknown and data variables (observations) $(T(\mathbf{s}_1), \ldots, T(\mathbf{s}_N), Y(\mathbf{s}_0))$ to be multivariate Gaussian. Since Y is corrupted by noise, we consider first the noise-free variables $(Y(\mathbf{s}_1), \ldots, Y(\mathbf{s}_N), Y(\mathbf{s}_0))$. Because the Ys are linearly related to the Ws, see Eq. (3.158), a relationship between the covariance of Y and W is as follows:

$$C_y = \phi_0 C_w \phi_0^T \tag{3.163}$$

with the extended design matrix now including the evaluation of the basis functions ϕ at the location to be estimated $\mathbf{s}_0$:

$$\phi_0 = \begin{pmatrix} \phi_1(\mathbf{s}_1) & \cdots & \phi_J(\mathbf{s}_1) \\ \vdots & & \vdots \\ \phi_1(\mathbf{s}_N) & \cdots & \phi_J(\mathbf{s}_N) \\ \phi_1(\mathbf{s}_0) & \cdots & \phi_J(\mathbf{s}_0) \end{pmatrix} \tag{3.164}$$

The extension to corrupted measurements then simply involves adding variance to the diagonal, except for the last row/column since this involves the true unknown (not corrupted with noise evidently):

$$(T(s_1), \ldots, T(s_N), Y(s_0)) \sim N(0, \phi_0 C_w \phi_0^T + E) \tag{3.165}$$

with

$$E = \begin{pmatrix} \sigma_\varepsilon^2 & \cdots & \cdots & 0 \\ \vdots & \ddots & & \vdots \\ \vdots & & \sigma_\varepsilon^2 & \vdots \\ 0 & \cdots & \cdots & 0 \end{pmatrix} \tag{3.166}$$

Again, because the multivariate normal distribution is now stated, the conditional mean and variance can be expressed as

$$\begin{aligned} E[Y_0 | t(\mathbf{s}_1), \ldots, t(\mathbf{s}_N)] &= \phi_0^T C_w \phi^T (\phi C_w \phi^T + \sigma_\varepsilon^2 I_N)^{-1} \mathbf{t} \\ \mathrm{var}(Y_0 | t(\mathbf{s}_1), \ldots, t(\mathbf{s}_N)) &= \phi_0^T C_w \phi_0^T \\ &\quad - \phi_0^T C_w \phi^T (\phi C_w \phi^T + \sigma_\varepsilon^2 I_N)^{-1} C_w \phi_0 \end{aligned} \tag{3.167}$$

Williams [1999] shows that expressions (3.155) and (3.167) are identical, meaning that the weight-space (W) approach and the function-space approach (Y) are equivalent. The difference lies in computational complexity. In the W-approach one needs to invert a $J \times J$ system (size of the approximation), while in the Y-approach the inversion is of an $N \times N$ matrix (size of data). Recall a similar duality in Section 3.4.4 between sample size space and model size space.

3.7.5.5. History of Applications of Gaussian Process Regression. Several communalities with different views of the same problem exist that invoke the Gaussian process of the unknown phenomenon being estimated (conditional expectations). Historically, these methods have been developed somewhat independently in a number of different areas (statistics, geostatistics, machine learning). Although the basic theory still goes back to Wiener and Kolmogorov, linear prediction is also well known in the geostatistics field [*Matheron*, 1970; *Journel and Huijbregts*, 1978; *Cressie*, 1993] as kriging. Kriging was established as the best linear unbiased estimator:

$$y(\mathbf{s}_0) = \sum_{n=1}^{N} w_n t(\mathbf{s}_n) = \mathbf{w}^T \mathbf{t} \tag{3.168}$$

In geostatistics, noise σ_ε^2 is modeled as the so-called nugget effect. The weights are derived based on a minimum estimation variance criterion, resulting in

$$\mathbf{w} = \mathbf{k}^T K^{-1} \tag{3.169}$$

This solution "identifies" the covariance, meaning that the covariance between the estimate $Y(\mathbf{s}_0)$ and the observations is the same as (identifies) the modeled covariance, based on the observations. The establishment of the kriging equation, initially, did not invoke any Gaussian process. However, simple kriging can be regarded as the conditional expectation of an unknown value of a Gaussian process given observations. Dual kriging [*Goovaerts*, 1997] is another form to express the same kriging, but now the unknown is written as a linear combination of covariance functions:

$$y(\mathbf{s}_0) = \sum_{n=1}^{N} w_n C(\mathbf{s}_0 - \mathbf{s}_n) \tag{3.170}$$

The weights are now established by means of identification with the observed values (the exact interpolator property):

$$\mathbf{w} = \mathbf{t}^T K^{-1} \tag{3.171}$$

Equation (3.156) looks like Eq. (3.170), but it is now written with covariance functions as basis functions. These covariance functions can be estimated from the observations, providing a method for inferring such basis functions. This is typically possible in 3D, but it becomes more difficult in higher dimensions because of emptiness of high-dimensional space and the limited amount of data for such inference. In such cases, inference can be made based on likelihood methods [*Diggle and Ribeiro*, 2007], rather than least-square fitting of covariance or variograms [*Cressie*, 1985].

3.7.5.6. Linear Inverse Problems and Kriging. The solution of linear inverse problems has many applications in the subsurface, either where the relationship between data variables and model variables is linear, or where the linear problem is solved iteratively as part of a larger nonlinear inversion. In this section, we define what a linear inverse problem is and how it is related to Gaussian process regression (kriging) [see *Hansen et al.*, 2006]. A linear inverse problems involves a linear relationship between data variables and model variables

$$\mathbf{d} = G\mathbf{m} \tag{3.172}$$

The aim is to invert (estimate/regress) the model variables from observed data. In this sense, this is an extension of Eq. (3.168) where one model variable is a linear combination of the observed data, but model and data variables are of the same type (although co-kriging can be used to extend the regression to any type, see *Goovaerts* [1997]).

The linear inverse problem can be solved within a Bayesian framework. Just as the case in Gaussian processes, we assume (a priori) that the model parameters follow a multivariate Gaussian distribution with some prior covariance C_m. If the models are defined on a grid with a number of cells N_{grid}, then the observed data (of size N_{data}) is simply

$$\mathbf{d}_{\text{obs}} = \left(d\left(\mathbf{s}_{(1)}\right),\ldots,d\left(\mathbf{s}_{(N_{\text{data}})}\right)\right) = \left(m\left(\mathbf{s}_{(1)}\right),\ldots,m\left(\mathbf{s}_{(N_{\text{data}})}\right)\right) \tag{3.173}$$

with $\mathbf{s}_{(n)}$, $n = 1, \ldots, N_{\text{data}}$ the locations where observations are available. As a result, the operator G simply contains ones and zeros as elements:

$$\begin{gathered} G_{n'n} = 1 \text{ if } \mathbf{s}_{(n')} = \mathbf{s}_{(n)}, \text{ zero else} \\ n = 1,\ldots,N_{\text{grid}}; n' = 1,\ldots,N_{\text{data}} \end{gathered} \tag{3.174}$$

G identifies grid locations with data locations. Equation (3.172), however, states that data variables can be forward modeled by any linear combination (not just ones and zeros) of model variables:

$$d_{n'} = \sum_{n=1}^{N_{\text{grid}}} g_{n'n} m(\mathbf{s}_n); \quad n' = 1,\ldots,N_{\text{data}} \tag{3.175}$$

In terms of physical modeling, this applies to a limited set of forward models in the subsurface such as pressure equations (as function of permeability) or tomography problems Arrival times are linear combinations of model velocities (which in itself is an approximation to the full physics in such problems). While traditionally such problems have been solved using least squares, such methods ignore any prior covariance and hence solutions tend to be too smooth (certainly smoother than the actual reality). The prior covariance on the model variables allows injecting any information about the spatial distribution of these properties as quantified by a Gaussian process. The solution to this problem are provided in *Tarantola* [1987, p. 66] and summarized here by listing the expression of the posterior mean $E[\mathbf{M}|\, d_{\text{obs}}]$ of the entire model $\mathbf{m}$ and the posterior covariance $C_{M|d_{\text{obs}}}$:

$$\begin{gathered} E[\mathbf{M}|\, d_{\text{obs}}] = E[\mathbf{M}] + C_M G^T\left(GC_M G^T + C_D\right)\left(d_{\text{obs}} - GE[\mathbf{M}]\right) \\ C_{M|d_{\text{obs}}} = C_M - C_M G^T\left(GC_M G^T + C_D\right)^{-1} GC_M \end{gathered} \tag{3.176}$$

Like Gaussian process regression, the posterior covariance is not a function of the observed data. C_D is the covariance of the data variables, which under a simple error model becomes

$$C_D = \begin{pmatrix} \sigma_\varepsilon^2 & \cdots & 0 \\ \vdots & \ddots & \vdots \\ 0 & \cdots & \sigma_\varepsilon^2 \end{pmatrix} \tag{3.177}$$

Conditioning to linear averages in geostatistics is also known as block kriging [*Journel and Huijbregts*, 1978]. The "blocks" do not need to be square or compact, the term refers to estimating mining blocks (averages) from point data, but the problem can be reversed to estimating points from block data. "Block data" simply refers to some linear averaging. Block kriging requires calculating covariances of the linear averages, covariances between the linear averages and the unknown, which are then plugged into Eq. (3.176). All these covariance can be calculated from C_M. The matrix C_D reflects the so-called nugget effect.

3.8. KERNEL METHODS

3.8.1. Introduction

Consider the following simple problem (see Figure 3.28). The aim is to cluster the two red points in one group and the two blue points in another group. To achieve this would require a complex discriminant function. A discriminant function is a function that divides space into two parts, to "discriminate" one part from the other. Obviously, this is useful for classification where any point in one half are then considered to belong to group "red" and the other to group "blue." Using highly nonlinear discriminant function can be problematic because (i) finding expressions for such function may not be trivial and (ii) overfitting may occur quite rapidly. This is shown in Figure 3.28. The discriminant function fits the problem of discriminating the blue points from the red points almost perfectly; however, that discrimination will have poor performance when applied to yet

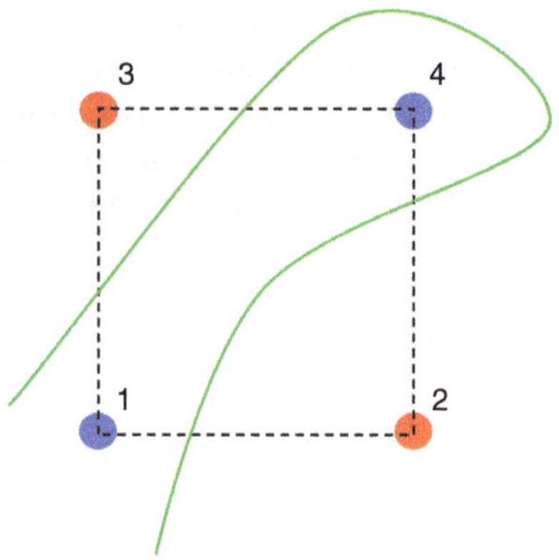

Distance d

	1	2	3	4
1	0	1	1	$\sqrt{2}$
2	1	0	$\sqrt{2}$	1
3	1	$\sqrt{2}$	0	1
4	$\sqrt{2}$	1	1	0

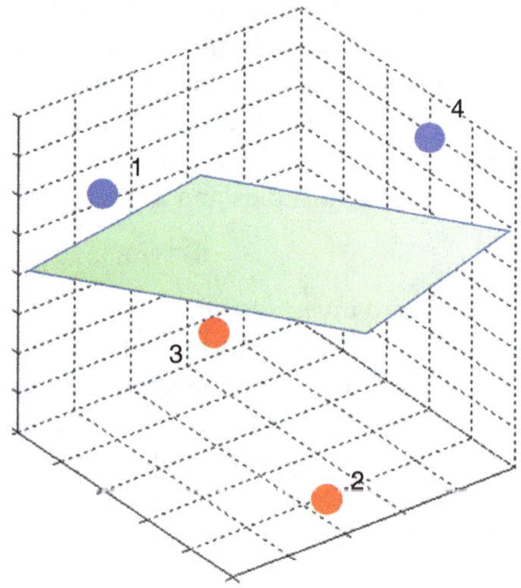

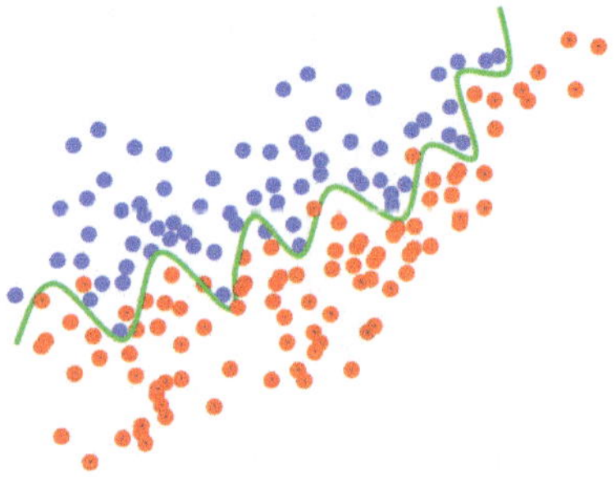

$k = 1 - \exp(-d)$

	1	2	3	4
1	0	0.86	0.86	0.94
2	0.86	0	0.94	0.86
3	0.86	0.94	0	0.86
4	0.94	0.86	0.86	0

Figure 3.28 A simple classification problem with a nontrivial solution. Linear classification is not possible in this XOR problem (exclusive or). Linear discriminant functions are desirable, in particular when going to higher dimension. The "trick" is to transform the distances to obtain a higher-dimensional space.

unseen blue/red points. This problem is obvious here, but it becomes difficult to diagnose in higher dimensions. An obvious solution is to split the data into a training set and a validation set and use such validation set to avoid overfitting. However, this would not mitigate the problem of what function to choose in the first place, and it may not be applicable when only few samples are available.

For these reasons, linear methods have their appeal, but clearly there is no linear solution to the problem of Figure 3.28, which uses 2D Cartesian space with Euclidean distances. How can we make a linear method work? Take now the example in Figure 3.29 with a complex curve in 2D. Consider now increasing the dimension by one and embedding that curve into a plane (a linear feature!) such that the projection back into 2D still results in the same curve. Figure 3.29 achieves this. One can imagine that if the curve is a helix, then further increasing the dimensions allows embedding the helix into a hyperplane (unfortunately, we cannot show that in a figure!)

The idea, therefore, is to change space using some transformation. However, we will not be transforming coordinates of a Cartesian axis system, as this again would call for complex multivariate transformation functions (and hence we are back to the same problem). To change space, we will change distances to create a new space. Recall that distances can be expressed as dot-products and vice versa (see Section 3.5.2). Now let us go back to our problem in Figure 3.28. Let us first calculate the distance between the four points, listed in the distance table; clearly, the two red points are far from each other than the red from the blue point and vice versa. Now we transform that distance table into a new distance table using the following simple equation:

$$k_{ij} = 1 - \exp(d_{ij}) \;\; i, j = 1, \ldots, 4 \qquad (3.178)$$

The exponential function of minus the distance makes objects that are far apart appear closer and objects that are closer further apart. That is exactly what we like to achieve, because it would move the two red points closer, the two blue point closer, and the red points further apart from the blue points. The problem is that this cannot be done in 2D. Why not? This can be explained by calculating the eigenvalues of the distance matrix d and of the matrix k, we find

$$\begin{aligned} d &: \lambda_1 = \lambda_2 = 1; \lambda_3 = \lambda_4 = 0 \\ k &: \lambda_1 = \lambda_2 = 0.44; \lambda_3 = 0.31; \lambda_4 = 0 \end{aligned} \qquad (3.179)$$

The distance table has only two positive eigenvalues. This means that we can only create a 1D or 2D projection from the distance table to Cartesian coordinates. The k-table, however, has three positive eigenvalues. This means that we can construct an orthogonal 3D Cartesian space, with its own (3D) Euclidean distance (despite the fact that we

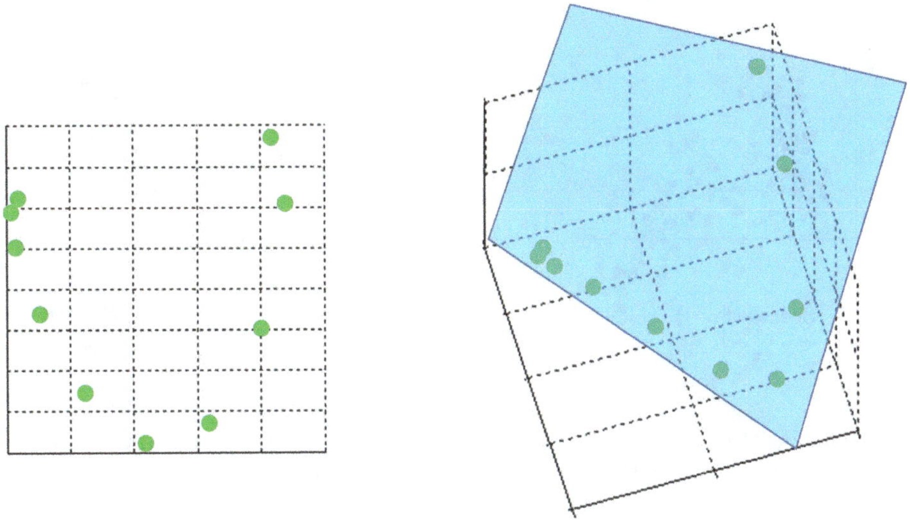

Figure 3.29 Embedding a nonlinear 2D function in a 3D plane, such that projection retrieves the nonlinear function.

only specified 2D distances). In fact, if we perform an MDS projection into this 3D space, we notice how the blue points moved up and the red down. A simple plane (a linear function!) now solves the classification problem, exactly what we set out to achieve.

The problem in Figure 3.28 involves a classification problem, but classification and regression problems are basically variations of the same theme as we discussed in Section 3.7. In classification problems, we seek a discriminant function. In regression, we seek also to estimate a function between predictors and predictants. In the both cases, we have to estimate a function. We saw in Section 3.7 that linear methods, such as Gaussian process regression, can be powerful, if the linear models are adequate to describe the actual complexity of data on which they are applied. In Figure 3.28, we observed how embedding the "data" into a new space, where linear operations apply more readily, is an interesting concept worthwhile pursuing. To understand a bit better how this works for regression problems, we consider two views of the same regression. Consider a data set

$$\left(\left(\mathbf{x}^{(1)},y^{(1)}\right),\ldots,\left(\mathbf{x}^{(L)},y^{(L)}\right)\right) \tag{3.180}$$

with $\mathbf{x}$ a vector of any dimension N of predictors $\mathbf{x}=(x_1, \ldots, x_N)$ (could include spatial location, so we exchange $\mathbf{x}$ and $\mathbf{s}$ in denoting predictors). A linear model is assumed

$$y(\mathbf{x},\mathbf{w})=\sum_{n=1}^{N} w_n x_n=\mathbf{w}^T\mathbf{x} \tag{3.181}$$

If we denote, as before,

$$\begin{aligned} X&=\left(\mathbf{x}^{(1)},\ldots,\mathbf{x}^{(L)}\right)\\ \mathbf{y}&=\left(y^{(1)},\ldots,y^{(L)}\right) \end{aligned} \tag{3.182}$$

Then the classical and primal solution for the least-square weights is (see Eq. (3.130))

$$\mathbf{w}=\left(X^TX\right)^{-1}X^T\mathbf{y} \tag{3.183}$$

An alternative form of presenting the same solution in a dual form is

$$\mathbf{w}=X^T\tilde{\mathbf{w}} \text{ with } \tilde{\mathbf{w}}=\left(XX^T\right)^{-1}\mathbf{y} \tag{3.184}$$

The primal form on X^TX and the dual form relies on XX^T. If this looks familiar, then refer to presentation of the data matrices in two ways: the space formed by the dimension of the vector $\mathbf{x}$, using covariances (X^TX), or the space formed by sample size with dot-products (XX^T). We discussed the advantage of working with XX^T over X^TX in the type of UQ problems we are dealing with. As a result, the presentation of the material that comes next, kernel mapping, can be highly effective in addressing certain UQ problems, in particular those that involve complex priors and nonlinear operations. Basically, we will see that it is easy to extend any linear statistical operation simply by changing dot-products (see Figure 3.30). This will invoke a transformation $\boldsymbol{\varphi}$ of the original Cartesian space; however, an explicit representation of $\boldsymbol{\varphi}$ (a possible high-dimensional function, which as we know, we would like to avoid) will not be required, only a dot-product (just a scalar!). The change in dot-product aims at embedding the data in a space in which linear operations apply more readily (see Figure 3.30).

3.8.2. Kernel-Based Mapping

Consider a general nonlinear mapping between two spaces

$$\boldsymbol{\varphi}:\mathbf{x}\in\mathbb{R}^N\mapsto\mathbf{x}\in\mathbb{R}^M \tag{3.185}$$

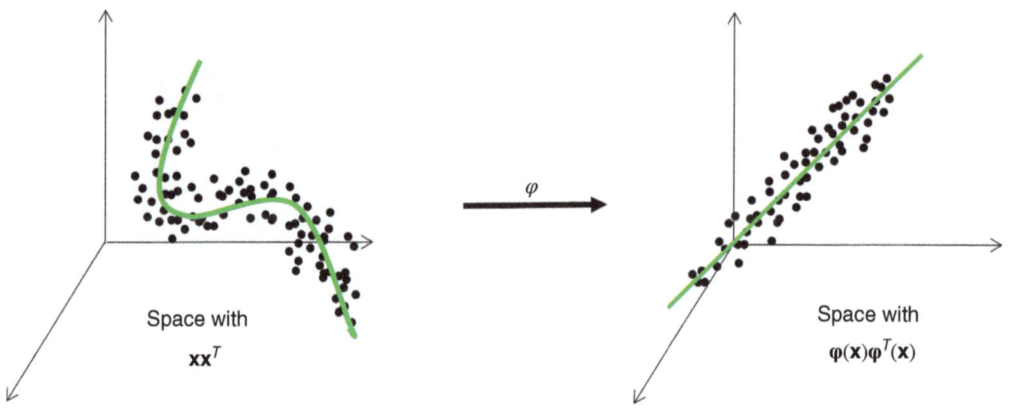

Figure 3.30 Changing dot-product to render linear modeling more appropriate. The space on the right is also termed the "feature space" in machine learning, features being quantified by the kernel function. Features in statistics are termed simply "predictors."

Both N and M could be very large (possibly even infinite) and hence a function $\boldsymbol{\varphi}$ that creates a space where linear methods more readily apply would be difficult to find. Instead, many such linear methods only require knowing a dot-product. In this new space, the dot-product between any two samples $\mathbf{x}^{(\ell)}$and $\mathbf{x}^{(\ell')}$ is

$$g_{\ell\ell'} = \boldsymbol{\varphi}^T\left(\mathbf{x}^{(\ell)}\right)\boldsymbol{\varphi}\left(\mathbf{x}^{(\ell')}\right) \quad (3.186)$$

where the matrix G consisting of elements $g_{\ell\ell'}$ is termed the Gram matrix (or kernel matrix). Then, when considering a new predictor $\mathbf{x}$, we write

$$k_\ell = \boldsymbol{\varphi}^T\left(\mathbf{x}^{(\ell)}\right)\boldsymbol{\varphi}(\mathbf{x}) \quad (3.187)$$

which leads to the definition of a kernel function:

$$k(\mathbf{x},\mathbf{x}') = \boldsymbol{\varphi}^T(\mathbf{x})\boldsymbol{\varphi}(\mathbf{x}') \quad (3.188)$$

Since a dot-product can be related to a distance, in that sense, the kernel matrix contains all the information needed to compute distances, and hence define a metric space. Such space has no orientation; therefore, the kernel matrix is rotationally invariant. Transformations such as rotation are, however, not really important when doing regression or performing classification. Therefore, the kernel matrix is viewed as the information bottleneck, filtering the necessary information to perform "learning" whether this is regression, classification, or orthogonal component analysis. Hence, the "a priori" insight that renders the problem more linear is provided by means of distances. The choice of the kernel function is, therefore, relevant to solve complex problems.

A kernel that will be used throughout this book is the radial basis function (RBF). The RBF kernel function is expressed as

$$k(\mathbf{x},\mathbf{x}') = \exp\left(-\frac{\|\mathbf{x}-\mathbf{x}'\|^2}{2\sigma^2}\right) \text{ with } \sigma > 0. \quad (3.189)$$

The performance of the RBF kernel depends highly on its bandwidth σ. If σ is small, the kernel matrix is close to the identity matrix ($K = I$); hence, all the $\mathbf{x}$ will tend to be very dissimilar. On the other hand, large values of σ makes the kernel matrix close to a constant matrix ($K = \mathbf{11}^T$), leading to all the $\mathbf{x}$ being very similar. Cross-validation techniques are very popular to estimate the kernel bandwidth in the case of supervised learning. However, for unsupervised learning (such as KPCA), the choice of the bandwidth remains an open question. We will rely on the rule of thumb of *Kwok and Tsang* [2004].

Figure 3.31 provides an example of what a kernel function does on mapping concentrations in feature space (kernel space), for the simple hydro case. The left plot is the MDS plot based on the Euclidean distance (or PCA with covariance) between the original concentration curves; one notices how for small distances the data is strongly clustered but spreads out for larger distances. This is common in classical metric spaces, a few large distances are easier to approximate in lower dimensions than a lot of small distances. The kernel transformation makes these distances more uniform (by increasing dimension). Figure 3.31 shows a nice arrangement of models along a curve-linear feature. Modeling in feature space is preferable over modeling in the original Cartesian space in these types of cases.

3.8.3. Kernel PCA

3.8.3.1. Method. Recall from Section 3.5.1 that PCA performs a projection onto orthogonal bases, derived from the covariance matrix (space of X^TX). This projection onto the n-th eigenvectors can be written as

$$\mathbf{u}_n^T\mathbf{x} = \sum_{\ell=1}^{L}\frac{v_n^{(\ell)}\left(\mathbf{x}^{(\ell)}\right)^T\mathbf{x}}{\sqrt{\lambda_n}} \quad n \le \text{rank}(X) \quad (3.190)$$

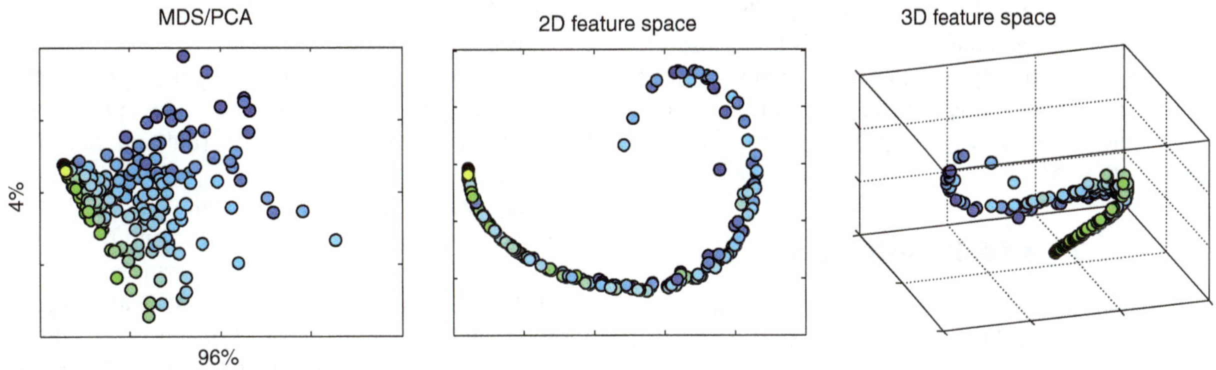

Figure 3.31 Comparison of classical MDS with MDS after kernel transformation. Notice, like in Figure 3.28, the impact of an increase in dimension.

Here we use the duality that exists in using X^TX or $XX^T$to calculate eigenvalues and eigenvectors **u** and **v**, respectively. We now apply the same kernel trick to turn PCA into kernel PCA (KPCA) by changing the dot-product $\mathbf{x}^T\mathbf{x}$ to $\boldsymbol{\varphi}(\mathbf{x})^T\boldsymbol{\varphi}(\mathbf{x})$ to obtain a projection (now in XX^Tspace) as

$$\sum_{\ell=1}^{L}\frac{v_n^{(\ell)}\boldsymbol{\varphi}^T\left(\mathbf{x}^{(\ell)}\right)\boldsymbol{\varphi}(\mathbf{x})}{\sqrt{\lambda_n}}=\sum_{\ell=1}^{L}\frac{v_n^{(\ell)}}{\sqrt{\lambda_n}}k\left(\mathbf{x}^{(\ell)},\mathbf{x}\right) \tag{3.191}$$

This leads to the following calculation for KPCA (compare with PCA in Section 3.5.1.2)

$$\begin{aligned}
&\text{Calculate } g_{\ell\ell'}=k\left(\mathbf{x}^{(\ell)},\mathbf{x}^{(\ell')}\right),\quad \ell,\ell'=1,\ldots,L\\
&\text{Center } g_{\ell\ell'}\leftarrow g_{\ell\ell'}-\frac{1}{L}g_{\ell\cdot}-\frac{1}{L}g_{\cdot\ell'}+\frac{1}{L^2}g_{\cdot\cdot}\\
&\text{Decompose } G=V^T\Lambda V\\
&\text{Project } \mathbf{y}=\sum_{\ell=1}^{L}\frac{v_n^{(\ell)}}{\sqrt{\lambda_n}}k\left(\mathbf{x}^{(\ell)},\mathbf{x}\right)
\end{aligned} \tag{3.192}$$

Note how the centering in the covariance is now replaced by the centering of the dot-product (see Section 3.5.2).

3.8.3.2. The Pre-image Problem. PCA is a bijective projection, meaning that one can uniquely recover the original vector, after projection and reconstruction. This is not the case for KPCA. This makes sense, intuitively: KPCA relies on distances (similarities) between vectors, and not on their exact location in a Cartesian setting. Therefore, reconstructing a new vector based on the original data is not unique. This is the case for UQ in the subsurface where $L \ll N$. For example, imagine knowing the distance between three vectors in 100-dimensional space and we are given a fourth vector for which only the distance to the previous three is specified. Reconstructing this fourth vector is a nonunique problem, many solutions exist. In computer science literature, this is termed a "pre-image problem" [*Schölkopf and Smola*, 2002]. Pre-image means seeking the "image" of an object, given the distances with other objects whose "images" are known.

The difficulty in the pre-image problem is that the mapping function $\boldsymbol{\varphi}$ into the feature space is unknown, nonlinear and nonunique, thus only approximate solutions can be generated. Consider a feature space expansion $\Psi=\sum_{\ell=1}^{L}\alpha^{(\ell)}\boldsymbol{\varphi}\left(\mathbf{x}^{(\ell)}\right)$ and denote $\mathbf{x}^*$ as its approximate pre-image. The pre-image problem attempts to minimize the squared distance in feature space:

$$\min_{\mathbf{x}^*}\|\Psi-\boldsymbol{\varphi}(\mathbf{x}^*)\|^2=\min_{\mathbf{x}^*}\left\|\sum_{\ell=1}^{L}\alpha^{(\ell)}\boldsymbol{\varphi}\left(\mathbf{x}^{(\ell)}\right)-\boldsymbol{\varphi}(\mathbf{x}^*)\right\|^2 \tag{3.193}$$

From the kernel trick, the minimization of Eq. (3.193) is equivalent to minimizing

$$\begin{aligned}
&\min_{\mathbf{x}^*} k(\mathbf{x}^*,\mathbf{x}^*)-2\sum_{\ell=1}^{L}\alpha^{(\ell)}k\left(\mathbf{x}^*,\mathbf{x}^{(\ell)}\right)\\
&\quad+\sum_{\ell'=1}^{L}\sum_{\ell=1}^{L}\alpha^{(\ell)}\alpha^{(\ell')}k\left(\mathbf{x}^{(\ell)},\mathbf{x}^{(\ell')}\right)
\end{aligned} \tag{3.194}$$

which formulates a nonlinear optimization problem that is only function of the kernel function and not of $\boldsymbol{\varphi}$. Gradient procedures can be used to minimize this expression. In the particular case of the RBF kernel, a fixed-point iterative approach can be used to find approximate pre-images [*Schölkopf and Smola*, 2002]. The following solution is then obtained:

$$\mathbf{x}_{t+1}^*=\frac{\sum_{\ell=1}^{L}\alpha^{(\ell)}\exp\left(-\left\|\mathbf{x}^{(\ell)}-\mathbf{x}_t^*\right\|^2/2\sigma^2\right)\mathbf{x}^{(\ell)}}{\sum_{l=1}^{L}\exp\left(-\left\|\mathbf{x}^{(\ell)}-\mathbf{x}_t^*\right\|^2/2\sigma^2\right)} \tag{3.195}$$

Unfortunately, the fixed-point iterative method suffers from encountering local minima and tends to be unstable. An interesting property of the fixed-point iterative method is that the resulting pre-image lies in the span of the available data, since the pre-image is simply a weighted sum of $\mathbf{x}^{(\ell)}$.

3.9. CLUSTER ANALYSIS

Cluster analysis is widely used to find hidden structures that may exist in data sets. The aim of clustering is to partition a set of L data points $\mathbf{x}^{(\ell)}$, $\ell = 1, \ldots, L$ into K groups or clusters, based on the similarity between data points: data within a cluster should be similar, and dissimilar to data in other clusters. An ideal cluster is a set of points that is compact and isolated [*Jain*, 2010]. Clustering is used in a variety of applications, such as data mining, image segmentation, and data compression. In subsurface modeling, clustering is sometimes applied to group subsurface models that are similar, given some defined measure of similarity, or for dimension reduction purposes.

3.9.1. *k*-Means

Among many available clustering algorithms, k-means clustering is probably one of the most widely used algorithms because of its simplicity and efficiency. The objective of k-means is to find the cluster configuration that minimizes the squared error over all K clusters:

$$J = \sum_{k=1}^{K} \sum_{\mathbf{x}^{(\ell)} \in c_k} \left\| \mathbf{x}^{(\ell)} - \boldsymbol{\mu}^{(k)} \right\|^2 \quad \boldsymbol{\mu}^{(k)} = \frac{\sum_{\mathbf{x}^{(\ell)} \in c_k} \mathbf{x}^{(\ell)}}{|S_k|} \tag{3.196}$$

with $\boldsymbol{\mu}^{(k)}$ the centroids of cluster c_k defined as the mean point of the cluster. $|S_k|$ the number of samples in the cluster c_k. The main steps of the k-means procedure are illustrated in Figure 3.32 and are as follows:

Step 1: Select randomly K centroids $\boldsymbol{\mu}^{(k)}$. These points need not correspond to any of the data points.

Step 2: Find the closest centroid $\boldsymbol{\mu}^{(k)}$ to the point $\mathbf{x}^{(\ell)}$ and assign point $\mathbf{x}^{(\ell)}$ to cluster c_k.

Step 3: Update centroids $\boldsymbol{\mu}^{(k)}$ using the equation above.

Step 4: Repeat Steps 2 and 3 until convergence, that is cluster configuration is stabilized (the squared error is minimized).

In the k-means procedure, the number of clusters K needs to be specified prior to clustering and remain fixed. Methods have been developed to determine the number of clusters (see Section 3.9.4). The k-means algorithm finds a local minima of Eq. (3.196); hence, the clustering results may differ with the choice of different initial clusters centers. This can be mitigated by running k-means multiple times with different initial randomly chosen centroids and selecting the partition with the smallest squared error.

3.9.2. *k*-Medoids

Instead of taking the mean values of the data within a cluster as centroids, k-medoids assigns the cluster center to the most central point of that cluster (referred to as the medoids, the point "in the middle"). Partitioning in k-medoids is still based on the minimization of the sum of the dissimilarities of the data and the cluster centers, but k-medoids does not require the use of a squared Euclidean distance in the calculation of the cost function (Eq. (3.196)). k-medoids algorithms may employ directly any dissimilarity distance matrix. k-medoids clustering is more robust to noise and outliers than k-means.

Among many algorithms for k-medoids clustering, partitioning around medoids (PAM) proposed by *Kaufman and Rousseeuw* [1990] is the most popular. PAM consists of first selecting randomly k-medoids and assigning each data point to the cluster with the closest medoid in terms of the dissimilarity measure. Then, iterating over the data points, each medoid and non-medoid are swapped. If the total cost of the configuration is decreased then the swap is preserved. The algorithm stops when no further permutations improve the quality of the clustering. The major

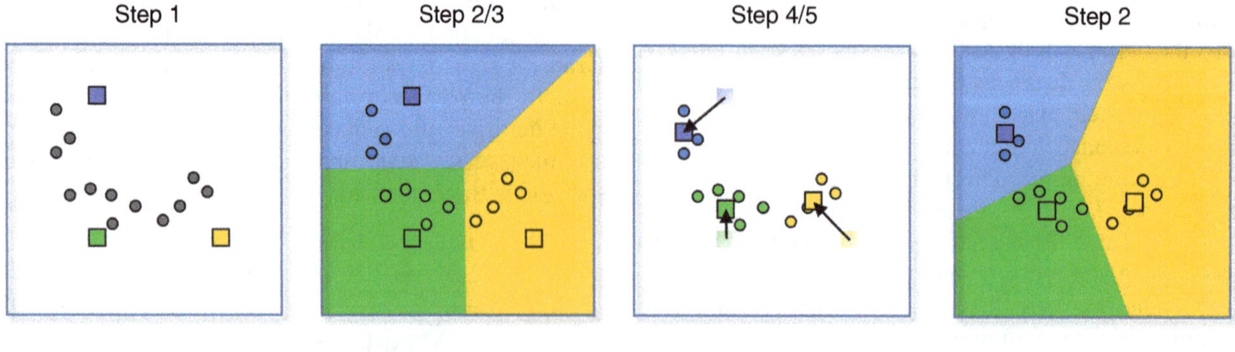

Figure 3.32 An example of partitioning. Image from *Caers* [2011].

drawback of PAM is its high computational cost, which makes k-medoids more costly than k-means. PAM becomes impractical in terms of CPU when the number of data points L or the number of clusters K is large. Its complexity is of the order $O^{K(L-K)^2}$.

3.9.3. Kernel Methods for Clustering

Data may contain nonlinear structures that cannot be easily separated by linear models such as k-means or k-medoids. An example of such a case was given in Figure 3.28. Linear methods such as k-means would perform poorly. One way to tackle this problem is to use kernel clustering (kernel k-means or kernel k-medoids). The use of kernels can, by increasing the dimension of the problem, result in increased linearity and separability. In kernel-based clustering, the data points are first mapped into a high-dimensional space (the feature space), and then clustering is applied in this space. For both kernel k-means and kernel k-medoids methods, the distance in feature space between a point and its centroid/medoid can be computed using only the kernel function. In the case of k-means, a pre-image problem must be solved to obtain the centroid coordinates in the original space.

3.9.4. Choosing the Number of Clusters

3.9.4.1. Silhouette Index. *Kaufman and Rousseeuw* [1990] proposed a "silhouette index" to determine the quality of the clustering. This index can also be used to find the optimal number of clusters K to use in the clustering algorithm. Let $a(\ell)$ be the average distance of a point $\mathbf{x}^{(\ell)}$ to all other points in the same cluster. It measures how well the point $\mathbf{x}^{(\ell)}$ is assigned to its cluster (the smaller, the better). Let $b(\ell)$ represent the minimum of the average distance between $\mathbf{x}^{(\ell)}$ and the points in different clusters:

$$b(\ell) = \min_k d\left(\mathbf{x}^{(\ell)}, c_k\right) \tag{3.197}$$

with $d(\mathbf{x}^{(\ell)}, c_k)$ the average distance between $\mathbf{x}^{(\ell)}$ and all points in c_k ($\mathbf{x}^{(\ell)}$does not belong to c_k). The silhouette index $s(\ell)$ is then defined as follows:

$$s(\ell) = \frac{b(\ell) - a(\ell)}{\max(a(\ell), b(\ell))} \tag{3.198}$$

If $s(\ell)$ is close to one, the data point $\mathbf{x}^{(\ell)}$ is well classified, whereas if $s(\ell)$ is close to zero, it is unclear whether $\mathbf{x}^{(\ell)}$ should belong to its assigned cluster or its neighboring cluster. A negative value is an indication that the data point $\mathbf{x}^{(\ell)}$ has been misclassified. The average value over all data points $\mathbf{x}^{(\ell)}$ is called the average silhouette index and can be evaluated for different number of clusters. The best clustering configuration is achieved when the average silhouette index is maximal. The silhouette index can be plot as a function of the number of clusters. This plot often has an "elbow" shape (and is sometimes referred to as elbow plot). The optimal number of clusters is obtained when the average silhouette index value bends "at the elbow," see for example Figure 3.33.

3.9.4.2. Davies–Bouldin Index. An alternative index to quantify cluster quality and optimal number of clusters is the Davies–Bouldin index [*Davies and Bouldin*, 1979]. It is defined as a function of the ratio of the within cluster scatter to the between cluster separation:

$$DB = \frac{1}{K}\sum_{i=1}^{K} \max_{j \neq i} \left\{ \frac{S_i + S_j}{M_{ij}} \right\} \tag{3.199}$$

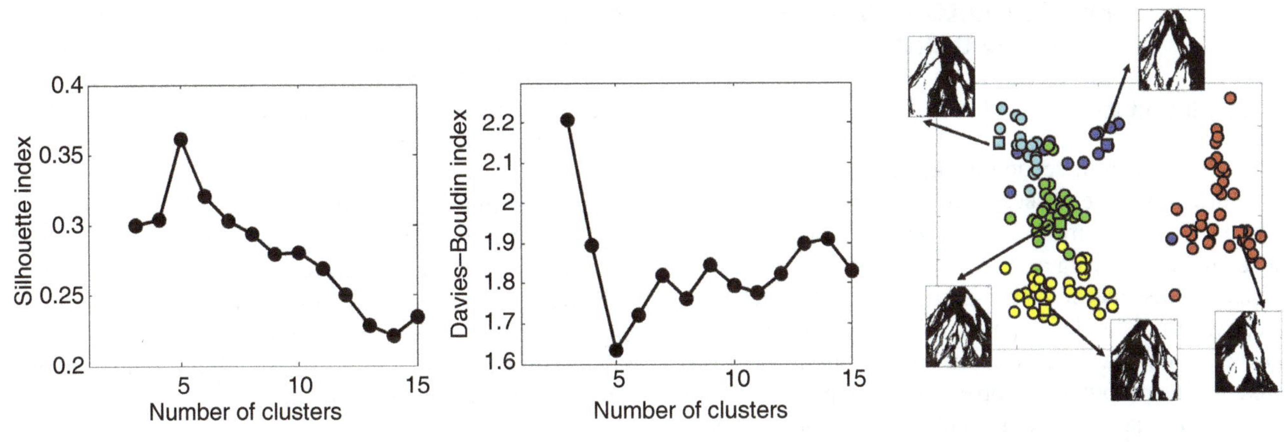

Figure 3.33 Choice of the optimal number of clusters using the silhouette index and Davies–Bouldin index. Five clusters are created using k-medoid, applied on the set of 136 overhead snapshot of a flume experiment. Note that the clustering is applied in high dimension (and not 2D).

where the intra-cluster (within) distance is defined as

$$S_i = \left(\frac{1}{N_{c_i}}\sum_{\ell=1}^{N_{c_i}} \left|\mathbf{x}^{(\ell)} - \boldsymbol{\mu}^{(i)}\right|^p\right)^{\frac{1}{p}} \quad (3.200)$$

with N_{c_i} the number of points assigned to cluster c_i and the cluster separation (inter-cluster distance) between cluster c_i and c_j as

$$M_{ij} = \left(\sum_{n=1}^{N} \left|\mu_n^{(i)} - \mu_n^{(j)}\right|^p\right)^{1/p} \quad (3.201)$$

Usually, $p = 2$. As opposed to the silhouette index, the optimum number of clusters is obtained by minimizing the index with respect to the number of clusters (Figure 3.33). The Davies–Bouldin index is described here for the k-means algorithm, but it could be estimated similarly for k-medoids, by replacing the centroids by the medoids.

3.9.5. Application

Clustering is applied to the set of 136 overhead snapshots of a flume experiment, based on the modified Hausdorff distance between snapshots. This data set will be further discussed in Chapter 5. Both the silhouette index and the Davies–Bouldin index suggest that the optimal number of clusters for this data set is 5. The k-medoid procedure was applied to identify five medoids, which correspond each to one of the snapshots. The five selected images can be considered as representative of the set of 136 images, as defined by the modified Hausdorff distance. This idea is very useful when dealing with a large number of spatial model realization in UQ. Using clustering, one can reduce this large set to a representative smaller set that has the same impact in terms of UQ [see *Scheidt and Caers*, 2009; *Scheidt and Caers*, 2013].

3.10. MONTE CARLO AND QUASI MONTE CARLO

3.10.1. Introduction

In this important section in the context of uncertainty quantification, we discuss various methods of Monte Carlo (MC) simulation. MC is used in many fields of science and engineering and literature is extensive. Here we focus on those methods that are relevant to UQ in subsurface systems. The development of MC goes back to John von Neumann who introduced random numbers generated by a computer to solve problems encountered in the development of atom bomb [*Von Neumann*, 1951]. MC refers to the famous casino in Monaco, where randomness is also used in a perhaps more joyful way than von Neumann's original application.

Broadly, MC uses random sampling to study properties of systems with components that behave in a random fashion [*Lemieux*, 2009]. The idea is simply to simulate on a computer the behavior of a system (in terms of outputs for example) by randomly generating the variables that control/model/describe the system (e.g., the inputs). Then, using the obtained results one can study the system, such as to perform a sensitivity analysis or apply any other statistical analysis or inference. This requires (i) the description of the "model," including a computer code implementing "the model," (ii) specification of distributions on the variables describing the system, (iii) generating samples from these distributions and evaluating the samples using the computer written code, and (iv) statistical analysis and learning from the obtained results. From a purely practical point of view, one needs to define (i) a mathematical model, (ii) a computer implementation of it, and (iii) a way of random sampling and propagating that sampling through the computer code; then observing and analyzing the output of that code, for whatever purpose is deemed relevant. This very broad description encapsulated many specific applications such as the following:

1. *Sampling from a known distribution:* A probability distribution is a mathematical model that can be implemented with a numerical model in a computer program. The inputs are pseudo-random numbers that are passed through this program to generate samples which can then be studied statistically. Underlying all MC methods, therefore, is the generation of these pseudo-random numbers. The pseudo-random number generator itself is a computer code that takes a deterministic input (the seed) and generates an (approximate) sequence of random numbers in the interval [0,1].

2. *Stochastic simulation:* This is generically described as the study of systems that contain stochastic components, broadly represented as $\mathbf{Y} = g(\mathbf{X})$. $\mathbf{X}$ denotes a set of random variables (the stochastic components) that needs to be simulated to generate outputs $\mathbf{Y}$. g is a "transfer function," or "forward model," or "system response model"; whatever flavor is used, it again represents some computer code (hence deterministic). Such computer code can be a simple function or be as complex as a numerical implementation of a partial differential equation that models a dynamic system (e.g., wave equations or flow and transport in porous media).

3. *MC integration:* A specific problem in MC simulation where the goal is to use random sampling to solve a deterministic problem (or problem that has no inherent stochastic component) of the kind:

$$\int_V f(\mathbf{x})d\mathbf{x} \quad (3.202)$$

MC integration solves this problem by generating random samples of **x**. For example, if $f(x) = x\,g(x)$ with $g(x)$ a positive function whose integral is unity, then MC integration boils down to estimating the mean of the random variable X.

3.10.2. Sampling from Known Distributions

3.10.2.1. Inversion. The method of inversion goes back to von Neumann. Inversion applies to the cdf, simply as follows, for a continuous distribution:

$$\begin{aligned} &u \leftarrow \text{rand}(\bullet) \\ &x = F^{-1}(u) \end{aligned} \tag{3.203}$$

Obviously, this method depends on the ability to calculate the inverse of the cumulative distribution F. For discrete distributions this becomes

$$\begin{aligned} &u \leftarrow \text{rand}(\bullet) \\ &x = \inf\{y : F(y) \geq u\} \end{aligned} \tag{3.204}$$

If an explicit expression for inf is not available, then this becomes a search problem.

3.10.2.2. Acceptance–Rejection. Sampling through inversion is usually only possible in univariate cases and with fully known cdfs of pdfs. In most cases, it will be impossible to directly sample from a distribution, in particular when we have only the pdf and then usually only up to some normalization constant (that we often cannot compute analytically). Since we cannot directly sample from the target pdf $f(x)$, we will first sample from another pdf $t(x)/T$ (i) that we know how to sample from and (ii) $t(x)$ majors $f(x)$ over the domain of interest, namely there exists some M such that $t(x)M \geq f(x)$. $t(x)$ itself is not a density, but

$$\int t(x)dx = T \tag{3.205}$$

The rejection–acceptance method works as follows:

$$\begin{aligned} &\text{Sample } y \text{ from } t(x)/M \\ &u \leftarrow \text{rand}(\bullet) \\ &\text{If } u \leq \frac{f(x)}{t(x)} \Rightarrow x = y \text{ else reject } y \end{aligned} \tag{3.206}$$

Figure 3.34 shows why this works in the case the majoring function is uniform. Each dot in this plot has coordinates

$$(y, u\,t(x)) \tag{3.207}$$

It makes intuitive sense that the values of y should be accepted based on the ratio between $t(x)$ and $f(x)$. In that way, the accepted values y will occur more frequently in the area where the ratio $t(x)/f(x)$ is close to unity.

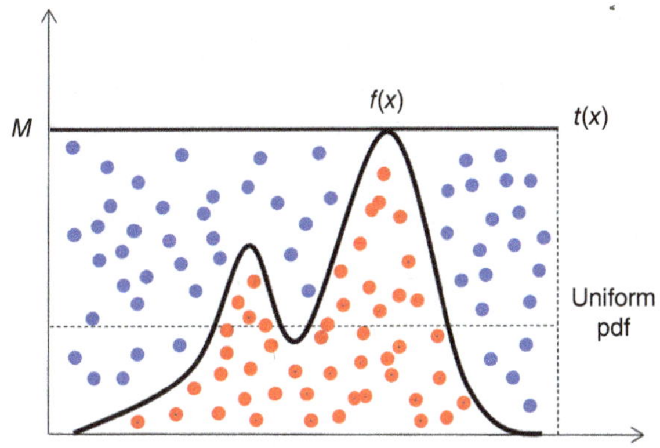

Figure 3.34 Sampling from a complex pdf by sampling from a uniform distribution and then accepting the red dots and rejecting the blue dots.

3.10.2.3. Compositions. Many applications of UQ in the subsurface require sampling from compositions. For example, lithologies such as shale and sand may exhibit different populations of petrophysical properties such as porosity and permeability. Hence, the total distribution of such properties may be expressed as

$$F(x) = \sum_{k=1}^{K} p_k F_k(x) \text{ with } \sum_{k=1}^{K} p_k = 1 \text{ and } p_k \geq 0\ \forall k \tag{3.208}$$

A mixture distribution with $K = \infty$ is termed a composition. Equation (3.208) can also be written in terms of densities. To sample from a mixture two draws are needed:

$$\begin{aligned} &u_1 \leftarrow \text{rand}(\bullet) \\ &i = \inf\left\{k' : \sum_{k=1}^{k'} p_k \geq u_1\right\} \\ &u_2 \leftarrow \text{rand}(\bullet) \\ &x = F_i^{-1}(u_2) \end{aligned} \tag{3.209}$$

3.10.2.4. Multivariate Gaussian. The Gaussian model is popular; hence, many ways exist to sample from it. The Gaussian model is applied in cases of not only multivariate modeling but also spatial modeling. In the latter case, the variables are the unknown random variables on a lattice or grid (random field). In the spatial case, one may already have measurements at certain locations (in geostatistics termed "hard data"). Without any hard data, the unconditional simulation of Gaussian field proceeds through, for example, an LU-decomposition (zero mean case):

$$C = LU \Rightarrow X = L^T\mathbf{y}, \quad \mathbf{y} \sim N(0,1) \tag{3.210}$$

C is the covariance matrix, which can be very large in the spatial case. To deal with these large matrices and also include the hard data, a sequential decomposition of the multivariate Gaussian into a chain of univariate Gaussians can be used:

$$\begin{aligned} f(X(\mathbf{s}_1),\ldots,X(\mathbf{s}_N)) = & f(X(\mathbf{s}_{(1)})) \times f(X(\mathbf{s}_{(1)})|X(\mathbf{s}_{(2)})) \\ & \times f(X(\mathbf{s}_{(3)})|X(\mathbf{s}_{(1)})X(\mathbf{s}_{(2)})) \times \cdots \\ & \times f(X(\mathbf{s}_{(N)})|X(\mathbf{s}_{(N-1)})\cdots X(\mathbf{s}_{(1)})) \end{aligned} \tag{3.211}$$

where the notation (i) refers to a random permutation of the grid locations. Should some hard observation $\mathbf{d}_{\text{obs}}$ be available (and those are standard Gaussian or transformed to standard Gaussian), then

$$\begin{aligned} f(X(\mathbf{s}_1),\ldots,X(\mathbf{s}_N)|\mathbf{d}_{\text{obs}}) = & f(X(\mathbf{s}_{(1)})|\mathbf{d}_{\text{obs}}) \\ & \times f(X(\mathbf{s}_{(1)})|X(\mathbf{s}_{(2)}),\mathbf{d}_{\text{obs}}) \times \cdots \\ & \times f(X(\mathbf{s}_{(N)})|X(\mathbf{s}_{(N-1)})\cdots X(\mathbf{s}_{(1)}),\mathbf{d}_{\text{obs}}) \end{aligned} \tag{3.212}$$

A practical implementation of this is termed sequential Gaussian simulation [*Goovaerts*, 1997] and will be used in several applications later.

3.10.2.5. Using MC to Approximate a Distribution. MC arose as a technique to solve integrals. This looks perhaps a bit odd, since random number generators are used to solve a deterministic problem. However, calculating integrals numerically is not a trivial problem, in particular in higher dimensions. The number of samples drawn will then determine the accuracy of that solution. One particular integral problem is the definition of the expectation of some function

$$\mu_h = E[h(X)] = \int h(x)f(x)dx \quad X \sim f(x) \tag{3.213}$$

Using MC, this integral is estimated simply by the arithmetic mean from L samples $\{x^{(1)}, \ldots, x^{(L)}\}$ drawn from $f(x)$

$$\hat{\mu}_h = \frac{1}{L}\sum_{\ell=1}^{L} h\left(x^{(\ell)}\right) \tag{3.214}$$

In UQ, we are interested not only in just estimating a mean or a function but also in an uncertainty statement on some function value. Such uncertainty statement would need to involve a pdf or some quantiles calculated from that pdf. With a limited set of samples, this distribution can be represented by an empirical cdf:

$$\hat{F}(h(x)) = \frac{1}{L}\sum_{\ell=1}^{L} 1_{h(x^{(\ell)}) \le h(x)} \tag{3.215}$$

While this cdf is discontinuous, a continuous approximation can be obtained by making suitable interpolations and extrapolations [*Deutsch and Journel*, 1992]. In similar vein, the quantiles for given percentile $0 < p < 1$ can be estimated as

$$\hat{q}_p = \inf\{h(x) : \hat{F}(x) \ge p\} \tag{3.216}$$

Generally, the estimate $\hat{F}$ is unbiased but $\hat{q}_p$ is biased because it relies on the estimate $\hat{F}$ and not the true F.

3.10.3. Variance Reduction Methods

3.10.3.1. Introduction. In estimating or approximating integrals such as Eq. (3.202) or approximating a distribution using MC sampling, accuracy in the estimates is increased as the number of samples increases. However, evaluating the function $h(x)$ in Eq. (3.214) (such as a forward model) may be CPU demanding; hence, the number of MC samples must be limited. This calls for the definition of a measure of accuracy in the estimates. One such measure of accuracy is the variance of the estimates which measures the degree of error caused by limited sampling. A high variance in the estimates means low accuracy and a low variance of the estimates is a sign of high accuracy.

When performing simple also termed *naïve Monte Carlo*, we know that the sampling variance of Eq. (3.214) (a measure of degree of error caused by limited sampling) is given by

$$\text{var}(\hat{\mu}_h) = \frac{\sigma^2}{L}, \quad \sigma^2 = \text{var}(h(X)) \tag{3.217}$$

This estimate is also unbiased by definition (see Section 3.13 for a formal definition of bias). In case of biased estimators, a better measure of accuracy is the mean squared error (MSE) which is defined as

$$\text{MSE}(\hat{\mu}_h) = \text{var}(\hat{\mu}_h) + \text{bias}^2(\hat{\mu}_h) \tag{3.218}$$

Variance reduction aims to improve on the variance of Eq. (3.217) produced by the (naïve) MC sampling. Most variance reduction sampling schemes achieve this by sampling X more strategically. Random sampling may cause accidental clustering of samples $x^{(\ell)}$ in the space defined by X; hence, such samples are "wasted." Stratification (spreading the samples) is one such method. Another approach is to oversample (bias) certain regions that affect the calculation of the integrals most. Some weighting scheme must then be introduced to account for the incurred bias. This method, known as importance sampling, can be very efficient in reducing variance, but its improper use may lead to such sampling to run astray and actually increase variance. In other words, efficiency is traded off for the risk to have even less efficiency than the naïve sampler.

3.10.3.2. Stratified Sampling. The idea behind stratification is simple: split the high-dimensional model space of $\mathbf{X}$ into mutually exclusive and exhaustive regions and spread samples equally over these regions (instead of random of the entire domain as in the naïve MC), see Figure 3.35. Consider V the domain in which $\mathbf{X}$ varies and consider a partition of that domain into regions V_k : $k = 1, \ldots, K$. Denote

$$p_k = P(\mathbf{X} \in V_k) \tag{3.219}$$

Then, the conditional density of $X \mid X \in V_k$ is

$$f(\mathbf{x}|\mathbf{x} \in V_k) = \frac{1}{p_k} f(\mathbf{x}) 1_{\mathbf{x} \in V_k} \tag{3.220}$$

Stratified sampling becomes practical when we know the size of each strata V_k and we know how to sample from $f(\mathbf{x}| \mathbf{x} \in V_k)$; the latter is usually the case when the components of $\mathbf{X}$ are independent or when $\mathbf{X}$ follows a multi-Gaussian distribution. For example, when the components of $\mathbf{X} = (X_1, \ldots, X_N)$ are independent, then we sample a number of samples $L_k : k = 1, \ldots, K$, per region, preferably $L_k \geq 2$ (to estimate sampling variance). The samples per each region are denoted as $\left(\mathbf{x}_k^{(1)}, \ldots, \mathbf{x}_k^{(L_k)}\right)$. The stratified (and unbiased) estimate of Eq. (3.214) becomes

$$\hat{\mu}_{h,\text{strat}} = \sum_{k=1}^{K} \frac{p_k}{L_k} \sum_{\ell=1}^{L_k} h\left(\mathbf{x}_k^{(\ell)}\right) \tag{3.221}$$

In the case the L_k are chosen proportional, namely $L_k = p_k L$ with L the total number of samples, then

$$\hat{\mu}_{h,\text{strat}} = \frac{1}{L} \sum_{k=1}^{K} \sum_{\ell=1}^{L_k} h\left(\mathbf{x}_k^{(\ell)}\right) \tag{3.222}$$

The variance of this estimate is

$$\text{var}\left(\hat{\mu}_{h,\text{strat}}\right) = \sum_{k=1}^{K} p_k^2 \frac{\sigma_k^2}{L_k} \quad \text{with} \quad \sigma_k^2 = \text{var}(\mathbf{X}|\mathbf{X} \in V_k) \tag{3.223}$$

From this it becomes clear that a good stratification scheme has small within strata variances. The optimal allocation of L_k depends on the sampling cost per each stratum. If these are the same then the optimal L_k is given as

$$L_k = \frac{L p_k \sigma_k}{\sum_{k'=1}^{K} L p_{k'} \sigma_{k'}} \tag{3.224}$$

In actual applications, however, the σ_k may not be known (a priori) or may have to be estimated. Under such circumstances a safe bet is to use proportional allocation. Because high-dimensional space becomes empty very rapidly, stratification becomes difficult to implement beyond dimension larger than 5. As we will see in Chapter 8, it is most useful when sampling a subset of the variables within a larger MC sampling scheme involving other methods of variance reduction such as importance sampling.

3.10.3.3. Latin Hypercube Sampling. The main issue with stratification is that a space becomes quite empty for dimensions larger than 5. For example, in any dimension N, a regular grid has only $L^{1/N} \ll L$ strata per each component of the random vector (of dimension N) with L samples.

In Latin hypercube sampling [LHS; *Mckay*, 1998], we consider more than one dimension at a time to avoid the $L^{1/N}$ problem. The method is quite straightforward and easily explained when $d = 2$ (see Figure 3.36). Instead of assigning a number of samples per each square, one now assigns a sample per each row *and* column combined.

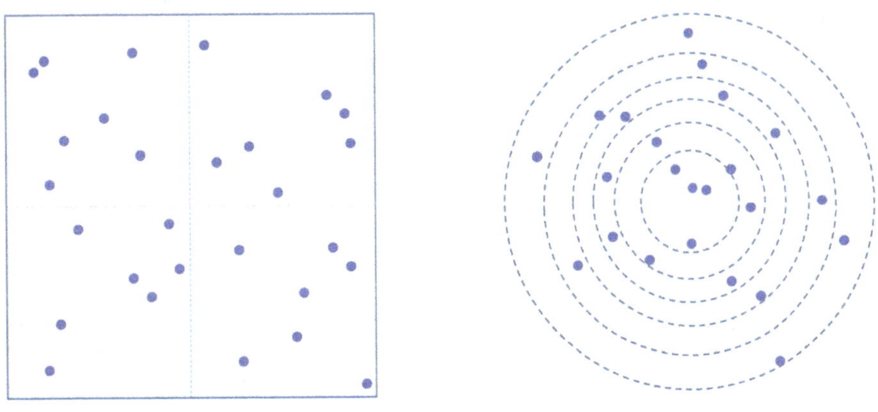

Figure 3.35 Sampling from a uniform distribution with seven samples per stratum. Sampling from a Gaussian distribution with four samples per stratum constructed from concentric circles.

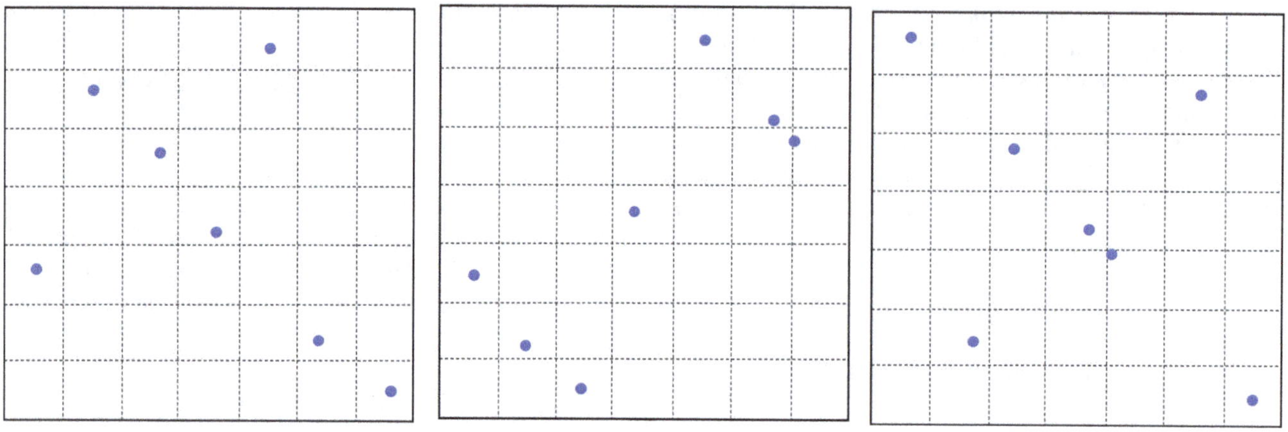

Figure 3.36 Three Latin hypercube samples in two dimensions.

The same idea is easily extended to higher dimensions. Note that in this context, the maximum number of combinations for a LHS with K divisions and N variables is $(K!)^{N-1}$. For example, a LHS with $K = 4$ divisions and dimension of $\boldsymbol{X}$ equal to 3 will have only 576 possible combinations of samples.

Theory shows that LHS can be much better than naïve MC, and moreover it is robust in the sense that the bound on the sampling variance is almost the naïve MC sampling variance for Eq. (3.214) (up to a factor of $n/(n-1)$ only). In other words, it cannot get worse and for most functions h performs much better.

3.10.3.4. Control Variates and Multilevel MC. In the method of control variates, the variance of the sampler is reduced (compare to naïve MC) by the introduction of another variable termed the control variable. The basic idea is for this additional variable to be correlated to the target, but that performing MC on this control variable has much less cost than for the target. Next to having samples of the target $h(\mathbf{x}_1), \ldots, h(\mathbf{x}_L)$ to compute the estimate Eq. (3.214), we now consider samples available of a control variable denoted by $(c_1, \ldots, c_L)$; hence, we have some estimate

$$\hat{\mu}_c = \frac{1}{L}\sum_{\ell=1}^{L} c_\ell \tag{3.225}$$

If $h(\boldsymbol{X})$ and C correlate, then this estimate will inform the estimate $\hat{\mu}_h$, whether it is too small or too large, for example by comparing $\hat{\mu}_c$ for a sample size of L with $\hat{\mu}_c$, assumed to be quite accurately known through extensive sample ($\gg L$). A control variate estimator is constructed as follows:

$$\hat{\mu}_{cv} = \frac{1}{L}\sum_{\ell=1}^{L}\left(h\left(\mathbf{x}^{(\ell)}\right) + \beta(\mu_c - c_\ell)\right) \tag{3.226}$$

A simple calculation shows that the optimal value of β, under the criterion of minimizing the estimation variance, is

$$\beta = \frac{\mathrm{cov}(h(\mathbf{X}), C)}{\mathrm{var}(C)} \tag{3.227}$$

which makes sense, since the better the correlation, the more the covariate approximates the unknown μ_h. The main problem is that estimating β requires knowing the mean μ_h, which defeats the very purpose of sampling. For that reason β is estimated using the sample $(h(\mathbf{x}^{(1)}), c_1), \ldots, (h(\mathbf{x}^{(L)}), c_L)$

$$\hat{\beta} = \frac{\sum_{\ell=1}^{L} h\left(\mathbf{x}^{(\ell)}\right)c_\ell + L\hat{\mu}_h\hat{\mu}_c}{(L-1)\,\sigma_c^2} \qquad \sigma_c^2 = \frac{1}{L-1}\sum_{\ell=1}^{L}(c_\ell - \hat{\mu}_c)^2 \tag{3.228}$$

Multilevel MC builds further on the idea of covariates. In multilevel MC, as the term suggests, we use more than one covariate (two levels: the target and a covariate). The goal again is to estimate $E[h(\mathbf{X})]$ of which samples $h_1(\mathbf{x}^{(\ell)})$, $\ell = 1, \ldots, L_1$ ($h_1 = h$) are available as well as samples $h_0(\mathbf{x}^{(\ell)})$, $\ell = 1, \ldots, L_0$ whose sampling is much less costly than h_1 ($L_0 > L_1$), then

$$E[h(\mathbf{X})] = E[h_1(\mathbf{X})] = E[h_0(\mathbf{X})] + E[h_1(\mathbf{X}) - h_0(\mathbf{X})] \tag{3.229}$$

This method again relies on the difference between the approximation and the target, which is estimated by

$$\hat{\mu}_{h,2L} = \frac{1}{L_0}\sum_{\ell=1}^{L_0} h_0\left(\mathbf{x}^{(\ell)}\right) + \frac{1}{L_1}\sum_{\ell=1}^{L_1}\left(h_1\left(\mathbf{x}^{(\ell)}\right) - h_0\left(\mathbf{x}^{(\ell)}\right)\right) \tag{3.230}$$

$2L$ refers to a two-level MC. The difference now with covariates is that $\beta = 1$ and the expected value of h_0 needs to be estimated (instead of assumed known with high accuracy). The question instead is what appropriate values for L_0 and L_1 are that minimize the estimation variance of $\hat{\mu}_{h,2L}$. To that end, consider the following costs and estimation variances:

$$\begin{aligned}&\text{cost}_0, \text{var}_0 : \text{the cost and variance for } h_0(\mathbf{X})\\&\text{cost}_1, \text{var}_1 : \text{the cost and variance for } h_1(\mathbf{X})\end{aligned} \tag{3.231}$$

Then based on Eq. (3.230), the total variance is

$$\text{var}_{\text{total}} = \frac{\text{var}_0}{L_0} + \frac{\text{var}_1}{L_1} \tag{3.232}$$

which is minimized for some fixed total cost by choosing

$$\frac{L_1}{L_0} = \frac{\sqrt{\dfrac{\text{var}_1}{\text{cost}_1}}}{\sqrt{\dfrac{\text{var}_0}{\text{cost}_0}}} \text{ or } L_1 \sim \sqrt{\frac{\text{var}_1}{\text{cost}_1}};\ L_0 \sim \sqrt{\frac{\text{var}_0}{\text{cost}_0}} \tag{3.233}$$

This makes sense, since one would like for each level (target and covariate) to run an amount of simulations that is proportional to the accuracy (estimation variance) and inversely proportional to the cost of doing so. The two-level MC can be extended to a multilevel MC with various covariates with different approximations of the target. Consider these levels as 0, 1, …, M with $h_M = h$ the target, then Eq. (3.230) can be generalized to

$$\begin{aligned}\hat{\mu}_{h,M} &= \frac{1}{L_0}\sum_{\ell=1}^{L_0} h_0\left(\mathbf{x}^{(\ell)}\right)\\&+\sum_{m=1}^{M}\frac{1}{L_m}\sum_{\ell=1}^{L_m}\left(h_m\left(\mathbf{x}^{(\ell)}\right) - h_{m-1}\left(\mathbf{x}^{(\ell)}\right)\right)\end{aligned} \tag{3.234}$$

Again, one chooses an amount of simulations based on cost and estimation variance

$$L_m \sim \sqrt{\frac{\text{var}_m}{\text{cost}_m}} m = 0, \ldots, M \tag{3.235}$$

3.10.3.5. Importance Sampling

3.10.3.5.1. Methodology. The previous two methods used the idea of correlated samples to steer sampling of $f(\mathbf{x})$ to areas of the sampling domain that lead most to variance reduction of a particular estimate, without inducing (much) bias. The importance sampler uses a different strategy. It still tries to steer the sampler to important areas of the domain but does so by means of another pdf $q(\mathbf{x})$ (instead of a covariate).

Because sampling is done from the wrong distribution, a correction will need to be made to obtain unbiased estimates. In that sense, importance sampling is more than a variance reduction technique of an estimate It can also be used to perform MC on a given distribution by means of sampling from another distribution. Since UQ relies on some form of MC, we will see that importance sampling has many applications in UQ (see Chapter 7). For that reason, it can also be seen as an alternative to rejection sampling and lies at the foundation of sequential MC, which is treated in the next section.

Consider again the estimation problem of (3.214) with samples distributed as $f(\mathbf{x})$. Consider now another distribution $q(\mathbf{x})$ and write the expectation of h as follows:

$$\mu_h = \int_V h(\mathbf{x})q(\mathbf{x})\frac{f(\mathbf{x})}{q(\mathbf{x})}d\mathbf{x} \tag{3.236}$$

One notices how the portion $h(\mathbf{x})q(\mathbf{x})$ in the integral leads to calculation of the expected value of $h(\mathbf{x})$ under $q(\mathbf{x})$, which is then corrected by a ratio of two pdfs $f(\mathbf{x})$ and $q(\mathbf{x})$. In importance sampling, we generate samples $\mathbf{x}^{(\ell)}$ from $q(\mathbf{x})$ and not $f(\mathbf{x})$, which then produces the estimate

$$\hat{\mu}_h = \frac{1}{L}\sum_{\ell=1}^{L} h\left(\mathbf{x}^{(\ell)}\right)\frac{f\left(\mathbf{x}^{(\ell)}\right)}{q\left(\mathbf{x}^{(\ell)}\right)} = \frac{1}{L}\sum_{\ell=1}^{L} w_\ell h\left(\mathbf{x}^{(\ell)}\right) \tag{3.237}$$

which requires the calculation of the ratio f/q. A sufficient condition for q is

$$f(E) = 0 \text{ such that } q(E) = 0 \text{ for that set } E \tag{3.238}$$

This basically entails that q must "cover" the range of f, being able to generate samples wherever f generates samples. The most critical question evidently is on good choices for q. Statistical theory provides some suggestions [*Lemieux*, 1997] based on general principles, but having domain knowledge (the specific application) on the target distribution may be more important. For example, where sampling of f has greatest impact on UQ and perhaps even the decisions made is probably more relevant to the context of this book. In that respect, any sampling to estimate Eq. (3.214) can also be used to approximate a distribution. Nevertheless, the choice of q will determine the amount of variance reduction achieved. Chapter 7 uses such domain knowledge to decide on a proposal distribution.

Consider a simple example in Figure 3.37 where the goal is to estimate $P(X > 4)$ where $X \sim N(0, 1)$ and hence the true exceedance probability $p_{\text{true}} = 3.1671 \times 10^{-5}$. When performing naïve MC with 100.000 samples, we get the estimate $\hat{p}_{\text{naive}} = 3.5000 \times 10^{-5}$. Consider now another pdf more centered around the target area of interest, $q(\mathbf{x}) \sim N(4, 2)$. When applying Eq. (3.237) we find that $\hat{p}_{\text{imp}} = 3.1676 \times 10^{-5}$.

3.10.3.5.2. Guideline for q. To study the sampling properties of importance sampling, and the impact of certain choices of q, recall that the variance of the MC sampler is

$$\text{var}\left(\hat{\mu}_{h,MC}\right) = \frac{1}{L}\left(E\left[h^2(\mathbf{X})\right] - \mu_h^2\right)\ \mathbf{X} \sim f(\mathbf{x}) \tag{3.239}$$

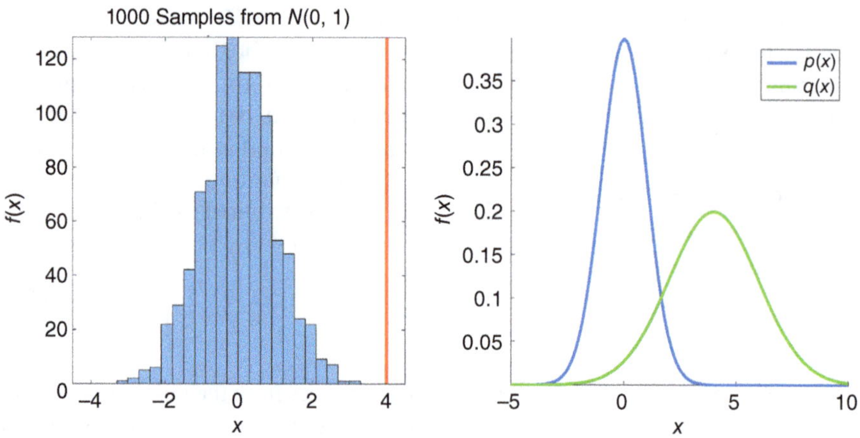

Figure 3.37 Using MC to estimate an exceedance probability from the blue pdf. Thousand samples provide basically no accuracy, while high accuracy can be obtained by using the green proposal distribution.

A simple calculation shows that

$$\mathrm{var}(\hat{\mu}_{h,MC}) = \frac{1}{L}\left(E\left[h^2(\mathbf{X})\frac{f(\mathbf{X})}{q(\mathbf{X})}\right] - \mu_h^2\right) \quad \mathbf{X} \sim f(\mathbf{x}) \tag{3.240}$$

Hence, importance sampling is more efficient than MC when

$$E\left[h^2(\mathbf{X})\frac{f(\mathbf{X})}{q(\mathbf{X})}\right] \le E\left[h^2(\mathbf{X})\right] \tag{3.241}$$

As a result, to be efficient, the ratio f/q should be small, making $\mathbf{x}$ more likely (according to that ratio) when $h(\mathbf{x})$ is larger. When $h(\mathbf{x})$ is small, then one should aim for a ratio larger than unity. Note that only when

$$\frac{f(\mathbf{x})}{q(\mathbf{x})} < 1 \quad \forall \mathbf{x} \quad \text{for which} \quad h(\mathbf{x}) \neq 0 \tag{3.242}$$

leads to variance reduction. This is also means that a poorly chosen q will lead to a variance increase. The more one knows about $h(\mathbf{x})$ and $f(\mathbf{x})$ the better a choice for q can be made, but again, this is very application specific. Getting insight into $g(\mathbf{x})$ and $f(\mathbf{x})$ in very high dimensions is not trivial. We will show in Chapter 7 that dimension reduction methods become very helpful in this regard.

q cannot be light-tailed compared to f, but a heavy tailed q often leads to inefficient sampling (close to rejection sampling). To study more quantitatively the importance sampling weights, a so-called effective sample size is introduced. When all weights are equal then the effective sample size is basically L. However, when weights become skew, then fewer samples have influence on the estimate of h. Denote as w_ℓ the weight given to a sample of q, namely $\mathbf{x}^{(\ell)}$, then an effective sample size L_{eff} can be defined as

$$L_{\text{eff}} = L\frac{\sum_{\ell=1}^{L} w_\ell}{\sum_{\ell=1}^{L} w_\ell^2} \tag{3.243}$$

Clearly, $L_{\text{eff}} \ll L$ when the weights are very skew. If L_{eff} is very small, then the importance sampler estimate may not be trusted and, worse, it may lead to an increase in variance. In the next section, on sequential MC, we will see how this problem can be alleviated.

3.11. SEQUENTIAL MC

3.11.1. Problem Formulation

Previously, we discussed methods to sample from univariate or higher-dimensional distributions. Here we consider a more specific problem, namely sampling from higher-dimensional distributions conditioned on observations. In particular, predicting, by means of MC, a future "signal" (realization, unknown, sample, event) from a past "signal" (typically denoted as "data"). In addition, this problem is formulated dynamically in time, not for a single static instance. The idea is to "assimilate" data as time progresses to make forecasts on some target future variable or event. Future observations will then become data as time progresses. Forecasting weather is an evident example [*Leeuwen and Jan*, 2009], but many other applications exists such as robotic navigation [*Dellaert et al.*, 2001], financial market analysis [*Aihara et al.*, 2009], visual object tracking [*Nummiaro et al.*, 2003], and so on. Data assimilation, data filtering, sequential MC (SMC), bootstrap filtering, particle filtering, and survival of the fittest sampling are all nomenclature to address basically the same problem.

Generally, SMC aims to dynamically predict an "unknown signal" from "observed data." Within the context of UQ, the data are modeled using data variables, but now occur at some discrete time events:

$$\mathbf{d}_{1:t} = \{\mathbf{d}_1,\ldots,\mathbf{d}_{t-1},\mathbf{d}_t\} \quad t \in N^+ \tag{3.244}$$

For example, one may repeat 3D seismic surveys to obtain a sequence termed 4D seismic (each 3D is compared with the base survey) or, in a groundwater setting, one may repeat hydraulic head or concentration measurements over time. The "unknown signal" is the unknown subsurface, modeled using model variables; here these model variables are updated in time (no longer a single static model)

$$\mathbf{m}_{1:t} = \{\mathbf{m}_1,\ldots,\mathbf{m}_{t-1},\mathbf{m}_t\} \tag{3.245}$$

Forward models are used to produce predictions

$$\mathbf{h}_{1:t} = \{\mathbf{h}_1,\ldots,\mathbf{h}_{t-1},\mathbf{h}_t\} \quad \mathbf{h}_{t'} = g_h(\mathbf{m}_{t'}) \tag{3.246}$$

Note that the forward model g_h is now explicitly written as a function of the time at which a prediction is made. Prior to having any data, hence at time $t = 0$, the initial or prior uncertainty on model and prediction are

$$\mathbf{m}_0 \sim f(\mathbf{m}_0); \quad \mathbf{h}_0 \sim f(\mathbf{h}_0) \quad \text{(no data yet)} \tag{3.247}$$

In addition, and this is an assumption used in most SMC, a Markov property is invoked whereby the unobserved signal is conditionally independent of all previous states $t-2$, …, 1 given state $t-1$. This basically means that knowing the previous state at $t-1$ is sufficient, and we do not need to include all the prior states $t-2$, …, 1. The probabilistic model is now fully specified by

$$\begin{gathered}\text{Prior}\, f(\mathbf{m}_0)\\ \text{Conditional}: f(\mathbf{m}_t|\mathbf{m}_{t-1},\mathbf{m}_{t-2},\ldots,\mathbf{m}_1) \cong f(\mathbf{m}_t|\mathbf{m}_{t-1})\\ \text{Conditional}: f(\mathbf{d}_t|\mathbf{m}_t,\mathbf{m}_{t-1},\ldots,\mathbf{m}_1) \cong f(\mathbf{d}_t|\mathbf{m}_t)\end{gathered} \tag{3.248}$$

The aim is to recursively estimate the following:

1. $f(\mathbf{m}_{0:t}|\,\mathbf{d}_{1:t})$: the full posterior distribution of all models, in time, given the time sequence of data.
2. $f(\mathbf{m}_t|\,\mathbf{d}_{1:t})$: the marginal for the present model given past data, also termed the filtering distribution.
3. $\int h(\mathbf{m}_{0:t})f(\mathbf{m}_{0:t}|\,\mathbf{d}_{1:t})d\mathbf{m}_{0:t}$: any expectation of some function of interest. This could be a mean, a covariance but also some summary statistic of the model, such as the prediction h deduced from it. More specifically, the marginal $\int h(\mathbf{m}_t)f(\mathbf{m}_t|\,\mathbf{d}_{1:t})d\mathbf{m}_t$.

In terms of the latter prediction, one could also aim to directly obtain $f(\mathbf{h}_{0:t}|\,\mathbf{d}_{1:t})$: the model is now a hidden variable.

Ways of obtaining samples from Eq. (3.248) will be the topic of Chapter 7. The solution to the above-formulated problem is obtained by the direct application of Bayes' rule:

$$f(\mathbf{m}_{0:t}|\mathbf{d}_{1:t}) = \frac{f(\mathbf{d}_{1:t}|\mathbf{m}_{0:t})f(\mathbf{m}_{0:t})}{\int f(\mathbf{d}_{1:t}|\mathbf{m}_{0:t})f(\mathbf{m}_{0:t})d\mathbf{m}_{0:t}} \tag{3.249}$$

Then, a recursive formula going from time t to $t+1$ is obtained as

$$f(\mathbf{m}_{0:t+1}|\mathbf{d}_{1:t+1}) = f(\mathbf{m}_{0:t}|\mathbf{d}_{1:t})\frac{f(\mathbf{d}_{t+1}|\mathbf{m}_{t+1})f(\mathbf{m}_{t+1}|\mathbf{m}_t)}{f(\mathbf{d}_{t+1}|\mathbf{d}_{1:t})} \tag{3.250}$$

In terms of the marginal, one therefore obtains the following recursion:

$$\begin{aligned}&\text{Predict}\, t \,\text{from}\, 1:t-1:\ f(\mathbf{m}_t|\mathbf{d}_{1:t-1})\\ &\quad = \int f(\mathbf{m}_t|\mathbf{m}_{t-1})f(\mathbf{m}_{t-1}|\mathbf{d}_{1:t-1})d\mathbf{m}_{t-1}\\ &\text{Update/assimilate data}\, \mathbf{d}_t:\ f(\mathbf{m}_t|\mathbf{d}_{1:t})\\ &\quad = \frac{f(\mathbf{m}_t|\mathbf{d}_t)f(\mathbf{m}_t|\mathbf{d}_{1:t-1})d\mathbf{m}_t}{\int f(\mathbf{m}_t|\mathbf{d}_t)f(\mathbf{m}_t|\mathbf{d}_{1:t-1})d\mathbf{m}_t}\end{aligned} \tag{3.251}$$

This iterative/recursive of updating models with data requires the solution of integrals over high-dimensional functions. From our previous discussion, see Section 3.10.2.5, we saw how MC can be used to estimate these integrals. This would be "perfect MC" sampling. However, such MC would be very inefficient, in particular if evaluation of either data forward models or prediction forward models are CPU demanding. Perfect MC ignores the sequential nature of these integrals (the next one depends of the previous one only because of the Markov property) and thereby ignoring important information to make such calculation more efficient. How this is achieved is discussed in the following section on a sequential sampling method termed SIR (sequential importance resampling).

3.11.2. Sequential Importance Resampling

Recall that importance sampling aims to calculate an expectation more accurately with lesser samples than a naïve MC. This was achieved by sampling from another distribution, then reweighting the samples to correct for that biased sampling. Importance sampling can equally be applied as a variance reduction method within sequential MC. In this context and using the notation in the context of the importance sampling method, see Eq. (3.237), we substitute as follows:

$$E[h(\mathbf{X})] \rightarrow E[h(\mathbf{m}_{0:t})] \tag{3.252}$$

$$f(\mathbf{x}) \rightarrow f(\mathbf{m}_{0:t}|\mathbf{d}_{1:t}) \tag{3.253}$$

$$q(\mathbf{x}) \rightarrow q(\mathbf{m}_{0:t}|\mathbf{d}_{1:t}) \tag{3.254}$$

$$\mathbf{x}^{(\ell)} \rightarrow \mathbf{m}^{(\ell)} \tag{3.255}$$

$$\Rightarrow \hat{\mu}_{h(\mathbf{m}_{0:t}),IS} = \frac{1}{L}\sum_{\ell=1}^{L} w\left(\mathbf{m}_{0:t}^{(\ell)}\right) h\left(\mathbf{m}_{0:t}^{(\ell)}\right);$$
$$w\left(\mathbf{m}_{0:t}^{(\ell)}\right) = \frac{f(\mathbf{m}_{0:t}|\mathbf{d}_{1:t})}{q(\mathbf{m}_{0:t}|\mathbf{d}_{1:t})} \tag{3.256}$$

which is estimated from L samples

$$\left\{\mathbf{m}_{0:t}^{(1)}, \ldots, \mathbf{m}_{0:t}^{(L)}\right\} \tag{3.257}$$

These samples are also termed "particles": think of points floating around in a very high dimensional space. The problem with this simple application of importance sampling within the context SMC is that the weights need to recalculated at each time t, since models get updates and hence weights need updating, regardless of how suitable a choice for q is taken. As time progresses, this becomes progressively more difficult from a computational point of view.

One way around this is to evoke again the Markov assumption and state that q at time t is conditionally depended on q at time $t-1$

$$q(\mathbf{m}_{0:t}|\mathbf{d}_{1:t}) = q(\mathbf{m}_{0:t-1}|\mathbf{d}_{1:t-1})q(\mathbf{m}_t|\mathbf{m}_{0:t-1}, \mathbf{d}_{1:t}) \tag{3.258}$$

For example, if q is simply the prior distribution $f(\mathbf{m})$ (which will certainly cover the target posterior distributions), then

$$q(\mathbf{m}_{0:t}|\mathbf{d}_{1:t}) = f(\mathbf{m}_0)\prod_{t'=1}^{t} f(\mathbf{m}_{t'}|\mathbf{m}_{t'-1}, \mathbf{d}_t) \tag{3.259}$$

The problem, however, with this approach is that as t progresses, the weights become very skewed, meaning that only very few samples (particles) get any weight (also termed degeneracy). This makes intuitive sense: as data becomes more constraining, the posterior distribution becomes narrower, and given a limited set of prior models, few of those prior models fall within the range of the posterior distribution. We discussed in Section 3.10.3.5 that such skewness is not desirable, as IS estimates may have very large variance, even larger than the MC sampler.

SIR, also termed bootstrap filtering, allows dealing with this problem by eliminating from the set of prior models/particles those models that have low weights in the importance sampler and targeting the generation of particles/models with higher weight. SIR employs a simple resampling idea. Since importance weights are normalized ($w \geq 0; \sum w = 1$), these weights form a discrete empirical distribution (see also Section 3.10.3.5) on the current particles/models. We can, therefore, replace the weighted estimate of Eq. (3.256) into an unweighted estimate by resampling particles/models from this discrete distribution

$$\hat{\mu}_{h,\mathrm{SIR}} = \frac{1}{L}\sum_{\ell=1}^{L} h\left(\tilde{\mathbf{m}}_{0:t}^{(\ell)}\right) \tag{3.260}$$

with $\tilde{\mathbf{m}}_{0:t}^{(\ell)}$ sampled according to the importance sampler weights $\mathbf{w}_{0:t}$. The advantage of SIR is that (i) it is easy to implement because it is very modular (generate model, calculate weights, resample models), hence no iterative inversion is needed and (ii) it is perfectly parallelizable since the forward models can be applied to all samples at the same time.

As additional observations become available, $\mathbf{d}_{t+1}$, the process is repeated. The proposal function is reestimated using the resampled particles from the previous time step, and Eq. (3.258) is rewritten as

$$q(\mathbf{m}_{0:t+1}|\mathbf{d}_{1:t+1}) = q(\mathbf{m}_{0:t}|\mathbf{d}_{1:t})q(\mathbf{m}_{t+1}|\mathbf{m}_{0:t}, \mathbf{d}_{1:t+1}) \tag{3.261}$$

This process is repeated each time new observations are gathered, providing an online estimate of $\mathbf{m}_t$ at any subsequent time step.

3.12. MARKOV CHAIN MC

3.12.1. Motivation

Markov chain MC (McMC) methods [*Geyer*, 2002] are used to sample iteratively from complicated distributions for which MC methods do not work anymore. Such situations arise, for example in Chapter 6 on inversion, where the aim is to infer model variables from data. While McMC methods apply to sampling from any distribution model, the interest in this book mostly lies in sampling from a posterior model within a Bayesian context (see Chapters 5 and 6); hence, the distribution to be sampled from is defined as

$$f(\mathbf{m}|\mathbf{d}) = \frac{f(\mathbf{d}|\mathbf{m})}{f(\mathbf{d})} f(\mathbf{m}) \simeq f(\mathbf{d}|\mathbf{m})f(\mathbf{m}) \tag{3.262}$$

A number of complications may arise when sampling from such posterior model:

1. The prior model $f(\mathbf{m})$ may not have an analytical expression, but rather is some computer algorithm that generates prior model realizations $\mathbf{m}^{(1)}$, $\mathbf{m}^{(2)}$, $\mathbf{m}^{(3)}$, …; hence, the density value $f(\mathbf{m}^{(\ell)})$ cannot be evaluated. This is not unusual in subsurface applications where complex computer codes generate very realistic representations of the subsurface. Basically this code maps random numbers into some subsurface model, without explicit knowledge of the underlying density.

2. The forward modeling code used to evaluate how well a model matches the data is again a complex computer algorithm to which we have little or no access.

The main point here is that in both (1) and (2) we have only access to the "black box" for generating prior models and for applying some response function on these prior model.

3.12.2. Random Walk

Since we cannot sample directly and independently from the posterior, we need to "walk around" the sample space iteratively and hope to find high-likelihood models. Jumping around randomly is not a good strategy. A better strategy would be to walk around models that have high likelihood, once we have found such a region. Unless the posterior is quite degenerate (random peaks), it is more plausible that higher likely models can be found in such region. This suggests walking around more like a drunk person (a random walk) than a grasshopper. The problem now is how to merge this random walking with sampling properly from a distribution. Intuitively, the random walk as a sampling method makes sense: walking around more frequently in high-likelihood areas than low-likelihood areas will generate samples from the posterior that reflects that frequency. The list of samples obtained along this random walk, however, does not constitute a sample from the posterior, because two consecutive samples along this random walk are not independent (and we would like i.i.d. samples). In fact, to make this walking relevant in terms of finding high-likelihood samples, they should be very dependent.

The random walk initiates a Markov chain of models $\mathbf{m}$ such that the next model is generated only knowing the current model. The key theoretical result is that if such chain of models is (i) irreducible and not transient, meaning essentially that any model can be "reached" from any other model through the walk, and (ii) is aperiodic (e.g., it does not get stuck in loops), then the Markov chain has a stationary distribution that equals the target distribution, here the posterior distribution. If this walking around is done "long enough" then the sample obtained when stopping the walk is a sample from the posterior distribution. More samples can be obtained by starting over, or waiting again long enough. These issues will be discussed in Section 3.12.5.

The McMC theory does, however, not state how such walking (iterating) should be performed but provides only the necessary and sufficient conditions for it. Hence, many methods have been designed to perform McMC sampling. Unless in very specific cases (e.g., Gaussian), there is also no theoretical result on when to stop (i.e., when "convergence" is reached). In that sense, McMC constitutes a family of approximate samplers.

3.12.3. Gibbs Sampling

The Gibbs sampler [*Geman and Geman*, 1984] is particularly useful for high-dimensional problems because it relies on dividing the model variables into "blocks":

$$\mathbf{m} = (\mathbf{m}_1, \mathbf{m}_2, \ldots, \mathbf{m}_K) \tag{3.263}$$

Each iteration of the Gibbs sampler consists of cycling through all these blocks $\mathbf{m}_k$ and drawing a new value for each block conditional on all the other blocks. These conditional distributions are, therefore, represented as

$$f(\mathbf{m}_k | \mathbf{m}_1, \mathbf{m}_2, \ldots, \mathbf{m}_{k-1}, \mathbf{m}_{k+1}, \ldots, \mathbf{m}_K, \mathbf{d}) \tag{3.264}$$

Notice how these conditional distributions are in fact partial posterior distribution in the model variables. Gibbs sampling is particularly useful in cases where the sequence of conditional distribution is available, such as in hierarchical Bayesian models [*Jaynes*, 2003], or in sequential simulation, often used in geostatistics [*Hansen et al.*, 2012].

3.12.4. Metropolis–Hastings Sampler

The Metropolis sampler [*Hastings*, 1970] adds an acceptance/rejection rule to the random walk and works as follows:

Generate an initial model from the prior: $\mathbf{m}^0 \sim f(\mathbf{m})$

For $k = 1, 2, \ldots$

1. Sample a new model from the proposal distribution $\mathbf{m}^* \sim f(\mathbf{m} | \mathbf{m}^{k-1})$

2. Calculate the ratio $\alpha = \frac{f(\mathbf{d}|\mathbf{m}^*)}{f(\mathbf{d}|\mathbf{m}^{k-1})}$

3. Set $\mathbf{m}^k = \begin{cases} \mathbf{m}^* & \text{with probability } \min(\alpha, 1) \\ \mathbf{m}^{k-1} & \text{else} \end{cases}$

For the Metropolis sampler, the proposal distribution must be symmetric $f(\mathbf{m}^i | \mathbf{m}^j) = f(\mathbf{m}^j | \mathbf{m}^i)$. The Metropolis–Hasting sampler, on the other hand, generalizes this to asymmetric proposal distributions, a much wider class of proposal distributions. Therefore, it requires the calculation of the following ratio which accounts for this asymmetry:

$$\alpha = \frac{f(\mathbf{d}|\mathbf{m}^*)}{f(\mathbf{d}|\mathbf{m}^{k-1})} \frac{f(\mathbf{m}^{k-1}|\mathbf{m}^*)}{f(\mathbf{m}^*|\mathbf{m}^{k-1})} \tag{3.265}$$

The use of an asymmetric proposal distribution often aids in increasing the speed of the random walk [*Gelman et al.*, 2004]. The ideal jumping distribution is the target distribution. In that case $\alpha = 1$, always; the sampler becomes a series of independent draws from the target distribution. This ideal case does not happen, but some desirable properties of jumping distributions are that (i) they are easy to sample from and (ii) they allow relatively large jumps and do not often get rejected. The Gibbs sampler and

Metropolis sampler are often used as basic building blocks in constructing more efficient samplers (see Section 3.12.6).

3.12.5. Assessing Convergence

MC by independent draws (i.i.d sampling) is simple: the resulting outcomes are samples from the target distribution (here the posterior distribution) and the empirical distribution approximates the target distribution. Sampling iteratively is more challenging because of the following reasons:

1. The iterations have to proceed long enough; otherwise, the sampler will reflect more the starting approximation rather than the actual distribution.
2. Since sampling is not done by independent draws, serial correlation exists between the samples that may cause inefficiency in the sampling. Even though at "convergence" (which is theoretically infinite but has good approximation for finite iterations), the correlated draws are samples from the target distribution, the amount of samples is effectively less than if one would sample by independent draws. The higher this correlation, the less the effective sample size is.

This requires methods for "monitoring" convergence. Note that such methods usually provide necessary indications/conditions that we are drawing from the target distribution, they are however not sufficient. The first stage of monitoring is with regard to the "burn-in" or "warm-up." Here we see significant changes in any properties calculated during in the chain. Any statistics that are being monitored display nonstationary variation. A common approach to monitoring is to start multiple chains and then study so-called "mixing" of the chains. In this mixing, we study two levels of variation in the chain: the variation of some quantity q (mean, variance, median, etc.) within a single chain and the variation between chains (similar to a variance analysis in cluster sampling). Consider again our target distribution $f(\mathbf{m}|\,\mathbf{d})$ and the monitoring of some estimated $\hat{q}$ (a scalar such as an estimate of the mean of the posterior distribution). Consider that we run m chains and that we are currently at iteration n. We compute

$$\begin{aligned} &B_{\hat{q}} \ : \text{the between variance of } \hat{q} \\ &W_{\hat{q}} \ : \text{the within variance of } \hat{q} \end{aligned} \tag{3.266}$$

$W_{\hat{q}}$ provides an estimate of the marginal posterior variance of q; however, this is an underestimate (has less variability) because the target distribution has not been fully sampled yet. One proposal, therefore, is to calculate a "corrected" estimate as

$$W_{\hat{q},\text{corr}} = W_{\hat{q}} + \frac{1}{n}B_{\hat{q}} \tag{3.267}$$

In the limit this will estimate the variance unbiasedly. In monitor convergence, we can therefore study the ratio

$$R = \frac{W_{\hat{q},\text{corr}}}{W_{\hat{q}}} \tag{3.268}$$

which should converge to 1 as $n \to \infty$.

3.12.6. Approximate Bayesian Computation

A "full" Bayesian method involves specifying models for the likelihood and prior and then to sample from the posterior. These models are full multivariate pdfs. In Chapter 6 we will discuss various applications of the full Bayesian approach. The problem with full Bayesian methods is that they often rely on certain model assumptions. One example is the overuse of the Gaussian distribution (by lack of anything better) or specification of likelihood by assuming independence in measurement errors. In that context, *Schoups and Vrugt* [2010] introduced a generalized likelihood function where residual errors are correlated, heteroscedastic, and non-Gaussian. However, within a geological context this may not be the main issue. Even if this allows for a more general likelihood function, it does not address how errors (model, data, epistemic, aleatory) can be separated. Full Bayes may require extensive computations to fully evaluate the likelihood model and/or normalization constant. An example of the latter is the unknown normalization constants in MRF models that may require McMC sampling just to evaluate this normalization constant [*Tjelmeland and Besag*, 1998; *Mariethoz and Caers*, 2015].

From a subsurface system application point of view, a full Bayesian approach may not be needed. Adhering rigorously to model specifications and performing rigorous sampling ignores the subjectivity and importance of the prior model. Why would one sample from a very subjectively chosen model very rigorously? This point will be strongly argued in Chapters 5, 6, and 7. The critical issue in UQ is the statement of a physically realistic and geologically plausible prior distribution. Hence, the main problem is not so much the sampling, but what exactly one is sampling from. We may not need to care too much about the "correct" posterior model and the "correct" sampling from it. Rather we would like for our posterior models to adhere to properties formulated in a physics-based prior and to reflect some field data.

A large family of methods that avoid likelihood pdfs, and also called "likelihood-free" methods [*Diggle and Graton*, 1984], are termed approximate Bayesian computation (ABC). This simple idea has many varieties, see ABCs method in *Beaumont et al.* [2002], *Sadegh and Vrugt* [2014], *Sadegh and Vrugt* [2013], and *Turner and Van Zandt* [2012], or extended rejection sampling in

Mosegaard [1995], or generalized likelihood uncertainty estimation in *Beven* [2009]. The basic method is as follows:

1. Draw **m** from the prior distribution $f(\mathbf{m})$, simulate the data from the forward model $\mathbf{d} = g_d(\mathbf{m})$.
2. Specify a distance and calculate $d(\mathbf{d}, \mathbf{d}_{\text{obs}})$.
3. Accept **m** if $d(\mathbf{d}, \mathbf{d}_{\text{obs}}) < \varepsilon$ for some specified threshold ε, otherwise reject.

This method accepts draws from the prior relative to the distance calculated from the observed data. There are three issues that need to be addressed: (i) specify an appropriate distance, note that **d** may be very high dimensional, hence comparison will be made based on some summary statistics evaluated on **d**, (ii) ε should be small, but not too small, and (iii) for certain prior models, the rejection rate may be very high.

What approximation is being made compared to a full Bayesian approach? Given the above distribution, the approximate distribution being sampled from is

$$f(\mathbf{m}|\mathbf{d}) = f(\mathbf{m})\int_d f(\mathbf{d}|\mathbf{m}) I(d(g_d(\mathbf{m}),\mathbf{d}_{\text{obs}}) < \varepsilon) d\mathbf{d} \quad (3.269)$$

with I an indicator operator (either 1 or 0). The actual (true) posterior distribution is obtained when $\varepsilon \to 0$ [*Beaumont et al.*, 2002; *Turner and Van Zandt*, 2012]. In the above algorithm

$$f(\mathbf{d}|\mathbf{m}) = \delta(\mathbf{d} - g(\mathbf{m})) \quad (3.270)$$

but could also be based on a measurement error model (see Chapter 6)

$$\mathbf{d} = g(\mathbf{m}) + \varepsilon \quad (3.271)$$

3.12.7. Multichain McMC

Traditional McMC methods work well for problems that are not too high dimensional (a few parameters, e.g., five or less). For high-dimensional and nonlinear problems, with multimodel distributions in such spaces, these methods can become difficult to apply, or have problems in terms of efficiency and convergence. One way to overcome these inefficiency problems is to create so-called adaptive chains, meaning that the proposal distribution changes during iteration.

The most common adaptive single chain methods are adaptive proposal [*Haario et al.*, 1999] and adaptive Metropolis [*Haario et al.*, 2001] methods. The proposal distribution here is multivariate Gaussian and the adaptation is made in the covariance matrix, by recalculating the covariance based on a set of samples along the chain as well as some consideration on the dimensionality of the problem. Although increase in efficiency is achieved in higher dimensions, the nature of the adaption makes it really only applicable for Gaussian-type distributions.

Multiple chain methods use multiple chains running in parallel and are known to out-perform single chain methods for complex posterior distributions, for example, that exhibit multimodality (e.g., [*Gilks et al.*, 1994; *Liu et al.*, 2000; *Craiu et al.*, 2009]). A multichain method popular in hydrology (as well applied to a variety of other problems) is the DiffeRential Evolution Adaptive Metropolis method [DREAM, *Vrugt et al.*, 2009; *Gupta et al.*, 2012; *Laloy and Vrugt*, 2012; *Vrugt*, 2016]. This method is based on an adaptive Metropolis sampler termed Differential Evolution Markov chain (DE-MC). DE-MC employs genetic algorithms (or any other differential evolution, [*Storn and Price*, 1997] to evolve a population of chains but using the Metropolis criterion (Eq. (3.265)) to evolve the population of chains [*Ter Braak*, 2006]. A traditional genetic algorithm (GA), (on its own) is not a sampler but an optimizer; hence, convergence is increased by combining GA and McMC, while still drawing from the posterior distribution. In DREAM, DE-MC is enhanced by using an adaptive randomized subspace sampling as well as other methods to get balance and ergodicity in the chain, leading to considerable improvement over other adaptive MCMC sampling approaches [*Vrugt*, 2016].

Another approach is to combine McMC with sequential MC. For example, *Andrieu et al.* [2010] propose a particle Markov chain Monte Carlo (PMCMC) methods, which relies on a combination of both McMC and SMC taking advantage of the strength of each. Here SMC algorithms are used to design efficient high-dimensional proposal distributions for MCMC algorithms.

3.13. THE BOOTSTRAP

3.13.1. Introduction

Even with the advent of new data scientific approaches such as machine learning, or computer vision, the basic approach to data analytics has not changed: (i) collect data, (ii) summarize data, and (iii) infer from data: statistical inference. In that regard, the bootstrap caused a revolution in statistical science since its inception [*Efron*, 1979; *Efron and Tibshirani*, 1994]. Statistical inference deals with the estimation of (population) parameter θ, in terms of estimates $\hat{\theta}$ and determining how accurate $\hat{\theta}$ is in terms of the true θ. A typical example is the estimate of the mean $\mu = E[X]$ of a random variable $X \sim F(x, \theta)$. An unbiased estimate of this expectation is the arithmetic average

$$\hat{\mu} = \frac{1}{L}\sum_{\ell=1}^{L} x^{(\ell)} \quad (3.272)$$

If the population variance σ is known, and $X \sim N(\mu, \sigma)$, then the sampling distribution of $\hat{\mu}$ (meaning how $\hat{\mu}$ varies in the population for that sample size) is

$$X \sim N\left(\hat{\mu}, \frac{\sigma}{\sqrt{L}}\right) \tag{3.273}$$

In case σ is not known, the standard error of the estimate can be calculated as

$$se_{\hat{F}} = \frac{s}{\sqrt{L}} \quad s = \sqrt{\frac{\sum_{\ell=1}^{L}\left(x^{(\ell)} - \hat{\mu}\right)^2}{L-1}} \tag{3.274}$$

The problem is that this simple setting does not easily extend to the general case for any random variable and any quantity of interest that needs to be estimated. Explicit expressions are usually not available.

The bootstrap instead provides an approximation of $se_{\hat{F}}$ by way of resampling. To that extent, we introduce the notion of a bootstrap sampling which samples from the original sample $(x^{(1)}, \ldots, x^{(L)})$ with replacement (basically putting all samples in a bag, taking a value, but putting it back too). We denote such sample as

$$\mathbf{x}_b = \left(x_b^{(1)}, \ldots, x_b^{(L)}\right) \tag{3.275}$$

The value $x_b^{(\ell)}$ can only be one of the original sample values. This resampling idea allows for values to be sampled multiple times, even if they are only found once in the original sample. Essentially, samples are drawn from the (discrete) empirical distribution constructed from the original sample. As many such new sample sets can be generated as desired

$$\mathbf{x}_b = \left(x_b^{(1)}, \ldots, x_b^{(L)}\right) \quad b = 1, \ldots, B \tag{3.276}$$

Using each such bootstrap sample, we recalculate the estimate as $\hat{\hat{\theta}}_b$ (the double hat indicating it is calculated under the bootstrap, the subscript, which bootstrap sample) and calculate the bootstrap standard error as

$$\widehat{se}_B(\hat{\theta}) = \sqrt{\frac{\sum_{b=1}^{B}\left(\hat{\hat{\theta}}_b - \hat{\hat{\theta}}\right)^2}{B-1}} \quad \hat{\hat{\theta}} = \frac{\sum_{b=1}^{B}\hat{\hat{\theta}}_b}{B} \tag{3.277}$$

Note, however, that θ does not have to be the mean, it could be any parameter. Figure 3.38 provides a summary of the procedure.

3.13.2. Nonparametric Bootstrap

3.13.2.1. One-Sample. In this section, we focus on the simple case of a single sample of a random variable $(x^{(1)}, \ldots, x^{(L)}) \sim F(x, \theta)$. This sample generates an empirical distribution $\hat{F}$ with emprical probability of $1/L$ on each sample $x^{(\ell)}$. The true parameter of a distribution can be written as a general function (e.g., an integral) of the true distribution

$$\theta = t(F) \tag{3.278}$$

Using the empirical distribution, one can now generate a so-called plugin estimate

$$\hat{\theta} = t(\hat{F}) \tag{3.279}$$

Nonparametric bootstrap provides a modular way of studying the sampling properties (e.g., bias, standard error, quantiles) of the estimate $\hat{\theta}$. Figure 3.39 provides an overview of this procedure.

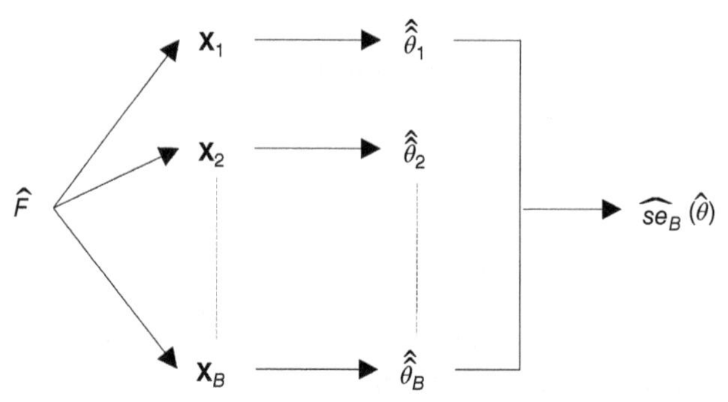

Figure 3.38 Summary of the bootstrap procedure.

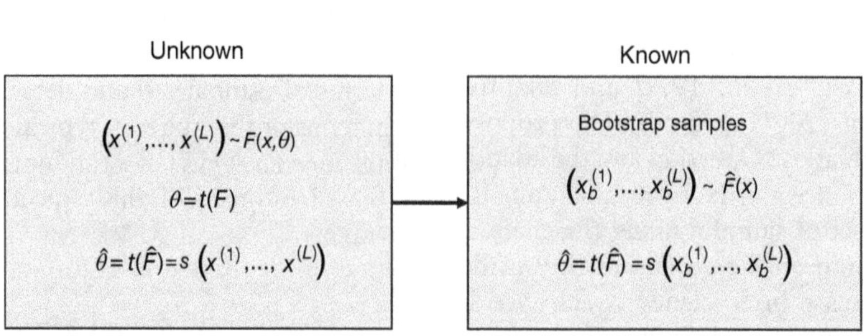

Figure 3.39 The one-sample bootstrap.

*3.13.2.2. **Bias Correction.*** A bias is a systematic error made by an estimator about an unknown quantity. Mathematically, this is represented as

$$\begin{aligned}\text{bias}_F &= E_F\left[\hat{\theta}\right]-\theta \\ &= E_F\left[t(\hat{F})\right]-t(F) \text{ plugin}\end{aligned} \tag{3.280}$$

The bootstrap can be used to estimate this bias as follows:

$$\begin{aligned}\text{bias}_F &= E_F\left[\hat{\theta}\right]-t(\hat{F}) \\ &= E_{\hat{F}}\left[\hat{\hat{\theta}}\right]-\hat{\theta}\end{aligned} \tag{3.281}$$

This bias estimate can then be used to correct the original estimate into a bias-corrected estimate:

$$\hat{\theta}_{\text{bias-corrected}} = 2\hat{\theta}-E_{\hat{F}}\left[\hat{\hat{\theta}}\right] \tag{3.282}$$

The main issue with bias-corrected estimators is that they may have higher variance. This is clear from Eq. (3.282) since the bias corrections have two quantities that need to be estimated (hence subject to estimation variance). In addition, the multiplication by two increases this effect.

*3.13.2.3. **Time Series.*** Bootstrap can also address more complex data structures than the simple one-sample model with single parameter θ. The idea depicted in Figure 3.39 can be extended to any probability model generated the "sample," whether these are time series, maps, 3D cubes of data, and so on. As illustration, consider a time series modeled as a first-order auto-regressive model

$$x(t)=\beta x(t-1)+\varepsilon(t) \tag{3.283}$$

given an observed time series $(x(1), \dots, x(T))$, β can be estimated as $\hat{\beta}$ through a simple least-square procedure (minimizing the squares error of residuals). In addition, it is assumed that the error follows a certain distribution F. Hence, the generating probability model for the data has two unknowns: (β, F). The empirical distribution of the residuals is

$$\left(\varepsilon(t)=x(t)-\hat{\beta}x(t-1),\frac{1}{T}\right),\quad t=1,\dots,T \tag{3.284}$$

which then allows generation of bootstrap samples x_b by the following recursion:

$$x_b(i)=\hat{\beta}x_b(i-1)+\varepsilon_b(i-1) \tag{3.285}$$

*3.13.2.4. **Regression.*** Another application of a more complex data structure offers itself in regression. For example, we may want to build a regression between model parameters and data, predictions and data, and so on in the context of UQ (see Chapter 7). Because we have limited samples to do so, we need to get some idea of confidence in that regression. In regression, the aim is to model the conditional expectation on some variable Y given obnservations (independent variables) $\mathbf{X}$, of dimension N. Consider the general situation of samples

$$\mathbf{z}^{(\ell)}=\left(\mathbf{x}^{(\ell)},y^{(\ell)}\right),\ell=1,\dots,L \tag{3.286}$$

The aim is to model the conditional expectation with a linear function

$$E[Y|\mathbf{X}]=\boldsymbol{\beta}^T\mathbf{x} \tag{3.287}$$

The probability structure for the y-values is expressed by means of an error model

$$y=\boldsymbol{\beta}^T\mathbf{x}+\varepsilon \tag{3.288}$$

The classical solution (see Eq. (3.130)) is

$$\hat{\boldsymbol{\beta}}=\left(X^TX\right)^{-1}X^T\mathbf{y}\quad \mathbf{y}=\left(y^{(1)},\dots,y^{(L)}\right) \tag{3.289}$$

To apply the bootstrap, we need to first state the probabilistic generating structure $P(\boldsymbol{\beta}, F)$. Here F is the distribution of the residuals ε, which is modeled empirically as $\left(\hat{\boldsymbol{\beta}},\hat{F}\right)$

$$\hat{F}:P(\varepsilon=\hat{\varepsilon}_\ell)=\frac{1}{L}\hat{\varepsilon}_\ell=y^{(\ell)}-\hat{\boldsymbol{\beta}}^T\mathbf{x}^{(\ell)} \tag{3.290}$$

This allows generating bootstrap samples as follows: estimate $\hat{\boldsymbol{\beta}}$, generate bootstrap sample of ε using Eq. (3.290), and then generate bootstrap samples of y using Eq. (3.288).

3.13.3. Bootstrap with Correlated Data

In geosciences one often deals with models, data, or predictions that are correlated in space and/or time. In such cases, one is equally interested in stating confidence on same statistical parameter whether it is the global mean of a spatial field or parameters of the spatial covariance function, modeling the observations.

A simple approach to deal with such correlated data is to first decorrelate [*Solow*, 1985] them, apply a standard one-sample nonparametric bootstrap, and then back-transform the uncorrelated bootstrap sample to a correlated one. Consider a random vector $\mathbf{X}=(X_1, \dots, X_N)$ whose elements are correlated (could be space, time, or just multivariate). Then the covariance matrix C_X (modeled from these observations) can be decomposed by means of a Cholesky decomposition:

$$C_X=LL^T \tag{3.291}$$

Decorrelated observations can now be generated as follows:

$$\mathbf{u}=L^{-1}\mathbf{x}\quad C_U=I_N \tag{3.292}$$

Generating any bootstrap sample $\mathbf{u}_b$ results in a boostrap sample $\mathbf{x}_b$

$$\mathbf{x}_b = L\mathbf{u}_b \quad b = 1,\ldots,B \tag{3.293}$$

from which any statistics (e.g., the global mean) can be calculated. Note that this model assumes Gaussianity; hence, some prior normal-score transformation may be required if the X_n are not Gaussian.

Consider now the specific case of a spatial model (a grid with unknown quantities) of which we have some partial observations (e.g., measurements at certain locations). The spatial model may not necessarily be a multi-Gaussian model with some covariance function. For example, the random field may be modeled with a Boolean model, a Markov random field, or some multiple-point geostatistical method (see Chapter 6). The idea of these methods is to generate posterior samples from some (implicit or explicit) posterior model of the unknown grid variables:

$$\left(x(\mathbf{s}_1),\ldots,x\left(\mathbf{s}_{N_{\text{grid}}}\right)\right) \tag{3.294}$$

from some limited set of observations on that grid $\left(x\left(\mathbf{s}_{(1)}\right),\ldots,x\left(\mathbf{s}_{(N_{\text{data}})}\right)\right)$. In a general method for spatial bootstrap, these conditional samples/realizations are resampled using the same sampling strategy (but different locations) to obtain resampled data sets:

$$\left(x_b\left(\mathbf{s}_{(1)}\right),\ldots,x_b\left(\mathbf{s}_{(N_{\text{data}})}\right)\right), b = 1,\ldots,B \tag{3.295}$$

Thus, resampling accounts for whatever correlation structure is assumed in generating the realizations. The bootstrap samples can then be used in any estimator:

$$t\left(x\left(\mathbf{s}_{(1)}\right),\ldots,x\left(\mathbf{s}_{(N_{\text{data}})}\right)\right) \to t\left(x_b\left(\mathbf{s}_{b,(1)}\right),\ldots,x_b\left(\mathbf{s}_{b,(N_{\text{data}})}\right)\right) \tag{3.296}$$

Consider the case of estimating the global mean of the domain and requiring some confidence on that estimate. The variability of this arithmetic estimator of that unknown mean is dependent on the correlation structure of the field (the pure random case would then be solved with the i.i.d bootstrap, but otherwise this method would yield incorrect confidence). This type of bootstrap allows accounting for that structure. Example applications of these ideas are presented in *Journel* [1994] and *Caumon et al.* [2009].

3.13.4. Bootstrap Confidence Intervals and Hypothesis Testing

In the application of UQ, it is often important to know if two sample sets follow the same distribution or a different distribution. Knowing whether two (empirical) distributions are different is, for example, relevant to the application of Bayes' rule. If the prior is not sufficiently different from the posterior, then clearly the data was not able to reduce uncertainty. Another application lies in sensitivity analysis: we would like to know if a variable/parameter is impacting the response or not. One way of testing this is to classify the response into two groups (positive/negative or high/low or reacting/non-reacting), then study the distribution of that variable in each group (see Chapter 5). If the distribution within the two groups is the same, then the parameter is not impacting the response; likewise, the degree of difference can inform how impacting that variable is with respect to other variables.

The problem is that typically we have a limited sample only. Either because the data is too expensive to acquire, or when computer models are involved, the amount of runs is limited. A hypothesis test is needed where the null hypothesis is defined as "no difference in the distributions exists" and the statistical evidence is used to test if it can be rejected (hypothetico-deductive reasoning, see Chapter 5). Because hypothesis testing requires sampling statistics, bootstrap is an ideal candidate for those tests that involve statistics whose sampling distributions are not known.

Here we present two different ways of addressing these hypotheses tests, each addressed with a different bootstrap method. First consider the null-hypothesis that two distributions are the same:

$$H_0 : F_1 = F_2 \tag{3.297}$$

with $F_1$and F_2 the two distributions in question. Next we define a test statistics, such as for example the difference in mean:

$$\theta = \mu(F_1) - \mu(F_2) \tag{3.298}$$

The difference is also estimated from the data as $\hat{\theta}$. Clearly, if $\hat{\theta}$ is significantly different from zero then the null-hypothesis should be rejected. To study this, we would like to generate bootstrap samples and use them to calculate bootstrap estimates $\hat{\hat{\theta}}_b$, which allows calculating the so-called "achieved significance level" (ASL):

$$\text{ASL} = P_{H_0}\left(\hat{\hat{\theta}}_b \geq \hat{\theta}\right) \tag{3.299}$$

The smaller ASL, the more evidence against the null-hypothesis. The problem now lies in how to resample (a question of the probability generating structure) to generate these bootstrap estimates. Indeed, H_0 leaves many possibilities for distributions $F_1 = F_2$ with a given test statistics (e.g., if the test statistics is the mean only, then many distributions can be constructed). To address this, a permutation test looks at every possible combination of creating two groups with the original sample values. For example, if we have L_1 samples in group 1 (F_1) and

L_2 samples in group 2 (F_2), then the amount of possible combinations is

$$\binom{L_1+L_2}{L_1} = \binom{L_1+L_2}{L_2} = \frac{(L_1+L_2)!}{L_1!L_2!} \tag{3.300}$$

and each of such a combination has probability $1/\binom{L_1+L_2}{L_1}$. A permutation test can be considered as a bootstrap *without* replacement. It selects L_1 samples randomly from the combined set of $L_1 + L_2$ samples, assigns it to group 1 and the remainder to group 2. As a result, samples in group 1 do not replicate in group 2 (that would be a bootstrap *with* replacement). When using a finite amount B of such bootstrap samples, then ASL in Eq. (3.299) can be approximated by

$$\text{ASL}_{perm} = \#\left(\hat{\theta}_b \geq \hat{\theta}\right) / \binom{L_1+L_2}{L_1} \tag{3.301}$$

A second and different way of addressing the same problem is to avoid the original null-hypothesis and directly test

$$H_0 : \theta = \theta_0 \tag{3.302}$$

The distributions are now simply the empirical distributions; hence, B bootstrap samples with replacement are drawn of size $L_1 + L_2$ of which the first L_1 are assigned to group one and the remainder to group 2. Unlike the permutation test, this way of testing does not assign probabilities to each sample (no explicit generating structure). This also means that the ASL obtained $\text{ASL}_{boot} = \#\left(\hat{\theta}_b \geq \hat{\theta}\right)/B$ has no interpretation of a probability as B goes to infinity.

The bootstrap sampling results from hypothesis testing can equally be used to construct confidence intervals on any statistic $\hat{\theta}$. The bootstrap estimates $\hat{\hat{\theta}}_b$ represent the empirical distribution from which any percentiles $\hat{F}^{-1}(\alpha)$ can be calculated. This simply means that for a given α, one finds amongst the B bootstrap samples the sample with rank $(\alpha \times B)/100$. This results in the confidence interval

$$\left[\hat{\theta}_{\alpha,Lo}, \hat{\theta}_{\alpha,Hi}\right] = \left(\hat{\hat{F}}_b(\alpha), \hat{\hat{F}}_b(1-\alpha)\right) \tag{3.303}$$

The link with hypothesis testing also becomes clear now. Consider the case again of testing difference

$$H_0 : \theta = 0 \Rightarrow \text{ASL} = \alpha \tag{3.304}$$

meaning that we can use confidence intervals to calculate ASL. Indeed, consider that $\hat{\theta} > 0$, if we chose α such that $\hat{\theta}_{\alpha,Lo} = 0$ then as a result

$$P_{\theta=0}\left(\hat{\hat{\theta}}_b \geq \hat{\theta}\right) = \alpha \tag{3.305}$$

REFERENCES

Aihara, S. I., A. Bagchi, and S. Saha (2009), On parameter estimation of stochastic volatility models from stock data using particle filter: Application to AEX index, *Int. J. Innov. Comput. I.*, *5*(1), 17–27.

Andrieu, C., A. Doucet, and R. Holenstein (2010), Particle Markov chain Monte Carlo methods, *J. R. Stat. Soc. Ser. B Stat. Methodol.*, *72*, 269–342, doi:10.1111/j.1467-9868.2009.00736.x

Beaumont, M. A., W. Zhang, and D. J. Balding (2002), Approximate Bayesian computation in population genetics, *Genetics*, *162*(4), 2025–2035.

Beven, K. (2009), *Environmental Modelling: An Uncertain Future?* Available from: http://books.google.dk/books/about/Environmental_Modelling.html?id=A_YXIAAACAAJ&pgis=1.

Bishop, C. M. (1995), *Neural Networks for Pattern Recognition*, Clarendon Press, pp. 482, doi:10.2307/2965437.

Breiman, L., J. H. Friedman, R. A. Olshen, and C. J. Stone (1984), *Classification and Regression Trees*, The Wadsworth Statisticsprobability Series, vol. *19*, doi:10.1371/journal.pone.0015807.

Caers, J. (2011), *Modeling Uncertainty in the Earth Sciences*. Wiley, Hoboken, NJ.

Caumon, G., P. Collon-Drouaillet, C. Le Carlier De Veslud, S. Viseur, and J. Sausse (2009), Surface-based 3D modeling of geological structures, *Math. Geosci.*, *41*(8), 927–945, doi:10.1007/s11004-009-9244-2.

Craiu, R. V., J. Rosenthal, and C. Yang (2009), Learn from thy neighbor: Parallel-chain and regional adaptive MCMC, *J. Am. Stat. Assoc.*, *104*(488), 1454–1466, doi:10.1198/jasa.2009.tm08393.

Cressie, N. A. C. (1985), Fitting variogram models by weighted least squares, *J. Int. Assoc. Math. Geol.*, *17*(5), 563–586, doi:10.1007/BF01032109.

Cressie, N. A. C. (1993), *Statistics for Spatial Data (Revised Edition)*. Wiley, New York, doi:10.2307/2533238.

Davies, D. L., and D. W. Bouldin (1979), A cluster separation measure, *IEEE Trans. Pattern Anal. Mach. Intell.*, *1*(2), 224–227, doi:10.1109/TPAMI.1979.4766909.

Dellaert, F., D. Fox, W. Burgard, and S. Thrun (2001), Monte Carlo localization for mobile robots, *Proceedings 1999 IEEE International Conference on Robotics and Automation (Cat. No.99CH36288C)*, February, 1322–1328, doi:10.1109/ROBOT.1999.772544.

Deutsch, C. V., and A. G. Journel (1992), *GSLIB: Geostatistical Software Library and User's Guide*. Oxford University Press, doi:10.1016/0098-3004(94)90041-8.

Diggle, P. J., and P. J. Ribeiro (2007), *Model-Based Geostatistics*, Springer Series in Statistics, vol. *1*, doi:10.1111/1467-9876.00113.

Diggle, P. J., and R. J. Gratton (1984), Monte Carlo methods of inference for implicit statistical models. *J. R. Stat. Soc. Ser. B Methodol.*, *46*, 193–227.

Dubuisson, M. P., and A. K. Jain (1994), A modified Hausdorff distance for object matching, *Proceedings of 12th International Conference on Pattern Recognition*, Jerusalem, Israel, vol. 1, 566–568, doi:10.1109/ICPR.1994.576361.

Efron, B. (1979), Bootstrap methods: Another look at the jackknife, *Ann. Stat.*, *7*(1), 1–26, doi:10.1214/aos/1176344552.

Efron, B., and R. J. Tibshirani (1994), *An Introduction to the Bootstrap*, CRC Press, Boca Raton, FL.

Feyen, L., and J. Caers (2006), Quantifying geological uncertainty for flow and transport modeling in multi-modal heterogeneous formations, *Adv. Water Resour.*, *29*(6), 912–929, doi:10.1016/j.advwatres.2005.08.002.

Gelman, A., J. B. Carlin, H. S. Stern, and D. B. Rubin (2004), *Bayesian Data Analysis*, Chapman Texts in Statistical Science Series, doi:10.1007/s13398-014-0173-7.2.

Geman, S., and D. Geman (1984), Stochastic relaxation, Gibbs distributions, and the Bayesian restoration of images, *IEEE Trans. Pattern Anal. Mach. Intell.*, *6*(6), 721–741, doi:10.1109/TPAMI.1984.4767596.

Geyer, C. J. (2002), *Introduction to Markov Chain Monte Carlo*, no. 1990, pp. 3–48.

Gilks, W. R., G. O. Roberts, and E. I. George (1994), Adaptive direction sampling, *J. R. Stat. Soc., Ser. D*, *43*(1), 179–189, doi:10.2307/2348942.

Gómez-Hernández, J. J, and X.-H. Wen (1998), To be or not to be multi-Gaussian? A reflection on stochastic hydrogeology, *Adv. Water Resour.*, *21*(1), 47–61, doi:10.1016/S0309-1708(96)00031-0.

Goovaerts, P. (1997), *Geostatistics for Natural Resources Evaluation*. Oxford University Press, New York.

Gupta, H. V., M. P. Clark, J. A Vrugt, G. Abramowitz, and M. Ye (2012), Towards a comprehensive assessment of model structural adequacy, *Water Resour. Res.*, doi:10.1029/2011WR011044.

Haario, H., E. Saksman, and J. Tamminen (1999), Adaptive proposal distribution for random walk metropolis algorithm, *Comput. Stat.*, *14*(3), 375, doi:10.1007/s001800050022.

Haario, H., E. Saksman, and J. Tamminen (2001), An adaptive metropolis algorithm, *Bernoulli*, *7*(2), 223, doi:10.2307/3318737.

Hansen, T. M., A. G. Journel, A. Tarantola, and K. Mosegaard (2006), Linear inverse Gaussian theory and geostatistics, *Geophysics*, *71*(6), R101, doi:10.1190/1.2345195.

Hansen, T. M., K. S. Cordua, and K. Mosegaard (2012), Inverse problems with non-trivial priors: Efficient solution through sequential Gibbs sampling, *Comput. Geosci.*, *16*(3), 593–611, doi:10.1007/s10596-011-9271-1.

Hastie, T., R. Tibshirani, and J. Friedman (2009), *The Elements of Statistical Learning Data Mining, Inference, and Prediction*, Second Edition, Springer Series in Statistics, doi:10.1007/978-0-387-84858-7.

Hastings, W. K. (1970), Monte Carlo sampling methods using Markov chains and their applications, *Biometrika*, *57*(1), 97–109, doi:10.1093/biomet/57.1.97.

Huttenlocher, D. P., G. A. Klanderman, and W. J. Rucklidge (1993), Comparing images using the Hausdorff distance, *IEEE Transactions on pattern analysis and machine intelligence*, *15*(9), 850–863.

Jain, A. K. (2010), Data clustering: 50 years beyond K-means, *Pattern Recogn. Lett.*, *31*(8), 651–66, doi:10.1016/j.patrec.2009.09.011.

Jaynes, E. T. (2003), Probability theory: The logic of science, *Math. Intell.*, *27*(2), 83–83, doi:10.1007/BF02985800.

Journel, A. G. (1994), Resampling from stochastic simulations, *Environ. Ecol. Stat.*, *1*(1), 63–91, doi:10.1007/BF00714200.

Journel, A. G., and C. V. Deutsch (1993), Entropy and spatial disorder, *Math. Geol.*, *25*(3), 329–355, doi:10.1007/BF00901422.

Journel, A. G., and C. J. Huijbregts (1978), *Mining Geostatistics*, Academic Press, London.

Kaufman, L., and P. J. Rousseeuw (1990), *Finding Groups in Data: An Introduction to Cluster Analysis*, Wiley, New York.

Kwok, J. T. Y., and I. W. H. Tsang (2004), The pre-image problem in Kernel methods, *IEEE Trans. Neural Netw.*, *15*(6), 1517–1525, doi:10.1109/TNN.2004.837781.

Laloy, E., and J. A. Vrugt (2012), High-dimensional posterior exploration of hydrologic models using multiple-try DREAM (ZS) and high-performance computing, *Water Resour. Res.*, *48*(1), W01526, doi:10.1029/2011WR010608.

Leeuwen, V., and P. Jan (2009), Particle filtering in geophysical systems, *Mon. Weather Rev. 137*, 4089–4114, doi:10.1175/2009MWR2835.1.

Lemieux, C. (2009). *Monte Carlo and Quasi-Monte Carlo Sampling*, Springer Science & Business Media, New York.

Liu, J. S., F. Liang, and W. H. Wong, (2000), The multiple-try method and local optimization in metropolis sampling, *J. Am. Stat. Assoc.*, *95*(449), 121–134.

Mariethoz, G., and J. Caers (2014), *Multiple-Point Geostatistics: Stochastic Modeling with Training Images*, Wiley Blackwell, Chichester.

Matheron, G. (1970), *La Théorie Des Variables Régionalisées*. Les Cahiers du Centre de morphologie mathématique de Fontainebleau, École Nationale Supérieure des Mines.

Mckay, D. J. C. (1998), Introduction to Monte Carlo methods, in *Learning in Graphical Models*, edited by M. I. Jordan, pp. 175–204, Kluwer Academic Press.

Mosegaard, K., and A. Tarantola (1995), Monte Carlo sampling of solutions to inverse problems, *J. Geophys. Res.*, *100*(B7), 12431–12447, doi:10.1029/94JB03097.

von Mises, R. (1964), *Mathematical Theory of Probability and Statistics*, Academic Press, New York.

Nummiaro, K., E. Koller-Meier, and L. Van Gool (2003), An adaptive color-based particle filter, *Image Vis. Comput.* *21*(1), 99–110, doi:10.1016/S0262-8856(02)00129-4.

Ramsay, J., and B. W. Silverman (2005), *Functional Data Analysis*, Springer Series in Statistics, Springer, New York.

Sadegh, M., and J. A. Vrugt (2013), Bridging the gap between GLUE and formal statistical approaches: approximate Bayesian computation, *Hydrol. Earth Syst. Sci.*, *17*(12), 4831–4850, doi:10.5194/hess-17-4831-2013.

Sadegh, M., and J. A. Vrugt (2014), Approximate Bayesian computation using Markov chain Monte Carlo simulation, *Water Resour. Res.*, *10*(2), 6767–6787, doi:10.1002/2014WR015386.Received.

Scheidt, C., and J. Caers (2009), Representing spatial uncertainty using distances and Kernels, *Math. Geosci.*, *41*(4), 397–419, doi:10.1007/s11004-008-9186-0.

Scheidt, C., and J. Caers (2013), Uncertainty quantification in reservoir performance using distances and kernel methods: Application to a west Africa deepwater turbidite reservoir, *SPE J.*, *14*(4), 680–692, doi:10.2118/118740-PA.

Schölkopf, B., and A. J. Smola (2002), *Learning with Kernels: Support Vector Machines, Regularization, Optimization, and Beyond*, MIT Press, Cambridge.

Schoups, G., and J. A. Vrugt (2010), A formal likelihood function for parameter and predictive inference of hydrologic models with correlated, heteroscedastic, and non-Gaussian errors, *Water Resour. Res.*, *46*, 1–17, doi:10.1029/2009WR008933.

Silverman, B. W. (1986), Density estimation for statistics and data analysis, *Chapman and Hall*, *37*(1), 1–22, doi:10.2307/2347507.

Solow, A. R. (1985), Bootstrapping correlated data, *J. Int. Assoc. Math. Geol.*, *17*(7), 769–775, doi:10.1007/BF01031616.

Storn, R., and K. Price (1997), Differential evolution: A simple and efficient heuristic for global optimization over continuous spaces, *J. Glob. Optim.*, *11*(4), 341–359, doi:10.1023/A:1008202821328.

Tarantola, A. (1987). *Inverse Problem Theory: Methods for Data Fitting and Model Parameter Estimation*, Elsevier, Amsterdam. Available from: http://www.scopus.com/inward/record.url?eid=2-s2.0-0023499373&partnerID=tZOtx3y1.

Ter Braak, C. J. F. (2006), A Markov chain Monte Carlo version of the genetic algorithm differential evolution: Easy Bayesian computing for real parameter spaces, *Stat. Comput.* *16*(3), 239–249, doi:10.1007/s11222-006-8769-1.

Tjelmeland, H., and J. Besag (1998), Markov random fields with higher-order interactions, *Scand. J. Stat.*, *25*, 415–433, doi:10.1111/1467-9469.00113.

Turner, B. M., and T. Van Zandt (2012), A tutorial on approximate Bayesian computation, *J. Math. Psychol.*, *56*(2), 69–85, doi:10.1016/j.jmp.2012.02.005.

Vapnik, V., and A. Lerner (1963), Pattern recognition using generalized portrait method, *Autom. Remote Control*, *24*, 774–780.

Von Neumann, J. (1951), Various techniques used in connection with random digits, Applied Math Series, Notes by G. E. Forsythe, National Bureau of Standards, 12, 36–38.

Vrugt, J. A. (2016), Markov chain Monte Carlo simulation using the DREAM software package: Theory, concepts, and MATLAB implementation, *Environ. Model. Softw.*, *75*, 273–316, doi:10.1016/j.envsoft.2015.08.013.

Vrugt, J. A, C. J. F. ter Braak, C. G. H. Diks, B. A. Robinson, J. M. Hyman, and D. Higdon (2009), Accelerating Markov chain Monte Carlo simulation by differential evolution with self-adaptive randomized subspace sampling, *Int. J. Nonlin. Sci. Num. Simul.*, *10*(3), 273–290, doi:10.1515/IJNSNS.2009.10.3.273.

Williams, C. (1999), Prediction with Gaussian processes, in *Learning in Graphical Models*, edited by M. I. Jordan, MIT Press, Cambridge, MA, pp. 599–621.

Zinn, B., and C. F. Harvey (2003), When good statistical models of aquifer heterogeneity go bad: a comparison of flow, dispersion, and mass transfer in connected and multivariate Gaussian hydraulic conductivity fields, *Water Resour. Res.*, *39*(3), 1–19, doi:10.1029/2001WR001146.

4

Sensitivity Analysis

4.1. INTRODUCTION

Prediction in subsurface systems often relies on the use of complex computer simulation models. These models may contain a large number of input parameters whose values are uncertain, thereby introducing uncertainty in the output of such computer models. Fundamental to any uncertainty quantification (UQ) is understanding the influence of each uncertain input parameter and how the input parameter uncertainty propagates to output (response, prediction) uncertainty. Sensitivity analysis (SA) is the study of how variation of input parameters impacts the uncertainty of a response of interest. With a good understanding of how input parameters influence the response, several actions/options are possible. SA can aid in reducing uncertainty in prediction by identifying high-impacting parameters, and hence which data to acquire to reduce uncertainty on said parameters. Additionally, one may wish to fix the values of the parameters that have little influence on the prediction, which may help in reducing the complexity of the problem, or even computational complexity. Finally, SA allows quantifying and understanding how the combined effect of parameters generates nonlinear variations in responses. In SA terminology, these are termed interactions. Parameter interaction occurs when the effect of a parameter value on the response of interest varies for different values of another parameter. Identifying the presence of interactions is important as it has consequences on the two previous points [*Saltelli et al.*, 2000]. In this chapter, we only consider interactions between two parameters (referred as two-way interactions) and ignore multi-way interactions (among three or more inputs). Multi-way interactions are difficult to calculate and interpret and are usually less significant than the lower-order sensitivity coefficients [*Zagayevskiy and Deutsch*, 2015].

Many reviews of SA exist (among many others, e.g., [*Saltelli et al.*, 2000; *Frey and Patil*, 2002; *Iooss and Lemaître*, 2014; *Song et al.*, 2015; *Borgonovo and Plischke*, 2016], etc.). The literature on SA uses different terminology, such as "influential," "sensitive," "important," "impacting," "effective," and "correlated" interchangeably. We will use the following terminology: a parameter is defined as influential when it has a significant impact on model response. When the parameter is influential, the common nomenclature in the SA literature is to define the response as sensitive to the parameter. The above definition requires a measure of a threshold of sensitivity beyond which a parameter is considered as influential. We will use the term *objective* measure of sensitivity to identify measures which compute the threshold, often using statistical techniques. However, many SA methods only provide a *subjective* measure of sensitivity, that is, they rank the parameters based on their influence on the response, but they require a subjective definition of the threshold by the user (often based upon visual inspection).

In this chapter, we do not offer an exhaustive list of the existing SA techniques, but we only discuss some of the state-of-the-art techniques that are well suited for the specificities of subsurface systems. Input parameters and model responses in subsurface modeling may have characteristics that require adaptation of existing SA techniques. In particular, input and output variables may be spatiotemporal (permeability, saturation maps, production as a function of time at many wells, etc.) and not univariate. In addition, uncertain parameters can be of any type, continuous, discrete, or nonnumerical, such as a depositional scenario. Not all SA methods are well suited for these general conditions. Finally, stochasticity in model responses may be present which introduces additional complications that render application of many SA methods problematic. Stochasticity in subsurface systems is often modeled through a set of alternative numerical models representing the spatial heterogeneity of the Earth. The set of alternative spatial models is generally

Quantifying Uncertainty in Subsurface Systems, Geophysical Monograph 236,
First Edition. Céline Scheidt, Lewis Li, and Jef Caers.

obtained using geostatistics and is sometimes referred to as spatial uncertainty or stochastic imaging [*Goovaerts*, 1997]. Spatial uncertainty introduces stochasticity in the model response. The implications of stochasticity will be further discussed in Section 4.5.

SA methods can be divided into two categories: local and global methods. Local SA evaluates sensitivity for a single (deterministic) set of input parameters. Sensitivity measures are usually defined by computing partial derivatives of the output function with respect to the parameters. The main advantage of local methods is their computational efficiency. Their major limitation is that they are only calculated for that single set; hence, they are only informative locally near the set of input parameters and not elsewhere in the parameter space. The main advantage of global SA (GSA) methods lies in their ability to assess the effect of inputs over the entire parameter space to evaluate sensitivity measures, and hence it is more appropriate for UQ. Global methods often require many model evaluations because of the need to cover sufficiently the parameter space, which is potentially vast in subsurface systems [*Saltelli et al.*, 2000]. Screening techniques, described next, attempt to minimize computational cost by selecting a reduced set of input parameters.

4.2. NOTATION AND APPLICATION EXAMPLE

We consider a system which takes N_p input parameters $\mathbf{x} = (x_1, \ldots, x_{N_p})$ and is used to predict a response $\mathbf{y}$. In the context of uncertainty in subsurface systems described in Chapter 3, $\mathbf{x}$ can be either the gridded model variables $\mathbf{m}_{\text{grid}}$ or the non-gridded model variables $\mathbf{p}$. The response of the system $\mathbf{y}$ can be the data variables $\mathbf{d}$ or the prediction variables $\mathbf{h}$. For many SA methods, $\mathbf{y}$ needs to be a scalar. $\mathbf{x}$ and $\mathbf{y}$ are outcomes of random variables $\mathbf{X} = (X_1, \ldots, X_{N_p})$ and $\mathbf{Y}$, respectively, and we assume that $\mathbf{Y}$ can be defined as a function of $\mathbf{X}$:

$$\mathbf{Y} = f(\mathbf{X}) = f(X_1, \ldots, X_{N_p}) \tag{4.1}$$

Because $\mathbf{X}$ is uncertain and thus characterized by a joint probability distribution, $\mathbf{Y}$ is also uncertain and can be characterized by its own joint probability distribution.

To illustrate some of the different SA methods presented in this chapter, we use a simplified version of the dense non-aqueous phase liquid (DNAPL) test case presented in Chapter 3. For clarity in presenting the results, the number of uncertain parameters was reduced from 14 to 6. The parameters that are kept are the mean and standard deviation of the hydraulic conductivity (K_{mean} and K_{sd}, respectively), the covariance model (K_{Cov}) and the covariance range (K_{range}) of the hydraulic conductivity, the river gradient (H_{rivGrad}), and the range of the Matèrn covariance model (H_{range}) used to define the boundary conditions. All other parameters are fixed to their mean value. Because most SA methods work with scalar responses, the influence of each parameter on the pollutant arrival time, a scalar, is assessed. In Sections 4.4.3 and 4.4.4, we analyze the pollutant arrival as a time-varying response instead of a scalar.

4.3. SCREENING TECHNIQUES

Screening techniques rely on simplified approaches to identify non-influential inputs of a computer model, while keeping the number of model evaluations small. They are often used for systems with many input parameters as a first screening, before using more advanced SA methods. Because screening techniques are computationally economical, only a subjective measure of sensitivity can be evaluated, that is, the parameters can only be ranked based on their importance and not defined as sensitive/insensitive based on some statistical criteria. Two of the most widely used screening techniques in subsurface modeling are presented next: the one-at-a-time (OAT) method and the Morris method.

4.3.1. OAT Method

The OAT approach [*Daniel*, 1973] is the most widely used screening method because of its simplicity. The idea is to vary the input parameters one by one, while keeping the other inputs fixed at a baseline ("nominal" value), and to monitor changes in the output. Tornado charts are a common tool in SA to visualize the effect of changing one parameter at a time. The Tornado chart consists of two OAT designs: changes in the response (output) when a parameter is changed from its nominal value to its lower and higher extremes are recorded. The corresponding variations of the response are plotted from the highest variation to the lowest in the Tornado chart (Figure 4.1). The nominal value for each parameter is usually taken as the midpoint between the two tested extremes. The number of model evaluations for the Tornado chart is therefore $2N_p + 1$, where N_p is the number of parameters. The low computational cost of the OAT is one of its main advantages.

If parameters are changed over small intervals, the OAT method can be considered as a local method. Because the OAT method only varies one parameter at a time, the effect of varying several parameters simultaneously is not investigated, and therefore interactions between parameters cannot be assessed. Care must, therefore, be taken to not remove a parameter using an OAT analysis, without quantifying interactions with other parameters. In addition, the OAT approach does not explore well the input parameter space; the results are dependent on the choice of the baseline values. The analysis is only strictly valid when the response can be modeled as a linear representation of the inputs. Moreover,

OAT approaches require numerical input parameters; hence, the impact of scenario-based parameters cannot be assessed by this technique, as one cannot define a "min" and "max" scenario. *Saltelli and Annoni* [2010] offer a critical review of the OAT method. Despite the OAT's limitations, several examples of its use can be found in the literature. In the context of the subsurface, and particularly in oil and gas applications, OAT is attractive as it can directly display the impact in terms of production volume or dollar amounts when a parameter is varied over its possible range of values (an example can be found in *Cavalcante et al.* [2015]). Other applications are, for example, in hydrological modeling [e.g., *Khalid et al.*, 2016] and ground water recharge [*Nolan et al.*, 2007].

To illustrate, the OAT approach was applied to the simplified DNAPL example. A fixed Gaussian covariance model is used, since the OAT method cannot handle scenario (nonnumerical) parameters. In addition, because OAT methods cannot be used in the presence of stochasticity in the response, both the boundary conditions and the hydraulic conductivity fields were generated using a Gaussian process (see Chapter 3 for more details) with a fixed random seed; there is no spatial uncertainty and hence no stochasticity. The resulting Tornado chart for the pollutant arrival time (in days) is shown in Figure 4.1.

Figure 4.1 shows that the parameter's hydraulic conductivity mean and standard deviation (K_{mean} and K_{sd}) and the river gradient ($H_{rivGrad}$) are influential on the arrival time. In particular, an increase of K_{mean} by two standard deviations results in contaminant arriving 3.5 days earlier, whereas a decrease of K_{mean} would result in late arrival time (+6 days). In addition, increasing the value of $H_{rivGrad}$ or decreasing the value of K_{sd} result in late arrival time of the contaminant. The effect of an increase or decrease of the parameter value is not necessarily symmetric.

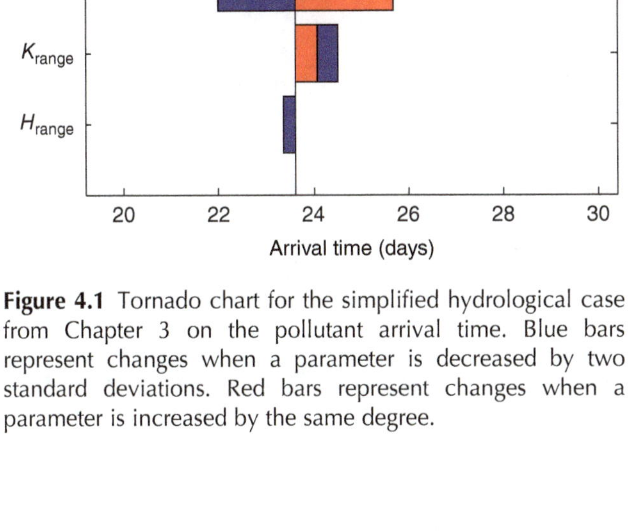

Figure 4.1 Tornado chart for the simplified hydrological case from Chapter 3 on the pollutant arrival time. Blue bars represent changes when a parameter is decreased by two standard deviations. Red bars represent changes when a parameter is increased by the same degree.

4.3.2. Morris Method

Morris [1991] extends the concept of OAT by deriving GSA measures from a set of local measures evaluated over the input parameter space. The Morris method is based on a repeated sampling of randomized OAT designs; hence, it is much more complete than the standard OAT method, but also more costly. It is well suited for models with a large number of input parameters (up to thousands, see an application in *Herman et al.* [2013]) and with relatively long computational times. Using the Morris method, the input parameters can be classified into three groups: (i) negligible effects, (ii) linear and additive effects, and (iii) nonlinear effects and/or interactions effects.

The Morris method works as follows. First, input parameters are scaled to the unit interval [0, 1] and discretized into q-levels, ($x_n \in \{0, 1/(q-1), 2/(q-1), \ldots, 1\}$), resulting in a q-dimensional lattice denoted Ω. In the Morris method, the elementary effect of the n-th parameter (EE_n) is evaluated by perturbing the n-th parameter of a point $\mathbf{x} = (x_1, \ldots, x_{N_p}) \in \Omega$ by a predefined increment Δ and keeping all other parameters constant

$$EE_n = \frac{f(x_1, \ldots, x_{n-1}, x_n + \Delta, \ldots, x_{N_p}) - f(x_1, \ldots, x_{n-1}, x_n, \ldots, x_{N_p})}{\Delta} \tag{4.2}$$

with Δ a multiple of values $1/(q-1)$ and $\mathbf{x}$ and $\mathbf{x} + \Delta$ belonging to the discretized input space Ω. q is often restricted to even values and $\Delta = q/(2(q-1))$ [*Morris*, 1991].

The Morris design is a strict OAT design as defined in the terminology proposed by *Daniel* [1973]. The design creates multiple trajectories through the parameter space as follows. For a given point $\mathbf{x}$, each trajectory contains a sequence of N_p perturbations, resulting in $N_p + 1$ model evaluations to evaluate the N_p elementary effects. The elementary effects depend on the point $\mathbf{x}$. To avoid this dependency, Morris suggests creating multiple trajectories by sampling a set of L randomly selected starting points to estimate a distribution of elementary effects for each parameter. The procedure is equivalent to performing a given number (L) of OAT designs. The total cost of the method is, therefore, $L(N_p + 1)$. The choice of q and L is critical in the Morris approach. High values of q create a large number of possible levels to explore, requiring a large number of trajectories to ensure that the space is well explored. Values such as $q = 4$ and $L = 10$ are often sufficient to provide meaningful results [*Wainwright et al.*, 2014]. More recent work proposes trajectories to maximize the coverage of the input space [*van Griensven et al.*, 2006; *Campolongo et al.*, 2007].

The distribution of the elementary effects of each parameter is analyzed to define sensitivities, using its mean μ_n and standard deviation σ_n, defined as

$$\mu_n = \frac{1}{L}\sum_{\ell=1}^{L} \mathrm{EE}_n^{(\ell)} \text{ and } \sigma_n^2 = \frac{1}{L-1}\sum_{\ell=1}^{L}\left(\mathrm{EE}_n^{(\ell)} - \mu_n\right)^2 \quad (4.3)$$

with $\mathrm{EE}_n^{(\ell)}$ the elementary effect of the n-th parameter (as defined in Eq. (4.2)) and of the ℓ-th randomly selected x_n.

The mean μ_n assesses the overall impact (influence) of the parameter on the response. A large mean indicates large changes in the response when perturbing a parameter (an influential parameter). The standard deviation or variance σ_n^2 is a measure of nonlinear and/or interaction effects. A large value indicates that the elementary effects depend highly on the choice of the sample point at which it is computed. On the other hand, a small σ_n^2 indicates similar values in elementary effects for different sample points, implying that the effect of the parameter is independent of other parameter values. In some instances, μ_n can be prone to type II errors (if the distribution contains both positive and negative values, they may cancel out). *Campolongo et al.* [2007] suggests replacing the mean of the elementary effects by the mean of the absolute value of the elementary effects:

$$\mu_n^* = \frac{1}{L}\sum_{\ell=1}^{L} \left|\mathrm{EE}_n^{(\ell)}\right| \quad (4.4)$$

A graphical representation of (μ_n^*, σ_n) is often used to interpret parameter sensitivity, as shown in Figure 4.2. *Saltelli et al.* [2008] suggests using three measures μ_n, μ_n^*, and σ_n to interpret influence of each parameter to get a better insight of the impact of each parameter.

The Morris method is applied to the hydrological case, using $q = 4$, $\Delta = 2/3$, and $L = 10$. The elementary effects were calculated on the arrival time (again, uncertainty in the covariance model was not considered). We observe in Figure 4.2 that the parameters K_{mean} and H_{rivGrad} have a large impact on the pollutant arrival time, with both a large μ_n^* and σ_n. The large value of σ_n indicates the presence of nonlinearity or interactions. Interestingly, K_{mean} has a μ_n close to zero, which indicates that the elementary effect for that parameter has a different sign depending on where in the parameter space it is computed. On the other hand, H_{rivGrad} has a negative μ_n of similar order as μ_n^*, which indicates that most elementary effects are negative. The parameters K_{range} and K_{sd} are less influential, but it still may have an impact on the arrival time, as well as interactions. However, H_{range} appears not to be influential.

The Morris method has been applied to various applications, but only recently in the geosciences, for example for carbon sequestration projects [*Wainwright et al.*, 2013; *Sarkarfarshi et al.*, 2014], oil reservoir applications [*Feraille and Busby*, 2009; *Gervais-Couplet et al.*, 2010], geothermal [*Finsterle et al.*, 2013], hydrogeology [*Wainwright et al.*, 2014; *Dessirier et al.*, 2015], flood inundation models [*Pappenberger et al.*, 2008], and hydrological studies [*Francos et al.*, 2003; *Herman et al.*, 2013] to list a few.

Screening techniques only provide subjective measures of sensitivity. Also, they do not evaluate or distinguish interactions from nonlinear effects. In addition, screening methods require continuity in the response; hence, they cannot be applied to stochastic models (i.e., in the presence of spatial uncertainty, as defined by a varying random seed in the computer simulation). The choice of the random seed may have an impact on the results, as

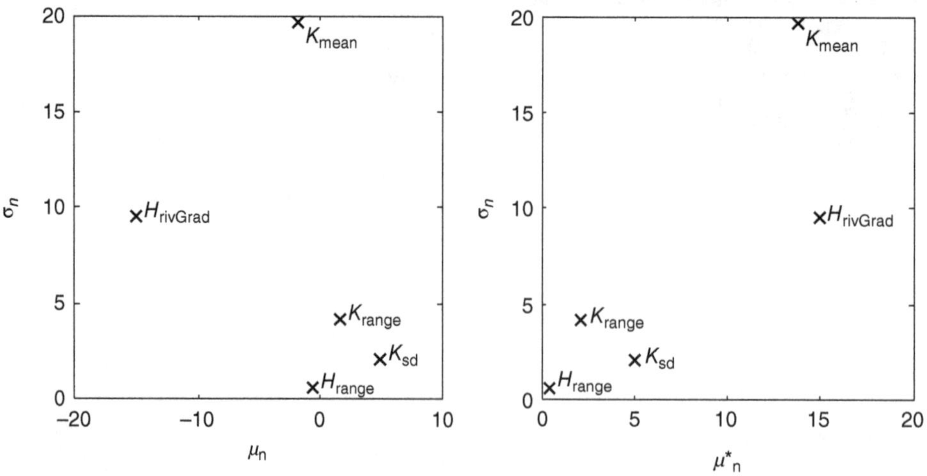

Figure 4.2 Morris method applied to the simplified hydrological case.

shown in Section 4.5. Nevertheless, the Morris method has recently gained attention in the geosciences community as it was shown to hold promise in identifying parameters that can be safely removed from the study, before applying the more complex, sampling-based techniques described next.

4.4. GSA METHODS

GSA methods are often based on some form of Monte Carlo sampling (see Section 3.10). All parameters are varied jointly, and hence their joint effect (interaction) on some desired response is evaluated. This global sampling provides a more rigorous assessment of parameter sensitivity, but at a much greater computational cost. GSA also considers the input parameter distribution, which was not the case for screening methods, and also allows for stochastic responses, which commonly occur in the geosciences.

Global sensitivity methods described in this section can be divided into four main categories based on their assumptions and how they determine sensitivities: linear regression-based SA, variance-based SA, regionalized/generalized SA [*Saltelli et al.*, 2000], and tree-based sensitivities. Linear regression-based SA relies on the assumption of a linear relationship between parameters and responses. Because they are relatively simple and efficient (low computational cost), linear-based SA are popular methods in UQ for subsurface systems. Variance-based SA methods do not require an assumption of linearity and rely on a decomposition of the variance of the response with respect to the parameters. These methods are popular as a research topic, but not yet widely applied in the geosciences because they suffer from high-computational demand and hence are yet impractical in many real-field applications. Many authors propose methods to approximate the model response using a surrogate model to make variance-based SA computationally feasible. Regionalized (or generalized) sensitivity analysis (RSA) methods are less used, but it can be quite powerful, especially with the latest developments that are tailored to handle many of the challenges when dealing with complex subsurface systems. Finally, tree-based SA has not been applied much in the context of subsurface modeling, but the latest development makes them potentially a valuable tool. We provide more details of each approach, including their advantages and drawbacks in the context of UQ for subsurface systems.

4.4.1. SA Based on Linear Models

4.4.1.1. Scatter Plots and Correlation Coefficient. One of the simplest SA methods is to analyze scatterplots of the response versus each input parameter, thereby visualizing the relationship between parameter and response. A strong correlation between an input parameter and a response indicates an influential parameter. Parameters are defined as influential/not influential by visual inspection, aided often by displaying the Pearson correlation coefficient. One of the drawbacks of scatterplots is that as many plots as the number of parameters are needed, which can be cumbersome when studying a large set of parameters. Scatterplots for the parameters of the synthetic case used through this chapter are shown in Figure 4.3 where a Latin hypercube sample (Section 3.10.3.3) with 200 runs was used.

The scatterplots show that the mean value of the hydraulic conductivity (K_{mean}) is correlated with the arrival time of the pollutant, with a Pearson correlation coefficient (ρ) of −0.83. A larger value of K_{mean} results generally in earlier arrival of the pollutant at the drinking well. The standard deviation of the hydraulic conductivity (K_{sd}) and the gradient of the river (H_{rivGrad}) also show some correlation with the arrival time. Very little influence is seen for the range of the covariance (K_{range}) and the covariance model (K_{Cov}) for the hydraulic conductivity and for the range of the covariance (H_{range}) used to define the boundary conditions.

One of the advantages of scatterplots is that they allow the identification of potentially complex dependencies (such as quadratic behavior) which can help to select an appropriate SA technique [*Frey and Patil*, 2002]. However, scatterplots are only visual tools, they do not provide a measure of sensitivity. The Pearson correlation coefficient is useful when a linear relationship between parameter and response occurs. The Pearson correlation coefficient is not meaningful for scenario-type of parameters (hence not shown in Figure 4.3).

4.4.1.2. Linear Regression Analysis. Regression analysis can also be used to assess the impact of the input parameters on the output response. Common practice in SA is to fit a linear regression model (also denoted as response surface) to the data. Consider a "main effect model," that is, a model that does not contain interactions:

$$y = \beta_0 + \sum_{n=1}^{N_p} \beta_n x_n + \varepsilon \tag{4.5}$$

Assuming that a linear model holds (the alternative is discussed in Section 4.4.1.3), the coefficients β_n can be used to determine the importance of each parameter x_n with respect to the response y. When x_n are independent, the absolute standardized regression coefficients can be taken as a measure of sensitivity:

$$\text{SRC}_n = \left| \beta_n \frac{\bar{s}_n}{s} \right| \tag{4.6}$$

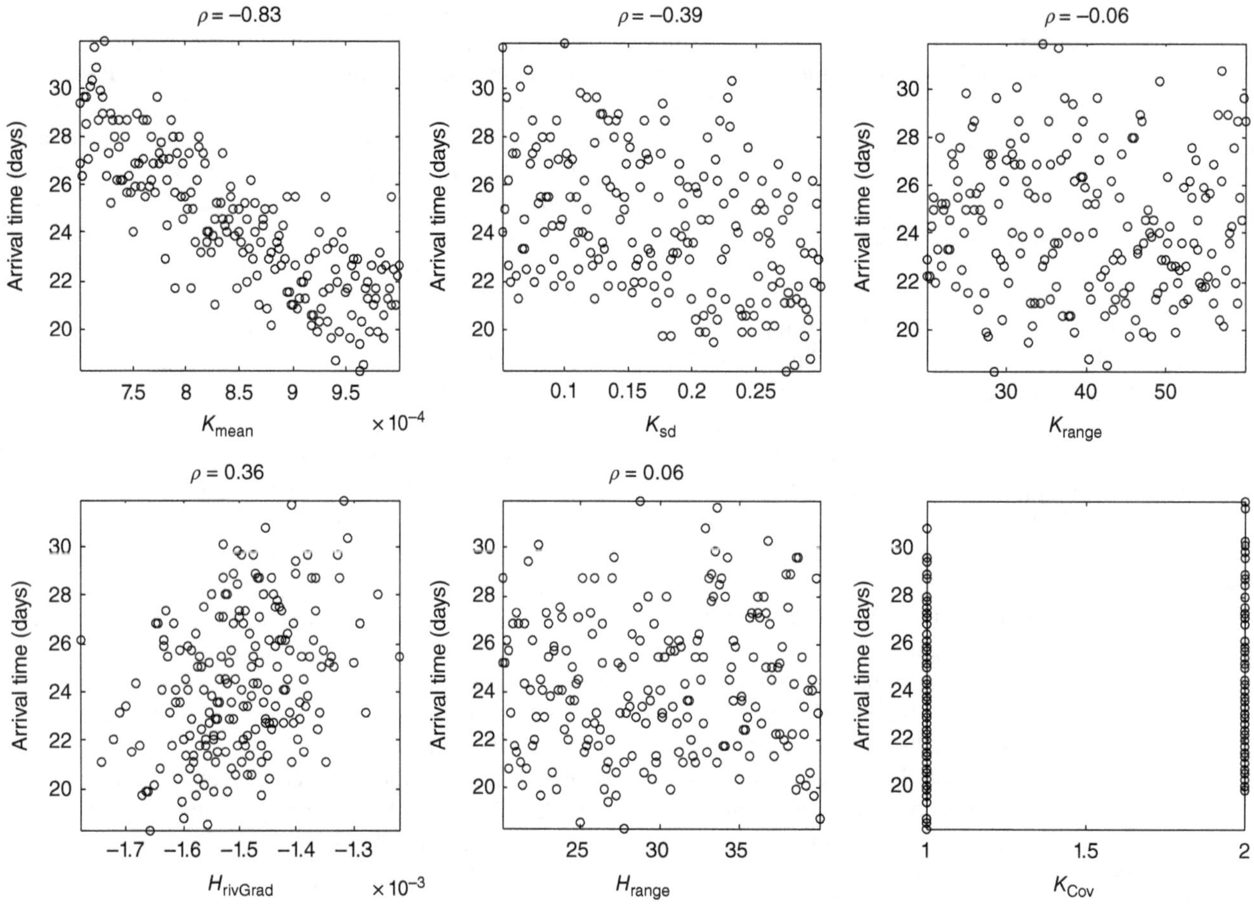

Figure 4.3 Scatterplots of parameters values versus the pollutant arrival time.

where $\bar{s}_n$ and s are estimated standard deviations for x_n and y. Calculating the SRC coefficients is equivalent to performing the regression analysis with the input and output variables normalized to mean of zero and variance of one [*Helton*, 1993]. A large SRC_n indicates a large change in the response for a unit change in the input parameter. Hence, the larger SRC_n is, the more important the input parameter x_n is. If the parameters are not independent (i.e., there is a correlation between the values of parameters x_n and $x_{n'}$), then the SRC is not suitable for measuring the importance of the input parameters [*Helton*, 1993]. Note that the coefficients β_n (hence SRC_n) are estimated based on the regression model, not on the data. As a consequence, if the model poorly fits the data, then the resulting sensitivity measure may be inaccurate.

The SRC_n are themselves uncertain because they are dependent on the sample used to fit the regression model. Hence, a statistical analysis of the significance level of each parameter is desirable through hypothesis testing on each of the coefficients of the model [*Draper and Smith*, 1981]. The null hypothesis is defined as

$$\begin{aligned} H_0 &: \beta_n = 0 \\ H_1 &: \beta_n \neq 0 \end{aligned} \tag{4.7}$$

If H_0 is rejected, then x_n is "significant." Otherwise, the input does not significantly impact the response and can be removed from the model. Under H_0 and assuming that the error ε in Eq. (4.5) follows a normal distribution, then the statistic

$$t_0 = \frac{\beta_n}{\sqrt{\sigma^2 c_{nn}}} \tag{4.8}$$

is a student variable with $L - m - 1$ degrees of freedom, with L being the total number of samples and m the number of terms in the model. The term c_{nn} is the n-th diagonal element of the data matrix $X^T X$ and $\hat{\sigma}$

$$\hat{\sigma} = \sqrt{\frac{\text{SS}_E}{L - m}} \tag{4.9}$$

where SS_E is the sum of the squares of the errors (refer to Section 3.7.2 on linear regression).

One then needs to compare t_0 to the critical value of the student t distribution with $L - m - 1$ degrees of freedom for a given significance level α (usually $\alpha = 0.05$). If $|t_0| > t_{\alpha/2}$, the null hypothesis is rejected and the corresponding term is deemed influential on the response. Otherwise, the term is considered to be non-influential. This test allows for the definition of an objective measure of sensitivity, as opposed to the coefficient SRC_n. Note that this test must be used with care, as the result depends on the form of Eq. (4.5) which is employed. Different results may be obtained if parameters are removed or added to Eq. (4.5). To overcome this drawback, methods such as stepwise regression can be used to automatically add statistically influential parameters [*Helton*, 1993]. In this approach, parameters are added to the regression model sequentially and each time the model is fit to the data. The coefficient of determination (R^2) [*Draper and Smith*, 1981] is calculated. R^2 measures the proportion of variance of the response that is explained by the model. The largest incremental change in R^2 is used to determine the most influential parameter. When the most influential parameter is determined, the process is repeated with the remaining parameters.

Even though most papers/books on regression analysis usually present the simple linear model, interactions terms can be added. Quadratic regression terms can be included as well [e.g., *Helton*, 1993; *Zagayevskiy and Deutsch*, 2015], but others find that linear and interaction terms are sufficient [*Dejean and Blanc*, 1999]. The hypothesis test described above is valid for Eq. (4.10); hence, the impact of interactions can be estimated as well. Note though that here interactions are modeled by means of a product of two parameters and, therefore, are assumed to have a symmetrical impact on the response.

$$y = \beta_0 + \sum_{n=1}^{N_p} \beta_n x_n + \sum_{n=1}^{N_p} \sum_{n'=1}^{N_p} \beta_{nn'} x_n x_{n'} + \varepsilon \qquad (4.10)$$

The regression method is popular for SA because it is straightforward and simple to apply. It has been used extensively in oil and gas applications [*Dejean and Blanc*, 1999; *White et al.*, 2001; *Zabalza-Mezghani et al.*, 2004; *Zagayevskiy and Deutsch*, 2015], hydrological modeling [e.g., *Manache and Melching*, 2004; *Muleta and Nicklow*, 2005; *Yang*, 2011], radioactive waste disposal [*Helton*, 1993], and stormflow models [*Gwo et al.*, 1996] to cite just a few. However, it should be used with care as it is not applicable when the relationship between the input and the output is nonlinear. Hence, before evaluating sensitivities based on the linear regression model, the quality of the regression model should be assessed. Classical measures include the evaluation of the coefficient of determination (R^2) and the predicted residual error sum of squares. In addition, if the residuals ε (in Eq. (4.5)) are not normally distributed or if the input parameters are not independent, the results of the regression analysis can only be interpreted subjectively, because the assumptions that are required to perform the t test are not valid.

Table 4.1 Linear regression analysis, without interactions.

Parameter	SRC	t-Value	p-Value
K_{mean}	0.84	−63.9	<0.001
K_{sd}	0.42	−32.8	<0.001
K_{range}	0.03	−2.3	0.02
H_{rivGrad}	0.38	29.3	<0.001
H_{range}	0.03	2.6	0.01

A linear regression model was applied on the DNAPL test case, using 100 parameter combinations obtained by LHS sampling. Because linear regression does not work well with scenario-type parameters, the analysis was performed using a (fixed) Gaussian covariance model. The standardized regression coefficients SRC_n and the t-statistics and corresponding p-values (see Section 3.13) for each parameter are displayed in Table 4.1. The estimated model is

$$\mathrm{ArrTime} = 67 - 27{,}913 \times K_{\mathrm{mean}} - 17.57 \times K_{\mathrm{sd}} - 0.008 \times K_{\mathrm{range}} + 10{,}906 \times H_{\mathrm{rivGrad}} + 0.017 \times H_{\mathrm{range}} \qquad (4.11)$$

We observe that the parameters K_{mean}, K_{sd}, and H_{rivGrad} are influential based on the linear model (small p-values, or $|t_0| > t_{\alpha/2} = 1.98$ for $\alpha = 0.05$). The parameters K_{range} and H_{range} are also influential, but far less than the other three parameters. The model quality is good with $R^2 = 0.98$.

The study was repeated using a linear regression model including interactions (Eq. (4.10)). Comparison of the t-statistics for both "main effect only" and "interaction" are presented in the Pareto plots in Figure 4.4. Interestingly, the influence of the main factors changes significantly between the two regression models, including H_{rivGrad} which is not influential when including interactions, but highly influential when considering only the main effects. This may be due to the importance of the interaction between H_{rivGrad} and H_{range}, which translates into a strong main effect when ignoring interactions.

4.4.1.3. Final Remarks on Linear Regression–Based SA.

All sensitivity measures described in this section assume that the linear model is valid for the case under consideration. If the linear hypothesis does not hold, one may attempt to apply the same techniques after applying a rank transformation to the response [*Saltelli et al.*, 2000]. Rank transformations may be useful since it linearizes nonlinear models and reduces the effect of long-tailed distributions [*Yang*, 2011]. Rank transformation though may not always fully linearize the problems, in particular with non-monotonic models.

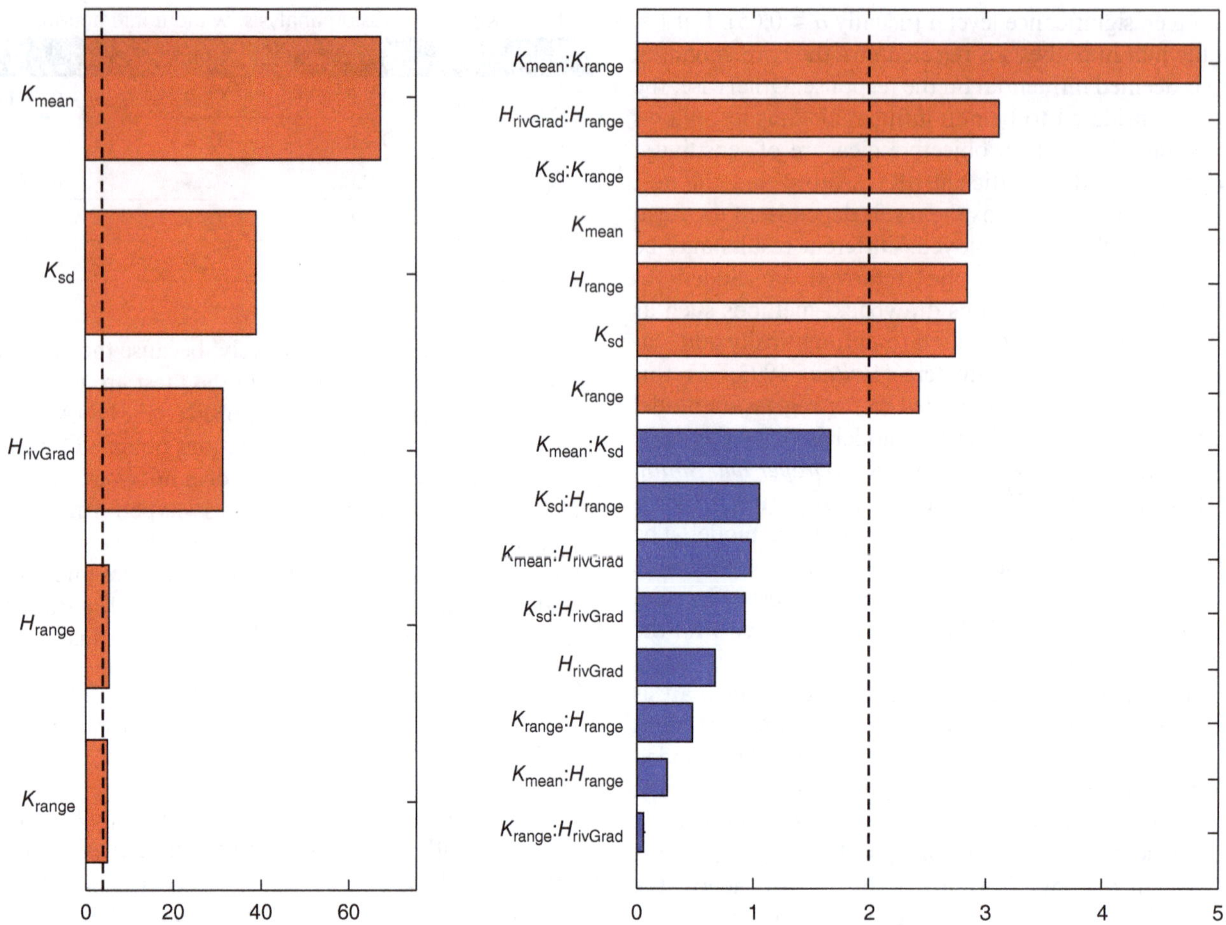

Figure 4.4 Pareto plots displaying the *t*-statistics for (left) regression model without interactions ($R^2 = 0.98$) and (right) regression model including interactions ($R^2 = 0.99$). Red bars show influential parameters. The black line represent the significance level of statistical test for $\alpha = 0.05$.

When using a regression model for SA, the response needs to be smoothly varying, which may not be the case in geoscience applications. Non-smooth or abrupt changes in the response occur when there is a stochastic component (spatial uncertainty), as well as discrete-valued input parameters, such as different geological interpretations. The impact of spatial uncertainty will be discussed in further details in Section 4.5.

One limitation of regression-based SA is their high computational cost. To increase their efficiency, experimental designs are often used instead of Monte Carlo sampling. Experimental design is a technique allowing to optimally define the number and value of input parameters to get the most information at the lowest cost (in terms of number of model evaluations), see Chapter 3. Example applications of experimental design with regression for petroleum applications can be found in *Damsleth et al.* [1992], *White et al.* [2001], *Dejean and Blanc* [1999] among others. For scatter plots or correlation coefficient analysis, a natural alternative to Monte Carlo sampling is the use of Latin hypercube designs [*Helton and Davis*, 2003].

4.4.2. Variance-Based Methods/Measures of Importance

Variance-based SA methods are quite popular in the geosciences (see References). Also referred to as measures of importance, variance-based SA attempts to evaluate the part of the total variance of the response Y that can be attributed to input parameter X_n:

$$\frac{\mathrm{Var}[E(Y|X_n = x_n)]}{\mathrm{Var}(Y)} \tag{4.12}$$

Among the variance-based global sensitivity methods, the Sobol' approach is the most popular. According to *Sobol'* [1993], any numerical computer model f, with associated parameters X and response $Y = f(\mathbf{X})$, can be expanded into summands of different dimensions:

$$f(\mathbf{x}) = f_0 + \sum_{n=1}^{N_p} f_n(x_n) + \sum_{1 \le n \le n' \le N_p} f_{n,n'}(x_n, x_{n'}) + \cdots \\ + f_{1,2,\ldots,N_p}(x_1, \ldots, x_{N_p}) \tag{4.13}$$

Assuming independence in the input parameters and orthogonality of the summands, the variance of each term in Eq. (4.13) can be obtained by means of Monte Carlo integration, leading to a decomposition of variance of Y similar to a functional ANOVA procedure:

$$\mathrm{Var}(Y) = \sum_{n=1}^{N_p} D_n + \sum_{1 \le n \le n' \le N_p} D_{n,n'} + \cdots + D_{1,2,\ldots,N_p} \tag{4.14}$$

where

$$D_n = \mathrm{Var}[E(Y|X_n)]$$
$$D_{n,n'} = \mathrm{Var}[E(Y|X_n, X_{n'})] - D_n - D_{n'} \tag{4.15}$$

and so on for higher-order interactions. The total variance of the response Y can be decomposed into partial variances, attributing variability of the response Y to each input parameter, including interactions. The Sobol' indices are obtained by normalizing each partial variance with respect to the total (unconditional) variance of Y as follows:

$$S_n = \frac{D_n}{\mathrm{Var}(Y)}, S_{n,n'} = \frac{D_{n,n'}}{\mathrm{Var}(Y)}, \ldots, S_{1,2,\ldots,N_p} = \frac{D_{1,2,\ldots,N_p}}{\mathrm{Var}(Y)} \tag{4.16}$$

The first-order Sobol' index S_n calculates the impact of the input parameter x_n by estimating the partial variance of Y explained by this parameter. It estimates by how much the variance of the response is reduced, on average, when the parameter x_n is fixed, that is, it measures the contribution of the parameter x_n to the total variance of the response. The second-order Sobol' index $S_{n,n'}$ measures the contribution of the interacting effects between x_n and $x_{n'}$ on the response variance. In Eqs. (4.15) and (4.16) the sum of all the terms should be equal to 1.

In addition to the indices defined in Eq. (4.16) the total effect index for a parameter x_n is defined as [*Homma and Saltelli*, 1996]:

$$S_{T_n} = S_n + \sum_{n<n'} S_{n,n'} + \sum_{n' \neq n, n'' \neq n, n' < n''} S_{n,n',n''} + \cdots \tag{4.17}$$

The total effect index represents the total contribution (including interactions) of a parameter x_n to the response variance; it is obtained by summing all first-order and higher-order effects involving the parameter x_n.

Evaluating this quantity would require estimating all $2^{N_p} - 1$ sensitivity indices, which is not possible with a reasonable number of model evaluations. The calculation of S_{T_n} can be obtained much more efficiently by evaluating the following equation [see *Saltelli et al.*, 2008 for details]:

$$S_{T_n} = \frac{\mathrm{Var}(Y) - \mathrm{Var}[E(Y|\mathbf{X}_{\sim n})]}{\mathrm{Var}(Y)} = \frac{E[\mathrm{Var}(Y|\mathbf{X}_{\sim n})]}{\mathrm{Var}(Y)} \tag{4.18}$$

with $\mathbf{X}_{\sim n} = (X_1, \ldots, X_{n-1}, X_{n+1}, \ldots, X_{N_p})$.

The smaller S_{T_n} is, the less x_n contributes to the variance. If $S_{T_n} = 0$, then x_n is a non-influential parameter and x_n can be fixed at any value within its range of uncertainty without significantly impacting the variance of Y [*Saltelli et al.*, 2008]. The difference between S_{T_n} and S_n represents the interaction effect of the parameter x_n with the other inputs. If $S_n = S_{T_n}$ for all n, there is no interaction effect.

Confidence intervals for the first-order and total effect indices can be computed using the bootstrap method (see Section 3.13). Sampling with replacement is performed to obtain a distribution of the S_n and S_{T_n}, from which confidence intervals can be derived [*Archer et al.*, 1997].

The main advantage of the Sobol' method is that the indices are defined for any type of response (nonlinear, non-monotonic) and any type of parameter. The Sobol' indices can also be applied for groups of parameters, and not just to each parameter separately [*Saltelli et al.*, 2008]. However, in its traditional formulation, the response must be univariate; if the response is, for example, a time series then the Sobol' indices need to be calculated at all desired times. *Gamboa et al.* [2014] propose a generalization of the Sobol' indices to determine sensitivities for any multidimensional responses, including time-varying responses or spatial maps. This new formulation has been applied to spatial outputs in the context of contamination [*De Lozzo and Marrel*, 2017] and cyclone-induced waves [*Rohmer et al.*, 2016]. The Sobol' method is computationally demanding because the indices are computed using a Monte Carlo procedure and a large sample is required to reach convergence. For details on how to compute the sensitivity indices, we refer to *Saltelli et al.* [2008], who shows that at least $m \times (N_p + 2)$ model evaluations are needed to compute the first-order and total effect indices, where m can vary from a few hundreds to a few thousands. Different algorithms exist to compute the sensitivity indices, for example *Glen and Isaacs* [2012] and *Wainwright et al.* [2014] use the Pearson correlation coefficient. Of particular interest, *Wainwright et al.* [2014] suggests an alternative method to compute the Sobol' total effect indices using nonparameteric regression. They additionally suggest that the Morris's mean |EE| can be used instead of S_{T_n} to rank parameter effects (including the interaction), reducing significantly the computational cost. However, this method does not allow to compute the Sobol' first-order effect.

Because of its high cost, the Sobol' method is often impractical for geoscience applications with high-dimensional inputs. Despite this high-computational demand, it has been applied in environmental modeling (see, e.g., [*Pappenberger et al.*, 2008; *Nossent et al.*, 2011]). Most studies based on Sobol' circumvent this problem by approximating the computer model by a surrogate model (meta-model) and use the surrogate model to compute the Sobol' indices at a much cheaper cost. For example, Gaussian process regression (kriging, Chapter 3) may provide such a surrogate model, others are polynomial chaos expansions (expressing output as a polynomial series). In many applications, surrogate models has been used to estimate the Sobol' indices, including water contaminant remediation [*Ciriello et al.*, 2013; *Luo and Lu*, 2014], CO_2 geological storage [*Rohmer*, 2014], oil and gas applications [*Touzani and Busby*, 2013; *Dai et al.*, 2014; *Sarma et al.*, 2015], landslide modeling [*Rohmer and Foerster*, 2011], and hydrogeology [*Marrel et al.*, 2012; *Oladyshkin et al.*, 2012] among many others.

Sobol' sensitivity indices were calculated on a modified version of the Libyan case presented in Chapter 1. The case is fully described in *Park et al.* [2016]. In total, 12 uncertain parameters are considered in this study, requiring a total of 14,000 simulations ($m = 1000$). The sampling procedure described in *Saltelli et al.* [2008] was applied and the sensitivity indices were computed using the approach of *Wainwright et al.* [2014]. The response considered was the total field water production at a given time. For more details about the case and the calculation of the indices, we refer to *Park et al.* [2016]. The sensitivity indices as a function of time are presented in Figure 4.5. Oil–water contact is by far the most influential parameter on the field water production, especially early in the simulation time. Because the first-order effect and the total effect have similar values, this model does not appear to contain many influential interactions. It is also worth noting that in some cases, the total effect appears slightly smaller than the first-order effect, which is due to numerical error. This numerical error may be reduced by increasing the sample size m.

An alternative variance-based method is the Fourier amplitude sensitivity test (FAST) method [*Cukier et al.*, 1978], which is more efficient than the Sobol' method while providing the same indices (S_n and S_{T_n}). The FAST method evaluates the contribution of a parameter x_n to the variance of the response based on the estimation of Fourier coefficients. In its original version, the FAST method only provides first-order sensitivity indices (S_n in Sobol'). *Saltelli et al.* [1999] present an extended FAST approach which can account for higher-order interactions. As for the Sobol' approach, FAST does not require any assumptions on the functional form of the model. One of the limitations of the FAST algorithm is its lack of reliability in the presence of uncertain discrete parameters, as well as its algorithmic complexity [*Frey and Patil*, 2002]. Example applications of FAST in geosciences includes groundwater modeling [*Fontaine et al.*, 1992] and hydrology [*Francos et al.*, 2003].

4.4.3. Generalized/Regionalized SA

4.4.3.1. Regionalized SA. The regionalized sensitivity analysis (RSA) method was originally developed by *Spear and Hornberger* [1980] to identify influential parameters in environmental systems. Spear and Hornberger refer to it as generalized SA, but the approach was later renamed to RSA [*Spear and Grieb*, 1994]. RSA is based on the principle of Monte Carlo filtering [*Rose et al.*, 1991; *Saltelli et al.*, 2004, pp. 151–191 for a review], which

Figure 4.5 Sobol' indices: (a) first-order effect and (b) total effect (including interactions) for a modified version of the Libyan case described in Chapter 1. Modified from *Park et al.* [2016].

consists of classifying each response generated from a Monte Carlo sampling of the uncertain input parameters into two classes ("behavior" or "non-behavior," as described by Spear and Hornberger). The definition of the behavioral classes usually requires the definition of a threshold: if the response is within the threshold, then the model is attributed to the category "behavior" B (acceptable models), otherwise to "non-behavior" $\bar{B}$ (unacceptable models). In the case of Spear and Hornberger, behavior is defined as the presence or absence of algae in the Peel Inlet of Western Australia. Parameter values leading to $\bar{B}$ are then analyzed in the Monte Carlo filtering procedure. In this instance, one is not interested in the variance of the response, but rather in which parameter values generated models found in class $\bar{B}$. Spear and Hornberger suggest to analyze the difference in input parameter distributions between the L_m samples that lead to the category B and the $L_{\bar{m}}$ samples that lead to class $\bar{B}$. The cumulative distribution of each parameter is calculated for each behavior and compared statistically: if the distribution for parameter x_n is different between the two classes, then x_n is defined as influential on the response. The comparison is done using a two-sampled Kolmogorov–Smirnov (KS) statistical test, where the null-hypothesis is defined as $H_0: f(x_n|B) = f(x_n|\bar{B})$. The KS test is based on the Smirnov statistic which is defined as the largest vertical distance between the two distribution functions (see Figure 4.6):

$$d_{L_m, L_{\bar{m}}} = \sup\left|F_{L_m}(x_n|B) - F_{L_{\bar{m}}}(x_n|\bar{B})\right| \tag{4.19}$$

Intuitively, parameters with larger $d_{L_m, L_{\bar{m}}}$ are more influential in discriminating the response between models in B and $\bar{B}$, implying more impact on the response. On the other hand, small values of $d_{L_m, L_{\bar{m}}}$ indicate a similar

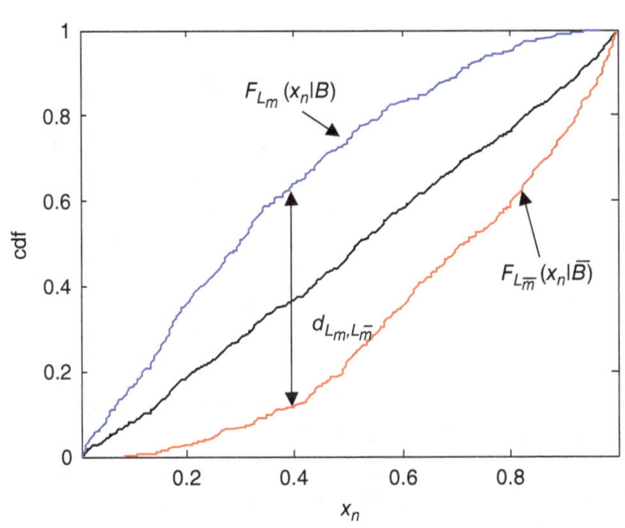

Figure 4.6 Schematic illustration of the KS test.

distribution of parameters between classes, hence little impact of the parameter on the response.

RSA and subsequent extensions [*Beven and Binley*, 1992; *Bastidas et al.*, 1999; *Pappenberger et al.*, 2008] have been extensively used in hydrology (e.g., [*Chang and Delleur*, 1992; *Lence and Takyi*, 1992; *Spear and Grieb*, 1994; *Tang et al.*, 2007], among others). More specifically, the extension based on generalized likelihood estimation (GLUE) due to *Beven and Binley* [1992] has been popular in hydrology [*Freer et al.*, 1996; *Hossain et al.*, 2004; *Ratto et al.*, 2007]. In the GLUE approach, the binary classification of the realizations into behavior/non-behavior is replaced by a likelihood weight (models in the non-behavior class have low likelihood).

Similar to variance-based methods, RSA has many global properties: (i) the entire range of values of the input parameters is considered and (ii) all parameters are varied simultaneously. Other advantages of RSA are that it is conceptually simple, model-independent and does not assume any functional form of the response (smoothness, linearity, etc.). However, because RSA is based on a KS test, it is only applicable to continuous parameters. In addition it cannot quantify interaction effects. This is one of the reasons why *Fenwick et al.* [2014] developed an extension of the basic RSA method of Spear and Hornberger, denoted as DGSA and described in the following.

4.4.3.2. Distance-Based Generalized SA. *Fenwick et al.* [2014] extend the principles of the RSA method in several aspects by proposing a distance-based generalized sensitivity analysis (DGSA). One extension of DGSA is to account for the possible high-dimensional responses of the computer models, which is typically the case in subsurface systems. They propose a distance-based classification procedure [*Scheidt and Caers*, 2009a, 2009b] to classify the responses (analogous to behaviors in RSA). This presents two advantages: (i) the response does not need to be univariate, as in the screening SA, linear-based methods, and traditional variance-based methods seen so far in this chapter and (ii) more than two classes can be constructed, which can be of interest because the modeler may wish to classify the response into more than two types of behavior, hence exploring more refined behavior. The definition of the sensitivity measure in DGSA does not rely on the KS test; instead, it uses the concept of distances between the prior cdfs and class-conditional cdfs. The advantage of using a distance measure over the KS test is that a distance can be computed for all types of input parameter distributions (continuous, discrete, scenario-based, etc.).

For each class c_k ($k = 1, \ldots, K$) and for a parameter x_n, the distance is calculated as the $L1$-norm between the prior distribution of x_n (denoted as $\hat{F}(x_n)$) and its class-conditional distribution (denoted as $\hat{F}(x_n|c_k)$):

$$\hat{d}_n^{(k)} = f_{\Delta\text{cdf}}\left(\hat{F}(x_n),\ \hat{F}(x_n|c_k)\right) \tag{4.20}$$

where $f_{\Delta\text{cdf}}$ is a distance between two cdfs, determined by evaluating the $L1$-norm. This distance represents the area between the curves.

To determine if the parameter x_n is influential on the response for a class k, a hypothesis test is formulated and evaluated using a resampling procedure. The resampling procedure estimates a distribution of distances $\hat{\hat{d}}_n^{(k)}$ that would occur if the input parameter x_n had no impact on the response (null hypothesis H_0). The observed distance is then compared to the resampling distribution. This estimated distribution under H_0 is obtained by repeatedly evaluating the $L1$-norm between the prior cdf of x_n and the cdf of a random sample of n_k parameters x_n from its prior distribution (n_k being the number of models in cluster k). The null hypothesis is then

$$\text{if } \exists k \text{ for which } \hat{d}_n^{(k)} \geq \hat{\hat{d}}_{n,\alpha}^{(k)}, \quad \text{then } H_0 \text{ is rejected}$$

where $\hat{\hat{d}}_{n,\alpha}^{(k)}$ is the alpha-percentile (usually, $\alpha = 0.95$) from the resampling procedure (see Figure 4.7).

A standardized measure of sensitivity of the parameter x_n on the response for class k can be expressed as

$$\hat{d}_n^{S(k)} = \frac{\hat{d}_n^{(k)}}{\hat{\hat{d}}_{n,\alpha}^{(k)}} \tag{4.21}$$

A parameter x_n is defined as influential if for at least one class the standardized measure of sensitivity is greater than 1. *Fenwick et al.* [2014] propose to visualize sensitivity measures using Pareto plots. Because sensitivity is defined based on classes, for each parameter, K bars

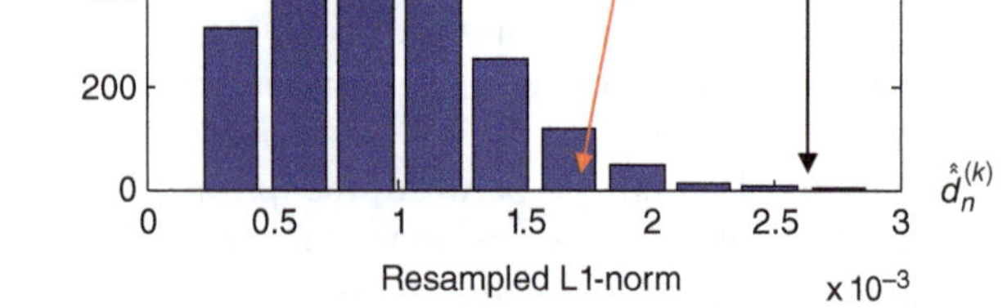

Figure 4.7 Histograms of L1-norm obtained by the resampling procedure. The observed L1-norm $\hat{d}_n^{(k)}$ and alpha-percentile $\hat{\hat{d}}_{n,0.95}^{(k)}$ are shown as well.

can be displayed (Figure 4.8a). For simplicity, an averaged sensitivity value can be calculated and displayed in the Pareto chart, as illustrated in Figure 4.8b:

$$s(x_n) = \frac{1}{K}\sum_{k=1}^{K} \hat{d}_n^{S(k)} \tag{4.22}$$

Note that for Figures 4.8–4.10, sensitivity on the full concentration curves as a function of time is performed, whereas in the previous examples the sensitivity was determined on a scalar: the pollutant arrival time. In addition, spatial uncertainty was included in the modeling exercise for the definition of both the boundary conditions and the hydraulic conductivities.

An alternative representation of DGSA results is to display $\widetilde{\text{ASL}} = 1 - \text{ASL}$ (the achieved significance level or ASL, see Section 3.13) obtained from the resampling procedure. For each observed $L1$-norm for parameter x_n and c_k, $\widetilde{\text{ASL}}\left(\hat{d}_n^{(k)}\right)$ can be estimated from the resampling procedure (see Eq. (3.299)):

$$\widetilde{\text{ASL}}\left(\hat{d}_n^{(k)}\right) = P\left(\hat{\hat{d}}_n^{(k)} \leq \hat{d}_n^{(k)}\right) \tag{4.23}$$

The global sensitivity measure $s^*(x_n)$ for x_n is then represented by the maximum $\widetilde{\text{ASL}}$ per class (see Figure 4.9):

$$s^*(x_n) = \max_k \widetilde{\text{ASL}}\left(\hat{d}_n^{(k)}\right) \tag{4.24}$$

The larger $s^*(x_n)$, the more influential the parameter is. Sensitivity can then be divided into three groups, based on the value of confidence of the hypothesis test:

1. High value: $s^*(x_n) > \alpha \rightarrow$ the parameter x_n is influential (critical).
2. Low value: $s^*(x_n) < \alpha \rightarrow$ the parameter x_n is non-influential.
3. $s^*(x_n) \approx \alpha \rightarrow$ the parameter x_n is important.

DGSA also estimates the impact of interacting parameters on the response. Contrary to SA based on a regression model which assumes symmetric interactions (interactions are modeled by the product of two parameters), DGSA models interactions through conditional densities. For example, two-way interactions are expressed using conditional densities, that is, $x_n \mid x_{n'}$ and $x_{n'} \mid x_n$ for interactions between parameters x_n and $x_{n'}$. In this manner, the approach is capable of evaluating asymmetric interactions. The principle for determining sensitivity for parameter interactions is similar to the main factors and works conceptually as follows: if no significant difference exists between the class-conditional distributions of a single parameter and the class-conditional distribution of the parameter additionally conditioned to a second parameter, then the two-way interaction is

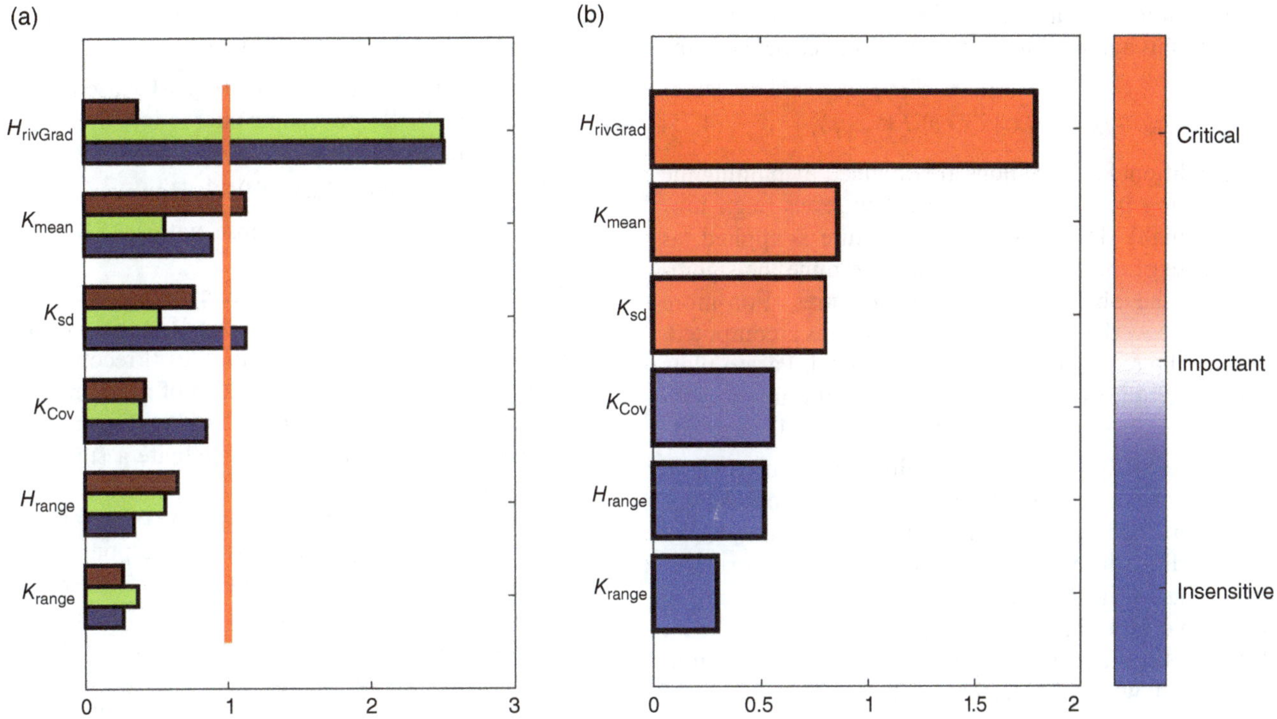

Figure 4.8 Pareto plots illustrating the sensitivity of each parameter on the pollutant concentration curves. (a) Standardized L1-norm per class (three classes used in this example). (b) Average standardized L1-norm.

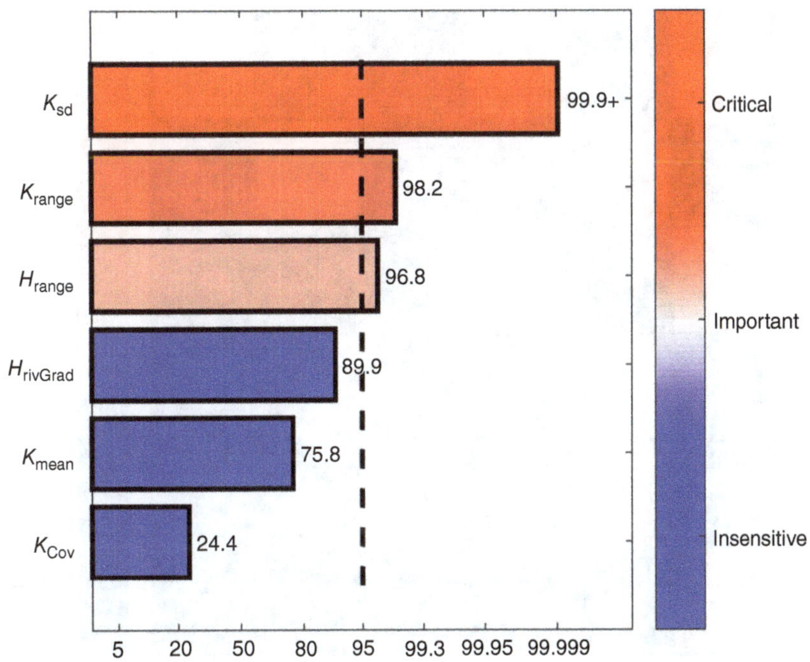

Figure 4.9 $\widetilde{\text{ASL}}$-based sensitivity values for each parameter (in percent) using $\alpha = 95\%$.

not influential. For a class c_k, the sensitivity measure for interaction is again a function of the distance between the two cdfs:

$$\hat{d}_{k,n|n'} = f_{\Delta\text{cdf}}\left(\hat{F}(x_n|x_{n'},c_k),\hat{F}(x_n,c_k)\right) \tag{4.25}$$

The conditional distributions are obtained by binning the conditioning parameter $x_{n'}$ into a few levels (e.g., low/medium/high). The resampling procedure is applied, as in the case of main sensitivities, to determine the significance of the observed sensitivity measures. For more details on how the sensitivity on interactions is computed, we refer the reader to *Fenwick et al.* [2014]. Pareto plots representing the average (per level of conditioning parameter and number of classes) of standardized L1-norm distances or $\widetilde{\text{ASL}}$-based sensitivity values can be used again to visualize the results. However, when the number of parameters is large, the high number of interactions makes it difficult to visualize the results in such a manner. An alternative visualization of two-way interactions using 2D bubble plots or H-plots is proposed in *Park et al.* [2016]. A third alternative is to use a table, as shown in Figure 4.10, where the diagonal values in the table show the $\widetilde{\text{ASL}}$-based sensitivity values for the one-way sensitivity, and the off-diagonal values are the $\widetilde{\text{ASL}}$-based sensitivity values for the two-way interactions (row|column).

Figure 4.10 shows a few interesting results. First, each parameter has at least one influential interaction with another parameter, which illustrates the high complexity of the case. In particular, the interaction $H_{\text{rivGrad}}|K_{\text{sd}}$ is critical to the variation of the response ($\widetilde{\text{ASL}}$-based sensitivity values of 98.9). Second, some interactions are asymmetric. For example, $K_{\text{Cov}}|K_{\text{mean}}$ is shown important, with a $\widetilde{\text{ASL}}$-based sensitivity values of 94.3. However, $K_{\text{mean}}|K_{\text{Cov}}$ is not influential on the response, with a $\widetilde{\text{ASL}}$-based sensitivity values of 57.7.

DGSA is well suited for geoscience applications and is best used when only a modest number of model evaluations can be afforded. *Fenwick et al.* [2014] recommend using Latin hypercube sampling instead of random sampling. As a rule of thumb, *Fenwick et al.* [2014] suggests to define the number of classes K and evaluate a sufficient number of models such that at least ten models are found in each class. DGSA can handle any type of parameter distributions as well as high-dimensional responses and does not require an assumption that the model response takes any particular functional form. DGSA was applied in the context of basin modeling [*Tong and Mukerji*, 2017], flood and drought hydrologic monitoring [*Chaney et al.*, 2015], and reservoir engineering [*Fenwick et al.*, 2014; *Park et al.*, 2016].

DGSA also works well with non-smooth responses, such as those subject to stochastic (spatial) uncertainty, or when discrete or scenario-based parameters exist. One of its main advantages is that asymmetric

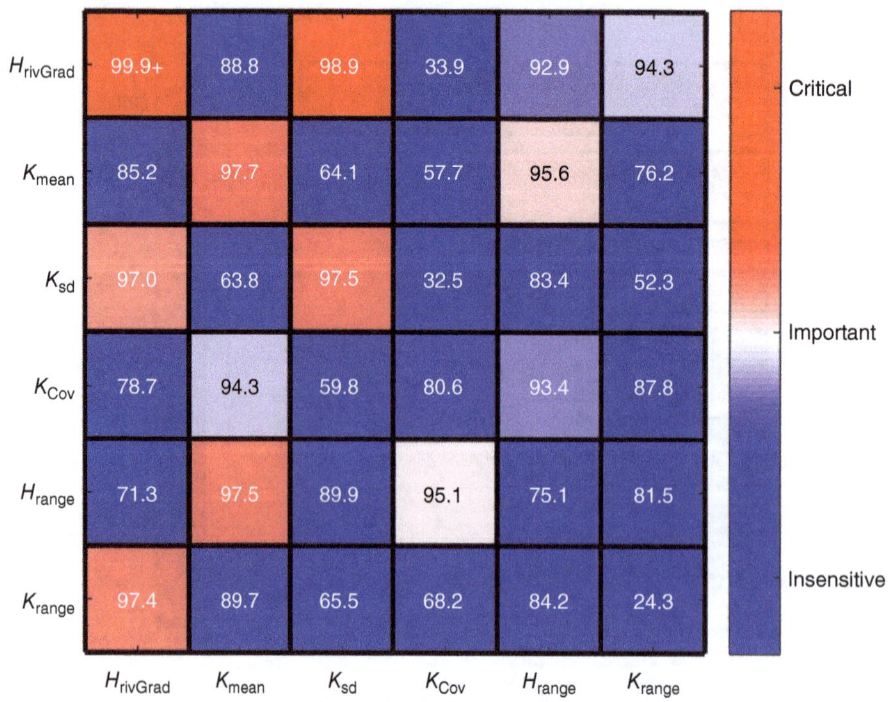

Figure 4.10 Table showing the $\widetilde{\text{ASL}}$-based sensitivity values for interactions (in percent).

interactions can be estimated (two-way interactions are not assumed to be symmetric). In theory, DGSA can be used to estimate interactions between more than two parameters, but this has not been explored and would require a large initial sample of models. Finally, because the responses are only used for classification, a proxy can be used instead of the time-consuming model evaluations. Note that DGSA does not require the proxy model itself to be accurate (only the classification of models must be accurate) which is a much less stringent requirement. One minor drawback with DGSA is that the results may lack in ease in interpretation, as the SA measure (standardized L1-norm or $\widetilde{\text{ASL}}$) may be harder to interpret than in OAT or Sobol' indices for example.

A comparison of the results obtained from DGSA and Sobol' on a case derived from a real oil field is proposed in *Park et al.* [2016]. The parameter ranking was similar for the two methods, for only a fraction of the cost for DGSA (14,000 for Sobol' vs. 1000 for DGSA).

4.4.4. Tree-Based SA

Classification and regression trees (CART) are also a powerful tool for SA [*Breiman et al.*, 1984]. As seen in Section 3.7.4, the overall impact of an input parameter x_n on the response y can be quantified using the CART variable importance procedure. In CART, the input parameter space is partitioned into smaller regions by recursive binary splitting. At each node of the tree, the splitting parameter is partitioned into two subbranches. The choice of the splitting parameter and the splitting point is based on the minimization of a cost function (often least square for continuous responses). For a tree with M terminal nodes (or regions R_m), the cost using the least-square formulation can be expressed as (see Eq. (3.138))

$$\text{Cost} = \sum_{\ell=1}^{L}\left(y^{(\ell)} - \hat{y}\right)^2, \text{ with } \hat{y} = \sum_{m=1}^{M}\hat{c}_m I(\mathbf{x}\in R_m) \quad (4.26)$$

with $\hat{c}_m$ the average of all responses in region R_m and I the indicator function.

The sensitivity of an input parameter can then be estimated by its overall contribution to the reduction of the cost function. In the CART literature, this is referred to as "relative importance." The relative importance of a parameter x_n for a regression tree containing M terminal nodes is defined as (Chapter 3)

$$I_n^2 = \sum_{j=1}^{M-1} i_j^2 I(v(j) = n) \quad (4.27)$$

where

1. $v(j)$ indicates the split variable at node j (e.g., if the j-th node splits variable x_n, then $v(j) = n$).

2. i_j^2 accounts for how much the cost is improved by splitting the tree at node j into two subbranches (right jR and left jL): $i_j^2 = \text{Cost}_j - \text{Cost}_{jL} - \text{Cost}_{jR}$, with Cost_j, Cost_{jL}, and Cost_{jR} the costs for the partition at node j and the left and right subbranches (as defined in Eq. (4.26)), respectively.

Recursive splitting of the regression tree allows for multi-way interactions to be accounted for, since all splits are conditional to the previous splits. Hence, the relative importance of an input parameter is a measure of sensitivity of that parameter, including all interactions.

In the classical approach, the regression tree is computed on a scalar response; hence, sensitivities can only be obtained for scalar responses. However, this can be extended to functional variable, by generalizing the definition of the cost function defined in Eq. (4.26). Assume the response of interest is a function of time $\mathbf{y}^{(\ell)} = y^{(\ell)}(t)$. Based on observations $(\mathbf{x}^{(\ell)}, y^{(\ell)}(t))$, the regression model can be generalized to

$$\begin{aligned}\hat{y}(t) &= \sum_{m=1}^{M}\hat{c}_m(t) I_m(\mathbf{x}\in R_m) \text{ with} \\ \hat{c}_m(t) &= \frac{1}{\#(\mathbf{y}^{(\ell)}(t)|\mathbf{x}^{(\ell)}\in R_m)}\sum_{\mathbf{y}^{(\ell)}(t)|\mathbf{x}^{(\ell)}\in R_m} y^{(\ell)}(t)\end{aligned} \quad (4.28)$$

The partitioning of the input space can then be obtained by generalizing the cost function to functional variables. Many such cost functions can be defined, only a few are presented here. For example, a cost function for a region R_m can be obtained by integrating the difference over all times, or by simply taking the L2-norm:

$$\text{Cost}_m = \sum_{y^{(\ell)}(t)|\mathbf{x}^{(\ell)}\in R_m}\int_T \left(y^{(\ell)}(t) - \hat{c}_m(t)\right)dt \quad (4.29)$$

$$\text{Cost}_m = \sum_{y^{(\ell)}(t)|\mathbf{x}^{(\ell)}\in R_m}\left\| y^{(\ell)}(t) - \hat{c}_m(t)\right\|_2 \quad (4.30)$$

The cost function can be further extended by using the concept of distances between two model responses (similarly to what was done in DGSA, see Section 4.4.3.2). Any dissimilarity/similarity distance between model responses can be used in tree-based techniques. Instead of using the average response $\hat{c}_m(t)$ in the calculation of the cost function (which may not always be an optimal criteria for physical variables), one can simply select the medoid (the model whose response dissimilarity to all other responses in the region is minimal) in each region and evaluate its distance to the responses of the models in the same region:

$$\text{Cost}_m = \sum_{y^{(\ell)}(t)|\mathbf{x}^{(\ell)}\in R_m} d\left(\mathbf{y}^{(\ell)}(t), \mathbf{y}_{\text{medoid}}(t)\right) \quad (4.31)$$

This corresponds to approximating the regression model using the medoid response of each region, as opposed to the average response of each region (classical procedure). Note that the use of distances between model

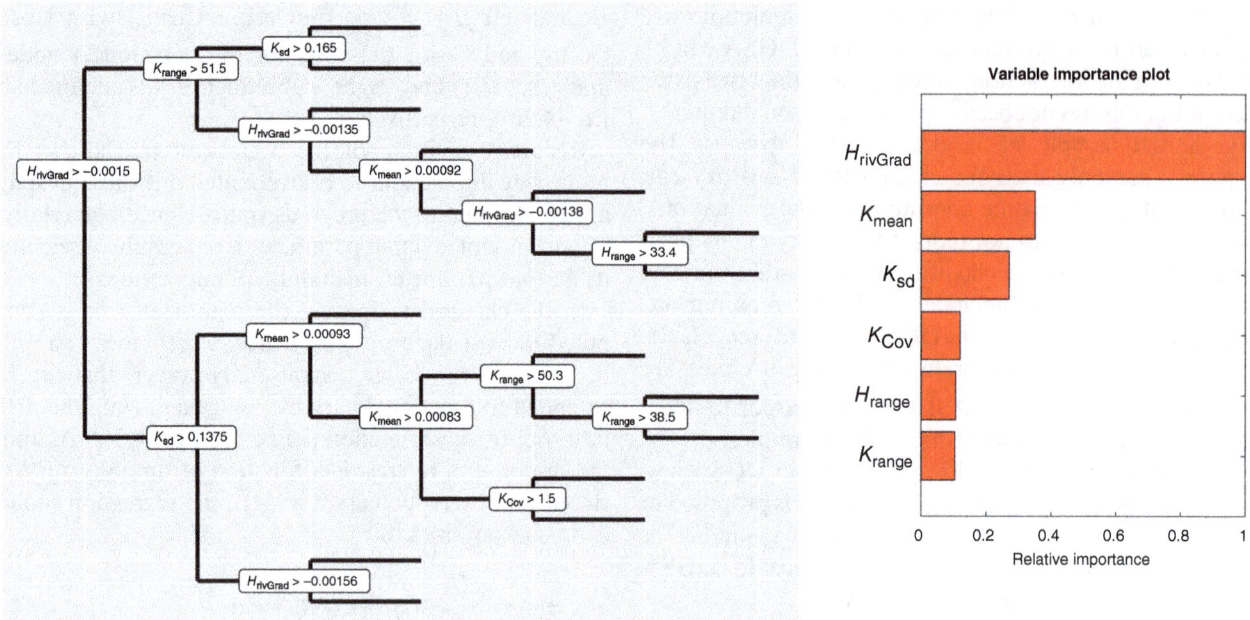

Figure 4.11 Application of CART-based SA on the simplified DNAPL example, using the cost of Eq. (4.31).

responses is not limited to time-varying data and can be applied to maps, 3D volume, and so on.

An example of tree-based relative importance for the DNAPL example is shown in Figure 4.11. Contrary to the application of CART for SA in Chapter 3, which uses the contaminant arrival time for response, here we use the entire contaminant production curve.

Even though the procedure is only described for a single regression tree (CART), variable importance can also be estimated using more complex, randomized tree, such as random forest (RF), see *Wei et al.* [2015] and *Breiman* [2001] for details. The idea behind RF is to construct multiple trees based on bootstrapped copies of the original data and a random sub-selection of K input parameters at each node of the tree. RF is quite popular because it reduces variance of single trees and thus improve predictions. In the case of RF, the variable importance can be estimated by simply averaging the relative importance (as defined in Eq. (4.27)) over all trees in the forest. An alternative approach proposed by *Breiman* [2001] is to compute variable importance based on the out-of-bag samples. This is often referred to as permutation importance and is described in *Wei et al.* [2015].

Currently, very few applications of SA based on regression trees to subsurface modeling exist. Examples of such includes *Mishra et al.* [2009], *Pappenberger et al.* [2006], *Spear and Grieb* [1994]. However, CART-based SA has many advantages that are well suited to the problem of subsurface modeling. There is no assumption on the functional form of the response. For example, high-dimensional model responses can be used as well as stochastic responses. Any type of input parameter with their own input distributions can be used (continuous, discrete and numerical, categorical, functional, maps/images). Finally, the resulting measure of sensitivity may not be easy to interpret (it is a function of the reduction in cost) and is rather subjective, that is, it is the modeler who decides which parameter are influential or non-influential.

4.5. QUANTIFYING IMPACT OF STOCHASTICITY IN MODELS

Most SA studies ignore stochasticity in the computer model, which for subsurface applications is often in the form of spatial uncertainty. The reason is either because the modeler believes that stochasticity is unimportant or the modeler is unaware that a stochastic component of the response may exist. Ignoring stochasticity may simply be a convenience, since it introduces discontinuities in the response and poses a problem for SA approaches relying on linear models. However, spatial uncertainty may have a major effect on the response, as illustrated in the contaminant concentrations profiles of the DNAPL example shown in Figure 4.12 and hence should not be neglected. Clearly in this example, when fixing the random seed for generating the spatial hydraulic conductivity and hydraulic heads at the boundaries, uncertainty in the contaminant arrival time is reduced compared to uncertainty obtained with varying random seed.

An additional complication in ignoring spatial uncertainty is that the results of classical SA approaches may be different for different choices of random seeds in the simulation. This is illustrated in Figures 4.13 and 4.14, where SA results for OAT and Morris methods are repeated three times each with different random seeds. One observes in Figure 4.13 that in two out of three analyses, K_{mean} and H_{rivGrad} parameters are the most influential. However, we observe that K_{sd} is the second most influential parameter in Figure 4.13 (left), but it the least influential parameter in Figure 4.13 (middle). Similar observations can be made when analyzing the results of the Morris method for three different random seeds (Figure 4.13). Even though the main two parameters, namely H_{rivGrad} and K_{mean}, are identified in all cases, μ_n^* and σ_n vary quite significantly. Differences are also observed for the other parameters.

These results indicate that performing SA using a single random seed may result in an incorrect assessment of parameter sensitivity, if spatial uncertainty is present. In addition, the use of linear regression to estimate sensitivity values in the presence of spatial uncertainty is not recommended, as most likely the stochasticity of the models will result in poor quality regression models, and hence the obtained sensitivities may not be reliable. For example, without spatial uncertainty, the main effect model of the arrival time had a coefficient of determination R^2 of 0.98. However, for the same case with spatial uncertainty, the R^2 value is only 0.43. If spatial uncertainty is part of the modeling procedure, we suggest using methods such as DGSA to estimate the parameters sensitivities.

An interesting problem is to quantitatively assess the impact of spatial uncertainty on the response. Some authors attempt to account for stochasticity in the model and estimate the influence of the stochastic component of the response. Of note is the work of *Iooss and Ribatet* [2009] and *Marrel et al.* [2012]. These authors expand on the joint modeling method proposed by *Zabalza-Mezghani et al.* [2004], who modeled the model response as a combination of a proxy of the mean response and a proxy for the variance (dispersion) around the mean. The dispersion component of the joint model is used to evaluate the Sobol' indices. *Rohmer* [2014] proposes to assign a categorical indicator to the set of stochastic realizations and to use meta-modeling techniques that can handle both continuous and categorical parameters to compute Sobol' sensitivity indices. *Park et al.* [2016] propose an approach to parameterize spatial uncertainty by ranking the models using KPCA+SOM. They then apply DGSA using the rank as input parameter. Other ranking techniques which classify the subsurface models in a one-dimensional ordering, such as Optimal Leaf Ordering [*Bar-Joseph et al.*, 2001], can be used as well. Note though that the rank obtained by either approach may be correlated to the value of other parameters. When this occurs, SA is a challenge since most SA techniques assume independence in the input parameters.

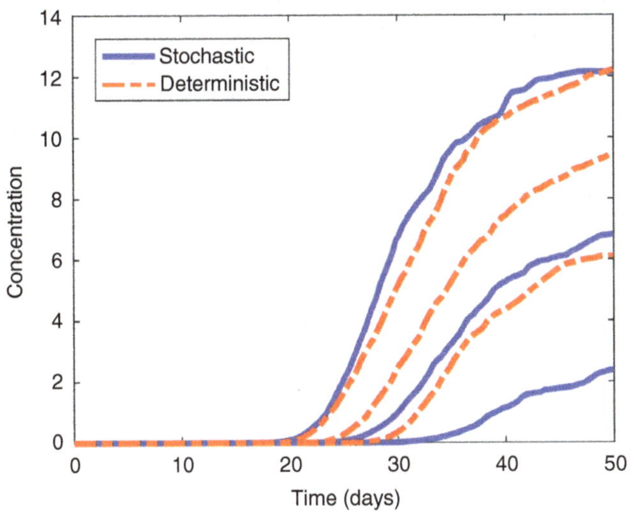

Figure 4.12 Uncertainty in contaminant concentration (P10, P50, and P90) with (solid blue lines) and without (dashed red lines) stochastic uncertainty in the construction of the hydraulic conductivities and boundary conditions.

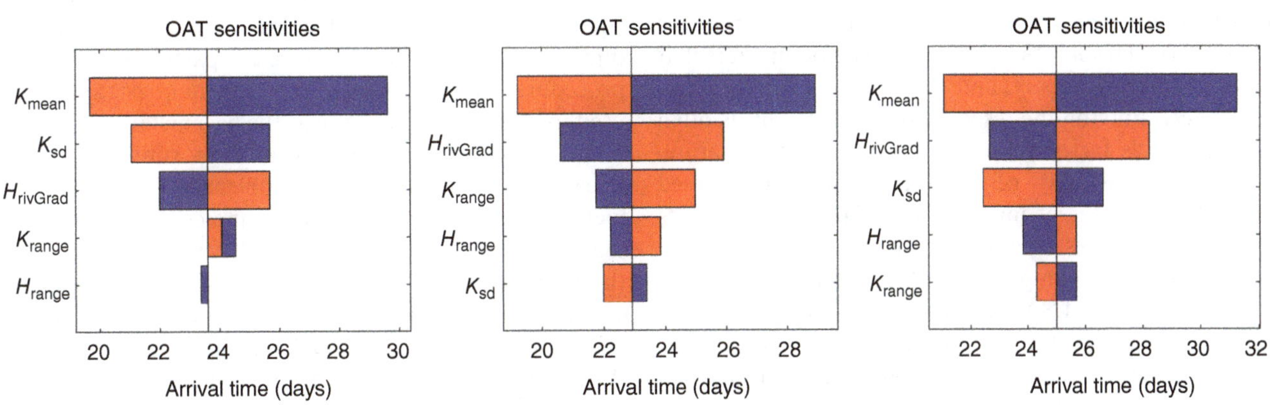

Figure 4.13 OAT results using three different random seeds.

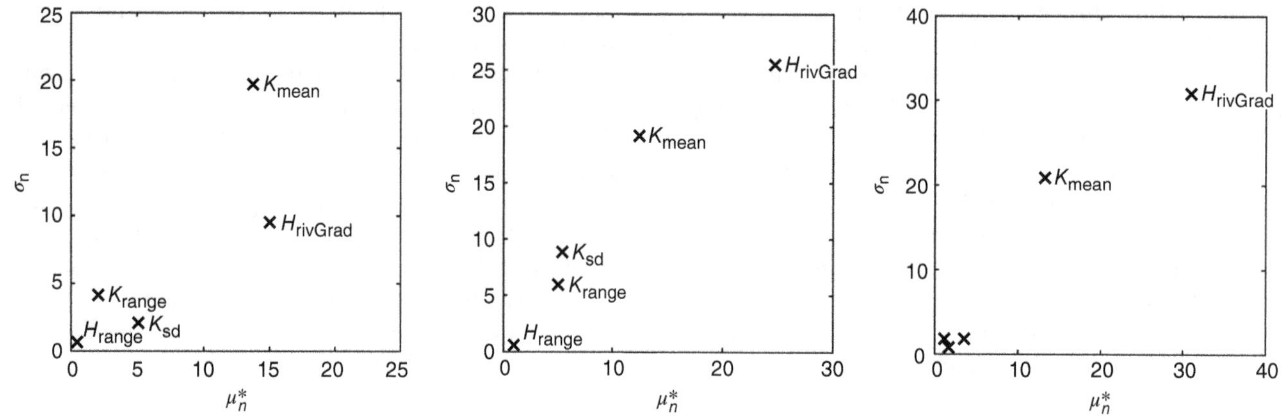

Figure 4.14 Morris method results using three different seeds.

4.6. SUMMARY

SA is a learning tool and is a critical step in UQ. The SA methods presented here allow modelers to obtain insight into how the model response varies with changes to the input in a structured, rigorous manner. SA is a fundamental step in many studies of uncertainty.

One of the main uses of SA is for model simplification. For example, the model may contain a large set of input parameters which may render any subsequent studies very cumbersome. In this case, the user may wish to reduce the number of parameters of the model. Using results from SA, the user can fix or remove non-influential parameters without significantly affecting the uncertainty of the response. Note, however, that non-influential parameters may have influential interactions with other parameters. Fixing such non-influential parameters without accounting for interactions may reduce uncertainty in the response artificially. Evidently, fixing an influential parameter to a certain value (often the mean value of the parameter distribution) may result in a reduction of uncertainty in the response.

SA provides insight into which parameters have the most influence on the response, allowing one to focus on additional resources to better estimate parameters that have the largest impact, and therefore potentially reducing uncertainty on the response of interest. This is often referred to as factor prioritization in the SA literature. Note that sensitivity of parameters may be affected strongly by ranges of inputs. Thus, if new data reduces the range of values a parameter may have, the influence of that parameter compared to others will likely diminish.

Although several SA techniques exist, there is no consensus on which methods are best applicable under what circumstances. Most studies compute sensitivities based on only one SA methodology, although different SA methods may rank parameters differently. Selecting an appropriate technique requires a clear statement of the objective of the analysis and the insights that the user wishes to obtain. In addition, which method to apply depends on the specific characteristics of the model under study, namely the number of parameters, the simulation time, the type of model (linearity, monotonicity, etc.), the type of input parameters (presence of discrete or scenario-based parameters), and the type of response (scalar vs. high-dimension). Finally, the choice of method depends on how to quantify model interactions, if at all desired. A summary of the main properties of the different SA methods discussed in this chapter is presented in Table 4.2.

Screening methods, such as OAT and Morris methods, are well suited for models that contain many input parameters and are computationally expensive, because they are relatively economical. They provide only subjective sensitivity measures, allowing to rank parameters based on their influence on the response and require a subjective judgment on which parameters are influential based upon visual inspection of the results. Linear regression methods were the first GSA techniques to be intensively employed and are still quite popular because of their mathematical rigor which provides objective sensitivity measures directly. These methods require no particular sampling procedure; Monte Carlo or experimental designs can be used for the analysis, as long as the parameters are uncorrelated. However, linear regression is not suitable for nonlinear and non-monotonic problems. Variance-based methods, and in particular Sobol' indices, are quite popular in subsurface applications, as they allow to apportion the proportion of variance in the response that can be explained by each input parameter. However, they require an extensive number of model evaluations, which makes them a challenge to use for computationally intensive problems without the use of proxy models. This poses the additional problem of the quality of the proxy model used

Table 4.2 Summary table of the SA techniques described in this chapter.

	OAT	Morris method	Regression	Sobol'	DSGA	Tree-based SA
Cost (amount of simulations)	Low N_p+1	Low $L(N_p+1)$	Low LHS, ED m	High quasi-random, LHS m (N_p+2)	Moderate LHS m	Moderate LHS m
Model assumption	Linear	Model free	Depends on regression model	Model free	Model free	Model free
Sensitivity measure	Subjective	Subjective	Objective	Subjective	Objective	Subjective
Interactions	No	Yes, qualitative	Depends on regression model, symmetric	Yes	Yes asymmetric	Yes
Discrete parameter	No	No	Yes	Yes	Yes	Yes
Stochasticity	No	No	No	With proxy only	Yes	Yes
Input distribution	No	No	Yes	Yes	Yes	Yes
High-dimensional response	No	No	No	Not in standard approaches	Yes	Yes

within variance-based analysis. In cases where a large number of model evaluations cannot be performed, an alternative could be to use the elementary effects evaluated using the Morris method, which have been shown to be a good proxy for the total sensitivity indices [*Wainwright et al.*, 2014]. In cases of a large number of parameters, one could also first apply the Morris method to screen out non-influential parameters and then do a variance-based analysis on the remaining subset of parameters. Note though that for variance-based methods, the sensitivity measure is still a subjective procedure, where the user must decide which parameters are influential. DGSA, the most recently developed GSA method described in this chapter, has the advantage of being computationally efficient compared to variance-based methods and able to handle any distribution in input parameters, and stochasticity in the model. Another advantage of DGSA lies in its ability to quantify asymmetric interactions. The resampling procedure additionally provides the user with an objective measure of parameter influence. Finally, tree-based SA looks very promising for subsurface applications because it presents similar advantages as DGSA, but the measure of sensitivity is subjective.

All SA methods discussed have the capability to reveal key sensitivities, but they rely on many hypotheses, which when violated may lead to erroneous interpretations. In particular, some of the SA methods assume a linear relationship between the response and the parameters, and if this condition is not met, then the obtained sensitivity values may be misleading. Rank transformation could be used in the presence of nonlinearity, but it requires monotonicity of the response to be useful. Because of the nature of the subsurface modeling, two or more parameters may have values which are correlated (such as porosity and permeability). This is a challenge for all SA techniques seen in this chapter. When parameters are correlated, the interpretation of the sensitivities for the correlated parameters is confounded, as well as the interactions with these parameters. To avoid this issue [*Iman and Helton*, 1988] propose the use of the partial correlation coefficient (PCC). The PCC measures the degree of linear relation between the input parameter x_i and the model output y after removing the linear effect of all the remaining parameters $x_{j,j\neq i}$.

In this chapter, we covered some of the SA techniques currently most applied in subsurface problems. The list is not an exhaustive set of all SA methods. Many other techniques exist and may in the future be of broader interest in applying to subsurface problems. One can note in particular the reliability methods (FORM and SORM, see *Saltelli et al.* [2000] for details or, *Jyrkama and Sykes* [2006] for application on groundwater recharge) and entropy-based measures (examples of applications include *Mishra et al.* [2009], *Pappenberger et al.* [2008], *Song et al.* [2015]). Entropy-based methods are attractive for delineating nonlinear and non-monotonic multivariate relationships compared to regression-based methods, and hence may be promising in applications to complex subsurface flow problems.

REFERENCES

Archer, G. E. B., A. Saltelli, and I. M. Sobol (1997), Sensitivity measures, anova-like techniques and the use of bootstrap, *J. Stat. Comput. Simul.*, *58*(2), 99–120.

Bar-Joseph, Z., D. K. Gifford, and T. S. Jaakkola (2001), Fast optimal leaf ordering for hierarchical clustering. *Bioinformatics*, *17*(Suppl 1), 22–29 Available from: http://bioinformatics.oxfordjournals.org/content/17/suppl_1/S22.abstract.

Bastidas, L., H. Gupta, and S. Sorooshian (1999), Sensitivity analysis of a land surface scheme using multicriteria methods, *J. Geophys. Res.*, *104*, 481–490, doi: http://onlinelibrary.wiley.com/doi/10.1029/1999JD900155.

Beven, K., and A. Binley (1992), The future of distributed models: Model calibration and uncertainty prediction, *Hydrol. Processes*, *6*, 279–298.

Borgonovo, E., and E. Plischke (2016), Sensitivity analysis: A review of recent advances, *Eur. J. Oper. Res.*, *248*(3), 869–887. Available from: http://linkinghub.elsevier.com/retrieve/pii/S0377221715005469.

Breiman, L. (2001), Random forests, *Mach. Learn.*, *45*(1), 5–32.

Breiman, L. et al. (1984), *Classification and Regression Trees*, Wadsworth, Belmont, CA.

Campolongo, F., J. Cariboni, and A. Saltelli (2007), An effective screening design for sensitivity analysis of large models, *Environ. Modell. Software*, *22*(10), 1509–1518.

Cavalcante Filho, J. S. A., Y. Xu, and K. Sepehrnoori (2015), Modeling fishbones using the embedded discrete fracture model formulation: Sensitivity analysis and history matching, *SPE Annual Technical Conference and Exhibition*, Houston, TX.

Chaney, N. W., et al. (2015), Flood and drought hydrologic monitoring: The role of model parameter uncertainty, *Hydrol. Earth Syst. Sci.*, *19*(7), 3239–3251.

Chang, F. J., and J. W. Delleur (1992), Systematic parameter estimation of watershed acidification model, *Hydrol. Processes*, *6*(1), 29–44, doi: 10.1002/hyp.3360060104.

Ciriello, V., et al. (2013), Polynomial chaos expansion for global sensitivity analysis applied to a model of radionuclide migration in a randomly heterogeneous aquifer, *Stochastic Environ. Res. Risk Assess.*, *27*(4), 945–954, doi: 10.1007/s00477-012-0616-7.

Cukier, R. I., H. B. Levine, and K. E. Shuler (1978), Nonlinear sensitivity analysis of multiparameter model systems, *J. Comput. Phys.*, *26*(1), 1–42.

Dai, C., H. Li, and D. Zhang (2014), Efficient and accurate global sensitivity analysis for reservoir simulations by use of probabilistic collocation method, *SPE J.*, *19*(4), 1–15.

Damsleth, E., A. Hage, and R. Volden (1992), Maximum information at minimum cost: A North Sea field development study with an experimental design, *J. Pet. Technol.*, *44*(12), 1350–1356.

Daniel, C. (1973), One-at-a-time plans. *J. Am. Stat. Assoc.*, *68*(342), 353–360.

De Lozzo, M., and A. Marrel (2017), Sensitivity analysis with dependence and variance-based measures for spatio-temporal numerical simulators, *Stochastic. Environ. Res. Risk Assess.*, *31*(6), 1437–1453.

Dejean, J. P., and G. Blanc (1999), Managing uncertainties on production predictions using integrated statistical methods. *SPE Annual Technical Conference and Exhibition*, Houston, TX, 3–6 October, pp. 1–15.

Dessirier, B., A. Frampton, and J. Jarsjo (2015), A global sensitivity analysis of two-phase flow between fractured crystalline rock and bentonite with application to spent nuclear fuel disposal, *J. Contam. Hydrol.*, *182*, 25–35, doi: http://dx.doi.org/10.1016/j.jconhyd.2015.07.006.

Draper, N. R., and H. Smith (1981), *Applied Regression Analysis*, Wiley, New York.

Fenwick, D., C. Scheidt, and J. Caers (2014), Quantifying asymmetric parameter interactions in sensitivity analysis: Application to reservoir modeling, *Math. Geosci.*, *46*(4), 493–511.

Feraille, M., and D. Busby (2009), Uncertainty management on a reservoir workflow. Paper IPTS 13768 presented at the International Petroleum Technology Conference, Doha, Qatar, (7–9 December), pp. 7–9.

Finsterle, S., et al. (2013), Microhole arrays for improved heat mining from enhanced geothermal systems, *Geothermics*, *47*, 104–115, doi: http://dx.doi.org/10.1016/j.geothermics.2013.03.001.

Fontaine, D. D., et al. (1992), The role of sensitivity analysis in groundwater risk modeling for pesticides, *Weed Technol.*, *6*(3), 716–724.

Francos, A., et al. (2003), Sensitivity analysis of distributed environmental simulation models: Understanding the model behaviour in hydrological studies at the catchment scale, *Reliab. Eng. Syst. Saf.*, *79*(2), 205–218.

Freer, J., K. Beven, and B. Ambroise (1996), Bayesian estimation of uncertainty in runoff production and the value of data: An application of the GLUE approach, *Water Resour. Res.*, *32*(7), 2161–2173.

Frey, H. C., and S. R. Patil (2002), Identification and review of sensitivity analysis methods, *Risk Anal.*, *22*(3), 553–578.

Gamboa, F., A. Janon, T. Klein, and A. Lagnoux (2014), Sensitivity analysis for multidimensional and functional outputs, *Electron. J. Stat.*, *8*, 575–603.

Gervais-Couplet, V., et al. (2010), Joint history matching of production and 4D-Seismic related data for a North Sea Field Case. *SPE Annual and Technical Conference*, Florence, Italy, pp. 1–18.

Glen, G., and K. Isaacs (2012), Estimating Sobol sensitivity indices using correlations, *Environ. Modell. Software*, *37*, 157–166, doi: http://dx.doi.org/10.1016/j.envsoft.2012.03.014.

Goovaerts, P. (1997), *Geostatistics for Natural Reources Evaluation*, Oxford University Press, New York.

van Griensven, A., et al. (2006), A global sensitivity analysis tool for the parameters of multi-variable catchment models, *J. Hydrol.*, *324*(1–4), 10–23.

Gwo, J. P., et al. (1996), Subsurface stormflow modeling with sensitivity analysis using a Latin hypercube sampling. *Ground Water*, *34*(5), 811–818.

Helton, J. C. (1993), Uncertainty and sensitivity analysis techniques for use in performance assessment for radioactive waste disposal, *Reliab. Eng. Syst. Saf.*, *42*(2–3), 327–367.

Helton, J. C., and F. J. Davis (2003), Latin hypercube sampling and the propagation of uncertainty in analyses of complex systems, *Reliab. Eng. Syst. Saf.*, *81*(1), 23–69.

Herman, J. D., et al. (2013), Technical note: Method of Morris effectively reduces the computational demands of global sensitivity analysis for distributed watershed models, *Hydrol. Earth Syst. Sci.*, *17*(7), 2893–2903.

Homma, T., and A. Saltelli (1996), Importance measures in global sensitivity analysis of nonlinear models, *Reliab. Eng. Syst. Saf.*, *52*, 1–17. Available from: http://www.sciencedirect.com/science/article/pii/0951832096000026.

Hossain, F., et al. (2004), Hydrological model sensitivity to parameter and radar rainfall estimation uncertainty, *Hydrol. Processes*, *18*(17), 3277–3291.

Iman, R. L., and J. C. Helton (1988), An investigation of uncertainty and sensitivity analysis techniques for computer models. *Risk Anal.*, *8*(1), 71–90.

Iooss, B., and P. Lemaître (2014), A review on global sensitivity analysis methods. *arXiv preprint arXiv:1404.2405*, (around 30), p. 23. Available from: http://arxiv.org/abs/1404.2405.

Iooss, B., and M. Ribatet (2009), Global sensitivity analysis of computer models with functional inputs. *Reliab. Eng. Syst. Saf.*, *94*(7), 1194–1204.

Jyrkama, M. I., and J. F. Sykes (2006), Sensitivity and uncertainty analysis of the recharge boundary condition, *Water Resour. Res.*, *42*(1), 1–11.

Khalid, K., et al. (2016), Application on one-at-a-time sensitivity analysis of semi-distributed hydrological model in tropical watershed, *IACSIT Int. J. Eng. Technol.*, *8*(2), 132–136.

Lence, B. J., and A. K. Takyi (1992), Data requirements for seasonal discharge programs: An application of a regionalized sensitivity analysis, *Water Resour. Res.*, *28*(7), 1781–1789.

Luo, J., and W. Lu (2014), Sobol' sensitivity analysis of NAPL-contaminated aquifer remediation process based on multiple surrogates, *Comput. Geosci.*, *67*, 110–116.

Manache, G., and C. S. Melching (2004), Sensitivity analysis of a water-quality model using Latin hypercube sampling, *J. Water Resour. Plan. Manag.*, *130*(3), 232–242.

Marrel, A., et al. (2012), Global sensitivity analysis of stochastic computer models with joint metamodels, *Stat. Comput.*, *22*(3), 833–847.

Mishra, S., N. Deeds, and G. Ruskauff (2009), Global sensitivity analysis techniques for probabilistic ground water modeling, *Ground Water*, *47*(5), 730–747.

Morris, M. D. (1991), Factorial sampling plans for preliminary computational experiments, *Technometrics*, *33*(2), 161–174.

Muleta, M. K., and J. W. Nicklow (2005), Sensitivity and uncertainty analysis coupled with automatic calibration for a distributed watershed model, *J. Hydrol.*, *306*(1–4), 127–145.

Nolan, B. T., et al. (2007), Factors influencing ground-water recharge in the eastern United States, *J. Hydrol.*, *332*(1–2), 187–205.

Nossent, J., P. Elsen, and W. Bauwens (2011), Sobol' sensitivity analysis of a complex environmental model, *Environ. Modell. Software*, *26*(12), 1515–1525. Available from: https://www.sciencedirect.com/science/article/pii/S1364815211001939.

Oladyshkin, S., F. P. J. de Barros, and W. Nowak (2012), Global sensitivity analysis: A flexible and efficient framework with an example from stochastic hydrogeology, *Adv. Water Resour.*, *37*, 10–22, doi: http://dx.doi.org/10.1016/j.advwatres.2011.11.001.

Pappenberger, F., I. Iorgulescu, and K. J. Beven (2006), Sensitivity analysis based on regional splits and regression trees (SARS-RT), *Environ. Modell. Software*, *21*(7), 976–990.

Pappenberger, F., et al. (2008), Multi-method global sensitivity analysis of flood inundation models, *Adv. Water Resour.*, *31*(1), 1–14, doi: http://linkinghub.elsevier.com/retrieve/pii/S0309170807000747.

Park, J., et al. (2016), DGSA: A Matlab toolbox for distance-based generalized sensitivity analysis of geoscientific computer experiments. *Comput. Geosci.*, *97*, 15–29, doi: http://dx.doi.org/10.1016/j.cageo.2016.08.021.

Ratto, M., et al. (2007), Uncertainty, sensitivity analysis and the role of data based mechanistic modeling in hydrology. *Hydrol. Earth Syst. Sci.*, *11*, 1249–1266.

Rohmer, J. (2014), Combining meta-modeling and categorical indicators for global sensitivity analysis of long-running flow simulators with spatially dependent inputs, *Comput. Geosci.*, *18*(2), 171–183.

Rohmer, J., and E. Foerster (2011), Global sensitivity analysis of large-scale numerical landslide models based on Gaussian-process meta-modeling, *Comput. Geosci.*, *37*(7), 917–927, doi: http://dx.doi.org/10.1016/j.cageo.2011.02.020

Rohmer, J., et al. (2016), Dynamic parameter sensitivity in numerical modelling of cyclone-induced waves: A multi-look approach using advanced meta-modelling techniques, *Nat. Hazards*, *84*(3), 1765–1792.

Rose, K. A., et al. (1991), Parameter sensitivities, Monte Carlo filtering, and model forecasting under uncertainty, *J. Forecast.*, *10*(1–2), 117–133.

Saltelli, A., and P. Annoni (2010), How to avoid a perfunctory sensitivity analysis, *Environ. Modell. Software*, *25*(12), 1508–1517, doi: http://dx.doi.org/10.1016/j.envsoft.2010.04.012

Saltelli, A., S. Tarantola, and K. P. S. Chan (1999), A quantitative model-independent method for global sensitivity analysis of model output, *Technometrics*, *41*(1), 39–56. Available from: http://amstat.tandfonline.com/doi/abs/10.1080/00401706.1999.10485594#.VsIgbvpqCso.mendeley.

Saltelli, A., K. Chan, and E. Scott (2000), *Sensitivity Analysis*, Wiley, New York.

Saltelli, A., et al. (2004), *Sensitivity Analysis in Practice: A Guide to Assessing Scientific Models*, Wiley, Chichester.

Saltelli, A., et al. (2008), *Global Sensitivity Analysis: The Primer*, Wiley, Chichester.

Sarkarfarshi, M., et al. (2014), Parametric sensitivity analysis for CO_2 geosequestration, *Int. J. Greenhouse Gas Control*, *23*, 61–71, doi: http://dx.doi.org/10.1016/j.ijggc.2014.02.003.

Sarma, P., et al. (2015), Identification of "big hitters" with global sensitivity analysis for improved decision making under uncertainty, paper presented at SPE Reservoir Simulation Symposium, Houston, TX.

Scheidt, C., and J. Caers (2009a), Representing spatial uncertainty using distances and kernels, *Math. Geosci.*, *41*(4), 397–419.

Scheidt, C., and J. Caers (2009b), Uncertainty quantification in reservoir performance using distances and kernel methods-application to a west Africa deepwater turbidite reservoir, *SPE J.*, *14*(4), 680–692.

Sobol, I. M. (1993), Sensitivity estimates for nonlinear mathematical models, *Math. Model. Comput. Exp.*, *1*, 407–414.

Song, X., et al. (2015), Global sensitivity analysis in hydrological modeling: Review of concepts, methods, theoretical framework, and applications. *J. Hydrol.*, *523*(225), 739–757, doi: http://dx.doi.org/10.1016/j.jhydrol.2015.02.013.

Spear, R. C., and T. M. Grieb (1994), Parameter uncertainty and interaction in complex environmental models, *Water Resour. Res.*, *30*(11), 3159–3169.

Spear, R. C., and G. M. Hornberger (1980), Eutrophication in peel inlet-II. Identification of critical uncertainties via generalized sensitivity analysis, *Water Res.*, *14*(1), 43–49.

Tang, Y., et al. (2007), Comparing sensitivity analysis methods to advance lumped watershed model identification and evaluation, *Hydrol. Earth Syst. Sci.*, *11*(2), 793–817. Available from: http://www.hydrol-earth-syst-sci.net/11/793/2007/.

Tong, Y., and T. Mukerji (2017), Generalized sensitivity analysis study in basin and petroleum system modeling, case study on Piceance Basin, Colorado, *J. Petrol. Sci. Eng.*, *149*, 772–781. Available from: http://linkinghub.elsevier.com/retrieve/pii/S092041051631021X.

Touzani, S., and D. Busby (2013), Smoothing spline analysis of variance approach for global sensitivity analysis of computer codes, *Reliab. Eng. Syst. Saf.*, *112*, 67–81, doi: http://dx.doi.org/10.1016/j.ress.2012.11.008.

Wainwright, H. M., et al. (2013), Modeling the performance of large-scale CO_2 storage systems: A comparison of different sensitivity analysis methods, *Int. J. Greenhouse Gas Control*, *17*, 189–205, doi: http://dx.doi.org/10.1016/j.ijggc.2013.05.007

Wainwright, H. M., et al. (2014), Making sense of global sensitivity analyses, *Comput. Geosci.*, *65*, 94–94, doi: http://dx.doi.org/10.1016/j.cageo.2013.06.006.

Wei, P., Z. Lu, and J. Song (2015), Variable importance analysis: A comprehensive review. *Reliab. Eng. Syst. Saf.*, *142*, 399–432. Available from: http://linkinghub.elsevier.com/retrieve/pii/S0951832015001672.

White, C. D., et al. (2001), Identifying and estimating significant geologic parameters with experimental design, *SPE J.*, *6*(3), 311–324.

Yang, J. (2011), Convergence and uncertainty analyses in Monte-Carlo based sensitivity analysis. *Environ. Modell. Softw.*, *26*(4), 444–457, doi: http://dx.doi.org/10.1016/j.envsoft.2010.10.007.

Zabalza-Mezghani, I., et al. (2004), Uncertainty management: From geological scenarios to production scheme optimization, *J. Petrol. Sci. Eng.*, *44*(1–2), 11–25.

Zagayevskiy, Y., and C. V. Deutsch (2015), A methodology for sensitivity analysis based on regression: applications to handle uncertainty in natural resources characterization. *Nat. Resour. Res.*, *24*(3), 239–274.

5

Bayesianism

5.1. INTRODUCTION

What really constitutes "Bayesian" modeling? Thomas Bayes did not write Bayes' rule in the form we often see it in textbooks. However, after a long time of being mostly ignored in history, his idea of using a "prior" distribution heralded a new way of scientific reasoning which can be broadly classified as Bayesianism. The aim of this chapter is to frame Bayesianism within the historical context of other forms of scientific reasoning, such as induction, deduction, falsification, intuitionism, and others. The application of Bayesianism is then discussed in the context of uncertainty quantification (UQ). This makes sense since quantifying uncertainty is about quantifying a lack of understanding or lack of knowledge. Science is all about creating knowledge. But then, what do we understand and what exactly is knowledge (the field of epistemology)? How can this ever be quantified with a consistent set of axioms and definitions, that is, if a mathematical approach is taken? Is such quantification unique? Is it rational at all to quantify uncertainty? Are we in agreement as to what Bayesianism really is?

These questions are not just practical questions toward engineering solutions, which is what most of this book is about, but aim at a deeper discussion around uncertainty. This discussion is philosophical, a discussion at the intersection of philosophy, science, and mathematics. In many scientific papers that address uncertainty in subsurface systems, or in any system for that matter, philosophical views are rarely touched upon. Many such publications would start with the "we take the Bayesian approach…," or "we take a fuzzy logic approach to…," and so on. But what entails making this decision? Quickly, papers become about algebra and calculus. Bayes or any other way of inferential reasoning is simply seen as a set of methodologies, technical tools, and computer programs. The emphasis lies on the beauty of the calculus, solving the puzzle, not on any desire of deeper understanding to what exactly one is quantifying. A pragmatic realist may state that in the end, the answer is provided by the computer codes, based on the developed calculus. Ultimately, everything is about bits and bytes and transistors amplifying or switching electronic signals, inputs and outputs. The debate is then which method is better, but such debate is only within the choices of the way of reasoning about uncertainty. That choice is rarely discussed. The paradigm is blindly accepted.

Our hope is that by being acquainted with debates about reasoning and being aware about the inherent subjectivity of making certain choices of reasoning, such as Bayesianism, will hopefully (i) allow gaining better insight into what such approaches require, what their limitations are and (ii) take things with a bit of grain of salt, perhaps, by adding some healthy self-criticism and skepticisms toward any practical method or technique. This book heavily relies on Bayes. Why? Bayes is like old medicine, we know how it works, what the side effects are, and it has been debated, tweaked, improved, and discussed since reverend Bayes' account was published by Price [*Bayes and Price*, 1763]. Let us say, it is the Prozac of UQ. It works when we know it will work, but we do not yet know what to do when it does not or what strong alternatives are, simply because these alternatives (e.g., possibility theory) have not been around that long, or been tested in practice to the extent that Bayesianism has.

Our discussion will start with a general overview of the scientific method and the philosophy of science. This by itself is a gigantic field; the purpose is not to be exhaustive but to lead readers to accessible books and publications, to provide some insight into that debate. For some readers this will be new, for the more mathematically (and philosophically) inclined this will be more common terrain. This discussion is useful in the sense that it will help introduce Bayesianism, as a way of inductive reasoning, compared to very different ways of reasoning. Bayes is popular but not accepted by all [*Earman*, 1992;

Quantifying Uncertainty in Subsurface Systems, Geophysical Monograph 236,
First Edition. Céline Scheidt, Lewis Li, and Jef Caers.

Klir, 1994; *Wang*, 2004; *Gelman*, 2008]. The philosophical discussion will then help in creating a better understanding of the Achilles heals of Bayesianism and hopefully will help the reader in designing a more thoughtful approach to UQ beyond the mere algorithmic and computer science aspects covered in the rest of this book.

5.2. A HISTORICAL PERSPECTIVE

In the philosophy of science, fundamental questions are posed such as the following: What is a "law of nature"? How much evidence and what kind of evidence should we use to confirm a hypothesis? Can we ever confirm hypotheses as truths? What is truth? Why do we appear to rely on inaccurate theories (e.g., Newtonian physics) in the light of clear evidence that they are false and should be falsified? How does science and the scientific method work? What is science and what is not (the demarcation problem)? Associated with the philosophy of science are concepts such as epistemology (study of knowledge), empiricism (the importance of evidence), induction and deduction, parsimony, falsification, paradigm, and so on, all which will be discussed in this chapter.

Aristotle (384–322 BC) is often considered to be the founder of both science and philosophy of science. His work covers many areas such as physics, astronomy, psychology, biology, and chemistry, mathematics, and epistemology. Attempting to not solely be Euro-centric, one should also mention the scientist and philosopher Ibn al-Haytham (Alhazen, 965–1040 AD), who could easily be called the inventor of the peer-review system, on which also this book is created. In the modern era, Galileo Galilei and Francis Bacon take over from the Greek philosophy of thought (rationality) over evidence (empiricism). Rationalism was continued by René Descartes. David Hume introduced the problem of induction. A synthesis of rationalism and empiricism was provided by Emanuel Kant. Logical positivism (Wittgenstein, Bertrand Russel, Carl Hempel) ruled much of the early twentieth century. For example Bertrand Russel attempted to reduce all of mathematics to logic (logicism). Any scientific theory then requires a method of verification using a logic calculus in conjunction with evidence, to prove such theory true of false. Karl Popper appeared on the scene as a reaction to this type of reasoning, replacing verifiability with falsifiability, meaning that for a method to be called scientific, it should be possible to construct an experiment or acquire evidence that can falsify it. More recently, Thomas Kuhn (and later Imre Lakatos) rejected the idea that one method dominates science. They see the evolution of science through structures, programs, and paradigms. Some philosophers, such as Feyerabend, go even further ("Against method"; [*Feyerabend*, 1993]) stating that no methodological rules really exist (or should exist).

The evolution of philosophy of science has relevance to UQ. This can be seen by simply replacing the concept of "theory" with "model," and observations/evidence with data. There is much to learn from how viewpoints toward scientific discovery differs, how they have changed, and how such change has affected our ways of quantifying uncertainty. One of the aims, therefore, of this chapter is to show that there is not really a single objective approach to UQ based on some laws or rules provided by a passive, single entity (the truth-bearing clairvoyant God!). UQ, just like other science disciplines, is dynamic and relies on interaction between data, models, and predictions and evolving views on how these components interact. It is likely that few methods covered in this book will not be used in 100 years; just consider the history of science as evidence.

The reader is referred to some excellent books on the topic. Those new to the field can consult the following:

1. Barker, G., and Kitcher, P. (2013), *Philosophy of Science: A New Introduction*, Oxford University Press.
2. Okasha, S. (2002), *Philosophy of Science: A Very Short Introduction*.
3. Chalmers, A. F. (2013), *What Is This Thing Called Science?*

Much of our treatment will follow the overviews provided in these books, the reasons, arguments, and counter-arguments, historical evolution, interwoven with personal experiences specific relevant to UQ, a topic not treated much, at least from the outset, in these books.

5.3. SCIENCE AS KNOWLEDGE DERIVED FROM FACTS, DATA, OR EXPERIENCE

Science has gained considerable credibility, including in everyday life because it is presented as "being derived from facts." It provides an air of authority, a truth that contrasts to the many uncertainties of daily life. This was basically the view with the birth of modern science in the seventeenth century. The philosophies that exalt this view are empiricism and positivism. Empiricism states that knowledge can only come from sensory experience. The common view was that (i) sensory experience produces facts to objective observers, (ii) facts are prior to theories, (iii) facts are the only reliable basis for knowledge.

Empiricism is still very much alive in the daily practice of data collection, model building, and UQ. In fact, many scientists find UQ inherently "too subjective" and of lesser standing than "data," physical theories, or numerical modeling. Many claim that decisions should be based merely on observations, not models. Our aim is here to present some serious doubt into this way of thinking.

Seeing is believing. "Data is objective, models are subjective." If facts are to be derived from sensory experience, mostly what we see, then consider Figure 5.1. Most readers see a panel of squares, perhaps from a nice armoire. Others (very few) see circles and perhaps will interpret this an abstract piece of art with interesting geometric pattern. Those who do not see circles at first, need to simply look longer, with different focusing of their retinas.

Hence, there seems to be more than meets the eyeball [*Hanson*, 1958]. Consider another example in Figure 5.2. What do you see? Most of us reading this book will

Figure 5.1 How many circles do you see?

recognize this as a section of a geophysical image (seismic, ground-penetrating radar (GPR), etc.). A well-trained geophysicist will observe potentially a "bright spot" which may indicate the presence of a gas (methane, carbon dioxide) in the subsurface formations. A sedimentologist may observe deltaic formations consisting of channel stacks. Hence, the experience in viewing an object is highly dependent on the interpretation of the viewer and not on the pure sensory light perceptions hitting one's retina. In fact, Figure 5.2 is a modern abstract work of art by Mark Bardford (1963) on display in the San Francisco Museum of Modern Art (September 2016).

Anyone can be trained to make interpretations and this is usually how education proceeds. Even pigeons can be trained to spot cancers as well as humans [*Levenson et al.*, 2015]. But this idea may also backfire. First off, the experts may not do better than random (*Financial Times*, 31 March 2013: "Monkey beats man on stock market picks", based on a study by the Cass Business School in London), or worse produce cognitive biases, as pointed out by a study of interpretation seismic images [*Bond et al.*, 2007].

First facts, then theory. Translated to our UQ realm as "first data, then models." Let us consider another example in Figure 5.3, now with an actual geophysical image and not a painting. A statement of fact would then be "this is a bright spot." Then, in the empiricist view, deduction, conclusions can be derived from it ("it contains gas"). However, what is relevant here is the person making this statement. A lay person will state as fact: "There are squiggly lines." This shows that any observable fact is influenced by knowledge ("the theory") of the object of study. Statements of facts are, therefore, not simply recordings of visual perceptions. Additionally, quite an amount of knowledge is needed to undertake the geophysical survey in the first place; hence, facts do not proceed theory. This is the case for the example here

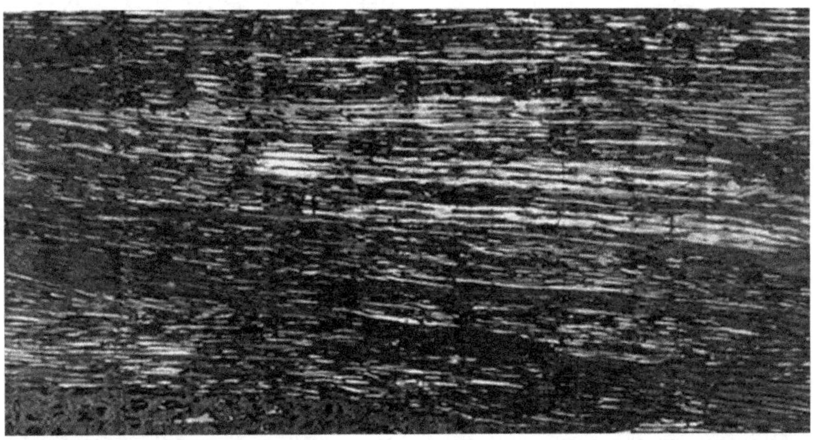

Figure 5.2 What do you see?

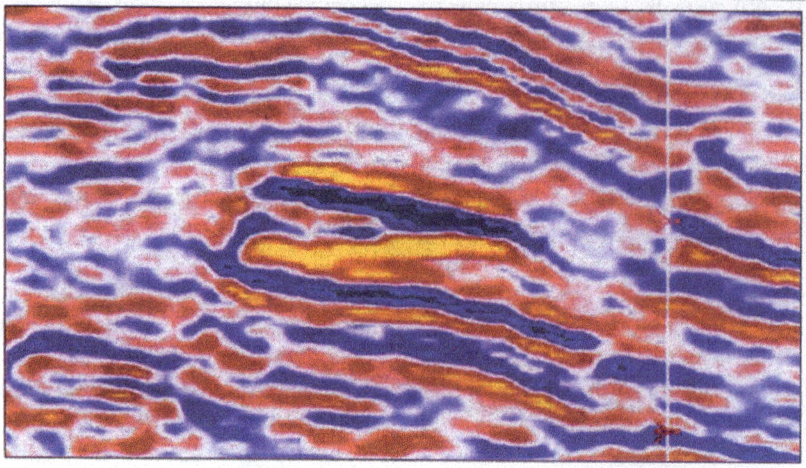

Figure 5.3 No art, just a geophysical image.

and is a reality for many scientific discoveries (we need to know where to look). A more nuanced view, therefore, is that data and models interact with each other.

Facts as basis for knowledge. "Data precedes the model." If facts depend on observers resulting in inherently subjective statements, then, can we trust data as a prerequisite to models (data precede models)? It is now clear that data does not come without a model itself, and hence if the wrong "data model" is used, then the data will be used to build incorrect models. "If I jump in the air and observe that I land on the same spot, then "obviously" the Earth is not moving under my feet." Clearly, the "data model" used here is lacking the concept (theory) of inertia. This again reinforces the idea that in modeling, and in particular UQ, data does not and should precede the model, or that one is subjective and the other somehow is not.

5.4. THE ROLE OF EXPERIMENTS: DATA

Progress in science is usually achieved by experimentation, the acquisition of information in a laboratory or field setting. Since "data" is central to UQ, we spend some time on what "data" are, what "experiments" aim to achieve, and what the pitfalls are in doing so.

First, the experiment is not without the "experimenter." Perceptual judgments may be unreliable, and hence such reliance needs to be minimized as much as possible. For example, in Figure 5.4, the uninformed observer may notice that the moon is larger when on the horizon, compared to higher up in the sky, which is merely an optical illusion (on which there still is no consensus as to why). Observations are, therefore, said to be both objective and fallible. Objective in the sense that they are shared (in public, presentations, papers, online) and subject to further tests (such as measuring the actual moon size by means of instruments, revealing the optical illusion). Often such progress happens when more advanced ways of testing or gathering data occur.

Believing that a certain amount and type of data will resolve all uncertainty and lead to determinism on which "objective" decisions can be based is an illusion because the real world involves many kinds of physical/chemical/biological process that cannot be captured by one way of experimentation. For example, performing a conservative tracer test, to reveal better hydraulic conductivity, may in fact be influenced by reactions in the subsurface taking place while doing such experiment. Hence, the hydraulic conductivity inferred through some modeling without geochemical reactions may provide a false sense of certainty about the information deduced from such experiment. In general, it is very difficult to isolate a specific target of investigation in the context of one type of experiment or data acquisition. A good example is in the interpretation of 4D geophysics (repeated geophysics). The idea of the repetition is to remove the influence of those properties that do not change over time, and therefore reveal only those that do change, for example, change in pressure, change in saturation, temperature, and so on. However, many processes may be at work at the same time, a change in pressure, saturation, rock compressibility, even porosity and permeability, geomechanical effects, and so on. Hence, someone interested in the movement of fluids (change in saturation) is left with a great deal of difficulty in unscrambling the time signature of geophysical sensing data. Furthermore, the inversion of data into a target of interest often ignores all these interacting effects. Therefore, it does not make sense to state that a pump test or a well test reveals permeability, it only reveals a pressure change under the conditions of the test and of

Figure 5.4 The harvest moon appearing gigantic as compared to the moon in the high sky. *Source:* https://commons.wikimedia.org/wiki/File:Harvest_Moon_over_looking_vineyards.jpg.

the site in question, and many of these conditions may remain unknown or uncertain.

Practitioners of UQ are often unaware of the difficulty in acquiring data, whether in the lab or the field. Those who perform UQ are often the "modelers" and their lab is the computer. The data are given to them, and they use it "as is." In Chapter 1, we discussed the use of ERT (electrical resistivity tomography) monitoring data in designing heat storage systems. Theoretically, this is a very worthy idea, but the practical implementation stands or falls with the experience of the experimenter. For example, if the injection and pumping wells are aligned on the ERT profile and the pumped water is used for re-injection, then an electric short-circuit is created rendering the collected data useless for monitoring the storage system [*Hermans et al.*, 2015]. This can also happen with perpendicular profiles when they are close to the injection or pumping well. In addition, problems may exist with the proper placement of sensors. Another confounding factor is noise. In the ERT case, and in most geophysical surveys, it is not just the presence of noise (random noise can be easily dealt with), but the presence of correlated noise whose correlation structure changes in time. Hence, such noise is nonstationary; no simple repetition occurs that allows to easily remove it, as is the case with random noise. Having an incorrect noise characterization or simply making the wrong assumptions may lead to meaningless results when such data is inverted. The problem often is that very little communication exists between the "experimenter" and the "modeler" (often two different persons).

A final issue that arises in experimentation is the possibility of a form of circular reasoning that may exist between an experimental setup and a computer model aiming to reproduce the experimental setup. If experiments are to be conducted to reveal something important about the subsurface (e.g., flow experiments in a lab), then often the results of such experiments are "validated" by a computer model. Is the physical/chemical/biological model implemented in the computer code derived from the experimental result, or are the computer models used to judge the adequacy of the result? Do theories vindicate experiments and do experiments vindicate the stated theory? To study these issues better, we introduce the notion of induction and deduction.

5.5. INDUCTION VERSUS DEDUCTION

Bayesianism is based on inductive logic [*Howson*, 1991; *Howson et al.*, 1993; *Chalmers*, 1999; *Jaynes*, 2003; *Gelman et al.*, 2004], although some argue that it is based both on induction and on deduction [*Gelman and Shalizi*, 2013]. Given the above consideration (and limitations) of experiments (in a scientific context) and data (in a UQ context), the question now arises on how to derive theories from these observations. Scientific experimentation, modeling, studies often rely on a logic to make certain claims. Induction and deductions are such kinds of logic. What such logic offers is a connection between premises and conclusions:

1. All deltaic systems contain clastic sands.
2. The subsurface system under study is deltaic.
3. The subsurface system contains clastic sands.

This logical deduction is obvious, but such logic only establishes a connection between premises 1 and 2 and the conclusion 3, it does not establish the truth of any of these statements. If that would be the case, then the following is also equally "logic":

1. All deltaic systems contain steel.
2. The subsurface system under study is deltaic.
3. The subsurface system contains steel.

The broader question, therefore, is if scientific theories can be derived from observations. The same question occurs in the context of UQ: can models be derived from data. Consider an experiment consisting of a set of n experiments:

Premises:

1. The reservoir rock is water-wet in sample 1.
2. The reservoir rock is water-wet in sample 2.
3. The reservoir rock is water-wet in sample 3.

...

20. The reservoir rock is water-wet in sample 20.

Conclusion: the reservoir is water-wet (and hence not oil-wet).

This simple idea is mimicked from Bertrand Russel's turkey argument (in his case it was a chicken). "I (the turkey) am fed at 9am" day after day, hence "I am always fed at 9am", until the day before Thanksgiving [*Chalmers*, 1999]. Another form of induction occurred in 1907: "But in all my experience, I have never been in any accident ... of any sort worth speaking about. I have seen but one vessel in distress in all my years at sea. I never saw a wreck and never have been wrecked nor was I ever in any predicament that threatened to end in disaster of any sort" [*E. J. Smith*, 1907, Captain, RMS Titanic].

Any model or theory derived from observations can never be proven in the sense as being derived from it.

This does not mean that induction (deriving models from observations) is completely useless. Some inductions are more warranted than others. Specifically, in the case when the observations set is "large," and performed under a "wide variety of conditions," although these qualitative statements depend clearly on the specific case. "When I swim with hungry sharks, I get bitten," this really needs to be asserted only once.

The second qualification (variety of conditions) requires some elaboration because we will return to it when discussing Bayesianism. The conditions that are being tested are important (the age of the driller, e.g., is not); hence, in doing so we rely on some prior knowledge of the particular model or theory being derived. Such prior knowledge will determine which factors will be studied, which are influencing the theory/model and which not. Hence, the question is how this "prior knowledge" itself is asserted by observations. One runs into the never-ending chain of what prior knowledge is used to derive prior knowledge. This point was made clear by David Hume, an eighteenth century Scottish philosopher [*Hume*, 1978, originally 1739]. Often, the principle of induction is argued because it has "worked" from experience. The reader need simply replace the example of the water-wet rocks with "Induction has worked in case j," and so on, to understand that induction is, in this way, "proven" by means of induction. The way out of this "mess" is to not make true/false statements, but to use induction in a probabilistic sense (probably true), a point we will return to when addressing Bayesianism.

5.6. FALSIFICATIONISM

5.6.1. A Reaction to Induction

Falsificationism, as championed by *Popper* [1959], appeared in the 1920s partly as a reaction to inductionism (and logical positivism). Popper claimed that science should not involve any induction (theories derived from observations). Instead, theories are seen as speculative or tentative, as created by the human intellect, usually to overcome limitations of previous theories. Once stated, such theories need to be tested rigorously with observations. Theories that are inconsistent with such observation should be rejected (falsified). The theories that survive are the best current theories. Hence, falsificationism has a time component and aims to describe progress in science, where new theories are born out of old ones by a process of falsification.

In terms of UQ, one can then see models not as true representations of actual reality but as hypotheses. One has as many hypotheses as models. Such hypothesis can be constrained by previous knowledge, but real field data should not be used to confirm a model (confirmation with data) but to falsify a model (reject, the model does not confirm with data).

A simple example illustrates the difference:

Induction:

Premise: All rock samples are sandstones.

Conclusion: The subsurface system contains only sandstone.

Falsification:

Premise: A sample has been observed that is shale.

Conclusion: The subsurface system does not consist just of sandstone.

The latter is clearly a logically valid deduction (true). Falsification, therefore, can only proceed with hypotheses that are falsifiable (this does not mean that one has the observations needed to falsify, but that such observation could exist). Some hypotheses are not falsifiable, for example, "the subsurface system consists of rock that are sandstone or not sandstone." This then raises the question of the degree of falsifiability of a hypothesis and the strength (precision) of the observation in falsifying. Not all hypotheses are equally falsifiable and not all observations should be treated on the same footing. A strong hypothesis is one that makes strong claims, and there is a difference between the following two statements:

1. Significant accumulation in the Mississippi delta requires the existence of a river system.
2. Significant accumulation in all deltas requires the existence of a river system.

Clearly 2 has more consequences than 1. Falsification, therefore, invites stating bold conjectures rather than safe conjectures.

The latter has considerable implication in UQ and model building. Inductionists tend to bet on one model, the best possible, best explaining most observations, within a static context, without the idea that the model they are building will evolve. Inductionists evolve their models, but that is not the outset of their viewpoint, there is always the hope that the best possible model will remain the best possible. The problem with this inductionist approach is that new observations that cannot be fit into the current model are used to "fix" the model with ad-hoc modification. A great example of this can be found in the largest oil reservoir in the world, namely the Ghawar field [see *Simmons*, 2013]. Before 2000, most modelers (geologists, geophysicist, engineers) did not consider fractures as being a driving heterogeneity for oil production. However, flow meter observations in wells indicated significant permeability. To account for this, the already existing models with already large permeabilities (1000–10,000 mD) were modified to as much as 200D (see Figure 5.5). While this dramatic increase in permeability in certain zones did lead to fitting the flow meter data

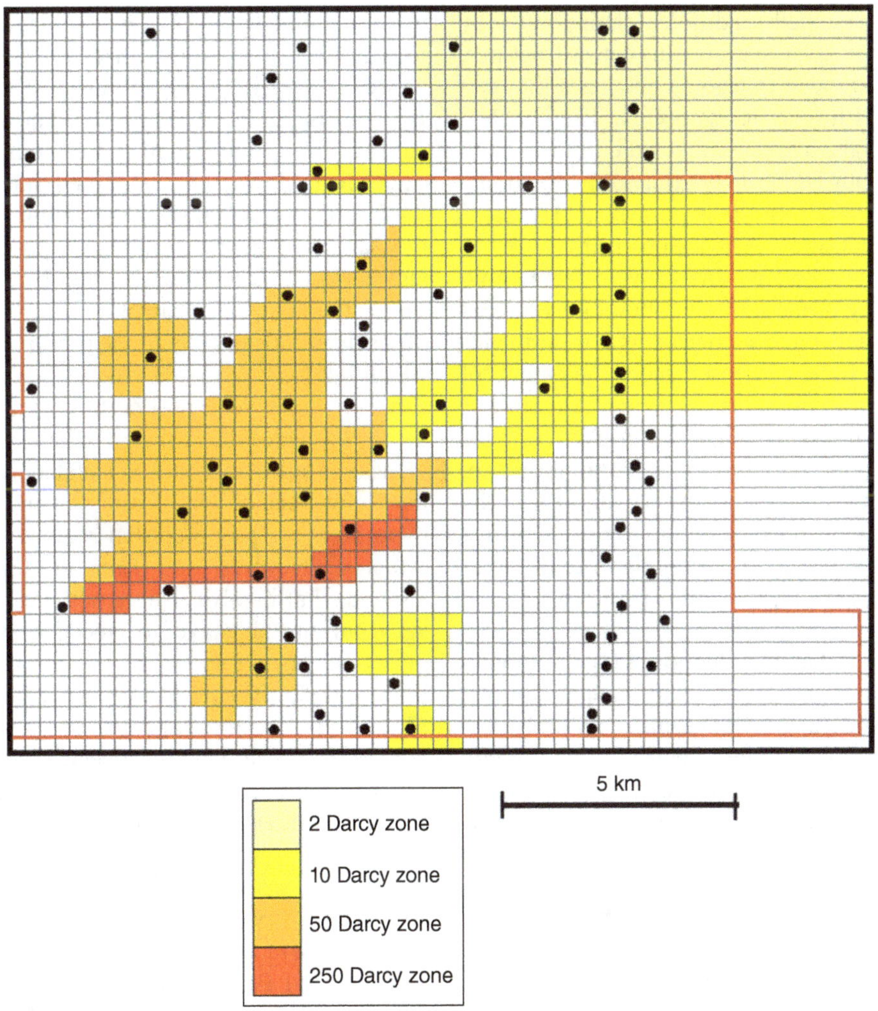

Figure 5.5 A reservoir modeled developed to reflect super permeability channels. Note the legend with permeability values [*Valle et al.*, 1993].

better, the ad-hoc modification cannot be properly tested with the current observations. It is just a fix to the model (the current "theory" of no fractures). Instead, a new test would be needed, such as new drilling to confirm or not the presence of a gigantic cave, that can explain such ridiculous permeability values. Today, all models built on the Ghawar field contain fractures.

Falsificationism does not use ad-hoc modification, because the ad-hoc modification cannot be falsified. In the Ghawar case, the very notion of fluid flow by means of an unrealistically large matrix permeability tells the falsificationist that bold alternative modifications to the theory are needed and not simple ad-hoc fixes, in the same sense that science does not progress by means of fixes. An alternative to the inductionist approach in Ghawar could be as follows: most fluid flow is caused by large permeability, except in some area where it is hypothesized that fractures are present despite the fact that we have not directly observed them. The falsificationist will now proceed by finding the most rigorous (new) test to test this hypothesis. This could consist of acquiring geomechanical studies of the system (something different than flow) or by means of geophysical data that aims to detect fractures. New hypotheses also need to lead to new tests that can falsify them. This is how progress occurs. The problem often is "time," a falsificationist takes the path of high risk, high gain, but time may run out on doing experiments that falsify certain hypotheses. "Failures" are often seen as that and not as lessons learned. In the modeling world, one often shies away from bold hypotheses (certainly if one wants to obtain government research funding!) and those modelers, as a group tends to gravitate toward some consensus under the banner of being good at "team-work." It is the view of the authors that such prohibits a realistic UQ. UQ needs to include bold hypotheses, model conjectures that are not the norm, or based on any majority vote, or by playing it safe, by being conservative. Uncertainty cannot be reduced by just great team-work, it will require equally rigorous observations (data) that can falsify any (preferably bold) hypothesis.

This does not mean that an inductionist type of modeling and falsification type of modeling cannot coexist. Inductionism leads to cautious conjectures and falsification to bold conjectures. Cautious conjectures may carry little risk, and hence, if they are falsified, then insignificant advance is made. Similarly, if bold conjectures cannot be falsified with new observations, significant advance is made. Important in all this is the nature of the background knowledge (recall, the prior knowledge), describing what is currently known about what is being studied. Any "bold" hypothesis is measured against such background knowledge. Likewise, the degree to which observations can falsify hypothesis need to be measured against such knowledge. This background knowledge changes over time (what was bold in 2000 needs no longer to be bold in 2015), and such change, as we will discuss, is explicitly accounted for in Bayesianism.

5.6.2. Falsificationism in Statistics

Schools of statistical inference are sometimes linked to the falsificationist views of science, in particular the work of Fischer, Neyman and Pearson; all well-known scientists in the field of (frequentists) statistics [*Fisher and Fisher*, 1915; *Fisher*, 1925; *Neyman and Pearson*, 1967; *Rao*, 1992; *Pearson et al.*, 1994; *Berger*, 2003; *Fallis*, 2013 for overviews and original papers]. Significance tests and confidence intervals p-values are associated with a hypothetico-deductive way of reasoning. Since these methods are pervasive in all areas of science, in particular in UQ, we present some discussion on their rationality together with opposing views of inductionism within this context.

Historically, Fisher can be seen as the founder of modern statistics. His work has a falsificationist foundation, steeped in statistical "objectivity" (lack of needed subjective assumption, which is the norm in Bayesian methods). The now well-known procedure starts with stating a null-hypothesis (a coin is fair), define an experiment (flipping), a stopping rule (e.g., number of flips), and a test-statistic (e.g., number of heads). Next the sampling distribution (each possible value of the test-statistic), assuming the null-hypothesis is true, is calculated. Then, we calculate a probability p that our experiment falls in an extreme group (e.g., 4 heads or less which has only probability of 1.2% for 20 flips). Then a convention is taken to reject (falsify) the hypothesis when the experiment falls in the extreme group, say, $p \leq 0.05$.

Fisher's test works only on isolated hypotheses, which is not how science progresses; often many competing hypotheses are proposed that require testing under some evidence. Neyman and Pearson developed statistical methods that involve rival hypotheses, but again reasoning from an "objective" perspective, without relying on priors or posteriors of Bayesian inductive reasoning. For example, in the case of two competing hypothesis H_1 and H_2, Neyman and Pearson reasoned that either hypotheses are accepted or rejected, leading to two kinds of errors (stating that one is false, while the other is false and vice versa), better known as type I and II errors. Neyman and Pearson improved on Fischer in defining better "low probability." In the coin example, a priori, any combination of 20 tosses have a probability of 2^{-20}, even under a fair coin, most tosses have small probability. Neyman and Pearson provide some more definition of this critical region (where hypotheses are rejected). If X is the random variable describing the outcome (e.g., a combination of tosses), then the outcome space is defined by the following inequality:

$$L(X) = \frac{P(X|H_1)}{P(X|H_2)} \leq \delta \quad P(L(X) \leq \delta | H_1) = \alpha \tag{5.1}$$

with δ depending on the significance level α and the nature of the hypothesis. This theorem known as the Fundamental Lemma [*Neyman and Pearson*, 1933] defines the most powerful test to reject H_1 in favor of H_2 at significance level α for a threshold δ. The interpretation of a likelihood ratio was provided by Bayesianists as the Bayes' factor (the evidential force of evidence). This was, however, not the original interpretation of Neyman and Pearson.

What then does a significance test tell us about the truth (or not) of a hypothesis? Since the reasoning here is in terms of falsification (and not induction), Neyman–Person interpretation is that if a hypothesis is rejected, then "one's actions should be guided by the assumption that it is false" [*Lindgren*, 1976]. Neyman and Pearson admit that significance test tells nothing about whether a hypothesis is true or not. However, they do attach the notion of "in the long run," interpreting the significance level as, for example, the number of times in 1000 times that the same test is being done. The problem here is that no testing can be done and will be done in exactly the same fashion, under the exact same circumstances. This idea would also invoke the notion that under a significance level of 0.05, a *true* hypothesis would be rejected with the probability of 0.05. The latter violates the very reason on which significance tests were formed: events with the probability p can never be proven to occur (that requires subjectivity!), let alone with the exact frequency of p.

The point here is to show that modern statistics need not be seen as purely falsificationist, a logical hypothetic-deductive way of reasoning. Reasoning in statistics comes with its own subjective notions of personal judgments (choosing which hypothesis, what significance level, stopping rules, critical regions, independence assumptions, Gaussian assumptions, etc.). This was in fact later acknowledged by Pearson himself [*Neyman and Pearson*, 1967, p. 277].

5.6.3. Limitations of Falsificationism

Falsificationism comes with its own limitations. Just as induction cannot be induced, falsificationism cannot be falsified, as a theory. This becomes clearer when considering real-world development of models or theories. The first problem is similar to the one discussed in using inductive and deductive logic. Logic only works if the premises are true; hence falsification, as a deductive logic cannot distinguish between a faulty observation and a faulty hypothesis. The hypotheses do not have to be false when inconsistent with observations, since observations can be false. This is an important problem in UQ that we will revisit later.

The real world involves considerably more complication than "the subsurface system is deltaic." Let us return to our example of monitoring heat storage using geophysics. A problem that is important in this context is to monitor whether the heat plume remains near the well and is compact, does not start to disperse, since then recovery of that heat becomes less efficient. A hypothesis could then be "the heat plume is compact," geophysical data can be used to falsify this by, for example, observing that the heat plume is indeed influenced by heterogeneity. Unfortunately, such data does not directly observe "temperature," instead it might measure electrical resistance, which is related to temperature and other factors, including survey design. Additionally, because monitoring is done from a distance of the plume (at the surface), the issue of limited resolution occurs (any "remote sensing" suffers from this limited resolution). This is then manifested in the inversions of the ERT data into temperature, since many inversion techniques result in smooth versions of actual reality (due to this limited resolution issue), from which the modeler may deduce that homogeneity of the plume is not falsified. Where now lies the error? In the instrumentation? In the instrumentation setup? In the initial and boundary conditions that are required to model the geophysics? In the assumptions about geological variability? In the smoothness constraint in the inversion process? Falsification does not provide a direct answer to this. In science, this problem is better known as the Duhem–Quine thesis after Pierre Duhem and Willard Quine [*Ariew*, 1984]. This thesis states that it is impossible to falsify a scientific hypothesis in isolation, because the observations required for such falsification themselves rely on additional assumptions (hypotheses) that cannot be falsified separately from the target hypothesis (or vice versa). Any particular statistical method that claims to do so ignores the physical reality of the problem.

A practical way to deal with this situation is to not consider just falsification but sensitivity to falsification. What impacts the falsification process? Sensitivity, even with limited or approximate physical models, provides more information that can lead to (i) changing the way data is acquired (the "value of information" in Chapter 3) and (ii) changing the way the physics of the problem (e.g., the observations) is modeled by focusing on what matters most toward testing the hypothesis (the target, see Chapter 1).

More broadly, falsification does not really follow the history of the scientific method. Most science has not been developed by means of bold hypotheses that are then falsified. Instead, theories that are falsified are carried through history; most notably, because observations that appear to falsify the theory can be explained by means of causes other than the theory that was the aim of falsification. This is quite common in modeling too: observations are used as claims that a specific physical model does not apply, only to discover later on that the physical model was correct but that data could be explained by some other factor (e.g., a biological reason, instead of a physical reason). Popper himself acknowledged this dogmatism (hanging onto models that have "falsified" to "some degree"). As we will see later, one of the problems in the application of probability (and Bayesianism) is that zero probability models are deemed "certain" to not occur. This may not reflect the actual reality that models falsified under such Popper–Bayes philosophy become "unfalsified" later by new discoveries and new data. Probability and "Bayesianism" are not at fault here, but the all-too common underestimation of uncertainties in many applications.

5.7. PARADIGMS

5.7.1. Thomas Kuhn

From the previous presentation, one may argue that both induction and falsification provide too much of a fragmented view of the development of scientific theory or methods that often do not agree with reality. Thomas Kuhn, in his book *The Structure of Scientific Revolution* [*Kuhn*, 1996] emphasizes the revolutionary character of scientific methods. During such revolution, one abandons one "theoretical" concept for another, which is incompatible with the previous one. In addition, the role of scientific communities is more clearly analyzed. Kuhn describes the following evolution of science

$$\text{Paradigm} \rightarrow \text{crises} \rightarrow \text{revolution} \rightarrow \text{new paradigm} \rightarrow \text{new crisis}$$

Such a single paradigm consists of certain (theoretical) assumptions, laws, methodologies, and applications adapted by members of a scientific community (e.g., evolution, plate tectonics, genetics, relativity theory). Probabilistic methods, or Bayesian methods, can be seen as such a paradigm: they rely on axioms of probability and the definition of a conditional probability, the use of prior information, subjective beliefs, maximum entropy, principle of indifference, algorithms of McMC, and so on. Researchers within this paradigm do not question the fundamentals of such paradigm, the fundamental laws or axioms. Activities within the paradigm are then puzzle-solving activities (e.g., studying convergence of a Markov chain) governed by the rules of the paradigm. Researchers within the paradigm do not criticize the paradigm. It is also typical that many researchers within that paradigm are unaware of the criticism on the paradigm or ignorant as to the exact nature of the paradigm, simply because it is a given: Who is really critical of the axioms of probability when developing Markov chain samplers? Or, who questions the notion of conditional probability when performing stochastic inversions? Puzzles that cannot be solved are deemed to be anomalies, often attributed to the lack of understanding of the community about how to solve the puzzle within the paradigm, rather than a question about the paradigm itself. Kuhn considers such unsolved issues as anomalies rather than what Popper would see as potential falsifications of the paradigm. The need for greater awareness and articulation of the assumptions of a single paradigm becomes necessary when the paradigm requires defending against offered alternatives.

Within the context of UQ, a few such alternative paradigms have emerged reflecting the concept of revolution as Kuhn describes. The most "traditional" of paradigms for quantifying uncertainty is by means of probability theory and its extension of Bayesian probability theory (the addition of a definition of conditioning). We provide here a summary account of the evolution of this paradigm, the criticism leveled, the counterarguments, and the alternatives proposed, in particular possibility theory.

5.7.2. Is Probability Theory the Only Paradigm for UQ?

5.7.2.1. The Axioms of Probability: Kolmogorov–Cox. The concept of numerical probability emerged in the mid-seventeenth century. A proper formalization was developed by *Kolmogoroff* [1950] based on classical measure theory. A comprehensive study of its foundations is offered in *Fine* [1973]. This topic is vast and of particular note are books by *Fine* [1973], *Feller* [2008], *Gnedenko et al.* [1962], *de Finetti et al.* [1975], *de Finetti* [1974, 1995], and *Jaynes* [2003]. Also of note is the work of *Shannon* [1948] on uncertainty-based information in probability. In other words, the concept of probability has been around for three centuries. What is probability? It is now generally agreed (the fundamentals of the paradigm) that the axioms of Kolmogorov as well as the Bayesian interpretation by *Cox* [1946] form the basis. Since most readers are unfamiliar with Cox theorem and its consequences for interpreting probability, we provide some high-level insight.

Cox works from a set of postulates, for example (we focus on just two of three postulates),

1. "A proposition p and its negation $\neg p$ is certain" or $\text{plaus}(p \cap \neg p) = 1$ which is also termed the logical principle of the excluded middle. plaus stands for plausibility.

2. Consider now two propositions p and q and the conjunction between them $p \cap q$. This postulate states that the plausibility of the conjunction is only function of the plausibility of p and the plausibility of q given that p is true. In other words,

$$\text{plaus}(p \vee q) = f(\text{plaus}(p), \text{plaus}(q|p)).$$

The traditional laws are recovered when setting plaus to be a probability measure or P, or stating as per Cox theorem "any measure of belief is isomorphic to a probability measure." This seems to suggest that probability is sufficient in dealing with uncertainty, nothing else is needed (due to this isomorphism). The consequence is that one can now perform calculations (a calculus) with "degrees of belief" (subjective probabilities, see Section 5.8.3) and even mix probabilities based on subjective belief with probabilities based on frequencies. The question is, therefore, whether these subjective probabilities are the only legitimate way of calculating uncertainty? For one, probability requires that either the fact is there or it is not there, nothing is left in the "middle." This then necessarily means that probability is ill-suited in cases where the excluded middle principle of logic does not apply. What are those cases?

5.7.2.2. Intuitionism. Probability theory is truth driven. An event occurs or does not occur. The truth will be revealed. From a hard scientific or engineering approach this seems perfectly fine, but it is not. A key figure in this criticism is the Dutch mathematician and philosopher Jan Brouwer. Brouwer founded the mathematical philosophy of intuitionism countering the then-prevailing formalism, in particular of David Hilbert as well as Bertrand Russell claiming that mathematics can be reduced to logic; the epistemological value of mathematical constructs lies in the fundamental nature of this logic.

In simplistic terms perhaps, intuitionists do not accept the law of excluded middle in logic. Intuitionism reasons from the point that science (in particular mathematics) is the result of the mental construction performed by humans rather than principles founded in the actual objective reality. Mathematics is not "truth," rather it constitutes applications of internally consistent methods used to realize more complex mental constructs, regardless of their possible independent existence in an objective reality. Intuition should be seen in the context of logic as the ability to acquire knowledge without proof or without understanding how the knowledge was acquired.

Classical logic states that existence can be proven by refuting nonexistence (the excluded middle principle). For the intuitionist, this is not valid; negation does not entail falseness (lack of existence), it entails that the statement is refuted (a counter example has been found). For an intuitionist, a proposition p is stronger than a statement of not (not p). Existence is a mental construction, not proof of nonexistence. One particular form and application of this kind of reasoning is fuzzy logic.

5.7.2.3. Fuzzy Logic. It is often argued that epistemic uncertainty (or knowledge) does not cover all uncertainty (or knowledge) relevant to science. One such particular form of uncertainty is "vagueness" which is borne out of the vagueness contained in language (note that other language-dependent uncertainties exists such as "context-driven"). This may seem rather trivial to someone in the hard sciences, but it should be acknowledged that most language constructs ("this is air," meaning 78% nitrogen, 21% oxygen, and <1% of argon, carbon dioxide, and other gases) are a purely theoretical construct, of which we still may not have incomplete understanding. The air that is outside is whatever that substance is, it does not need human constructs, unless humans use if for calculations, which are themselves constructs. Unfortunately, (possibly flawed) human constructs is all that we can rely on.

The binary statements "this is air" and "this is not air" again are theoretical human constructs. Setting that aside, most of the concepts of vagueness are used in cases with unclear borders. Science typically works with classification systems ("this is a deltaic deposit," "this is a fluvial deposit"), but such are again man-made constructs. Nature does not decide to "be fluvial," it expresses itself through laws of physics, which are still not fully understood.

A neat example presents itself in the September 2016 edition of EOS: "What is magma?" Most would think this is a solved problem, but it is not, mostly due to vagueness in language and the ensuing ambiguity and difference in interpretation by even experts. A new definition is offered by the authors: "*Magma*: naturally occurring, fully or partially molten rock material generated within a planetary body, consisting of melt with or without crystals and gas bubbles and containing a high enough proportion of melt to be capable of intrusion and extrusion."

Vague statements ("this may be a deltaic deposit") are difficult to capture with probabilities (it is not impossible but quite tedious and construed). A problem occurs in setting demarcations. For example, in air pollution, one measures air quality using various indicators such PM2.5, meaning particles which pass through a size-selective inlet with a 50% efficiency cut off at 2.5 μm aerodynamic diameter. Then standards are set, using a cutoff to determine what is "healthy" (a green color), what is "not so healthy" (orange color), and unhealthy (a red color). Hence, if the particular matter changes by one single particle, then the air goes suddenly from "healthy" to "not so healthy" (from green to orange).

In several questions of UQ, both epistemic and vagueness-based uncertainty may occur. Often, vagueness uncertainty exists at a higher-level description of the system, while epistemic uncertainty may then deal with questions of estimation because of limited data within the system. For example, policy makers in the environmental sciences may set goals that are vague, such as "should not exceed critical levels." Such a vague statement then needs to be passed down to scientist who is required to quantify risk of attaining such levels by means of data and numerical models, where epistemic uncertainty comes into play. In that sense, there is no need to be rigorously accurate, for example, according to a very specific threshold, given the above argument about such thresholds and classification systems.

Does probability easily apply to vagueness statements? Consider a proposition "the air is borderline unhealthy." The rule of the excluded middle no longer applies because we cannot say that the air is either not unhealthy or unhealthy. Probabilities no longer sum to one. It has, therefore, been argued that the propositional logic of probability theory needs to be replaced with another logic: fuzzy logic (although other logics have been proposed such as intuitionistic, trivalent logic, we will limit the discussion to this one alternative).

Fuzzy logic relies on fuzzy set theory [*Zadeh*, 1965, 1975, 2004]. An example of fuzzy set A such as "deltaic" is said to be characterized by a membership function $\mu_{\text{deltaic}}(u)$ representing the degree of membership given some information u on the deposit under study, for example $\mu_{\text{deltaic}}(\text{deposit}) = 0.8$ for a deposit with info u under study. Probabilists often claim that such membership function is nothing more than a conditional probability $P(A|u)$ in disguise [*Loginov*, 1966]. The link is made using the following mental construction. Imagine 1000 geologists looking at the same limited info u and then voting whether the deposit is "deltaic" or "fluvial." Let us assume these are the two options available. $\mu_{\text{deltaic}}(\text{deposit}) = 0.832$ means that 832 geologists picked "deltaic" and hence a vote picked at random has 83.2% chance of being deltaic. However, the conditional probability comes with its limitations as it attempts to cast a very precise answer into what is still a very vague concept. What really is "deltaic"? Deltaic is simply a classification made by humans to describe a certain type of depositional system subject to certain geological processes acting on it. The result is a subsurface configuration, termed architecture of clastic sediments. In modeling subsurface systems, geologists do not directly observe the processes (the deltaic system) but only the record of it. However, there is still no full agreement as to what is "deltaic" or when "deltaic" ends and "fluvial" starts as we go more upstream? (Recall our discussion on "magma.") What are the processes actually happening and how all this gets turned into a subsurface system? Additionally, geologist may provide interpretations based on personal experiences, different education (schools of thought about "deltaic"), and different education levels. What then does 0.832 really mean? What is the meaning of the difference between 0.832 and 0.831? Is this due to education? Misunderstanding or disagreement on the classification? Lack of data provided? It clearly should be a mix of all this, but probability does not allow an easy discrimination. We find ourselves again with a Duhem–Quine problem.

Fuzzy logic does not take the binary route of voting up or down, but it allows a grading in the vote of each member, meaning that it allows for more gradual transition between the two classes for each vote. Each person takes the evidence at his/her value and makes a judgment based on their confidence, education level: I do not really know, hence 50/50, I am pretty certain, hence 90/10. (More advanced readers in probability theory may see now a mixture models of probability stated based on the evidence what the u is. However, because of the overlapping nature of how evidence is regarded by each voter, these prior probabilities are no longer uniform.)

5.7.2.4. The Dogma of Precision. Clearly, probability theory (randomness) does not work well when the event itself is not clearly defined, subject to discussion. Probability theory does not support the concept of a fuzzy event; hence, such information (however vague and incomplete) becomes difficult and nonintuitive to account for. Probability theory does not provide a system for computing with fuzzy probabilities expressed as likely, unlikely, and not very likely. Subjective probability theory relies on elicitation rather than estimation of a fuzzy system. It cannot address questions of the nature, "What is the probability that the depositional system *may* be deltaic"? One should question, under all this vagueness and ambiguity what really the meaning of the digit "2" or "3" is in $P(A|u) = 0.832$. The typical reply of probabilists to possibilists is to "just be more precise" and the problem is solved. But this would ignore a particular form of lack of understanding, which goes to the very nature of UQ. The precision required does not agree with the vagueness in concepts.

The advantage and disadvantage of the application of probability to UQ is that it requires, dogmatically, precision. It is an advantage in the sense that it attempts to render subjectivity into quantification, that the rules are very well understood, the methods deeply practiced, because of the nature of the rigor of the theory, the community (after 300 years of practice) is vast. But, this rigor does not always jive with reality. Reality is more complex than "Navier stokes" or "deltaic," so we apply rigor to concepts (or even models) that probably deviate considerable from the actual processes occurring in nature. Probabilists often call this "structural" error (yet another classification and often ambiguous concept, because it has many different interpretations) but provide no means of determining what exactly this is and how it should be precisely estimated, as is required by their theories. It is left as a "research question," but can this question be truly answered within probability theory itself? For the same reasons, probabilistic method (in particular Bayesian, see next) are computationally very demanding, exactly because of this dogmatic quest for precision.

5.7.2.5. Possibility Theory: Alternative or Compliment?.

Possibility theory has been popularized by *Zadeh* [1978] and also by *Dubois and Prade* [1990]. The original notion goes back further to the economist *Shackle* [1962] studying uncertainty based on degrees of potential surprise of events. Shackle also introduces the notion of conditional possibility (as opposed to conditional probability). Just as probability theory, possibility theory has axioms. Consider Ω to be a finite set, with subsets A and B that are not necessarily disjoint:

axiom 1: $\text{pos}(\emptyset) = 0$ (Ω is exhaustive)

axiom 2: $\text{pos}(\Omega) = 1$ (no contradiction)

axiom 3: $\text{pos}(A \cup B) = \max(\text{pos}(A), \text{pos}(B))$ ("additivity")

A noticeable difference with probability theory is that addition is replaced with "max" and the subsets for axiom 3 need not be disjoint. Additionally, probability theory uses a single measure, the probability, whereas possibility theory uses two concepts, the possibility and the necessity of the event. This necessity, another measure, is defined as

$$\text{nec}(A) = 1 - \text{pos}(\bar{A}) \tag{5.2}$$

If the complement of an event is impossible, then the event is necessary. $\text{nec}(A) = 0$ means that A is unnecessary. One should not be "surprised" if A does not occur, it says nothing about $\text{pos}(A)$. $\text{nec}(A) = 1$ means that A is certainly true, which implies $\text{pos}(A) = 1$. Hence, nec carries a degree of surprise, $\text{nec}(A) = 0.1$ a little bit surprised, and $\text{nec}(A) = 0.9$ very surprised if A is not true. Possibility also allows for indeterminacy (which is not allowed in epistemic uncertainty), this is captured by $\text{nec}(A) = 0$, $\text{pos}(A) = 1$.

Logically then

$$\text{nec}(A \cap B) = \min(\text{nec}(A), \text{nec}(B)) \tag{5.3}$$

Possibility does not follow the rule of the excluded middle because

$$\text{pos}(A) + \text{pos}(\bar{A}) \geq 1 \tag{5.4}$$

An example goes as follows. Consider a reservoir. It either contains oil (A) or contains no oil ($\bar{A}$) (something we like to know!). $\text{pos}(A) = 0.5$ means that I am willing to bet that the reservoir contains oil as long as the odds are even *or better*. I would not bet that it contains oil. Hence, this describes a degree of belief very different from subjective probabilities.

Possibilities are sometimes called "imprecise probabilities" [*Hand and Walley*, 1993] or are interpreted that way. "Imprecise" need not be negative, as discussed above, it has its own advantages, in particular in terms of computation. In probability theory, information is used to update degrees of belief. This is based on Bayes' rule whose philosophy will be studied more closely in the next section. A counterpart to Bayes' rule exists in possibility theory, but because of the imprecision of possibilities over probabilities, no unique way exists to update possibilities into new possibility, given new (vague) information. Recall that Bayes' rule relies on the product (corresponding to a conjunction in classical logical)

$$P(A|B) = \frac{P(B|A)}{P(B)} P(A) \tag{5.5}$$

Consider first the counterpart of the probability density function $f_X(x)$ in possibility theory, namely the possibility distribution $\pi_X(x)$. Unlike probability densities, which could be inferred from data, possibility distributions are always specified by users, and hence take simple form (constant, triangular) functions. Densities express likelihoods, a ratio of the densities assessed in two outcomes denotes how much more (or less) likely one outcome is over the other. A possibility distribution simply states how possible an outcome x is. Hence, a possibility distribution is always equal or less than unity (not the case for a density). Also, note that $P(X = x) = 0$, always, if X is a continuous variable, while $\text{pos}(X = x)$ is not zero everywhere. Like a joint probability distribution, we can define a joint possibility distribution $\pi_{X,Y}(x, y)$ and conditional possibility distributions $\pi_{X|Y}(x|y)$. The objective now is to infer $\pi_{X|Y}(x|y)$ from $\pi_{Y|X}(y|x)$ and $\pi_X(x)$.

As mentioned above, probability theory relies on logical conjunction (see Figure 5.6). This conjunction has the following properties:

$a \cap b = b \cap a$ (commutativity)

if $a \leq a'$ and $b \leq b'$ then $a \cap b \leq a' \cap b'$ (monotonicity)

$(a \cap b) \cap c = a \cap (b \cap c)$ (associativity)

$a \cap 1 = a$ (neutrality)

Possibility theory, as it is based on fuzzy sets rather than random sets, relies on an extension of the conjunction operation. This new conjunction is termed a triangular norm (T-norm) [*Jenei and Fodor*, 1998; *Klement et al.*, 2004; *Höhle*, 2003] because it follows the following four properties:

1. $T(a,b) = T(b,a)$ (commutativity)
2. if $a \leq a'$ and $b \leq b'$ then $T(a,b) = T(a',b')$ (monotonicity)
3. $T(a,T(b,c)) = T(T(a,b),c)$ (associativity)
4. $T(a,1) = a$ (neutrality)

a	b	$a \cap b$
0	0	0
1	0	0
0	1	0
1	1	1

a	b	$T_1(a,b)$	$T_2(a,b)$	$T_3(a,b)$
0.1	0.1	0.1	0.01	0.053
0.9	0.1	0.1	0.09	0.098
0.1	0.1	0.1	0.09	0.098
0.9	0.9	0.9	0.81	0.82

$T_1(a, b) = \min(a, b)$; $T_2(a, b) = a.b$; $T_2(a, b) = \dfrac{ab}{a+b-ab}$

Figure 5.6 Example of *t*-norms for conjunction operations.

Recall that Cox relied on the postulate that plaus($p \cap q$) = f(plaus(p), plaus($q|p$)). Similarly, possibility theory relies on

$$\pi_{Y|X}(y|x) = T\left(\pi_X(x), \pi_{Y|X}(y|x)\right) = T\left(\pi_Y(x), \pi_{X|Y}(x|y)\right) \tag{5.6}$$

For example, for the minimum triangular norms we get

$$\pi_{X|Y}(x|y) = \begin{cases} 1 & \text{if } \pi(x) = \min\{\pi(x), \pi_{Y|X}(x|y)\} \\ \min\{\pi(x), \pi_{Y|X}(x|y)\} & \text{if } \pi(x) > \min\{\pi(x), \pi_{Y|X}(x|y)\} \end{cases} \tag{5.7}$$

and for the product triangular norm, we get something that looks Bayesian

$$\pi_{X|Y}(x|y) = \frac{\pi_{Y|X}(x|y)\pi(x)}{\pi(y)} \tag{5.8}$$

5.8. BAYESIANISM

5.8.1. Thomas Bayes

Hopefully, the reader has appreciated the scant overview of a long history of scientific tradition as well as philosophical approaches. UQ today often has a Bayesian flavor. What does this mean? Most researcher simply invoke Bayes' rule, as a theorem within probability theory. They work within the paradigm. But what is really the paradigm of Bayesianism? It can be seen as a simple set of methodologies, but it can also be regarded as a philosophical approach of doing science, in the same sense as empiricism, positivism, falsificationism, or inductionism. The reverend Bayes' would perhaps be somewhat surprised by the scientific revolution and main stream acceptance of the philosophy based on his rule.

Thomas Bayes was a statistician, philosopher, and reverend. Bayes presented a solution to the problem of inverse probability in "An Essay towards solving a Problem in the Doctrine of Chances." This essay was read one year after his death, by Richard Price for the Royal Society of London. Bayes' theorem remained in the background until it was reprinted in 1958, and even then it took a few more decades before an entire new approach to scientific reasoning, Bayesianism was created [*Earman*, 1992; *Howson et al.*, 1993].

Prior to Bayes' most works on chance were focused on direct inference, such as the number of replications needed to calculate a desired level of probability (how many flips of the coin are needed to assure 50/50 chance?). Bayes' treated the problem of inverse probability: "given the number of times an unknown event has happened and failed, required: the chance that the probability of its happening in a single chance lies between any two degrees of probability that can be named" (see the Biometrika publication of Bayes' essay). Bayes' essay has essentially four parts. Part 1 consists of a definition of probability and some basic calculation which are now known as the axioms of probability. Part 2 uses these calculations in a chance event related to a perfectly leveled billiard table (see Figure 5.7). Part 3 consists of using the equations obtained from the analysis of the billiard problem to his problem of inverse probability. Part 4 consists of more numerical studies and applications.

Bayes', in his essay, was not concerned with induction and the role of probability in it. Price, however, in the preface to the essay did express a wish that the work would in fact lead to a more rational approach to induction than was then currently available. What is perhaps less known is that "Bayes' theorem" in the form that we now know it was never written by Bayes'. However, it does occur in the solution to his particular problem. As mentioned above, Bayes' was interested in a chance event with unknown

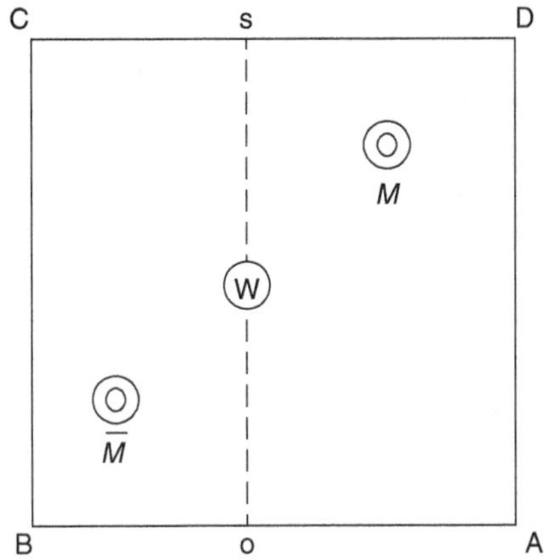

Figure 5.7 Bayes' billiard table: "to be so made and leveled that if either of the ball O and W thrown upon it, there shall be the same probability that it rests upon any one equal part of the plane as another" [*Bayes and Price*, 1763].

probability (such as in the billiard table problem), given a number of trials. If M counts the number of times that an event occurs in n trials, then the solution is given through the binomial distribution:

$$P(p_1 \leq p \leq p_2 | M = m) = \frac{\int_{p_1}^{p_2} \binom{n}{m} p^m (1-p)^{n-m} P(dp)}{\int_0^1 \binom{n}{m} p^m (1-p)^{n-m} P(dp)} \quad (5.9)$$

where $P(dp)$ is the prior distribution over p. Bayes' insight here is to "suppose the chance is the same that it (p) should lie between any two equi-different degrees." $P(dp) = dp$, in other words the prior is uniform, leading to

$$P(p_1 \leq p \leq p_2 | M = m) = \frac{(n+1)!}{m!(n-m)!} \int_{p_1}^{p_2} \binom{n}{m} p^m (1-p)^{n-m} dp \quad (5.10)$$

Why uniform? Bayes' does not reason from the current principle of indifference (which can be debated, see later), but rather from an operation characterization of an event concerning the probability which we know absolutely nothing about prior to the trials. The use of prior distributions, however, was one of the key insights of Bayes' that very much lives on.

5.8.2. Rationality for Bayesianism

Bayesians can be regarded more as relativists than absolutists (such as Popper). They believe in prediction based on imperfect theories. For example, they will take an umbrella on their weekend, if their ensemble Kalman filter prediction of the weather at their trip location puts a high (posterior) probability of rain in 3 days. Even if the laws involved are imperfect and probably can be falsified (many weather prediction are completely wrong!), they rely on continued learning from future information and adjustments. Instead of relying on Popper's zero probability (rejected or not), they rely more on an inductive inference yielding nonzero probabilities.

If we now take the general scientific perspective (and not the limited topic of UQ), then Bayesians see science progress by hypotheses, theories, and evidence offered toward these hypotheses all quantified using probabilities. In this general scientific context, we may therefore state hypothesis H, gather evidence E, with $P(H|E)$ the probability of the hypothesis in light of the evidence, $P(E|H)$ the probability that the evidence occurs when the hypothesis is true, $P(H)$ the probability of the hypothesis without any evidence, and $P(E)$ the probability of the evidence, without stating that any hypothesis is true:

$$P(H|E) = \frac{P(E|H)}{P(E)} P(H) \quad (5.11)$$

$P(H)$ is also termed the prior probability and $P(H|E)$ the posterior probability. We provided some discussion on a logical way of explaining this theorem [*Cox*, 1946] and the subsequent studies that showed this was not quite as logical as it seems [*Halpern*, 1995, 2011]. Few people today know that Bayesian probability has six axioms [*Dupré and Tiplery*, 2009]. Despite these perhaps rather technical difficulties, a simple logic underlies this rule. Bayes' theorem states that the extent to which some evidence supports a hypothesis is proportional to the degree to which the evidence is predicted by the hypothesis. If the evidence is very likely ("sandstone has lower acoustic impedance than shale) then the hypothesis ("acoustic impedance depends on mineral composition") is not supported significantly when indeed we measure that "sandstone has lower acoustic impedance than shale." If, however, the evidence is deemed very unlikely, for example ("shale has higher acoustic impedance than sandstone"), then the hypothesis of another theorem ("acoustic impedance depends not only on mineralization, but also fluid content") will be highly confirmed (have high posterior probability).

Another interesting concept is how Bayes' deals with multiple evidences of the same impact on the hypothesis. Clearly, more evidence leads to an increase in probability of a hypothesis supported by that evidence. But evidences of the same impact will have a diminishing effect. Consider that a hypothesis has equal probability as some alternative hypothesis:

$$P(H) = 0.5$$

Now consider multiple evidence sources such that

$$P(H|E_1) = 0.8; P(H|E_2) = 0.8; P(H|E_3) = 0.8$$

Then, according to a model of conditional independence and Bayes' theorem [*Bordley*, 1982; *Genest and Zidek*, 1986; *Journel*, 2002; *Clemen and Winkler*, 2007]:

$$P(H|E_2, E_1) = 0.94; P(H|E_3, E_2, E_1) = 0.98$$

Compounding evidence leads to increasing probability of the hypothesis, but it will never reach unity, unless some evidence states $P(H|E_j) = 1$.

5.8.3. Objective Versus Subjective Probabilities

In the early days of the development of Bayesian approaches, several general principles were stated under which researchers "should" operate, resulting in an "objective" approach to the problem of inference, in the sense that everyone is following that same logic. One such principle is the principle of maximum entropy [*Jaynes*, 1957], of which the principle of indifference (Laplace) is a special case. Subjectivists do not see probabilities as objective (leading to prescribing zero probabilities to well-confirmed ideas). Rather, subjectivists [*Howson et al.*, 1993] see Bayes' as an objective theory of inference. Objective is the sense that *given* prior probabilities and evidence, posterior probabilities are calculated. In that sense, subjective Bayesian make no claim on the nature of the propositions on which inference is being made (in that sense, they are also deductive).

One interesting application of reasoning this way occurs when disagreement occurs on the same model. Consider modeler A (the conformist) who assigns a high probability to some relatively well-accepted modeling hypothesis and low probability to some rare (unexpected) evidence. Consider modeler B (the skeptic) who assigns low probability to the norm and hence high probability to any unexpected evidence. As a consequence, when the unexpected evidence occurs and hence is confirmed $P(E|H) = 1$, then the posterior of each is proportional to $1/P(E)$. Modeler A is forced to increase their prior more than the Modeler B. Some Bayesians, therefore, state that the prior is not that important as continued new evidence is offered. The prior will be "washed out" by cumulating new evidence. This is only true for certain highly idealized situations. It is more likely that two modelers will offer two hypotheses; hence, evidence needs to be evaluated against each other. However, there is always a risk that neither model can be confirmed, regardless how much evidence is offered; hence, the prior model space is incomplete, which is the exact problem of the objectivist Bayes'. Neither objective nor subjective Bayes' addresses this problem.

5.8.4. Bayes with Ad-Hoc Modifications

Returning now to the example of Figure 5.5, Bayesian theory, if properly applied, allows for assessing these ad-hoc model modifications. Consider that a certain modeling assumption H is prevailing in multiphase flow: "oil flow occurs in rock with permeability of 10–10000 md" (H), now this modeling assumption is modified ad hoc to "oil flow occurs in rock with permeability of 10–10000 md and 100–200D)" ($H \cap$ ad hoc). However, this ad-hoc modification, under H, has very low probability, $P(\text{ad hoc}) \simeq 0$ and hence $P(H \cap \text{ad hoc}) \simeq 0$. The problem in reality is that those making the ad-hoc modification often do not use Bayesianism, hence never assess or use the prior $P(\text{ad hoc})$.

5.8.5. Criticism of Bayesianism

What is critical to Bayesianism is the concept of "background knowledge." Probabilities are calculated assuming some commonly assumed background knowledge. Recall that theories cannot be isolated and independently tested. This "background" consists of all the available assumptions tangent to the hypothesis at hand. The problem with using Eq. (5.11) is that such "background knowledge" (BK) is taken implicit:

$$P_{BK_0}(H|E) \simeq P_{BK_0}(E|H) P_{BK_0}(H) \rightarrow P_{BK_1}(H) \quad (5.12)$$

where 0 indicates time $t = 0$. The posterior then includes the "new knowledge" which is included in the new background knowledge at the next stage $t = 1$. A problem occurs when applying this to the real world: What is this "background knowledge"? In reality, the prior and likelihood are not determined by the same person. For example, in our application, the prior may be given by a geologist, the likelihood by a data scientist. It is unlikely that they have the same "background knowledge" (or even agree on it). A more "honest" way of conveying this issue is to make the background knowledge explicit. Suppose that $BK^{(1)}$is the background knowledge of person 1, who deals with evidence (the data scientist) then

$$P\left(H|E \cap BK^{(1)}\right) \simeq P\left(E \cap BK^{(1)}|H\right) P\left(H|BK^{(1)}\right) \quad (5.13)$$

Suppose $BK^{(2)}$ is person 2 (geologist) who provides the "prior," meaning the person provides background knowledge on his/her own, without evidence. Then, the new posterior can be written as

$$P\left(H|E \cap BK^{(1)} \cap BK^{(2)}\right) \simeq P\left(E \cap BK^{(2)}|H\right) P\left(H|BK^{(2)}\right) P\left(H|BK^{(1)}\right) \quad (5.14)$$

assuming, however, there is no overlap between background knowledge. In practice, the issue that different components of the "system" (model) are constructed by different modelers with different background knowledge is ignored. Even if one would be aware of this issue, it would be difficult to implement in practice. The ideal

Bayesian approach rarely occurs. No single person understands all the detailed aspects of the scientific modeling study at hand. A problem then occurs with dogmatism. The study in Figure 5.5 illustrates this. Hypotheses that are given very high probability (no fractures) will remain high, particularly in the absence of strong evidence (low to medium $P(E)$). Bayes' rule will keep assigning very high probabilities to such hypotheses, particularly due to the dogmatic belief of the modeler or the prevailing leading idea of what is going on. This is not the problem of Bayes', but its common (faulty) application. Bayes' itself cannot address this.

More common is to select a prior based on general principles or mathematical convenience, for example using a maximum entropy principle. Under such principle, complete ignorance results in choosing a uniform distribution. In all other cases, one should pick the distribution that makes the least claims, from whatever information is currently available, on the hypothesis being studied. The problem here is not so much the ascribing of uniform probabilities but providing a statement of what all the possibilities are (on which then uniform probabilities are assigned). Who chooses these theories/models/hypotheses? Are those the only ones?

The limitation, therefore, of Bayesianism is that no judgment is leveled to the stated prior probabilities. Hence, any Bayesian analysis is as strong as the analysis of the prior. In subsurface modeling, this prior is dominated by the geological understanding of the system. Such geological understanding and its background knowledge is vast, but qualitative. Later we will provide some ideas on how to make quantitative "geological priors."

5.8.6. Deductive Testing of Inductive Bayesianism

The leading paradigm of Bayesianism is to subscribe to an inductive form of reasoning: learning from data. Increasing evidence will lead to increasing probabilities of certain theories, models, or hypothesis. As discussed in the previous section, one of the main issues lies in the statement of a prior distribution, the initial universe of possibilities. Bayesianism assumes that a truth exists, that such truth is generated by a probability model, and also than any data/evidence are generated from this model. The main issue occurs when the truth is not even within the support (the range/span) generated by this (prior) probability model. The truth is then not part of this initial universe. What happens then? The same goes when the error distribution on the data is chosen too optimistic, in which case the truth may be rejected. Can we verify this? Diagnose this? Figure out whether the problem lies with the data or the model? Given the complexity of models, priors, and data in the real world, this issue may in fact go undiagnosed if one stops the analysis with the generation of the posterior distribution. *Gelman and Shalizi* [2013] discuss how mis-specified prior models (the truth is not in the prior) may result in either no solution, multi-model solutions to problems that are unimodal or complete nonsense.

Work by *Mayo* [1996] looks at these issues. Mayo attempted to frame tests within classical hypothesis testing. Recall that classical statistics relies on a deductive form of hypothesis testing, which is very similar in flavor to Popper's falsification. In similar vein, some form of model testing can be performed posterior to the generation of the posterior. Note that Bayesian model averaging [*Henriksen et al.*, 2012; *Refsgaard et al.*, 2012; *Rings et al.*, 2012; *Tsai and Elshall*, 2013; *Brunetti et al.*, 2017] and model selection are not tests of the posterior, rather they are consequences of the posterior distribution, yet untested! Classical checks are whether posterior models match data, but these are checks based on likelihood (misfit) only.

Consider a more elaborate testing framework (see for *Gelman et al.*, 1996). These formal tests rely on generating replicates of the data given some model hypothesis and parameters are the truth. Take a simple example of a model hypothesis with two faults (H=two faults) and the parameters $\boldsymbol{\theta}$ representing those faults (e.g., dip, azimuth, length, etc.). In Chapter 3, we discussed a bootstrap (Monte Carlo)-based determination of achieved significance level (ASL) as

$$\mathrm{ASL}(\boldsymbol{\theta}) = P\left(S\left(\mathbf{d}_{\mathrm{rep}}\right) \geq S(\mathbf{d}_{\mathrm{obs}}) \mid H, \boldsymbol{\theta}\right) \tag{5.15}$$

Here we consider calculating some summary statistic of the data as represented by the function S. This summary statistic could be based on some dimension reduction method, for example, a first or second principal component score. The uncertainty on $\boldsymbol{\theta}$ is provided by its posterior distribution; hence, we can sample various $\boldsymbol{\theta}$ from the posterior. Therefore, we first sample $\mathbf{d}_{\mathrm{rep}}$ from the following distribution (averaging out over posterior in $\boldsymbol{\theta}$)

$$P\left(\mathbf{d}_{\mathrm{rep}} \mid H, \mathbf{d}_{\mathrm{obs}}\right) = \int P\left(\mathbf{d}_{\mathrm{rep}} \mid H, \boldsymbol{\theta}\right) P(\boldsymbol{\theta} \mid H, \mathbf{d}_{\mathrm{obs}})\, d\boldsymbol{\theta} \tag{5.16}$$

and calculate average ASL over the posterior distribution. Analytically, this equals to

$$\mathrm{ASL} = \int \mathrm{ASL}(\boldsymbol{\theta}) P(\boldsymbol{\theta} \mid H, \mathbf{d}_{\mathrm{obs}}) d\boldsymbol{\theta} \tag{5.17}$$

or for given limited sample $\boldsymbol{\theta}^{(\ell)}$, $\ell = 1, \ldots, L \sim P(\boldsymbol{\theta} \mid H, \mathbf{d}_{\mathrm{obs}})$

$$\mathrm{ASL} = \frac{1}{L} \sum_{\ell=1}^{L} \mathrm{ASL}\left(\boldsymbol{\theta}^{(\ell)}\right) \tag{5.18}$$

(Chapter 8 provides such example) These tests are not used to determine whether a model is true, or even should be falsified but whether discrepancies exist between model and data. The nature of the functions S defines the "severity" of the tests [*Mayo*, 1996]. Numerous complex functions will allow for a more severe testing of the prior modeling hypothesis. We can learn how the model fails by generating several of these summary statistics, each representing different elements of the data (a low, a middle, and some extreme cases, etc.)

Within this framework of deductive tests, the prior is no longer treated as "absolute truth," rather the prior becomes a modeling assumption that is "testable" given the data. However, some may disagree on this point: Why should the data be any better than the prior? In the next section, we will try to get out of this trap, by basing priors on physical processes, with the hope that such priors are more realistic representations of the universe of variability, rather than simply relying on statistical methods that are devoid of physics.

5.9. BAYESIANISM IN GEOLOGICAL SCIENCES

5.9.1. Introduction

In the study of subsurface systems, one of the leading uncertainties in most cases is due to the geological variability of the depositional system with which one has to deal with in engineering applications. We discussed in Chapter 1 that many uncertainties are present that requires handling, such as boundary conditions, initial conditions, physical laws, chemical reactions, and so on. Some of these can be estimated directly from data (such as boundary conditions and initial conditions), others are physical laws that we keep refining through experimentation and modeling, such as multiphase flow. Here we discuss specifically the use of Bayesianism in dealing with uncertainty related to the geological system. Evidently, the science involved here are the various geological sciences that matter for the kind of studies in this book such as sedimentary geology, carbonate geology, geochemistry, and structural geology.

We discussed previously how the prior is mostly used as a smoother, regularizer, a mathematical construct to conveniently solve the Bayesian puzzle of getting to the posterior (and ways to sample from it). Most of the prior formulation used has very little to do with "background knowledge" of geological systems, simply because most Bayesians are not geologists, nor do they understand necessarily well the language and science of geology. We will discuss these kinds of techniques in more detail in Chapter 6. A principled approach to defining geologically founded prior is still in its infancy, and in the following section we provide some conceptual thoughts.

5.9.2. What Is the Nature of Geological Priors?

5.9.2.1. Constructing Priors from Geological Field Work. In a typical subsurface system, the model variables are parameterized in a certain way, for example with a grid, or a set of objects with certain lengths, widths dips, azimuths, and so on. What is the prior distribution of these model variables? Since we are dealing with a geological system, for example a delta, a fluvial, or a turbidite system, a common approach is to do geological field work. This entails measuring and interpreting the observed geological structures, on outcrops, and creating a history of their genesis, with an emphasis on generating (an often qualitative) understanding of the processes that generated the system. The geological literature contains a vast amount of such studies.

To gather all this information and render it relevant for UQ, geological databases based on classification systems have been compiled (mostly by the oil industry). Analog databases, for example, on proportions, paleo-direction, morphologies, and architecture of geological bodies or geological rules of association [*Eschard and Doligez*, 2002; *Gibling*, 2006] for various geological environments [FAKT: *Colombera et al.*, 2012; CarbDB: *Jung and Aigner*, 2012; WODAD: *Kenter and Harris*, 2006; *Pyrcz et al.*, 2008] have been constructed. Such relational databases employ a classification system based on geological reasoning. For example, the FAKTS database classifies existing studies, whether literature-derived or field-derived from modern or ancient river systems, according to controlling factors, such as climate, and context-descriptive characteristics, such as river patterns. The database can, therefore, be queried on both architectural features and boundary conditions to provide the analogs for modeling subsurface systems. The nature of the classification is often hierarchical. The uncertain style or classification, is often termed "geological scenario" [*Martinius and Naess*, 2005] and variations within that style.

While such approach appears to gather information, it leaves the question of whether the collection of such information and the extraction of parameters values to state prior distribution produce realistic priors (enough variance, limited bias) for what is actually in the subsurface. Why?

1. Objects and dimensions in the field are only apparent. An outcrop is only a 2D section of a 3D system. This invokes stereological problems in the sense that structural characteristics (e.g., shape, size, texture) of 2D outcrops are only apparent properties of the 3D subsurface. These apparent properties can drastically change depending on the position/orientation of the survey. Furthermore, interpreted 2D outcrops of the subsurface may be biased because large structures are more frequently observed

than small structures [*Lantuejoul*, 2013]. The same issue occurs when doing 2D geophysical surveys to interpret 3D geometries [*Davies and Sambrook Smith*, 2006]. For example, quantitative characterization of 2D GPR imaging [e.g., *Bristow and Jol*, 2003] ignore uncertainty on the 3D subsurface characteristics resulting from the stereological issue.

2. The database is purely geometric in nature. It records the end-result of deposition not the process of deposition. In that sense, it does not include any physics underlying the processes that took place and therefore may not capture the complexity of geological processes fully to provide a "complete" prior. For that reason, the database may aggregate information that should not be aggregated, simply because each case represents different geological processes, accidently creating similar geometry. For modeling, this may appear irrelevant (who cares about the process), yet it is highly relevant. Geologists reason based on geological processes, not just the final geometries; hence, this "knowledge" should be part of a prior model construction. Clearly, priors should not ignore important background knowledge, such as process understanding.

The main limitation is that this pure parameterization-based view (the geometries, dimensions) lacks physical reasoning, hence ignore important prior information. The next section provides some insight into this problem and suggests a solution.

5.9.2.2. Constructing Priors from Laboratory Experiments. As discussed earlier in this book, natural depositional systems are subject to large variability whose very nature is not fully understood. For example, channelized transport systems (fan, rivers, delta, etc.) reconfigure themselves more or less continually in time, and in a manner often difficult to predict. The configurations of natural deposits in the subsurface are thus uncertain. The quest for quantifying prior uncertainty necessitates understanding the sedimentary systems by means of physical principles, not just information principles (such as the principle of indifference). Quantifying prior uncertainty, thus, requires stating all configurations of architectures of the system deemed *physically* possible and at what frequency (a probability density) they occur. This probability density need not be Gaussian or uniform. Hence, the question arises: What is this probability density for geological systems, and how does one represent it in a form that can be used for actual predictions using Bayesianism?

The problem in reality is that we observe geological processes over a very short time span (50 years of satellite data and ground observations), while the deposition of the relevant geological systems we work with in this book may span 100,000 years or more. For that reason, the only way to study such system is either by computer models or by laboratory experiments. These computer models solve a set of partial differential equations (PDEs) that describe sediment transport, compaction, diagenesis, erosion, dissolution, and so on [*Koltermann and Gorelick*, 1992; *Gabrovsek and Dreybrodt*, 2010; *Nicholas et al.*, 2013]. The main issue here is that PDEs offers a limited representation of actual physical process and require calibration with actual geological observations (such as erosion rules), boundary conditions, and source terms. Often their long computing times limit their usefulness for constructing complete priors.

For that reason, laboratory experiments are increasingly used to study geological deposition, simply because physics occurs naturally, and not through an artificial computer code. Next we provide some insight into how laboratory experiments work and how they can be used to create realistic analogs of depositional systems.

5.9.2.3. Experimenting the Prior. We consider a delta constructed in an experimental sedimentary basin subject to constant external boundary conditions (i.e., sediment flux, water discharge, subsidence rates), see Figure 5.8. The dataset used is a subset of the data collected during an experiment conducted in 2010 [*Wang et al.*, 2011]. Basin dimensions were 4.2 m long, 2.8 m wide, and 0.65 m deep. The sediment consisted of a mix of 70% quartz sand and 30% anthracite coal sand. These experiments are used for a variety of reasons. One of them is to study the relationship between the surface processes and the subsurface deposition. An intriguing aspect of these experiments is that much of the natural variability is not due to forcing (e.g., uplift, changing sediment source) but due to the internal dynamics of the system itself, that is, it is autogenic. In fact, it is not known if the autogenic behavior of natural channels is chaotic [*Lanzoni and Seminara*, 2006], meaning one cannot predict with certainty the detailed configuration of even a single meandering channel very far into the future. This then has immediate impact on uncertainty in the subsurface in the sense that configuration of deposits in the subsurface cannot be predicted with certainty away from wells. The experiment, therefore, investigates uncertainty related to the dynamics of the system, our lack of physical understanding (and not some parameter uncertainty or observational error). All this is a bit unnerving, since this very fundamental uncertainty is *never* included in any subsurface UQ. At best, one employs a Gaussian prior, or some geometric prior extracted from observation databases, as discussed above. The following are the fundamental questions:

1. Can we use these experiments to construct a realistic prior, capturing uncertainty related to the physical processes of the system?

Figure 5.8 Flume experiment of a delta with low Froude number performed by John Martin, Ben Sheets, Chris Paola and Michael Kelberer. *Source:* https://www.esci.umn.edu/orgs/seds/Sedi_Research.htm.

2. Can a statistical prior model represent (mimic) such variability?

To address these questions and provide some insight (not an answer quite yet!), we run the experiment under constant forcing for long enough to provide many different realizations of the autogenic variability, a situation that would be practically impossible to find in the field. The autogenic variability in these systems is due to temporal and spatial variability in the feedback between flow and sediment transport, weaving the internal fabric of the final subsurface system.

Under fixed boundary conditions, the observed variability in deposition is therefore the result of only the autogenic (intrinsic) variability in the transport system. The dataset we use here is based on a set of 136 time-lapse overhead photographs that capture the dynamics of flow over the delta approximately every minute. Figure 5.9 shows representative images from this database. This set of images represents a little more than 2 h of experimental run time. Figure 5.9b shows the binary (wet-dry) images for the same set, which will be used in the investigation.

The availability of a large reference set of images of the sedimentary system enables testing any statistical prior by allowing a comparison of the variability of the resulting realizations, since all possible configurations of the system are known. In addition, the physics are naturally contained in the experiment (photographs are the result of the physical depositional processes). A final benefit is that a physical analysis of the prior model can be performed, which aids in understanding what depositional patterns should be in the prior for more sophisticated cases.

5.9.2.4. Reproducing Physical Variability with Statistical Models. To attempt to reproduce physical variability, we employ a geostatistical method termed multiple-point geostatistics (see Chapter 6 for a more elaborate discussion). MPS methods have grown popular in the last decade because of their ability to introduce geological realism in modeling via the training image (TI) [*Mariethoz and Caers*, 2014]. Similar to any geostatistics procedure, MPS allows for the construction of a set of stochastic realizations of the subsurface. TIs, along with trends (usually modeled using probability maps or auxiliary variables), constitute the prior model as defined in the traditional Bayesian framework. The choice of the initial set of TIs has a large influence on the stated uncertainty, and hence a careful selection must be done to avoid artificially reducing uncertainty from the start.

It is unlikely that all possible naturally occurring patterns can be contained in one single TI within the MPS framework (although this is still the norm; similarly it is the norm to choose for a multi-Gaussian model by default). To represent realistic uncertainty, realizations should be generated from multiple TIs. The set of all these realizations then constitutes a wide prior uncertainty model. The choice of the TIs brings a new set of questions: How many TIs should one use and which ones should be selected? Ideally, the TIs should be generated in such a way that natural variability of the system under study is represented (fluvial, deltaic, turbidite, etc.), so that that all natural patterns are covered in the possibly infinite set of geostatistical realizations. *Scheidt et al.* [2016] use methods of computer vision to select a set of representative TIs. On such computer vision method evaluates a rate

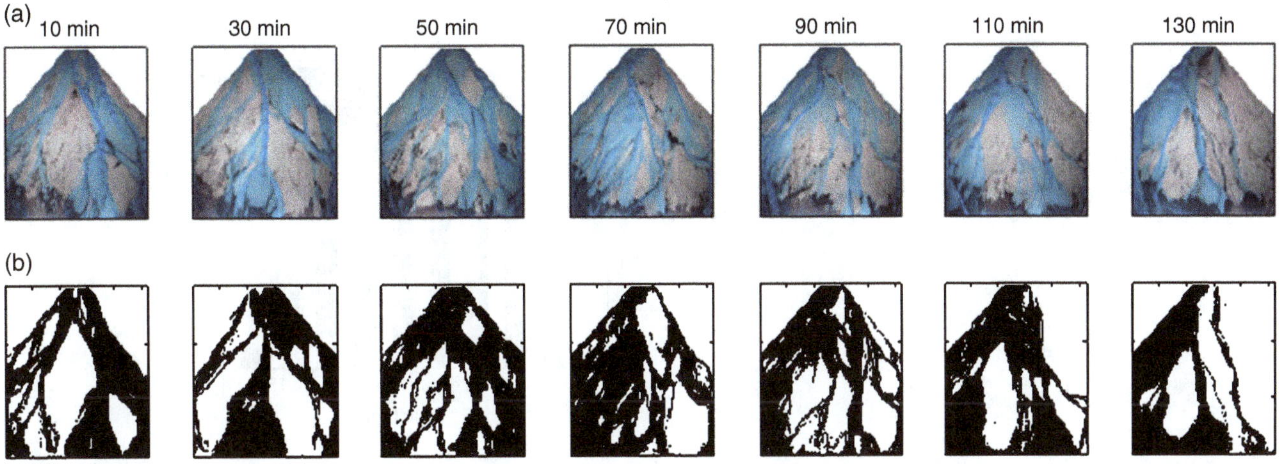

Figure 5.9 Examples of a few photographs images of the flume experiment for different time. Flow is from top to bottom. (a) Photographs of the experiments. The blue pixels indicate locations where flow moves over the surface. The black sediment is coal which is the mobile fraction of the sediment mixture, and the tan sediment is sand. (b) Binary representation of the photographs. Black represents wet (flow) pixels and white represents dry (no flow) pixels.

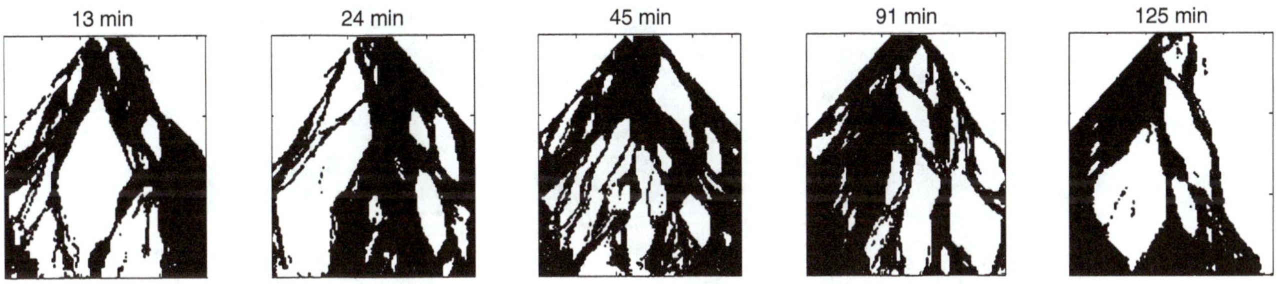

Figure 5.10 Selected images by clustering based on the modified Hausdorff distance. The value at the top of the image represents the time in minutes of the experiment.

of change between images in time and the TIs are selected in periods of relative temporal pattern stability (see Figure 5.10).

The TI set shown in Figure 5.10 displays patterns consistent with previous physical interpretations of the fundamental modes of this type of delta system: a highly channelized, incisional mode; a poorly channelized, depositional mode; and an intermediate mode. This suggests that some clues to the selection of appropriate TIs lie in the physical properties of the images from the experiment.

With a set of TIs available, multiple geostatistical realization per each TI can be generating (basically a hierarchical model of realizations). These realizations can now be compared with the natural variability generated in the laboratory experiments, to verify whether such set of realizations can at all reproduce natural variability. *Scheidt et al.* [2016] calculate the modified Hausdorff distance (MHD), described in Chapter 2, between any two geostatistical realizations and also between any two overhead shots. A QQ-plot of the distribution of the MHD between all the binary snapshots of the experiment and MPS models is shown in Figure 5.11a, showing similarity in distribution.

The result is encouraging but also point out a mostly ignored question of what a complete prior geological entails, that the default choices (one TI, one Boolean model, one multi-Gaussian distribution) make very little sense when dealing with realistic subsurface heterogeneity. The broader question remains on how such a prior should be constructed from physical principles and how statistical models, such as geostatistics, should be employed in Bayesianism when applied to geological systems. This fundamental question remains unresolved and certainly under-researched.

5.9.2.5. Field Application. The above flume experiments have helped in understanding the nature of a geological prior, at least for deltaic-type deposits. Knowledge accumulated from these experiments will create scientific

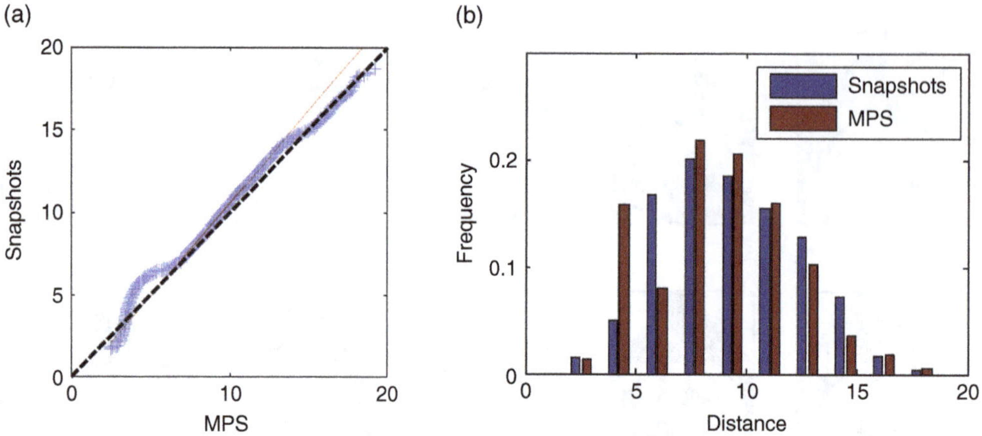

Figure 5.11 (a) QQ-plot of the MHD distances between the 136 images from the experiment and 136 images generated using direct sampling (DS). (b) Comparison of the variability, as defined by MHD, between generated realizations per each training image (TI) (red) and the images from the experiment (blue) closest (in MHD) to the selected TI.

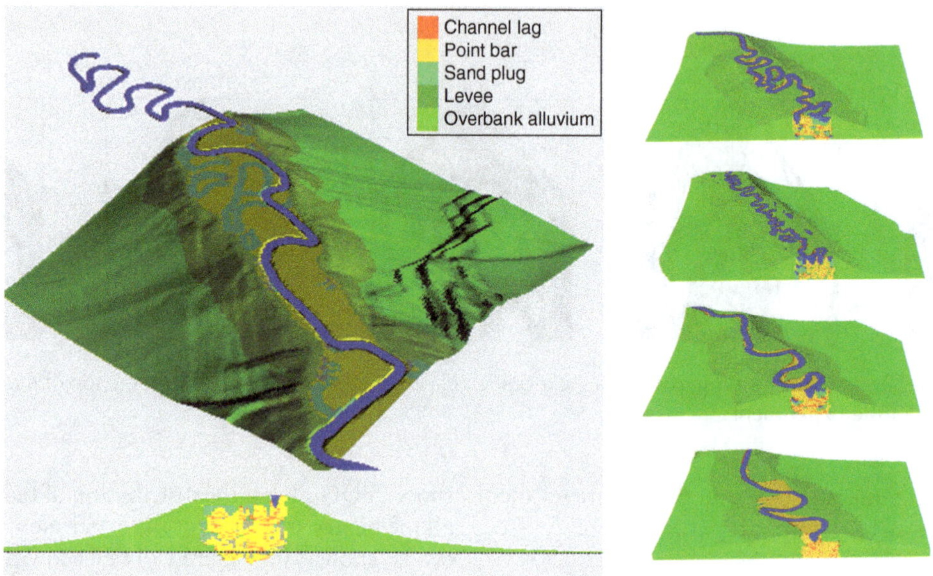

Figure 5.12 Example of a FLUMY model with several realizations of the prior generated from FLUMY with uncertain input parameters.

understanding on the fundamental processes involved in the genesis of these deposits and thereby help to understand better the range of variability of the generated stratigraphic sequences.

It is unlikely, however, that laboratory experiments will be of direct use in actual applications, since they take considerable time and effort to set up. In addition, there is a question of how to scale to the real world. It is more likely in the near future that computer models, built from such understanding, will be used in actual practice. In Chapter 6, we discuss various such computer models for depositional systems (process-based, process-mimicking. etc.).

We consider here one such computer model, FLUMY [*Cojan et al.*, 2005], which is used to model meandering channels, see Figure 5.12. FLUMY uses a combination of physical and stochastic process models to create realistic geometries. It is not an object-based model, which would focus on the end result, but it actually creates the depositional system. The input parameters are, therefore, a combination of physical parameters as well as geometrical parameters describing the evolution of the deposition.

Consider a simple application to an actual reservoir system (Courtesy of ENI). Based on geological understanding generated from well data and seismic, modelers are asked to input the following FLUMY parameters: channel width, depth and sinuosity (geometric), and two aggradation parameters: (i) decrease of the alluvium thickness away from the channel and (ii) maximum thickness deposited on levees during an overbank flood. More parameters exist, but these are kept fixed for this simple application.

The prior belief now consists of (i) assuming the FLUMY model as a hypothesis that describes variability in the depositional system and (ii) prior distributions of the five parameters. After generating 1000s of FLUMY models (see Figure 5.12) we can run the same analysis as done for the flume experiment to extract modes in the system that can be used as TIs for further geostatistical modeling.

Another approach is to define a certain desired model response (data or prediction or both) and to perform a sensitivity analysis on which of the FLUMY parameters are most impacting the response (e.g., using global sensitivity analysis in Chapter 4).

5.9.3. Moving Forward

Eventually, philosophical principles will need to be translated into workable practices, ultimately into data acquisition, computer codes, and actual decisions. A summary of some important observations and perhaps also personal opinions based on this chapter are as follows:

1. *Data acquisition, modeling, and predictions "collaborate"*; going from data to models to prediction ignores the important interactions that take place between these components. Models can be used, prior to actual data acquisition, to understand what role they will play in modeling and ultimately in the decision-making process. The often classical route of first gathering data and then creating models may be completely inefficient if the data has no or minor impact on any decision. This should be studied beforehand and hence requires building models of the data, not just of the subsurface.

2. *Prior model generation is critical to Bayesian approaches* in the subsurface and statistical principles of indifference are very crude approximation of realistic geological priors. Uniform and multi-Gaussian distributions have been clearly falsified by many case studies [*Gómez-Hernández and Wen*, 1998; *Zinn and Harvey*, 2003; *Feyen and Caers*, 2006]. They may lead to completely erroneous predictions when used in subsurface applications. One can draw an analogy here with Newtonian physics: it has been falsified but it is still around, meaning it can be useful to make many predictions. The same goes with multi-Gaussian-type assumptions. Such choices are logical for an "agent" that has limited knowledge and hence (rightfully) uses the principle of indifference. More informed agents will, however, use more realistic prior distributions. The point, therefore, is to use more informed agents (geologists) into the quantification of prior. The use of such agent would make use of the vast geological (physical) understanding that has been generated about over many decades.

3. *Falsification of the prior*. It now seems logical to propose workflows of UQ that have both the inductions and deduction flavors. Falsification should be part of any a-priori application of Bayesianism and also on the posterior results. We will propose several ways of falsifying realistic geological priors with data, prior to application of Bayes' rule for the applications in Chapter 8. Such approaches will rely on forms of sensitivity analysis as well as developing geological scenarios that are tested against data. The point here is not to state rigorous probabilities on scenarios but to eliminate scenarios from the pool of possibilities because they have been falsified. The most important aspect of geological priors are not the probabilities given to scenarios but the generation of a suitable set of representative scenarios to represent the geological process taking place. This was illustrated in the flume experiment study.

4. *Falsification of the posterior*. The posterior is the result of the prior model choice, the likelihood model choice and all of the auxiliary assumptions and choices made (dimension reduction method, sampler choices, convergence assessment, etc.). Acceptance of the posterior "as is" would follow the pure inductionist approach. Just as the prior, it would be good practice to attempt to falsify the posterior. This can be done in several ways, particularly using hypothetico-deductive analysis, such as the significance tests introduced in this chapter. Chapter 8 will illustrate this practice.

REFERENCES

Ariew, R. (1984), The Duhem thesis, *Br. J. Philos. Sci.*, *35*(4), 313–325.

Bayes, M., and M. Price (1763), An essay towards solving a problem in the doctrine of chances. By the late Rev. Mr. Bayes, F. R. S. communicated by Mr. Price, in a letter to John Canton, A. M. F. R. S, *Phil. Trans. R. Soc. London*, *53*, 370–418.

Berger, J. O. (2003), Could Fisher, Jeffreys and Neyman have agreed on testing? *Stat. Sci.*, *18*(1), 1–32.

Bond, C. E., et al. (2007), What do you think this is? "Conceptual uncertainty" in geoscience interpretation. *GSA Today*, *17*(11), 4–10.

Bordley, R. F. (1982), A multiplicative formula for aggregating probability assessments, *Manag. Sci.*, *28*(10), 1137–1148.

C. S. Bristow, and H. M. Jol (Eds.) (2003), *Ground Penetrating Radar in Sediments*, Special Publication, *211*, Geological Society Publishing House, Bath, ME, 330pp.

Brunetti, C., N. Linde, and J. A. Vrugt (2017), Bayesian model selection in hydrogeophysics: Application to conceptual subsurface models of the South Oyster Bacterial Transport Site, Virginia, USA, *Adv. Water Resour.*, *102*, 127–141.

Chalmers, A. F. (1999), *What Is This Thing Called Science?* Third Edition, Open University Press. Available from: http://www.amazon.com/dp/0872204529.

Clemen, R. T., and R. L. Winkler (2007), Aggregating probability distributions, *Adv. Dec. Anal.*, *120*(919), 154–176.

Cojan, I., et al. (2005), Process-based reservoir modelling in the example of meandering channel, in *Geostatistics Banff 2004*, edited by O. Leuangthong and C. V. Deutsch, pp. 611–619, Springer, The Netherlands.

Colombera, L., et al. (2012), A database approach for constraining stochastic simulations of the sedimentary heterogeneity of fluvial reservoirs. *AAPG Bull.*, *96*(11), 2143–2166.

Cox, R. T. (1946), Probability, frequency and reasonable expectation, *Am. J. Phy.*, *14*(1), 1. Available from: http://scitation.aip.org/content/aapt/journal/ajp/14/1/10.1119/1.1990764.

Davies, N. S., and G. H. Sambrook Smith (2006), Signatures of quaternary fluvial response, Upper River Trent, Staffordshire, U.K.: A synthesis of outcrop, documentary, and GPR data. *Geomorph. N.F.*, *50*, 347–374.

Dubois, D., and H. Prade (1990), The logical view of conditioning and its application to possibility and evidence theories, *Int. J. Approx. Reason.*, *4*(1), 23–46.

Dupré, M. J. and F. J. Tiplery (2009), New axioms for rigorous Bayesian probability, *Bayesian Anal.*, *4*(3), 599–606.

Earman, J. (1992), *Bayes or Bust? A Critical Examination of Bayesian Confirmation Theory*, MIT Press, Cambridge, MA.

Eschard, R., and B. Doligez (2002), Using quantitative outcrop databases as a guide for geological reservoir modelling, in *Geostatistics Rio 2000*, edited by M. Armstrong, C. Bettini, N. Champigny, A. Galli and A. Remacre, pp. 7–17, Springer, Dordrecht.

Fallis, A. (2013), *Fisher, Neyman and the Creation of Classical Statistics*, Springer, New York.

Feller, W. (2008), *An Introduction to Probability Theory and Its Applications*, Second Edition, vol. *2*, Wiley, New York, p. xxiv–669..

Feyen, L., and J. Caers (2006), Quantifying geological uncertainty for flow and transport modeling in multi-modal heterogeneous formations, *Adv. Water Resour.*, *29*(6), 912–929.

Feyerabend, P. (1993), Against method, *Pool-108-46-235-15. Nycmny.Fios. …*, *3*(3), 280.

Fine, A. (1973), Probability and the interpretation of quantum mechanics, *Br. J. Philos. Sci.*, *24*(1), 1–37.

de Finetti, B. (1974), The value of studying subjective evaluations of probability, in *The Concept of Probability in Psychological Experiments*, edited by C. A. S. Staël Von Holstein, pp. 1–14, Springer, Dordrecht.

de Finetti, B. (1995), The logic of probability, *Philos. Stud.*, *77*(1), 181–190.

de Finetti, B., A. Machí and A. Smith (1975), *Theory of Probability: A Critical Introductory Treatment*. Available from: http://books.google.com/books?id=Q44uAAAAIAAJ.

Fisher, R. A. (1925), *Statistical Methods for Research Workers*. Available from: http://psychclassics.yorku.ca/Fisher/Methods.

Fisher, R. A., and R. A. Fisher (1915), Frequency distribution of the values of the correlation coefficient in samples from an indefinitely large population, *Biometrika*, *10*(4), 507–521. Available from: http://biomet.oxfordjournals.org/cgi/reprint/10/4/507.pdf.

Gabrovsek, F., and W. Dreybrodt (2010), Karstification in unconfined limestone aquifers by mixing of phreatic water with surface water from a local input: A model, *J. Hydrol.*, *386*(1–4), 130–141.

Gelman, A. (2008), Objections to Bayesian statistics rejoinder, *Bayesian Anal.*, *3*(3), 467–477.

Gelman, A. and C. R. Shalizi (2013), Philosophy and the practice of {Bayesian} statistics, *Br. J. Math. Stat. Psychol.*, *66*(1), 8–38.

Gelman, A., et al. (1996), Posterior predictive assessment of model fitness via realized discrepancies, *Stat. Sin.*, *6*(4), 733–807. Available from: http://www3.stat.sinica.edu.tw/statistica/j6n4/j6n41/j6n41.htm.

Gelman, A., et al. (2004), *Bayesian Data Analysis*, Chapman & Hall, Boca Raton, FL.

Genest, C., and J. V. Zidek (1986), Combining probability distributions: A critique and an annotated bibliography, *Stat. Sci.*, *1*(1), 114–135. Available from: http://projecteuclid.org/euclid.ss/1177013825.

Gibling, M. R. (2006), Width and thickness of fluvial channel bodies and valley fills in the geological record: A literature compilation and classification, *J. Sediment. Res.*, *76*(5), 731–770.

Gnedenko, B. V., I. Aleksandr, and A. Khinchin (1962), *An Elementary Introduction to the Theory of Probability*, Dover Publications, New York.

Gómez-Hernández, J. J., and X. H. Wen (1998), To be or not to be multi-Gaussian? A reflection on stochastic hydrogeology, *Adv. Water Resour.*, *21*(1), 47–61.

Halpern, J. Y. (1995), A logical approach to reasoning about uncertainty: A tutorial, in *Discourse, Interaction, and Communication*, edited by K. Lehrer, S. Cohen, and X. Arrazola, pp.141–155, Springer, Dordrecht.

Halpern, J. Y. (2011), A counter example to theorems of Cox and Fine, *arXiv preprint arXiv:1105.5450*, *10*, 67–85. http://arxiv.org/abs/1105.5450.

Hand, D. J., and P. Walley (1993), Statistical reasoning with imprecise probabilities, *Appl. Stat.*, *42*(1), 237.

Hanson, N. R. (1958), Patterns of discovery, *Philos. Rev.*, *69*(2), 247–252.

Henriksen, H. J., et al. (2012), Use of Bayesian belief networks for dealing with ambiguity in integrated groundwater management, *Integr. Environ. Assess. Manag.*, *8*(3), 430–444.

Hermans, T., et al. (2015), Quantitative temperature monitoring of a heat tracing experiment using cross-borehole ERT, *Geothermics*, *53*, 14–26.

Höhle, U. (2003), Metamathematics of fuzzy logic, *Fuzzy Sets Syst.*, *133*(3), 411–412.

Howson, C. (1991), The "old evidence" problem, *Br. J. Philos. Sci.*, *42*(4), 547–555.

Howson, C., P. Urbach, and B. Gower (1993), Scientific reasoning: The Bayesian approach. *orton.catie.ac.cr*.

Hume, D. (1978), A Treatise of Human Nature (1739). British Moralists, 1650–1800.

Jaynes, E. T. (1957), Information theory and statistical mechanics, *Phys. Rev.*, *106*(4), 181–218.

Jaynes, E. T. (2003), Probability theory: The logic of science, *Math. Intell.*, *27*(2), 83–83.

Jenei, S., and J. C. Fodor (1998), On continuous triangular norms, *Fuzzy Sets Syst.*, *100*(1–3), 273–282. Available from: http://www.sciencedirect.com/science/article/pii/S0165011497000638/pdf?md5=ef4265149e771493d1917f70c288f249&pid=1-s2.0-S0165011497000638-main.pdf.

Journel, A. G. (2002), Combining knowledge from diverse sources: An alternative to traditional data independence hypotheses, *Math. Geol.*, *34*(5), 573–596.

Jung, A., and T. Aigner, (2012), Carbonate geobodies: Hierarchical classification and database – A new workflow for 3D reservoir modelling, *J. Pet. Geol.*, *35*, 49–65. doi:10.1111/j.1747-5457.2012.00518.x

Kenter, J. A. M., and P. M. Harris (2006), Web-based Outcrop Digital Analog Database (WODAD): Archiving carbonate platform margins, *AAPG International Conference*, Australia, November, pp. 5–8.

Klement, E. P., R. Mesiar, and E. Pap (2004), Triangular norms. Position paper I: Basic analytical and algebraic properties. *Fuzzy Sets Syst.*, *143*(1), 5–26.

Klir, G. J. (1994), On the alleged superiority of probabilistic representation of uncertainty, *IEEE Tran. Fuzzy Syst.*, *2*(1), 27–31.

Kolmogoroff, A. N. (1950), *Foundations of the Theory of Probability*, Chelsea Publishing Company, New York.

Koltermann, C., and S. Gorelick (1992), Paleoclimatic signature in terrestrial flood deposits, *Science*, *256*(5065), 1775–1782.

Kuhn, T. S. (1996), *The Structure of Scientific Revolution*, University of Chicago Press, Chicago, IL.

Lantuéjoul, C. (2013), *Geostatistical Simulation: Models and Algorithms*. Springer Science & Business Media, Berlin.

Lanzoni, S., and G. Seminara (2006), On the nature of meander instability, *J. Geophys. Res. Earth Surf.*, *111*(4).

Levenson, R. M., et al. (2015), Pigeons spot cancer as well as human experts, *PLOS ONE*. Available from: http://www.sciencemag.org/news/2015/11/pigeons-spot-cancer-well-human-experts.

Lindgren, B. (1976), *Statistical Theory*, MacMillan, New York.

Loginov, V. J. (1966), Probability treatment of Zadeh membership function and their use in pattern recognition, *Eng. Cybernet.*, *20*, 68–69.

Mariethoz, G., and J. Caers (2014), *Multiple-Point Geostatistics: Stochastic Modeling with Training Images*, Wiley Blackwell, Chichester.

Martinius, A. W., and A. Naess (2005), Uncertainty analysis of fluvial outcrop data for stochastic reservoir modelling, *Pet. Geosci.*, *11*(3), 203–214.

Mayo, D. (1996), *Error and the Growth of Experimental Knowledge*, Chicago University Press, Chicago, IL.

Neyman, J., and E. S. Pearson (1933), On the problem of the most efficient tests of statistical hypotheses, *Philos. Trans. R. Soc. A. Math., Phys. Eng. Sci.*, *231*(694–706), 289–337.

Neyman, J., and E. S. Pearson (1967), *Joint Statistical Papers*, University of California Press, Berkeley, CA.

Nicholas, A. P., et al. (2013), Numerical simulation of bar and island morphodynamics in anabranching megarivers, *J. Geophys. Res. Earth Surf.*, *118*(4), 2019–2044.

Pearson, K., R. A. Fisher, and H. F. Inman (1994), Karl Pearson and R. A. Fisher on statistical tests: A 1935 exchange from nature, *Am. Stat.*, *48*(1), 2–11. Available from: http://www.jstor.org/stable/2685077.

Popper, K. R. (1959), *The Logic of Scientific Discovery*, Hutchinson, London.

Pyrcz, M. J., J. B. Boisvert, and C. V. Deutsch (2008), A library of training images for fluvial and deepwater reservoirs and associated code. *Comput. Geosci.*, *34*, 542–560. doi:10.1016/j.cageo.2007.05.015

Rao, C. R. (1992), R. A. Fisher: The founder of modern statistics, *Stat. Sci.*, *7*(1), 34–48. Available from: http://www.jstor.org/stable/2245989.

Refsgaard, J. C., et al. (2012), Review of strategies for handling geological uncertainty in groundwater flow and transport modeling, *Adv. Water Resour.*, *36*, 36–50. doi:http://dx.doi.org/10.1016/j.advwatres.2011.04.006.

Rings, J., et al. (2012), Bayesian model averaging using particle filtering and Gaussian mixture modeling: Theory, concepts, and simulation experiments, *Water Resour. Res.*, *48*(5). doi: http://doi.wiley.com/10.1029/2011WR011607.

Scheidt, C., et al. (2016), Quantifying natural delta variability using a multiple-point geostatistics prior uncertainty model, *J. Geophys. Res. Earth Surf.*, *121*, 1800–1818.

Shackle, G. L. S. (1962), The stages of economic growth, *Pol. Stud.*, *10*(1), 65–67.

Shannon, C. E. (1948), A mathematical theory of communication, *Bell Syst. Tech. J.*, *27*, 379–423. Available from: http://cm.bell-labs.com/cm/ms/what/shannonday/shannon1948.pdf.

Simmons, M. (2013), *Twilight in the Desert: The Coming Saudi Oil Shock and the World Economy*, Wiley, Hoboken, NJ.

Tsai, F. T. C., and A. S. Elshall (2013), Hierarchical Bayesian model averaging for hydrostratigraphic modeling: Uncertainty segregation and comparative evaluation, *Water Resour. Res.*, *49*(9), 5520–5536.

Valle, A., A. Pham, P. T. Hsueh, and J. Faulhaber (1993), Development and use of a finely gridded window model for a reservoir containing super permeable channels, *Society of Petroleum Engineers 8th Middle East Oil Show*, Bahrain, vol. *2*, pp. 305–320.

Wang, P. (2004), The limitation of Bayesianism, *Artif. Intell.*, *158*(1), 97–106.

Wang, Y., K. M. Straub, and E. A. Hajek (2011), Scale-dependent compensational stacking: An estimate of autogenic time scales in channelized sedimentary deposits, *Geology*, *39*(9), 811–814.

Zadeh, L. A. (1965), Fuzzy Sets, *Inf. Control*, *8*(3), 338–353.

Zadeh, L. A. (1975), Fuzzy logic and approximate reasoning, *Synthese*, *30*(3–4), 407–428.

Zadeh, L. A. (1978), Fuzzy sets as a basis for a theory of possibility, *Fuzzy Set. Syst.*, *1*(1), 3–28.

Zadeh, L. A. (2004), Fuzzy logic systems: Origin, concepts, and trends, *Science*, 16–18.

Zinn, B., and C. F. Harvey (2003), When good statistical models of aquifer heterogeneity go bad: A comparison of flow, dispersion, and mass transfer in connected and multivariate Gaussian hydraulic conductivity fields. *Water Resour. Res.*, *39*(3), 1–19.

6

Geological Priors and Inversion

6.1. INTRODUCTION

Inverse problems are ubiquitous in the Earth Sciences. When dealing with data and modeling some form of "inversion" usually applies. Applications range from the local kilometer scale (mostly the topic of this book) to the global Earth scale (such as in global seismology). Here we deal with inversion within the context of uncertainty quantification (UQ). Inversion can be an important part of UQ, but UQ requires more than inversion. UQ requires considering many components not just data and model but also the prediction, the decision to be made, the a-priori geological information at hand, and so on. Inversion here will be understood as the narrower set of methodologies that infer models from data, the relation between the two being modeled by a physical model (in the sense of not a pure mathematical or statistical model).

The chapter proceeds by first providing a theory of inversion in its most general form, mostly based on the work of Tarantola and his school of thought [*Tarantola*, 1987; *Mosegaard*, 1995]. This theory provides a high-level description. Any practical implementation of inverse modeling will require a model parameterization. Model variables and data variables must be chosen to represent the description of the system as a whole. This choice is important and not unique, in particular since models can be very large and hence a good model parameterization is important to render the inverse solution computationally manageable. We will discuss not only the common explicit grid model but also how to parameterize more complex subsurface structures such as faults and horizons. We will also provide some alternatives to the grid model and present implicit models. We will compare several forms of inversion, from the deterministic inversion that is still quite common, to the fully stochastic (Bayesian) inversion relying on Markov chain Monte Carlo (McMC). Next some specifics are discussed regarding the inversion of geophysical data that often requires the addition of statistical rock physics in the forward model and the use of dynamic observations with the ensemble Kalman filter. The latter involve all time-varying observations such as time-lapse geophysics, well test, pump test, tracer test, and so on. The literature on these subjects is vast; our focus lies on the integration of geological information into the inverse modeling and the context within UQ.

6.2. THE GENERAL DISCRETE INVERSE PROBLEM

6.2.1. Introduction

In this section, we will review a general formulation of the inverse problem, and not yet focus on a specific method to parameterize it or to present inverse solution methodologies. The aim here is to show that many of the standard formulations (in particular the Bayesian/Gaussian) are derived from a more general formulation. Under the usual inverse modeling paradigm these assumptions are taken for granted, not necessarily without consequences.

Tarantola and Valette [1982] provided one of the first comprehensive formulation of "the inverse problem." It is different from the usual Bayes' formulation. They (TV) state conditions necessary for such formulation:

1. The formulation should address linear to "strongly" nonlinear problems. We will later quantify what "strong" means exactly.

2. The formulation must address both overdetermined and underdetermined problems.

3. The formulation must be invariant to any change in parameterization. In other words, solving an inverse problem with frequency or period should yield the same solution (uncertainty), because a change of variables does not add any new information to the problem.

Quantifying Uncertainty in Subsurface Systems, Geophysical Monograph 236,
First Edition. Céline Scheidt, Lewis Li, and Jef Caers.

4. The formulation must be general enough to allow for "errors in theory" and "errors in observations."

5. The formulation must allow for any a priori information. Although this notion in TV is somewhat differently formulated than in Bayesianism.

TV follows the general notion in probability that any *state of information* on a parameter can be defined by a probability density function. Hence, solving inverse problems requires combining measurements with prior states of information on model parameters and with states of information on the physical correlation between data and model. They consider inverse problems within the broader approach to scientific discovery by considering three major steps:

1. Parameterization of the physical system: stating the minimal set of model parameters whose values completely characterize the system.
2. Forward modeling: discovery of physical laws that allow making predictions on the results of measurements.
3. Inverse modeling: infer actual values of the model.

From a philosophical point of view, TV adhere to notions of probability theory and hence some form of Bayesianism, although Tarantola also prescribed to falsificationism [*Tarantola*, 2006] after having formulated his more inductionist-inspired theory.

TV limit themselves to a "discrete" inverse problem, meaning that model and data variables are discretized and represented through a finite list of parameters. In other words, they do not treat problems that involve functions. This is mainly of technical concern, functions can easily be represented with discrete coefficients after some appropriate decomposition (e.g., wavelet, Fourier, Discrete Cosine Transform).

6.2.2. Representation of Physical Variables in Probability Theory

6.2.2.1. Model and Data Space. Many applications in probability theory deal with variables that are not based on physics (here meant to include chemistry), such as number of people, income, IQ, crime level, and so on, which are represented by events (sets) or random variables. Such variables are often represented by Cartesian axis systems in which linear (matrix) operations apply. The Cartesian axis system (e.g., income vs. IQ) represents the parameterization of the problem.

In our type of inverse problems, we deal with model variables and data variables that are often physical in nature. As discussed in Chapter 2, the model variables and data variables are represented by

$$\mathbf{m} = (m_1, m_2, \ldots) \text{ and } \mathbf{d} = (d_1, d_2, \ldots) \tag{6.1}$$

as part of an abstract space of points or manifold Ω_m for the model space and Ω_d for the data space. Recall that model and data are treated in the same manner, namely data are considered to be measurements of observable parameters.

6.2.2.2. Metric Densities. Physical variables (compressibility, resistivity, frequency) cannot be treated on the same footing as nonphysical ones. Such variables often have an inverse (bulk modulus, conductivity, period). Consider a simple model composed of two physical variables, resistivity and period, $\{\rho, T\}$, and two model realizations $\{\rho_1, T_1\}$ and $\{\rho_2, T_2\}$, then the Euclidean distance is

$$d_{\rho T} = \sqrt{(\rho_1 - \rho_2)^2 + (T_1 - T_2)^2} \tag{6.2}$$

Consider now the same Euclidean distance for conductivity and frequency $\{\kappa, f\}$

$$d_{\kappa f} = \sqrt{(\kappa_1 - \kappa_2)^2 + (f_1 - f_2)^2} \tag{6.3}$$

Clearly, $d_{\kappa f} \neq d_{\rho T}$. This is a problem; the two realizations represent the exact same physical system, just expressed differently, yet their parametrization in a Cartesian axis system leads to different distances. A difference in distance logically means a difference in "density." If one would consider now 100 realizations, say from a uniform distribution, and calculate the distances in the $\{\rho, T\}$ representation and make a multi-dimensional scaling (MDS) plot of these, then that MDS plot is different if one uses the $\{\kappa, f\}$ representation. The empirical density calculated over these two MDS plots is different, see Figure 6.1a. This matter is not without consequence. A modeler working with $\{\rho, T\}$ will get different results than a modeler working with $\{\kappa, f\}$, and if these prior distributions are used in a UQ study, both the modelers will produce different posterior distributions.

Clearly, a uniform Cartesian space is not a good choice for physical parameters. In this case, a better choice would be to define the following distance:

$$d_{\rho T} = \sqrt{\left(\log\left(\frac{\rho_1}{\rho_2}\right)\right)^2 + \left(\log\left(\frac{T_1}{T_2}\right)\right)^2} \tag{6.4}$$

Now $d_{\kappa f} = d_{\rho T}$, see Figure 6.1b, and as a result, the prior densities are invariant to a change in variable. *Jeffreys* [1946] provides a general framework for assigning (prior) distributions such that the distributions are invariant to (certain) transformations. To study this issue a bit deeper, consider a transformation of coordinates $\mathbf{x}$ (e.g., model or data variables)

$$\mathbf{x}^* = t(\mathbf{x}) \tag{6.5}$$

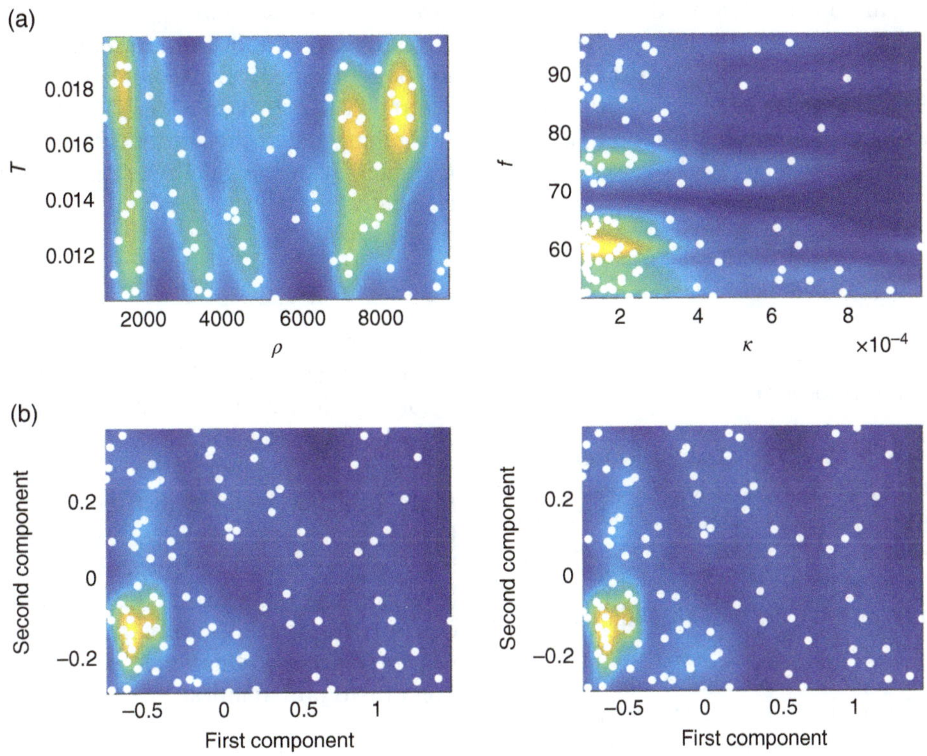

Figure 6.1 (a) Prior densities depend on the parameterization. (b) Prior densities are invariant to the parameterization.

Consider that **X** is a random vector with density $f(\mathbf{x})$. Per definition of the density function

$$\int_A f(\mathbf{x}^*)d\mathbf{x}^* = \int_A f(\mathbf{x})d\mathbf{x} \tag{6.6}$$

hence, per elementary property of integrals

$$f(\mathbf{x}^*) = f(\mathbf{x})\left|\frac{\partial \mathbf{x}}{\partial \mathbf{x}^*}\right| \tag{6.7}$$

The above example shows that each parameterization comes with a different "density." This essentially means that the volume associated with some event A (a set basically) is not uniform. Consider $V(A)$ to be the volume with the set A, then define $v(\mathbf{x})$ as the volume density, or

$$dV(\mathbf{x}) = v(\mathbf{x})d\mathbf{x} \Rightarrow \int_A v(\mathbf{x})d\mathbf{x} = V(A) \tag{6.8}$$

then given some total volume V

$$\mu(\mathbf{x}) = \frac{v(\mathbf{x})}{V} \tag{6.9}$$

which is termed the "homogenous probability density." When dealing with Jeffrey's parameters we have that an elementary volume decreases proportional to its magnitude [*Jeffreys*, 1946]

$$dV(\rho, T) = \frac{d\rho}{\rho}\frac{dT}{T} \tag{6.10}$$

which entails the homogenous probability density

$$\mu(\rho, T) = \frac{1}{\rho T} \tag{6.11}$$

Logically, then $\{f, \kappa\}$ has density of similar shape

$$\mu^*(f, \kappa) = \frac{1}{f\kappa} \tag{6.12}$$

Note that the functional form $1/x$ corresponds to the lognormal distribution when taken in the limit to infinity. Jeffrey's parameters often exhibit lognormal distributions.

In the statistical literature, Eq. (6.9) is often termed the "non-informative" distribution [*Jaynes*, 2003], a rather poor choice of terminology, as this density *informs* the metric being used. While Tarantola terms this the homogeneous density, a more appropriate term may be to use "metric density" as it reflects the metric used to measure the distance in the space of models and space of data parameters. In fact, given a distance, the general representation of a metric density is obtained from differential geometry [see e.g., *Sternbergh*, 1999] as

$$f_{\text{metric}}(\mathbf{x}) = \sqrt{\det(D(x_1, x_2, \ldots))} \tag{6.13}$$

with $D(x_1, x_2, \ldots)$ the metric tensor defined on the manifold Ω_x (e.g., a Euclidean metric tensor is the identity matrix for an n-dimensional manifold). A metric tensor is the derivative of the distance function defined on that space. From now on we will use the notation f_{metric} for metric density and

$$P_{\text{metric}}(A) = \int_A f_{\text{metric}}(\mathbf{x})d\mathbf{x} \tag{6.14}$$

6.2.3. Conjunction of Information

6.2.3.1. Information Sources: Experimentation and Theory. Tarantola makes an analogy between how research in physics works and how inverse problems are formulated [*Tarantola*, 1987]. Hc argucs that two realms exist: (i) "the realm of experimentation," conducting of experiments in lab/field and (ii) "the realm of theorization," the creation of theories and models of the real world. In Chapter 5 we discussed the various views (induction, deduction, paradigm) on how these worlds interact. Tarantola views each as providing information about the unknown world. These two information streams need to be conjoined into one single information model. Since he relies on probability theory, information is represented by probability density distributions. One of the major differences with Bayes' theory is a quest for symmetry (not atypical for a physicist) between the theoretical world and the empirical world. In classical Bayes' theory, at least technically, the hypothesis comes first (the prior) and hypotheses are then assessed probabilistically with evidence. The prior basically always existed, the data is then specific to the case studied (recall that *Gelman and Shalizi* [2013] question the classical induction notion of Bayes' theory as was discussed in Chapter 5). If data variables represent the observable world and model variables represent the unobservable world, then both sources (experimentation and theorization) provide information on each, represented by two probability density functions:

1. $f_{\text{prior}}(\mathbf{d}, \mathbf{m})$: any experimental information available on both the data variables and the model variables (observable and unobservable), prior to formulating theories. Although there is a Bayesian flavor here, Bayes' theory would consider $f_{\text{prior}}(\mathbf{d})$ and $f_{\text{prior}}(\mathbf{m})$ separately.
2. $f_{\text{theory}}(\mathbf{d}, \mathbf{m})$: any association between the data and model variables as speculated through theories. For example, such theories can be represented by partial differential equations.

Note that these are pdfs, not deterministic functions, so any uncertainties associated with both are part of the formulation.

6.2.3.2. Conjunction. $f_{\text{prior}}(\mathbf{d}, \mathbf{m})$ and $f_{\text{theory}}(\mathbf{d}, \mathbf{m})$ are considered probabilistic information sources that need to be conjoined. Because they express uncertainty, Tarantola calls on fuzzy logic ("and" operation between two vague statements) to address this problem. Recall (Chapter 5) that logical operations require a "neutral" element. In classical logic, a neutral element for a proposition is "1," since

$$p \cap 1 = p \tag{6.15}$$

In probability theory, the "neutral element" is termed a "non-informative distribution," and it models complete ignorance; hence, it will not affect any conjunction of probabilistic information (it does not add any information). The neutral element in this context is not necessarily the uniform distribution, since we already know something about physical variables. This knowledge is injected through the parameterization (e.g., deciding to model wave propagation with frequencies or periods). This information needs to be accounted for. To understand this better, consider now the conjunction of two probabilities:

$$P_1(A) = \int_A f_1(\mathbf{x})d\mathbf{x} \quad P_2(A) = \int_A f_2(\mathbf{x})d\mathbf{x} \tag{6.16}$$

The conjunction of two probabilities becomes

$$(P_1 \cap P_2)(A) = \int_A (f_1 \cap f_2)(\mathbf{x})d\mathbf{x} \tag{6.17}$$

The neutral element here is then, using the metric density,

$$P_{\text{metric}}(A) = \int_A f_{\text{metric}}(\mathbf{x})d\mathbf{x} \tag{6.18}$$

which imposes the conditions

$$(P_n \cap P_{\text{metric}})(A) = P_n(A) \quad n = 1,2 \tag{6.19}$$

From Eqs. (6.17) and (6.18), Tarantola deduces that the following is a conjunction of probabilistic information with metric density as a neutral element:

$$(f_1 \cap f_2)(\mathbf{x}) \simeq \frac{f_1(\mathbf{x})f_2(\mathbf{x})}{f_{\text{metric}}(\mathbf{x})} \tag{6.20}$$

Consider now $\mathbf{x} = (\mathbf{d}, \mathbf{m})$, the conjunction of information $f_{\text{prior}}(\mathbf{d}, \mathbf{m})$ and $f_{\text{theory}}(\mathbf{d}, \mathbf{m})$ in the context of inverse problems is defined as

$$f(\mathbf{d}, \mathbf{m}) \simeq \frac{f_{\text{theory}}(\mathbf{d}, \mathbf{m})f_{\text{prior}}(\mathbf{d}, \mathbf{m})}{f_{\text{metric}}(\mathbf{d}, \mathbf{m})} \tag{6.21}$$

This is a general formulation of the discrete inverse problem. Note the symmetry in the formulation in terms of $(\mathbf{d}, \mathbf{m})$. It reflects that any choice of parameterization should not distinguish **d** and **m** as being any different in the sense they are all physical variables representing the system. Before specifying a narrower formulation that will eventually be used to sample inverse solutions, we provide a discussion on how conditional probabilities are special form of conjunction (the reader may also refer to Chapter 5).

6.2.3.3. Conditional Distributions as Conjunctions. As was discussed in Chapter 5, probabilities cannot handle vague information and quantify the probability of it raining Sunday given that it *may* rain Saturday. Instead, in probability theory, either something is true or false; there is no middle. How can we formulate a conditioning such as $P(A|B)$ with a conjunction? Consider A and B to be two (overlapping) sets on some manifold Ω. Define

$$f_{\text{metric},A}(\mathbf{x}) = \begin{cases} f_{\text{metric}}(\mathbf{x}) \; \textit{if} \; \mathbf{x} \in A \\ 0 \end{cases} \tag{6.22}$$

as the metric density limited by A. Then for any other event, we define

$$P_{\text{metric},A}(B) = \int_B f_{\text{metric},A}(\mathbf{x}) d\mathbf{x} \tag{6.23}$$

Figure 6.2 describes the situation in 1D.

Tarantola defines a conditional probability as a conjunction

$$P(A|B) = (P \cap P_{\text{metric},A})(B) \tag{6.24}$$

in other words, a conjunction between the probability model over A with the metric density (the neutral element evaluated in B). The conjunction is more general in the sense that B must be a set and can be a fuzzy set (no hard bounds). This also shows that conditional probabilities need not be derived from the more "primitive" definition:

$$P(A|B) = \frac{P(A \cap B)}{P(B)} \tag{6.25}$$

6.2.4. More Limited Formulations

$$f_{\text{theory}}(\mathbf{d}, \mathbf{m})$$

We will derive the usual (and more limited) formulation of the inverse problem, by simplifying each term in Eq. (6.21). To get to the classical Bayesian formulation, we state that

$$f_{\text{theory}}(\mathbf{d}, \mathbf{m}) = f_{\text{forward}}(\mathbf{d}|\mathbf{m}) f_{\text{metric}}(\mathbf{m}) \tag{6.26}$$

which requires assuming

$$f_{\text{metric}}(\mathbf{d}, \mathbf{m}) = f_{\text{metric}}(\mathbf{d}) f_{\text{metric}}(\mathbf{m}) \tag{6.27}$$

and assuming that the physical relationship between **d** and **m** can be expressed using a forward model $g(\mathbf{m})$. Often one assumes error on this relationship, for example

$$f_{\text{forward}}(\mathbf{d}|\mathbf{m}) \simeq \exp\left((\mathbf{d} - g(\mathbf{m}))^T C_{\text{theory}}^{-1} (\mathbf{d} - g(\mathbf{m})) \right) \tag{6.28}$$

This error is "model" error, reflecting the uncertain relationship between variables due to the fact that the actual physical relationship remains unknown and is simply being approximated. The model error is here limited to a covariance (because of the choice of the multivariate Gaussian).

$$f_{\text{prior}}(\mathbf{d}, \mathbf{m})$$

$$f_{\text{prior}}(\mathbf{d})$$

First, we assume that the prior on model parameters and data parameters are independent (recall that this prior is different from the Bayesian prior):

$$f_{\text{prior}}(\mathbf{d}, \mathbf{m}) = f_{\text{prior}}(\mathbf{d}) f_{\text{prior}}(\mathbf{m}) \tag{6.29}$$

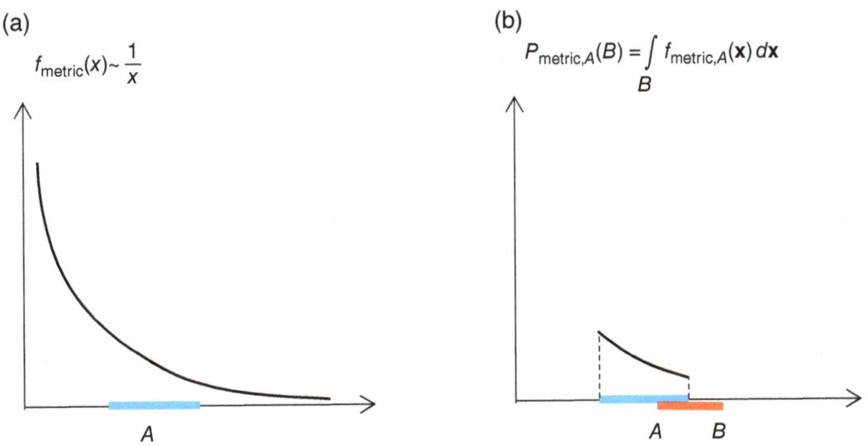

Figure 6.2 (a) Metric density of which the integral need not exist. (b) The integral always exists.

Consider first $f_{\text{prior}}(\mathbf{d})$. If measurements were perfect then this term simply identifies the observations

$$f_{\text{prior}}(\mathbf{d}) = \delta(\mathbf{d} - \mathbf{d}_{\text{observed}}) \tag{6.30}$$

In reality, the true observables $\mathbf{d}$ are not observed, for a variety of reasons, most importantly because we have experimental setups that are limited; $\mathbf{d}$ is not observed but something else related to $\mathbf{d}$ is. This situation is extremely common and several examples were given in Chapter 5. Consider, therefore, the general situation where $\mathbf{d}$ is being "probed" so that we only get $\mathbf{d}_{\text{probed}}$ instead of $\mathbf{d}$. Then, the joint distribution $f(\mathbf{d}, \mathbf{d}_{\text{probed}})$ can be written as

$$f(\mathbf{d}, \mathbf{d}_{\text{probed}}) = f(\mathbf{d}_{\text{probed}} | \mathbf{d}) f_{\text{metric}}(\mathbf{d}) \tag{6.31}$$

modeling that the output of instruments are some function of real physical variables being measured/probed. A common assumption (a choice) is to use an additive error model

$$\mathbf{d}_{\text{probed}} = \mathbf{d} + \boldsymbol{\varepsilon} \tag{6.32}$$

in particular, assume the error is not a function of what is being measured (homoscedasticity):

$$f_{\text{prior}}(\mathbf{d}) = f(\mathbf{d}_{\text{probed}} | \mathbf{d}) = f(\mathbf{d}_{\text{probed}} - \mathbf{d}) = f(\boldsymbol{\varepsilon}) \tag{6.33}$$

This is then evaluated in $\mathbf{d}_{\text{probed}} = \mathbf{d}_{\text{observed}}$, the actual observed measurements. A common assumption is to use a multivariate Gaussian distribution:

$$f_{\text{prior}}(\mathbf{d}) \simeq \exp\left((\mathbf{d} - \mathbf{d}_{\text{observed}})^T C_\varepsilon^{-1} (\mathbf{d} - \mathbf{d}_{\text{observed}})\right) \tag{6.34}$$

with C_ε^{-1} the covariance of the error vector $\boldsymbol{\varepsilon}$. Often this covariance is taken as diagonal simply because the off-diagonal elements are hard to assess (repeated measurements are rarely available).

$$f_{\text{prior}}(\mathbf{m})$$

A very common assumption here is to assume a multivariate Gaussian prior. In other words, the model $\mathbf{m}$ is considered to be a Gaussian process, when such model has a space component, which is the norm in the subsurface.

6.2.4.1. Simplified Formulation. Given Eqs. (6.26), (6.27), and (6.29), we can now state a simplified formulation of the general inverse problem:

$$f(\mathbf{d}, \mathbf{m}) \simeq f_{\text{prior}}(\mathbf{m}) \frac{f_{\text{prior}}(\mathbf{d}) f_{\text{forward}}(\mathbf{d} | \mathbf{m})}{f_{\text{metric}}(\mathbf{d})} \tag{6.35}$$

and as a result

$$f(\mathbf{m}) \simeq f_{\text{prior}}(\mathbf{m}) \int_{\mathbf{d}} \frac{f_{\text{prior}}(\mathbf{d}) f_{\text{forward}}(\mathbf{d} | \mathbf{m})}{f_{\text{metric}}(\mathbf{d})} d\mathbf{d} \tag{6.36}$$

A simplified form can now be retrieved by equating the integral part as a likelihood model:

$$L(\mathbf{m}) = \int_{\mathbf{d}} \frac{f_{\text{prior}}(\mathbf{d}) f_{\text{forward}}(\mathbf{d} | \mathbf{m})}{f_{\text{metric}}(\mathbf{d})} d\mathbf{d} \tag{6.37}$$

Note that Tarantola uses a slightly different notation for the posterior, $f(\mathbf{m})$, and not $f(\mathbf{m}| \mathbf{d})$, which is used in the Bayesian formulation.

6.3. PRIOR MODEL PARAMETERIZATION

6.3.1. The Prior Movie

It is easy to write mathematically the posterior $f(\mathbf{m})$, then develop an elegant theory and derive equations. Reality, however, begs for an actual statement of the prior (see Chapter 5) and for computer implementations. Apart from the multivariate Gaussian prior, very few explicit spatial prior models are available (and for that reason, the wrong reason, the Gaussian is very popular). Most prior models and their parametrizations are intricately linked with the computer algorithms that generate them. This is not a bad idea at all. A modern trend in geological sciences is to simulate depositional systems of all kind with numerical simulation models. These computer codes have inputs that are uncertain and sometime have built-in (aleatory) randomness. The prior model can then simply be represented by the (possibly infinite) realizations created by the computer code. *Tarantola* [1987] calls this the "prior movie" (a long and possibly boring one!). Here we provide an overview of these "movies," what they are and how they are created, who the director is. We focus specifically on the spatial model parameterization simply because the subsurface is a 3D system. Hence, simple univariate models (e.g., N layers with unknown properties) are often unrealistic prior model representations.

6.3.2. Gridding

Many prior model parameterizations involve some form of grid. A grid is basically a discretization of what in reality is more like a continuous function. The choice of grid is dependent on many factors such as computational demand, need to resolve details, purpose of the study, errors in numerical solutions, and so on. We will not be providing a treatment on this topic and the reader may refer to *Mallet* [2014] or related books on the topic. Although it is common to use a simple 3D Cartesian grid with a regular grid or a regular mesh, this simple grid is not preferable when complex geological structures are present, in particular faults, complex layering and complex geological architecture. Here we provide a broad overview of model parametrizations aiming to represent

3D variability, in particular the subsurface geological system, since this is central to the topics treated in this book.

In many subsurface applications, such as oil/gas, geothermal, even aquifers, faulting has a major impact on fluid flow, and hence, uncertainty of such faulting needs to be taken into account. Section 6.5.1 covers the issue of fault uncertainty. The composition of modeled faults and layers is termed the "structural model." Structural models are often built from geophysical data, as reflectors inform internal layering of the system due to contrasting petrophysical properties. Faults appear on geophysical images because of an off-set in reflectors, due to a fault throw. Because faults tend to be vertical structures, they are only indirectly visible by means of such throw. A structural model is often built from point interpretation from geophysical data (see Figure 6.3). One of the major difficulties in parametrizing the subsurface model (and hence any model of uncertainty) lies in the gridding of complex structures. In addition, the various properties need to follow the stratigraphy, including the offset due to faulting. This means that such properties cannot be directly modeled in real space (see Figure 6.4). Some form of mapping needs to be constructed to a depositional space, where any effect of faulting is removed (see Section 6.5.1). Such mapping can be geometric [*Mallet*, 2004, 2014; *Caumon et al.*, 2009] or based on an actual unfaulting of the system. The latter requires an enormous amount of additional information such as stress, strain, faulting history, and hence, the former is often preferred.

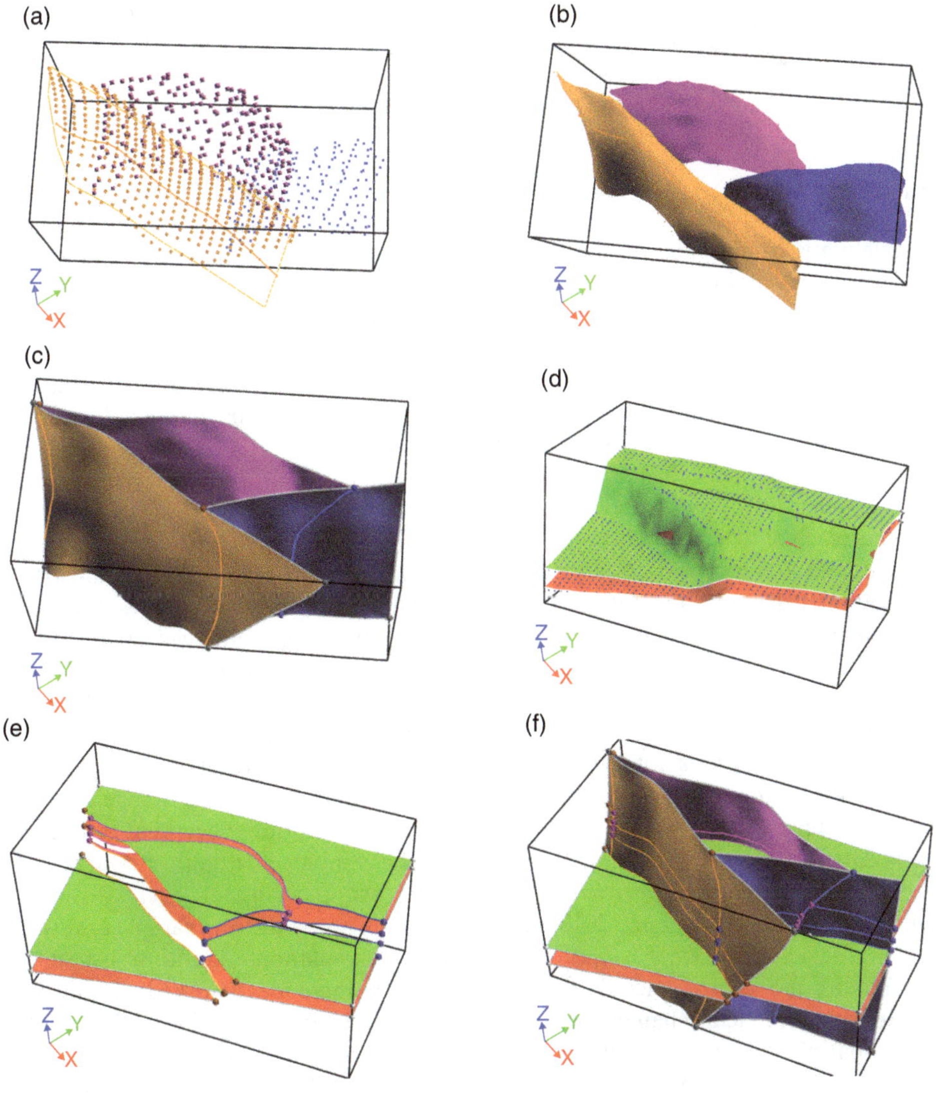

Figure 6.3 (a) Generation of point sets interpreted from a geophysical image of the subsurface, (b) initial fault surface, (c) generating proper intersection, (d) creation of the point set for layering, (e) construction of layer surfaces, and (f) putting it all together [*Caers*, 2011].

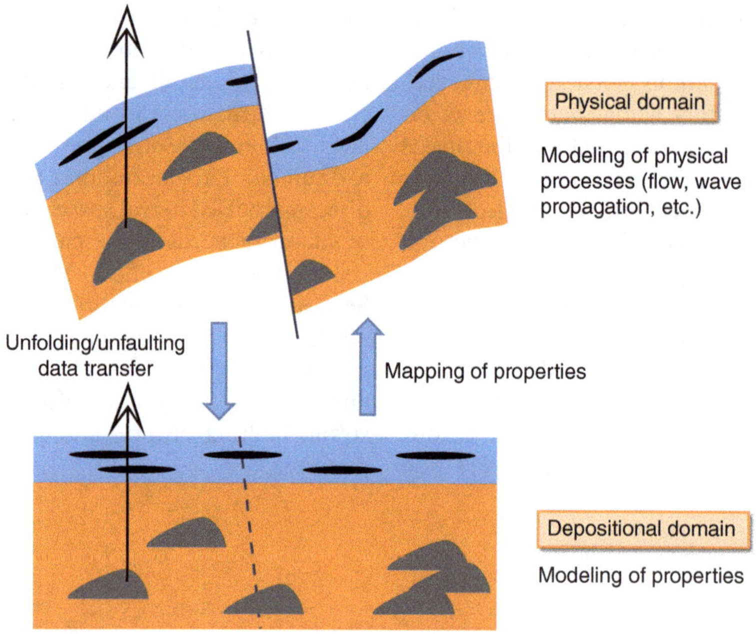

Figure 6.4 Assigning properties in a stratigraphic grid by creating a "depositional domain" [*Caers*, 2011].

6.3.3. Process-Based Prior Model

Process-based modeling is gaining increased attention in generating realistic geological variability through the explicit modeling of the physical processes of deposition, erosion, growth, and diagenesis [e.g., *Koltermann and Gorelick*, 1992; *Gabrovsek and Dreybrodt*, 2010; *Nicholas et al.*, 2013]. Many computer codes are now available. These process models aim not just at creating a stratigraphy but also to model the time-varying depositional process [*Paola*, 2000]. Such models require many input parameters (each of which are uncertain such as boundary conditions and source terms). Because of the considerable computational demand and the nature of forward modeling, constraining such models to complex 3D data such as multiple wells, geophysics, or other data is not possible currently, except for very limited representations [e.g., *Karssenberg et al.*, 2001]

One effort around these limitations is not to model the physics of deposition but to mimic the physics using so-called pseudo-genetic methods that generate structure-imitating model realizations [*Jussel et al.*, 1994; *Scheibe and Freyberg*, 1995; *Koltermann and Gorelick*, 1996; *Deutsch and Tran*, 2002; *Ramanathan et al.*, 2010; *Guin et al.*, 2014].

For instance, a fracture model may include fracture growth and interactions that mimic mechanical processes [*Davy et al.*, 2013]. Similarly, the processes of channel evolution through time (e.g., sedimentation, avulsion) can be accounted for while simulating the objects [*Pyrcz*, 2003; *Zhang et al.*, 2009]. Such ideas have also been used to develop 3D models of karst networks [*Borghi et al.*, 2012; *Rongier et al.*, 2014] by accounting for preexisting geology, fracturing, and phases of karstification without solving the flow, transport, and calcite dissolution equations. Because such models can be generated in a matter of seconds (compared to hours or days for process models), some limited constraining to data can be achieved [*Michael et al.*, 2010; *Bertoncello et al.*, 2013].

6.3.4. Geostatistics

6.3.4.1. Multiple-Point Geostatistics: A Prior Movie Based on Training Images. A new class of structure imitating approaches emerged 20 years ago [*Guardino and Srivastava*, 1993; *Mariethoz and Caers*, 2015]. It uses a training image that represents a fully informed description of how the subsurface may look like, but with the locations of different repeating structures being unknown. The concept of a training image can be seen as a vehicle to convey the prior conceptual geological knowledge that is to be combined with other sources of information (e.g., boreholes, outcrop, etc.) via a simulation algorithm. Figure 6.5 provides an example of this idea. The first successful simulation algorithm [snesim, *Strebelle*, 2002] based on these ideas worked with high-order conditional statistics or multiple-point statistics (MPS) derived from the training image. The *snesim* algorithm was restricted to categorical images with a small number of categories. The concept of a training image opened up a whole set of possible simulation methods. Indeed, why not use techniques derived from pattern recognition, texture synthesis, and machine learning algorithms? A large variety of methods to generate realizations for the prior movie based on training images have been developed [*Arpat and Caers*, 2007; *Honarkhah and Caers*, 2010; *Tahmasebi et al.*, 2012; *Huang et al.*, 2013; *Straubhaar et al.*, 2013; *Mahmud et al.*, 2014; *Mariethoz and Lefebvre*, 2014]. The generation of training images can be challenging [*Chugunova and Hu*, 2008; *Comunian et al.*, 2012; *Comunian et al.*, 2014]. Common approaches include using a process-based, an object-based method [*Deutsch and Tran*, 2002; *Pyrcz et al.*, 2009] or outcrop data [*Huysmans and Dassargues*, 2009]. Another approach is to model Markov random fields based on these training images [*Kindermann and Snell*, 1980; *Tjelmeland and Besag*, 1998; *Mariethoz and Caers*, 2015]. While such methods rely on consistency offered by probability theory, the challenging parameter inference of the Markov random field (MRF) model and computational burden in simulating such model using McMC makes them difficult to apply in actual applications.

What is relevant for UQ (see our discussion in Chapter 4) is the selection of training images that represent a realistic prior for the geological depositional under study (fluvial, deltaic, etc.). As illustrated in Chapter 4, one single training image rarely, if ever, represents realistic prior geological uncertainty. In that sense, an a-priori rich set of training images (100s) can be proposed and an attempt made to falsify them by means of data [*Park et al.*, 2013; *Scheidt et al.*, 2015]. An example of such falsification with geophysical data is presented in Chapter 7.

6.3.4.2. Variogram-Based Geostatistics. Variogram-based approaches are widely used, but they are often insufficient to capture the complexity of geological structures. Sequential indicator simulations [SIS; *Goovaerts*, 1997] or transition probability-based techniques, such as T-Progs [*Carle and Fogg*, 1996; *Jones et al.*, 2003], were remarkable advances in the 1990s and they are still among the most popular techniques to model geological heterogeneity. Unfortunately, they cannot properly reproduce curvilinear features, such as

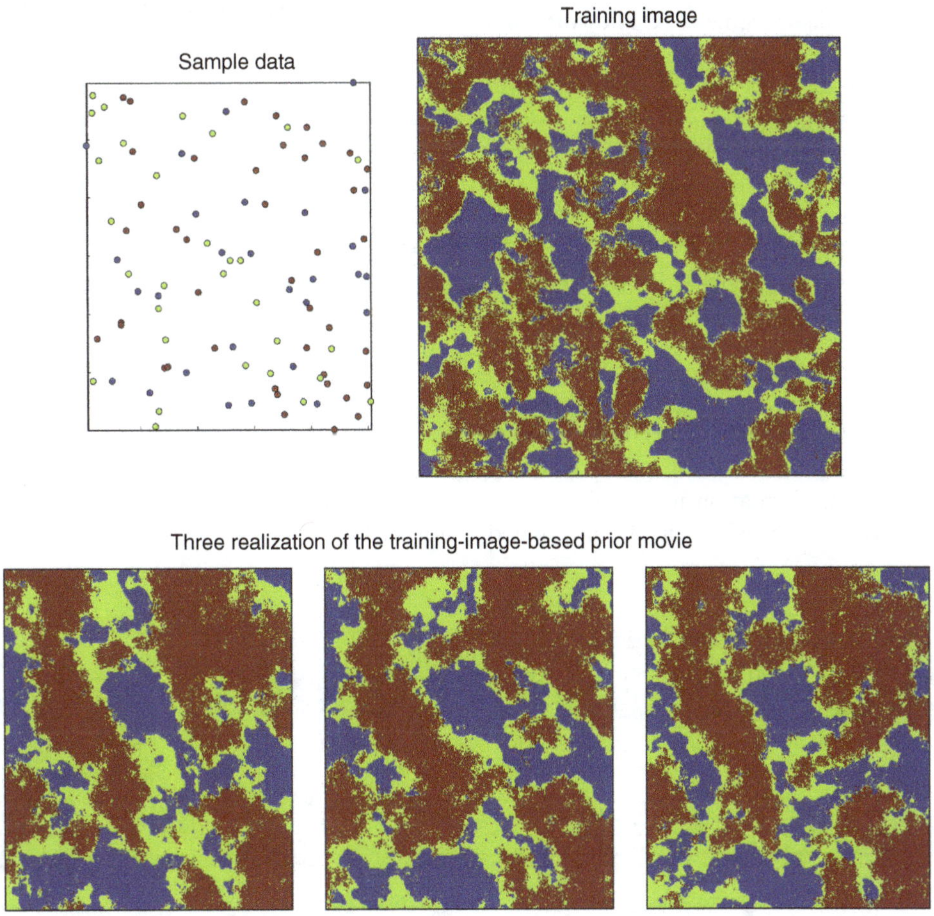

Figure 6.5 MPS: generating realizations from training images, constrained to hard data. Here three "snapshots" of the prior movie are shown.

channels [*Strebelle*, 2002] or more complex structures, and they do not include conceptual geological information beyond simple transitional constraints on the dimension and relations between structures. They are also limited in simulating realistic subsurface hydraulic connectivity, which often has considerable impact on fluid flow.

A method of increasing popularity is the truncated pluri-Gaussian approach [*Dowd et al.*, 2003; *Mariethoz et al.*, 2009] With pluri-Gaussian methods, it is possible, for example, to impose channels to be surrounded by levees, which in turn are surrounded by a flood plain. As compared to SIS or T-Progs, the inference of the underlying variogram is more complex.

6.3.5. Non-Grid-Based Prior Movies

6.3.5.1. Object-Based Priors. Object-based methods allow for the direct modeling of geometries, their spatial distribution and interactions. In that sense, they do not need the definition of a grid. From a practical point of view, such methods are efficient in terms of memory and computational demand. In addition, they are framed within a rigorous probabilistic framework, allowing for a more theoretical treatment than, for example, MPS methods, whose properties can only be studied with simulations.

In particular, Boolean methods allow for a consistent treatment of observations. Such treatment is required since larger objects tend to be more visible in data, for example from wells or geophysics, than smaller objects. In other words, the data provides a biased apparent geometry. Any method that conditions object models to ad-hoc fixes will, therefore, create object models inconsistent with the prior (violating Bayes' rule) even though they perfectly match the data. This is in particular the case with 2D geophysics and data from wells.

An example of this idea is shown in Figure 6.6. A 2D GPR survey detects scour formation in a braided river system. Any direct use of dimensions of 2D interpreted scour features would be biased toward larger objects. Object-based methods allow for the following:

1. The formulation of a prior distribution of geometries, spatial distribution and rules, for example from analog data or general understanding of the system. This is the prior movie.

2. A fast sampling by means of McMC methods (see Chapter 3) that generates posterior models in accordance with prior geometries and rules and matching any data. The sampling is fast because of the absence of a grid. Such McMC sampling relies on perturbations that are in accordance with the prior by generating (birthing) new objects and removing (dying) objects according to some point process model (e.g., Poisson, Strauss [*Strauss*, 1975]). The algorithm below is an example of McMC sampling in the case of a Poisson process with intensity λ. This type of sampling uses a Barker kernel [*Illian et al.*, 2008; *Lantuejoul*, 2013] to make sure that the birth and death process of points eventually results in a number of points, randomly distributed in space, where the amount of points follows a Poisson distribution.

Metropolis-Hasting-Green Simulation of a Poisson process.

```
n ← current number of objects
draw u = {+1, -1,0} with probabilities
p₊₁ ← 0.5 · min(1,λ/(n + 1))
birth probability
p₋₁ ← 0.5 · min(1,n/(n+λ))
death probability
p₀ ← 1 - p₊₁ - p₋₁
no change probability
if u == +1 do
      add 1 object
else if u == -1 do
      remove 1 object
else if u == 0 do
      do nothing
end if
```

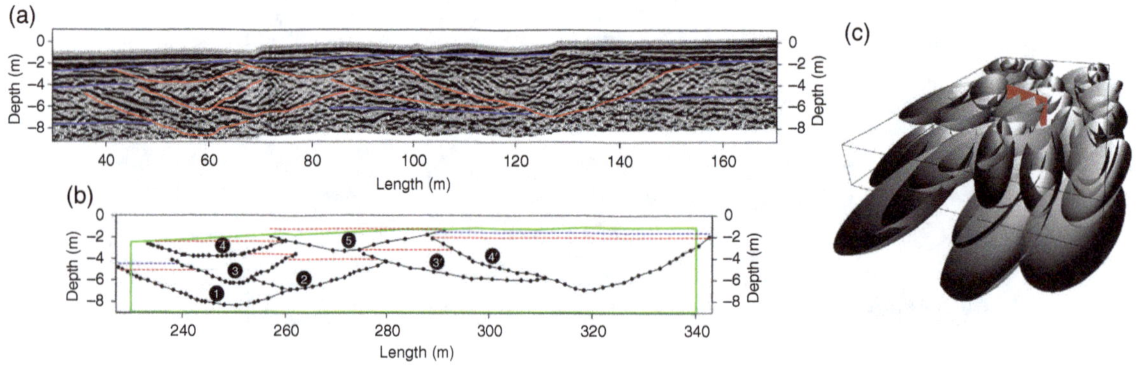

Figure 6.6 (a) A 2D GPR section with interpreted scour features. (b) A graphical presentation of a 2D GPR. (c) A 3D object model with 3D scour objects constrained to the interpreted section (red plane).

6.3.5.2. Level-Set Representation of Surfaces. Level sets have gained attention in the modeling of subsurface heterogeneity, in particular in cases with strong contrasts such as channel boundaries, layer boundaries, or faults [*Dorn et al.*, 2000; *Zhao et al.*, 2001; *Dorn and Lesselier*, 2006; *Frank et al.*, 2007; *Caumon et al.*, 2009; *Iglesias and McLaughlin*, 2011]. Level sets alleviate the difficulty arising when explicitly gridding complex surfaces, instead representing them by implicit functions, potentially saving considerable memory and possibly computation time. Note that other implicit methods, such a kriging and radial basis interpolation, do not need a grid either; they only require distances (or distance functions).

A level-set $\Pi(F, c)$ of an implicit function F and a scalar c is defined such that

$$\Pi(F(x,y,z),c) = \{(x,y,z)|F(x,y,z) = c\} \tag{6.38}$$

where $F(x, y, z)$ is any dimensional implicit function, with (x, y, z) as coordinates, and a scalar representing any iso-contour value [*Osher and Fedkiw*, 2002]. $F(x, y, z)$ determines the geometry of the manifolds defined as iso-contours of Π. For the 2D case, the level set is called the level curve. For the 3D case the level set is referred to as the level surface. In many practical applications, the implicit function $F(x, y, z)$ is defined as the signed distance function, representing a distance from a specific surface. Perturbations of a surface in this implicit representation entail simply recalculating the level-surface function for a different iso-contour value of c [*Frank et al.*, 2007]. Thus, perturbing a surface in 3D is done by calculating an implicit function at a different iso-contour.

A common problem in dealing with surfaces is to model the interaction between such surfaces and the rules that exist in doing so, such as a younger fault truncated by an older fault. Using explicit triangularized surface models, this may become challenging, in particular when procedures (such as perturbations or Monte Carlo simulations) need to be automated, which is often a requirement in inversion and UQ in general.

The level-set methodology defines surface terminations as a Boolean operation [*Osher and Fedkiw*, 2002]. Consider two different level-sets, with different level-set functions $\Pi(F_1, c_1)$ and $\Pi(F_2, c_2)$, each representing surfaces by taking a specific contour level $c_1 = c'_1$ and $c_2 = c'_2$ (see Figure 6.7). Consider also that the surface of level set 1 is older, meaning that the surface of level set 2 needs to be truncated. To do so, we first need to know on which side surface 2 lies with respect to surface 1 (see

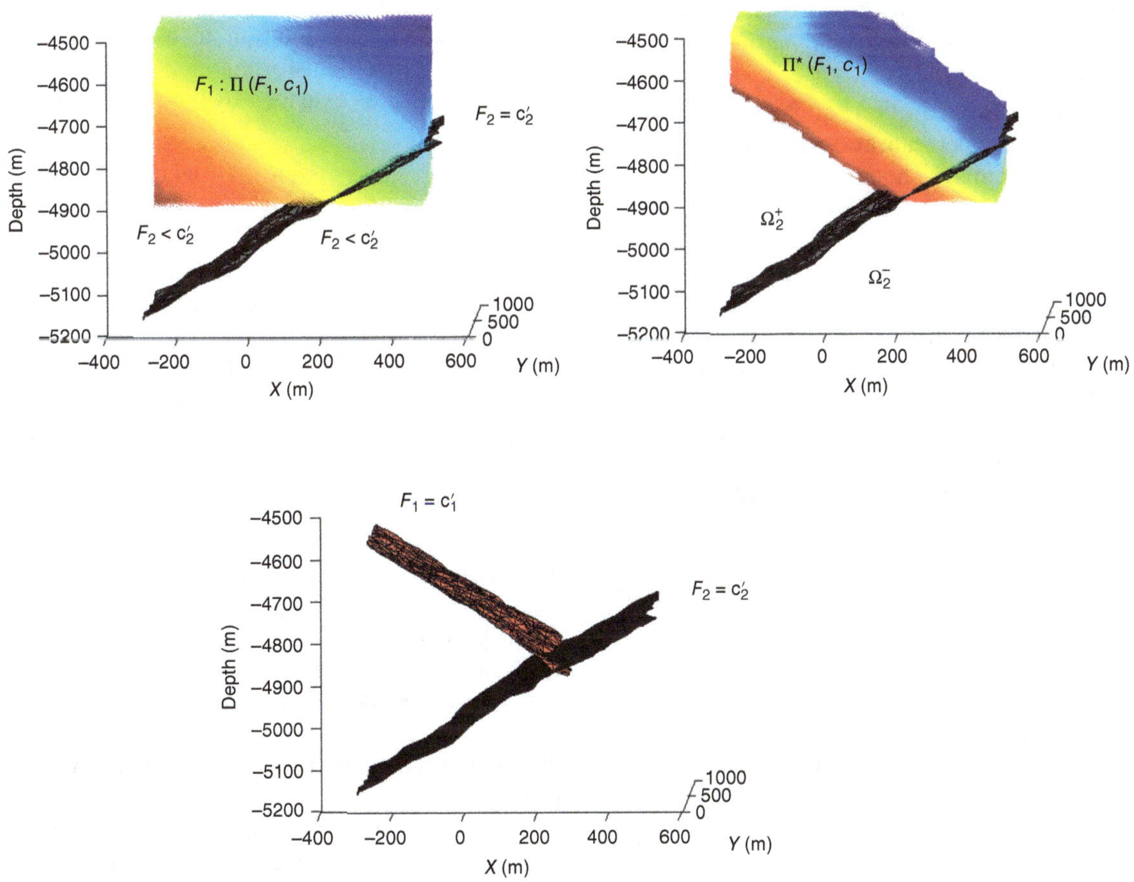

Figure 6.7 Intersection of two surfaces as constructed by a Boolean operation on two level-set functions.

Figure 6.7). The surface defined by $\Pi(F_2, c_2 = c'_2)$ divides space in two regions, Ω_2^+ and Ω_2^-. The convention here is that the plus sign is where the distance function is positive, the minus sign negative (or minus a positive distance). Consider now a new level set, $\Pi^*(F, c)$, which represents the truncated surface of fault 2. Then, the level surface for the truncated surface is written using the difference operator from constructive solid geometry [*Voelcker and Requicha*, 1977] as follows:

$$\Pi^*(F_1, c_1) = \Pi(F_1, c_1) - \Pi(F_2, c_2 | c_2 < c'_2) \tag{6.39}$$

The truncated surface is obtained by calculating this level surface at iso-contour $c_1 = c'_1$.

6.3.6. Dimension Reduction and Model Expansion from a Limited Prior Movie

*6.3.6.1. **General Purpose.*** Most of the above-mentioned algorithms for generating a prior movie allow for a possible infinite set of model realizations, an infinitely long prior movie (if the grid has infinitesimally small grid cells). Many of these algorithms do not rely on any explicit model formulation, in particular those relying on training images. The difficulty in dealing with such prior movies is that such algorithms generate possibly large realizations, whose dimension is equal to the number of grid cells (see Chapter 1). Such models are difficult to handle in inverse problems. To alleviate this problem, methods of dimension reduction can be applied. First, a limited set of prior model realizations is generated (a short movie or some summary shots from that movie, like a trailer), and then these realizations are used to build a reduced dimensional representation of the model variation. Model expansion (expanding the short movie) is done by fitting a probability model in reduced dimensional space. Methods of bijective dimension reduction (reduction and construction are unique, see Chapter 3), such as principal component analysis (PCA), are particularly useful here. However, PCA is limited in what it can represent.

In general terms, the aim is to replace a high-dimensional model realization $\mathbf{m}$ with a reduced dimension model $\mathbf{m}^*$ such that $\dim(\mathbf{m}^*) \ll \dim(\mathbf{m})$ and either sampling or optimization of $\mathbf{m}^*$is more efficient than with $\mathbf{m}$.

*6.3.6.2. **Kernel Methods for Dimension Reduction.*** A successful parameterization would require obtaining a mapping between some small set of parameters $\mathbf{m}^*$ and the model realization $\mathbf{m}$

$$\mathbf{m} = \mathbf{T}(\mathbf{m}^*) \tag{6.40}$$

where $\mathbf{T}$ is estimated from a limited sample of model realizations $\mathbf{m}^{(1)}, \mathbf{m}^{(2)}, \ldots, \mathbf{m}^{(L)}$. A linear mapping entails PCA, which is calculated from the experimental covariance

$$C = \frac{1}{L}\sum_{\ell=1}^{L} \mathbf{m}^{(\ell)} \left[\mathbf{m}^{(\ell)}\right]^T \tag{6.41}$$

The size of the covariance matrix is $N \times N$, where N is the dimension of $\mathbf{m}$. The problem here is the possible large size of C, which would render the eigenvalue decomposition impractical. To solve this problem, we rely on the duality between eigenvalues of covariance matrices $N \times N$ and dot-product matrix B of size $L \times L$.

$$b_{\ell\ell'} = \left[\mathbf{m}^{(\ell)}\right]^T \mathbf{m}^{(\ell')} \tag{6.42}$$

Relationships between eigenvalues and eigenvectors of C and B are given in Chapter 3. The eigenvalue decomposition can be used to determine a discrete Karhunen–Loeve (KL) expansion of the model $\mathbf{m}$, which for covariance-based models is classically known as

$$\mathbf{m} = V\Lambda^{1/2}\mathbf{m}^* \tag{6.43}$$

$\mathbf{m}^*$ is a vector of uncorrelated random variables. If $\mathbf{m}^*$ is Gaussian, then so is $\mathbf{m}$. At best, the linear PCA can only provide a Gaussian-type parameterization of a general prior movie.

To extend the formulation to a more general prior movie, the above dual formulation between covariance and dot-product is extended by means of kernels

$$k_{\ell\ell'} = \left[\boldsymbol{\varphi}\left(\mathbf{m}^{(\ell)}\right)\right]^T \boldsymbol{\varphi}\left(\mathbf{m}^{(\ell')}\right) \tag{6.44}$$

where $\boldsymbol{\varphi}$ is some unspecified multivariate (multi-point) transformation of $\mathbf{m}$. We refer to Chapter 3, for an introduction to kernel methods, in particular kernel PCA, or KPCA.

The same KL expansion can be applied after kernel mapping, resulting in an expansion of the transformed realization $\boldsymbol{\varphi}(\mathbf{m})$ and hence a parametrization of that realization in feature space:

$$\boldsymbol{\varphi}(\mathbf{m}) = (\mathbf{m}^*)^T \mathbf{T}(K, \mathbf{m}^*) \tag{6.45}$$

The mapping $\mathbf{T}$ is based on an eigen-decomposition of K [see *Sarma et al.*, 2008; *Vo and Durlofsky*, 2016 for details]. What is needed, however, is a parametrization of $\mathbf{m}$, not of $\boldsymbol{\varphi}(\mathbf{m})$. The problem is that $\boldsymbol{\varphi}^{-1}$ is not known, only the dot-product. In Chapter 3, this problem was introduced as the pre-image problem, which itself is an ill-posed problem: many $\mathbf{m}$ can be mapped onto the same $\boldsymbol{\varphi}(\mathbf{m})$.

Figure 6.8 illustrates this problem. A large set of realizations with discrete structures is generated of which two are shown. The KL expansion (PCA) and the reconstruction of a new realization result in a smoothed

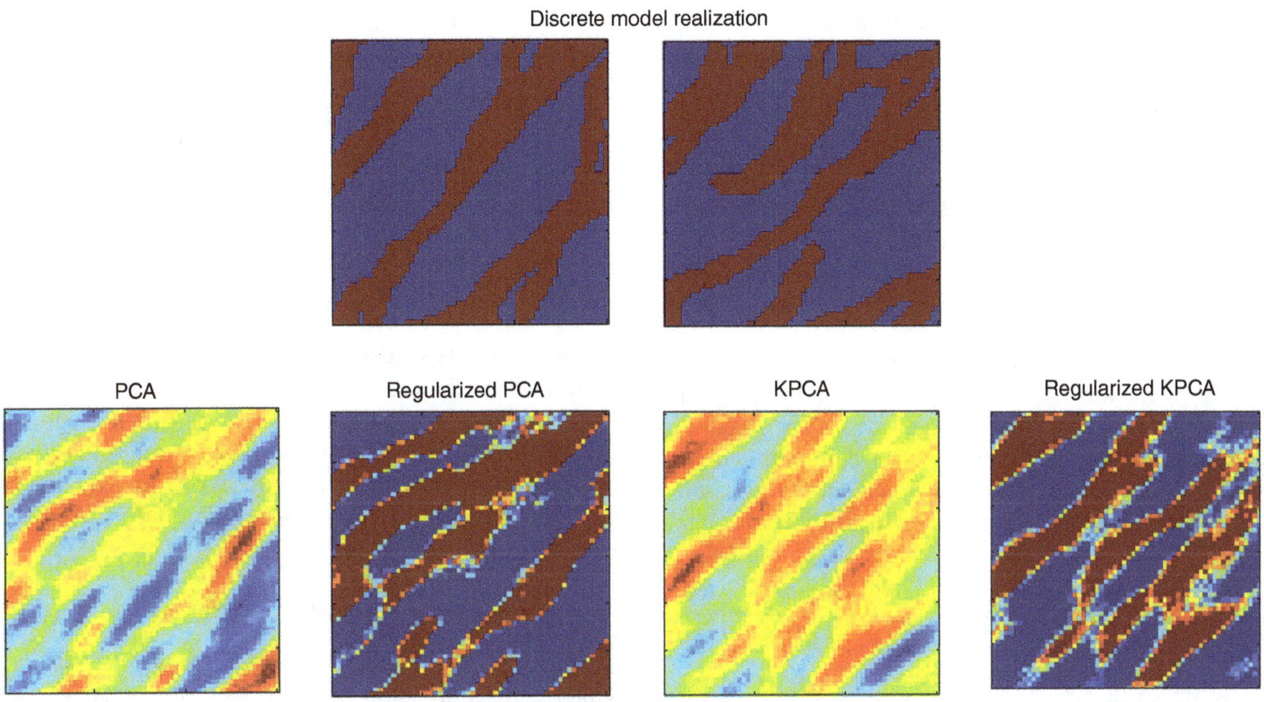

Figure 6.8 Discrete model realization and expansion of the sampling space using various dimension reduction methods and their regularizations to generate discreteness in the reconstruction [*Vo and Durlofsky*, 2016].

reconstruction, as does the KPCA with a fixed-point iteration (Chapter 3). *Vo and Durlofsky* [2014, 2016] add a regularization to the PCA reconstruction (a linear transform) and the KPCA reconstruction (pre-image problem) fixed-point algorithm. This results in generating new realizations that contain discrete structures instead of smooth ones. The problem of "smoothing" discrete spatial structures is an issue in most techniques relying on some form of dimension reduction/compression whether one uses wavelets [*Khaninezhad et al.*, 2012; *Scheidt et al.*, 2015] or other decomposition techniques.

6.4. DETERMINISTIC INVERSION

6.4.1. Linear Least Squares: General Formulation

To get to a treatment of deterministic inversion, we will start from a stochastic model. Deterministic cases can always be written as special cases of stochastic models; it just requires substituting a Dirac function. All deterministic solutions can be derived from the general formulation of Eq. (6.21).

In deterministic inversion [*Menke*, 2012], we often seek one solution; hence, what is relevant is the uniqueness of that solution. Inverse problems rarely have unique solutions; the key is to turn a nonunique into a unique one. Two approaches exist. The first approach is to acknowledge that the problem is nonunique and work out a stochastic solution (a posterior pdf), then take some model of that solution, for example, a maximum a posteriori model or maximum likelihood model. The second approach is to consider inversion as an optimization problem with a nonunique solution, then change the function to be optimized such that a unique solution exists. This requires imposing some additional "characteristics" on the solution (e.g., a smoothness or a distance from a base model).

The general linear solution is directly derived from Tarantola's formulation by making the following assumptions:

1. The forward model is linear and exact: $\mathbf{d} = G\mathbf{m} \to f_{\text{forward}}(\mathbf{d}|\,\mathbf{m}) = \delta(\mathbf{d} - G\mathbf{m})$.
2. The variables are Cartesian (hence not Jeffrey's parameters): $f_{\text{metric}}(\mathbf{d}) = cte$.
3. The prior on the model parameters is Gaussian with mean $\mathbf{m}_{\text{prior}}$ and covariance C_m.
4. The prior on the data parameters is Gaussian with mean $\mathbf{d}_{\text{obs}}$ and covariance C_d.

The posterior can now be written as

$$f(\mathbf{m}) \simeq \exp\left(-\frac{1}{2}(G\mathbf{m}-\mathbf{d}_{\text{obs}})^T C_d^{-1}(G\mathbf{m}-\mathbf{d}_{\text{obs}}) - \frac{1}{2}(\mathbf{m}-\mathbf{m}_{\text{prior}})^T C_m^{-1}(\mathbf{m}-\mathbf{m}_{\text{prior}})\right) \quad (6.46)$$

According to Chapter 3, the maximum a-posterior solution (also the mean of the posterior) equals

$$\mathbf{m}_{\text{MAP}} = \mathbf{m}_{\text{prior}} + C_m G^T \left(G C_m G^T + C_d\right)^{-1} \left(\mathbf{d}_{\text{obs}} - G\mathbf{m}_{\text{prior}}\right) \tag{6.47}$$

One can attach to this single solution a measure of "accuracy" by stating the posterior covariance:

$$\hat{C}_m = C_m - C_m G^T \left(G C_m G^T + C_d\right)^{-1} G C_m \tag{6.48}$$

The eigenvalues of the posterior covariance $\hat{C}_m$ indicates how ill-posed the solution is, for example, by comparing the largest with the smallest eigenvalue. The maximum a-posterior solution should not be confused with the maximum likelihood solution, which simply maximizes $f_{\text{prior},d}(G\mathbf{m})$ (or minimize minus its logarithm) to obtain

$$\mathbf{m}_{ML} = \left(G^T C_m^{-1} G\right)^{-1} G C_m^{-1} \mathbf{d}_{\text{obs}} \tag{6.49}$$

6.4.2. Regularization

6.4.2.1. Theory. A different view of the same problem starts from a purely deterministic formulation of the problem. In such formulation, inverse modeling requires generating models that match data. In that sense, consider for a linear forward model (see Section 6.4.5 for nonlinear models) the following optimization problem:

$$\hat{\mathbf{m}} = \min_{\mathbf{m}} \|G\mathbf{m} - \mathbf{d}_{\text{obs}}\|_2^2 = \min_{\mathbf{m}} (G\mathbf{m} - \mathbf{d}_{\text{obs}})^T (G\mathbf{m} - \mathbf{d}_{\text{obs}}) \tag{6.50}$$

The solution of this least square problem is provided in Chapter 3:

$$\hat{\mathbf{m}} = \left(G^T G\right)^{-1} G^T \mathbf{d}_{\text{obs}} \tag{6.51}$$

The problem is that the system is often underdetermined (many solutions exists) or that the matrix $(G^T G)^{-1}$ is close to singular. One way around this is to use SVD (Chapter 3). Another way is to "stabilize" the formulation by adding a so-called regularization term in the misfit function [*Tikhonov and Arsenin*, 1977]:

$$O(\mathbf{m}) = \|G\mathbf{m} - \mathbf{d}_{\text{obs}}\|_2^2 + \alpha \|\mathbf{m} - \mathbf{m}_0\|_2^2 \tag{6.52}$$

with $\mathbf{m}_0$ some reference, sometimes called a priori model, although this should not necessarily be interpreted in a Bayesian sense. The minimization of $O(\mathbf{m})$ leads to a regularization of the generalized least-square solution (Chapter 2):

$$\hat{\mathbf{m}} = \left(G^T G + \alpha I\right)^{-1} \left(G^T \mathbf{d}_{\text{obs}} + \alpha \mathbf{m}_0\right) \tag{6.53}$$

See also *Levenberg* [1944] and *Marquardt* [1963]. One can also introduce weight matrices (the equivalent of covariance matrixes in the Bayesian approach), W_m for the model variables and W_d for the data variables

$$O(\mathbf{m}) = \left\| W_d^T (G\mathbf{m} - \mathbf{d}_{\text{obs}})\right\|_2^2 + \alpha \left\| W_m^T (\mathbf{m} - \mathbf{m}_0)\right\|_2^2 \tag{6.54}$$

to obtain

$$\hat{\mathbf{m}} = \left(G^T W_d W_d^T G + \alpha W_m^T W_m\right)^{-1} \left(G^T W_d W_d^T \mathbf{d}_{\text{obs}} + \alpha W_m^T W_m \mathbf{m}_0\right) \tag{6.55}$$

Equation (6.55) can also be seen as a form of ridge regression. Various choices can be made in terms of the weight matrices. One is to choose them to be (often diagonal) covariance matrices. This will impose certain smoothness constraints in the solution as modeled by the covariance matrix (a kriging-like solution, see Chapter 3). Another simple option is to optimize

$$O(\mathbf{m}) = \|(G\mathbf{m} - \mathbf{d}_{\text{obs}})\|_2^2 + \alpha \left\| W_m^T (\mathbf{m} - \mathbf{m}_0)\right\|_2^2 \tag{6.56}$$

and have W_m to contain first-order difference (gradients), which allows modeling the sharpness of the boundaries occurring in the inverse solution [*Menke*, 2012].

6.4.3. Resolution

Approaches of deterministic inversion often introduce a "resolution matrix": How much of the model can the data resolve? This resolution model going back to [*Backus and Gilbert*, 1970] can also be framed within a stochastic context. To understand this, consider the true "earth" $\mathbf{m}_{\text{true}}$ and the exact data extracted from it. Consider now the change in difference from the prior model:

$$\hat{\mathbf{m}} - \mathbf{m}_{\text{prior}} = R\left(\mathbf{m}_{\text{true}} - \mathbf{m}_{\text{prior}}\right) \tag{6.57}$$

This basically states how the difference between the true earth and the prior is filtered into a difference between some solution and the prior. If R is the identity matrix then the data identifies the true earth. Any deviation (e.g., as measured by eigenvalues) is a measure of the loss of information on the true earth due to this filtering operation.

If we take $\hat{\mathbf{m}}$ as the MAP solution then one can derive [*Tarantola*, 1987] that the resolution matrix takes the Wfollowing expression:

$$R = I - \hat{C}_m C_m^{-1} \tag{6.58}$$

Another way of stating this is that

$$\hat{C}_m = (I - R) C_m \tag{6.59}$$

In other words, if R is close to the identity matrix then data resolves the model. This form also quantifies what is resolved by the data and what is resolved by the a-priori information (at least under the above assumptions of Gaussianity, linearity, etc.):

$$\mathrm{tr}(I) = \mathrm{tr}(R) + \mathrm{tr}\left(\hat{C}_m C_m^{-1}\right) \tag{6.60}$$

which is loosely interpreted as

No.total parameters = no.resolved by data + no.resolved by prior

In most subsurface setting, the resolution of data will decrease with depth. An example of this will be presented in Section 6.4.7.

6.4.4. Kalman Filter

An interesting property of linear inverse problems presents itself when the data vector is divided into mutually exclusive parts:

$$\mathbf{d} = (\mathbf{d}_1, \mathbf{d}_2, \mathbf{d}_3 \ldots) \text{ and hence } \mathbf{d}_{\mathrm{obs}} = (\mathbf{d}_{\mathrm{obs},1}, \mathbf{d}_{\mathrm{obs},2}, \mathbf{d}_{\mathrm{obs},3} \ldots) \tag{6.61}$$

with the data prior covariance matrix written as block matrices (no cross-correlation between the data blocks)

$$C_d = \begin{pmatrix} C_{d_1} & 0 & 0 & \cdots \\ 0 & C_{d_2} & 0 & \cdots \\ 0 & 0 & C_{d_3} & \cdots \\ \vdots & \vdots & \vdots & \ddots \end{pmatrix} \text{ and } G = \begin{pmatrix} G_1 \\ G_2 \\ G_3 \\ \vdots \end{pmatrix} \tag{6.62}$$

For example, these blocks of data could be blocks that present themselves consecutively in time. It can be shown that for the linear/Gaussian case, solving the full inverse problem with $\mathbf{d}$ is the same as solving consecutive updating problems as follows:

$$\begin{aligned} \mathbf{m}_{t+1} &= \mathbf{m}_t + C_t G_t^T \left(G_{t+1} C_t G_{t+1}^T + C_{d_{t+1}}\right)^{-1} \left(\mathbf{d}_{\mathrm{obs},t+1} - G_{t+1}\mathbf{m}_t\right) \\ C_{t+1} &= C_t - C_t G_t^T \left(G_{t+1} C_t G_{t+1}^T + C_{d_{t+1}}\right)^{-1} G_{t+1} C_t \end{aligned} \tag{6.63}$$

Meaning that data set $\mathbf{d}_{t+1}$ is used to update the model $\mathbf{m}_{t+1}$ from model $\mathbf{m}_t$. In other words, the "memory" of consecutive data is perfectly and consistently integrated in the linear/Gaussian case. Since most cases are not linear nor Gaussian, the question arises on how to perform updates in such case and if such "memory" property exists. This will be treated in Section 6.7.

6.4.5. Nonlinear Inversion

When the forward model is nonlinear and hence of the general form

$$\mathbf{d} = g(\mathbf{m}) \tag{6.64}$$

then no simple analytical solution as in the linear case is available, even when assuming Gaussianity. In case the forward model is only mildly nonlinear, meaning that near the inverse solution a linear approximation of the form

$$g(\mathbf{m}) \simeq g(\mathbf{m}_0) + G(\mathbf{m} - \mathbf{m}_0) \tag{6.65}$$

exists, for some $\mathbf{m}_0$ not far from the solution. The matrix G contains as entries the first derivative of the forward model with regard to the model parameters.

$$[G]_{ij} = \frac{\partial g_i}{\partial m_j} \quad i = 1, \ldots, \dim(\mathbf{d});\ j = 1, \ldots, \dim(\mathbf{m}) \tag{6.66}$$

$\mathbf{m}_0$ can be some reference model or prior model. Then the posterior mean (or MAP) becomes

$$\mathbf{m}_{\mathrm{MAP}} = \mathbf{m}_0 + C_m G^T \left(G C_m G^T + C_d\right)^{-1} \left(\mathbf{d}_{\mathrm{obs}} - g(\mathbf{m}_0)\right) \tag{6.67}$$

Another case presents itself when the solution can be linearized near the maximum likelihood point. This means that the posterior is Gaussian at that point. However, now no single-step analytical form is available and hence iterative descent methods (e.g., quasi-Newton, conjugate gradient) must be employed to get to the maximum likelihood point. Once that point has been reached, the usual MAP and posterior covariance equations apply.

6.4.6. Conceptual Overview of Various Model Assumptions

Figure 6.9 provides a summary of the various combinations of modeling assumptions presented earlier. The axes here are the data variables and the model variables. These 1D axes represent a high-dimensional manifold. The cases that are presented (from top to bottom, left to right) are as follows:

1. A linear forward model with Gaussian assumptions on model variables.
2. A forward model that can be linearized near the solution with Gaussian assumptions on model variables.
3. A forward model where the solution can be linearized near the maximum likelihood point with Gaussian assumptions on model variables.
4. A nonlinear forward model with Gaussian assumptions on model variables.
5. A nonlinear forward model with non-Gaussian modeling assumptions.

Clearly, the last case is more challenging; no simple expression or gradient optimization method will lead to the complex, skewed, and multi-model posterior distribution $f(\mathbf{m})$.

6.4.7. Illustration of Deterministic Inversion

6.4.7.1. Field Data. We will now illustrate some of the concepts presented in the previous section with a real field case. The case concerns an area in the Flemish Nature

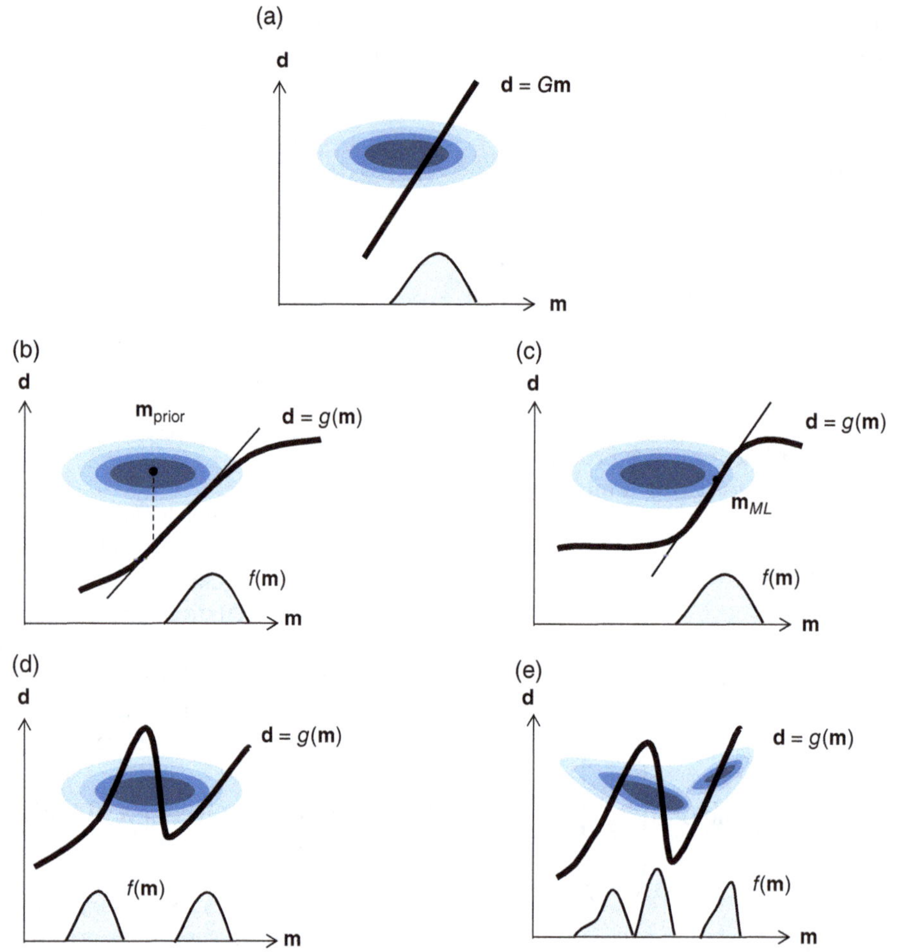

Figure 6.9 Cases of inverse problems (adapted from *Tarantola* [1987]). The ellipses are multivariate Gaussian distributions. (a) The linear case, (b) approximation of the linear case by expansion in the mean, (c) expansion in the maximum likelihood point, (d) a nonlinear problem with multiple solutions, and (e) nonlinear and non-Gaussian case.

Reserve "The Westhoek" (coastal area in northwest Belgium). The study aims at monitoring the intrusion of seawater into a freshwater aquifer within a dune system due to the construction of sea inlets. The depositional system of the dunes is about 30 m in thickness and mainly consists of sand with interconnecting semipermeable clay lenses. Because of their stratigraphy, these lenses enhance horizontal flow and diminish vertical flow. To investigate seawater intrusion, ERT data was collected (see Chapter 1 for an introduction to ERT). The data acquisition geometry consists for 72 electrodes with a spacing of 3 m and a dipole-dipole array. Individual reciprocal error estimates were used to weigh the data during inversion, the global noise level was estimated to be 5%. In addition, EM conductivity logs were available at the site (see Figure 6.10).

6.4.7.2. Regularized Inversions. We will illustrate methods of regularization: (i) regularization with smoothness constraint, (ii) regularization with a reference model, (iii) regularization with structural inversion, and (iv) regularization with a geostatistical constraint. First, the solution of the inverse problem with smoothness constraint is based on the minimization of the following function [*Kemna*, 2000]:

$$O(\mathbf{m}) = \left\| W_d^T(g(\mathbf{m}) - \mathbf{d}_{\text{obs}})\right\|_2^2 + \alpha \left\| W_m^T \mathbf{m}\right\|_2^2 \tag{6.68}$$

with the matrix

$$W_d = \operatorname{diag}\left(\frac{1}{\varepsilon_1}, \frac{1}{\varepsilon_2}, \cdots\right) \tag{6.69}$$

and ε_i the error variances of the measurement estimated from the measurement reciprocal error. W_m is a matrix evaluating the first-order roughness of $\mathbf{m}$. The problem is solved using an iterative Gauss–Newton scheme. The iteration process starts with a homogeneous initial guess $\mathbf{m}_0$, as the mean apparent resistivity of the data. Figure 6.10a shows the resulting inversion. The cumulative sensitivity is a by-product of the inversion and

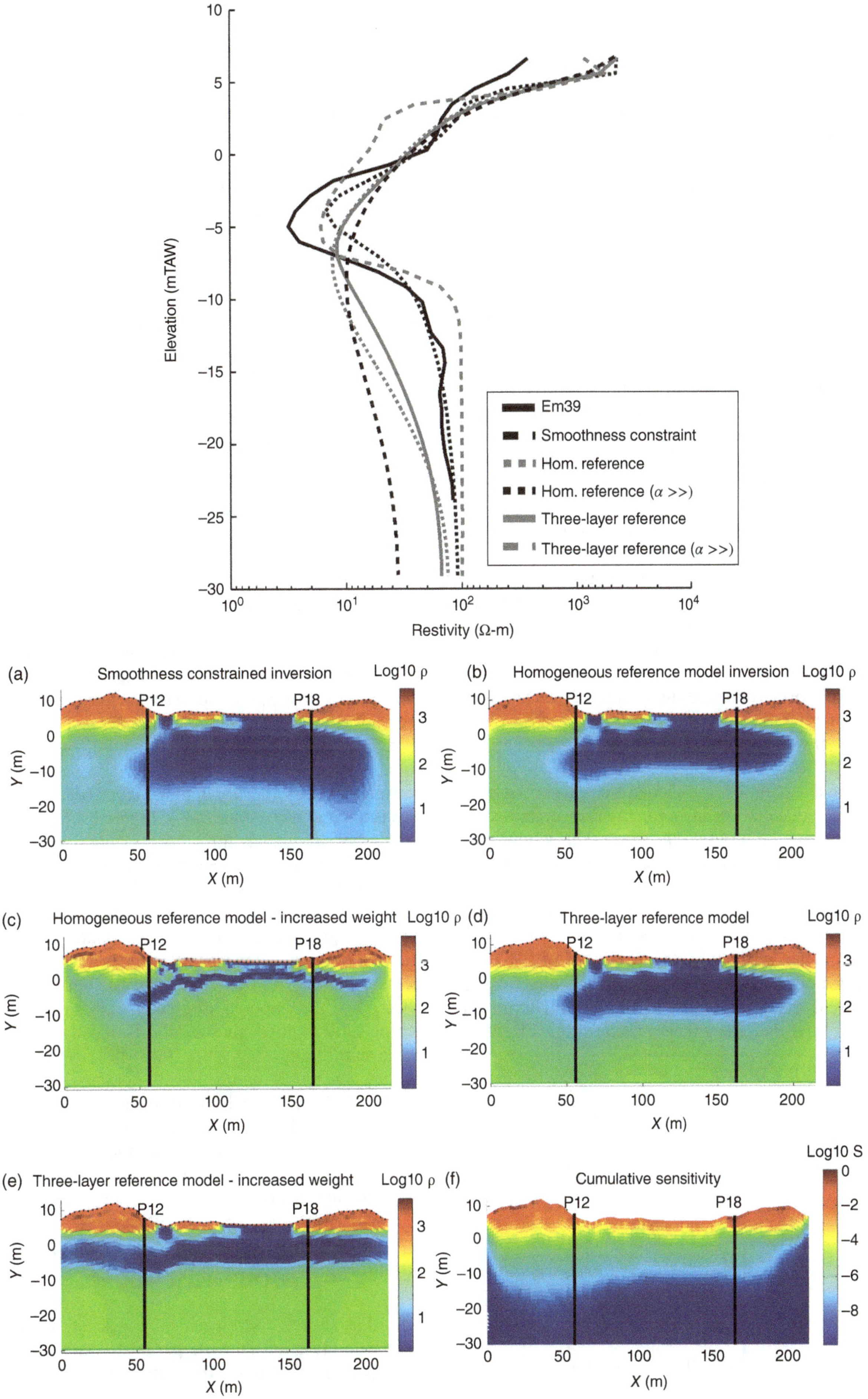

Figure 6.10 Well-log of P12 with measured resistivity (Em39), plus the various inverse solutions at that location. Inversion (a) with smoothness constraints, (b) using a homogenous reference model, (c) using a heavily weighted homogenous reference model, (d) using a three-layer reference model, and (e) using a heavily weighted three-layer reference model. (f) Cumulative sensitivity [*Hermans et al.*, 2012].

function of J^TJ with J the Jacobian evaluated in the solution (a Hessian). Figure 6.10f shows that sensitivity decreases with depth.

A second way of regularization is to use a reference model:

$$O(\mathbf{m}) = \| W_d^T(g(\mathbf{m}) - \mathbf{d}_{\text{obs}})\|_2^2 + \alpha \| W_m^T(\mathbf{m} - \mathbf{m}_{\text{ref}})\|_2^2 \quad (6.70)$$

where α is called the closeness factor, weighting the importance of the reference model during the inversion process. This factor is often chosen arbitrarily. We investigate two cases, one where the reference model is a homogenous model and one where the reference model is a three-layer model. In each case, the influence of α is evaluated by taking a small value (little impact of the reference) and a larger value (more impact, heavily weighted toward the reference). The influence of the reference model is therefore felt in two ways: (i) the value of α and (ii) the sensitivity of the inversion, the reference model tends to dominate the solution more where sensitivity to data is low. To make a comparison between the various results in Figure 6.10, the model results were evaluated at the location where the well-log is available. Overall the inversion involving a reference model seems to improve over the smooth-constrained inversion. The choice of reference model and α are subjective choices and one may not always have sufficient information to constrain them.

In regularization by so-called structural inversion, the aim is to reduce the penalty for rapid changes across a boundary (one lithology to the next). This would reflect any prior information about existence of boundaries. The formulation of the inverse problem remains exactly the same as for the smoothness constraint inversion, except that now the constraints are formulated on the gradient. This gradient is a function of $W_m^T W_m$. The constraints can be imposed in various direction; for example, if boundaries exist in the vertical more than the horizontal then the gradient is constrained by

$$W_m^T W_m = \beta_x W_x^T W_x + \beta_z W_z^T W_z \quad (6.71)$$

The modeler may then choose the ratio β_x/β_z for each element of the model based on some prior information [*Menke*, 2012]. The W's are first-order difference matrices:

$$W = \begin{pmatrix} -1 & 1 & 0 & \cdots & 0 \\ 0 & -1 & 1 & \cdots & 0 \\ 0 & 0 & -1 & \cdots & 0 \\ \vdots & \vdots & \vdots & \ddots & \vdots \\ 0 & 0 & 0 & \cdots & 0 \end{pmatrix} \quad (6.72)$$

Figure 6.11 shows that the use of structural inversion does not improve on the smoothness inversion. The problem in this specific case is that not only lithology causes contrast but also the saltwater lenses that intruded into the zone of study. Because of enhanced horizontal flow, the saltwater sits on top of the clay, making them difficult to discriminate with a prior model based on boundaries and contrasts.

The latter result then suggests that information about the spatial distribution of the model variables could improve the results. Such prior spatial information can be captured by means of a spatial covariance model. A regularization based on such covariance is then

$$O(\mathbf{m}) = W_m^T \|(g(\mathbf{m}) - \mathbf{d}_{\text{obs}})\|_2^2 + \alpha \| C_m^{-1/2}(\mathbf{m} - \mathbf{m}_0)\|_2^2 \quad (6.73)$$

with C_m the covariance matrix that is calculated from estimating or modeling the spatial covariance. The problem now is to estimate this spatial covariance. In most cases, only a few wells are available; hence, only a vertical variogram or covariance can be confidently deduced. To get more information on the horizontal component, it is not uncommon to calculate the variogram on the smoothed inversion and borrow on the ratio of horizontal and vertical range however, this may not be without bias, see *Mukerji et al.* [1997]. In this specific case, a Gaussian variogram with vertical range equal to 8.4 m was estimated from wells. The ratio between vertical and horizontal ranges (the anisotropy) was estimated to be equal to 4. Figure 6.12 shows some results, in particular the improvement over previous methods in terms of the comparison with the well-log.

Evidently, when looking at Figures 6.10–6.12, all solutions look smooth and are unique regardless of the type of regularization. The Hessian matrix J^TJ in the solution provides some idea of confidence near that (best guess) solution, but this should not be confused with an UQ. Also, the imposed prior constraints are not very geologically inspired, they tend to be mostly mathematical in nature imposing some properties on the resulting images, such that uniqueness is obtained. Uniqueness trumps geological realism in deterministic inversion. "Prior" should be taken with a grain of salt and should not be confused for the actual injection of rich geological information into the posterior results. This is the topic of the following section.

6.5. BAYESIAN INVERSION WITH GEOLOGICAL PRIORS

In this section, we specifically discuss methods of stochastic inversion that use parameterization based on geological information. In Section 6.3, we discussed

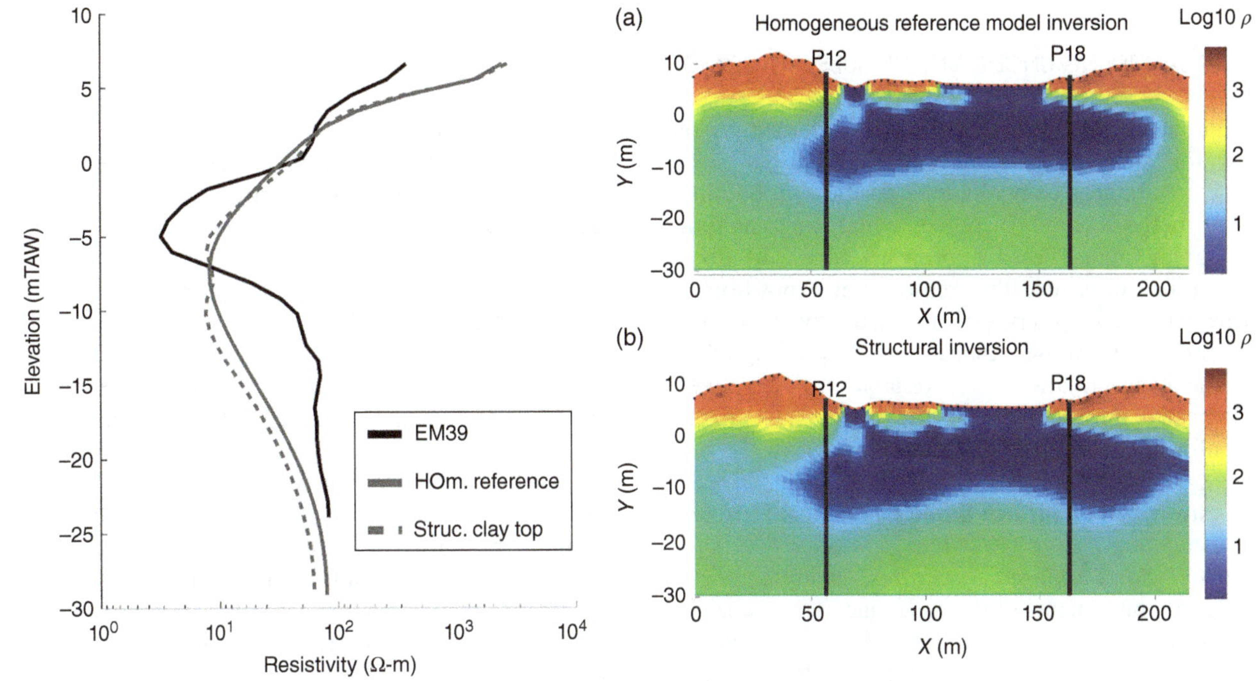

Figure 6.11 Comparison between structural inversion and a homogeneous reference inversion.

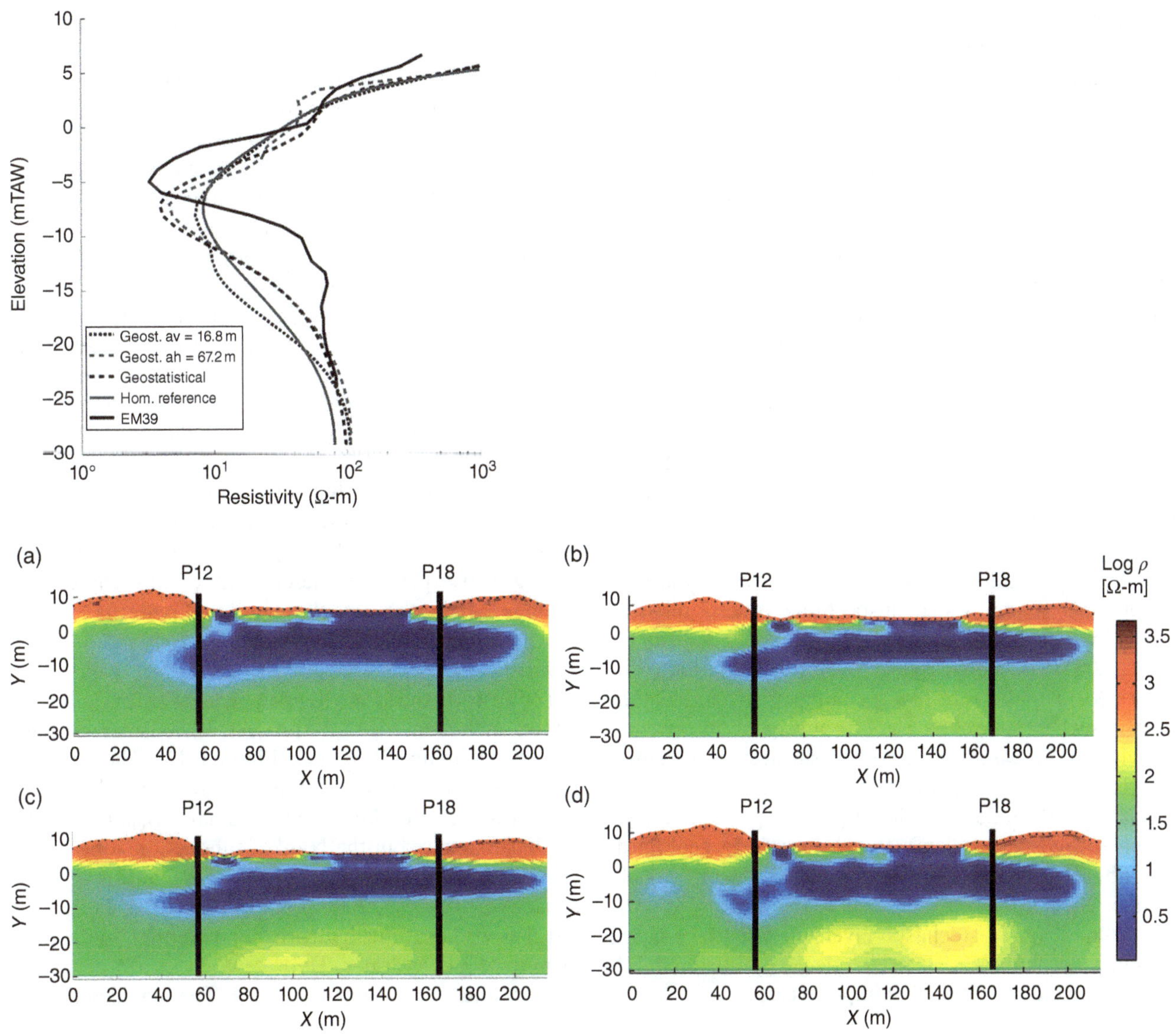

Figure 6.12 Comparison with the well-log: (a) Smoothness based on inversion, (b) imposing a spatial covariance, (c) multiplying the horizontal range by two, and (d) multiplying the vertical range by two.

several methods of prior model parameterization that allow for inclusion of geological information beyond simple covariances or two-point transition probabilities. These methods are (i) process-mimicking methods, (ii) training image-based approaches, and (iii) object-based approaches. For that reason, our exposition will not be exhaustive in terms of stochastic inversion, but it will focus on those application where significant geological information is available to develop a geologically informed prior (as opposed to, e.g., a non-informative prior or a Gaussian prior).

6.5.1. Inversion of Surface-Based Geological Structures

6.5.1.1. Surface-Based Structures. Many important geological components of the subsurface are represented by surfaces, not by rectangular or other cell-based grid. Examples are fault surfaces, horizons that separate distinct stratigraphic units, salt-domes, thin geological features such as shale drapes, or architectural boundaries such as channel boundaries. Such surfaces can be created from point interpretations on a geophysical image, or they could simply be defined through an object-based model. These surfaces are often triangularized from objects on which operations can be performed.

The difficulty of modeling (and inverting) with surface-based geological structures is that such surfaces are subject to geological rules that are not as easy to impose as, for example, grid-based properties [*Caumon et al.*, 2009]. Any automated CAD methods run the risk of generating interaction between surfaces that are inconsistent with general geological principle. Since a geological surface forms a boundary between two volumes of rocks, it needs to have two well-defined sides, and hence, such volumes should not overlap. Layers should not be leaking in the presence of faults. This means that a surface should terminate at a fault and not in the middle of nowhere. Faults, however, may terminate in the middle of nowhere.

This means that any surface-based model should undergo some reality checks. Some of these are manual, simply based on visual inspection, others rely on algorithms such as surface curvature analysis [*Thibert et al.*, 2005; *Groshong*, 2006]. Mesh quality needs to be inspected to make sure that any numerical models that involve such structures have good convergence properties. Checks can be based on individual surfaces or the entire structural model. The latter can go as far as restoring the structural model into the original depositional state (itself an inverse problem), thereby verifying geological plausibility [*Maerten and Maerten*, 2006; *Moretti*, 2008; *Mallet*, 2014].

The challenge in stochastic inversion of surface-based geological structures is (i) to include general understanding of the tectonic setting that imposes certain rules, (ii) to construct models that follow these rules, (iii) to provide consistency and quality checks of the generated models, and (iv) to design automatic perturbation methods that preserve these three criteria.

6.5.1.2. Example: Inversion of Faults Networks from Partial Fault Interpretations

6.5.1.2.1. Geological Priors for Faults. Faults provide important components to the subsurface system and may impact considerably the design of engineering operations, such as oil/gas production, geothermal energy production, or CO_2 sequestration. Locations of faults are uncertain for many reasons: (i) inaccurate geophysical imaging (see Section 6.6.2), (ii) imprecise representation of vertical surfaces in geophysical images, (iii) sparseness of well data and "luck" in cutting vertical structures, and (iv) interpretation subjectivity. Figure 6.13 provides a general overview of the evolution of fault modeling in the later part of the twentieth century. Because of the advance in computer graphics allowing the digital representation of complex surfaces, fault modeling has evolved from simple disc-like object modeled with a Poisson process to complex data structures that allow for both geometrical and topological perturbations. The current state is such that most geometric information can be incorporated, but that (i) the physics of tectonic process is largely ignored and that (ii) generating multiple models or perturbing models (needed for McMC) is tedious if not impossible. This brings up a fundamental question about modeling subsurface fault networks under uncertainty: How can one address the uncertainty in fault networks while accounting for knowledge pertaining to underlying tectonic processes, in the same sense that property models should integrate information about depositional process? This is a challenging question because the effect of tectonic settings on fault network patterns cannot be analytically expressed. Incorporating incomplete fault interpretations and tectonic concepts rigorously is another challenge because most structural geologists use qualitative descriptions of different tectonic settings and the resulting fault patterns [*Aydin and Schultz*, 1990; *Nieto-Samaniego and Alaniz-Alvarez*, 1997].

To illustrate the idea of a geological prior for faults that allows understanding of the process, we consider a specific case, the Kumano Basin (see Figure 6.14). The Kumano Basin and the Nankai Trough are located on the southwest coast of Japan. The Kumano Basin is a fore-arc basin situated in the Nankai subduction zone. It has been studied by many authors for its tsunami-generating capacity [*Aoki et al.*, 2000; *Gulick et al.*, 2010; *Tsuji et al.*, 2012] as well as for its hydrocarbon potential, in the form of natural gas hydrates. Hydrate deposits are present in clusters and thin seams in the area. The old accretionary

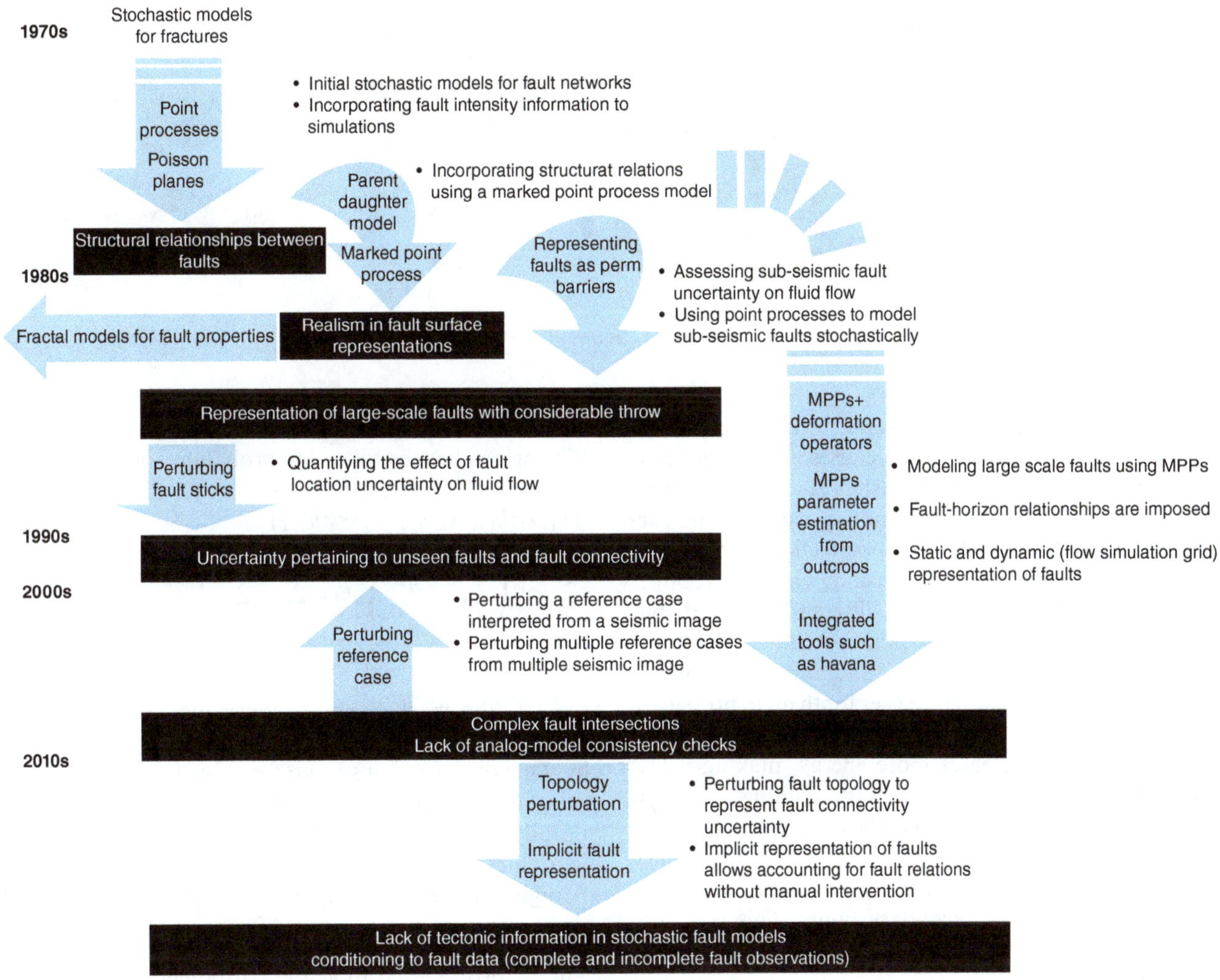

Figure 6.13 Overview of the advances in stochastic modeling of faults. The reader is referred to a history of fault modeling including the following: *Priest and Hudson* [1976], D. Veneziano [unpublished data, 1978], *Andrews* [1980], *Chilès* [1988], *Thore et al.* [2002], *Hollund et al.* [2002], *Cherpeau et al.* [2012] *Caumon et al.* [2004]; *Cherpeau et al.* [2010], *Lajaunie et al.* [1997] *Hesthammer and Fossen* [2000]. MPP = marked point process.

prism (zone of interest in Figure 6.14) is underneath the Kumano Basin. The old accretionary prism is poorly imaged because of attenuation of seismic waves. The seismic image is too blurry to interpret fault surfaces with confidence. Lateral continuity of fault markers is also compromised in the seismic image. Because of the effects of overlying hydrate deposits and a water surface, seismic artifact multiples exists that resemble fault surfaces.

The classical approach for an interpreter would be to extend the clearly visible partial faults (perhaps add a few) and use some truncation rules based on general tectonic understanding of the system. However, such a way of working leads usually to a deterministic fault model or some minor variations of that deterministic model, which may be completely incorrect. Therefore, the questions are (i) how to generate many fault network models adhering to the partially visible faults and (ii) account for geological understanding of the various ways faults could have been generated, in other words, the geological prior of faults. Obviously, this is a Bayesian problem, but the question is on the prior (and it is not multi-Gaussian, for sure). Because of the wealth of understanding of tectonic processes, this prior should entail more than a simple set of parameters (dip, angle, length) with some prior pdf, such as uniform or triangular distribution. This prior distribution needs to reflect the geological understanding of the tectonics that created the system (see Chapter 5: the prior needs to integrate the understanding of the geological depositional system). This understanding is "prior information." The question is how to quantify it. Ignoring it, because it cannot be easily quantified, leads to unrealistic posterior uncertainty. Let us, therefore, conceptually formulate what this understanding is as it relates to this setting.

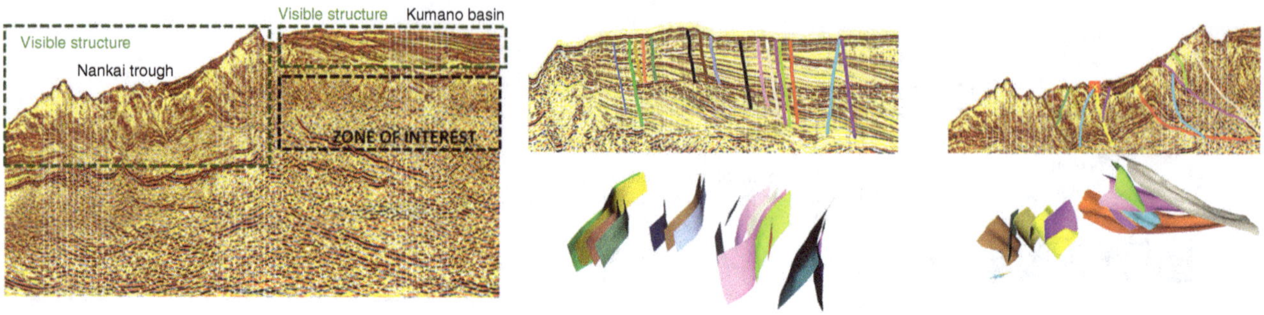

Figure 6.14 Kumano Basin, consisting of two areas with clearly interpretable faulting (normal faulting and Y-faulting) and a zone of interest with poor seismic imaging.

Fault networks are formed during a tectonic process where change of stress direction causes secondary faulting. Primary faults occur in clusters due to weakening of the brittle host rock. Such faulting weakens the surrounding formations such that imperfections are sheared due to stress. When the stress direction changes, younger (secondary) faults propagate at a different angle and abut (terminate) against the older (primary) faults. The effect of weakening due to primary faulting also impacts secondary faulting, making it easier for them to propagate closer to primary faults. However, weakening due to secondary faulting releases more energy, making it less likely to generate further secondary faulting. The end result of this physical process can be conceptualized qualitatively with a model of attraction between primary faults, attraction between primary and secondary faults and repulsion between secondary faults. This is a very relevant prior information and has impact on predictions (as shown later).

We use this conceptual model based on the fault interpretations from the Nankai Trough to model uncertainty of fault patterns of the old accretionary prism. The reasoning behind this modeling is that it is our subjective belief that the old accretionary prism experienced similar stresses acting on the modern Nankai Trough. Y-faults and nested splays are the main fault patterns interpreted in the Nankai Trough. A conceptual model involving Y-faults and nested splays caused by stress reversal could explain this observation. The question is now how to turn these concepts into math.

6.5.1.2.2. From Geological Prior to Marked Point Process Prior. Having specified, at least conceptually, how physical processes create fault patterns, a further quantification is needed into an actual mathematical representation of all this prior information, qualitative and quantitative. Because faults are objects of certain dimension that occur in space, a marked point process prior seems a logical choice. A marked point process requires the definition of the point process as well as the "marks," which in this case are fault objects.

More specifically, we use a Strauss process [*Strauss*, 1975], which has the following probability density model:

$$f(\{(\mathbf{x}_1;h_1),(\mathbf{x}_2;h_2),\ldots,(\mathbf{x}_n;h_n)\}) \simeq \exp\left(\sum_{r=1}^{N_h} -\gamma_r n - \sum_{q=1}^{N_h} \beta_{r,q} \sum_{i:h_r=r} \sum_{k:h_q=q} \mathbf{1}\left(\|\mathbf{x}_i-\mathbf{x}_k\| \le \tau_{r,q}\right)\right) \quad (6.74)$$

$(\mathbf{x}_i; h_i)$ is the marked point, consisting of the location (point) and a mark indicator that defines the fault hierarchy. One of the reasons to go for a MPP is that location (point) and hierarchy (mark) are clearly related (they are not independent). N_h here is the number of distinct fault hierarchies where $h_k = 1$ refers to the oldest faults that are highest in the hierarchy. One notices that this density function has a Poisson part $\exp\left(\sum_{r=1}^{N_h} -\gamma_r n\right)$. The coefficient γ_r, also referred to as the self-potential, defines the density of faults that belong to the hierarchy $h_k = r$.

The second component $\exp\left(\sum_{q=1}^{N_h} \beta_{r,q} \sum_{i:h_r=r} \sum_{k:h_q=q} \mathbf{1}\left(\|\mathbf{x}_i-\mathbf{x}_k\| \le \tau_{r,q}\right)\right)$ is an addition/subtraction to the location density function for a fault object due to its interactions with other fault objects. In particular, $\beta_{r,q}$, the pair-potential, models attraction or repulsion between faults with hierarchies between two fault families. $\mathbf{1}(\|\mathbf{x}_i - \mathbf{x}_k\| \le \tau_{r,q})$ denotes the number of fault pairs that are at most a distance $\tau_{r,q}$ apart. Each fault object at location $\mathbf{x}_k$ affects the location density function by $\exp(-\gamma_r n)$. Fault object pairs (with $h_r = r$ and $h_q = q$) that are at most a distance $\tau_{r,q}$ or closer affect the location density function with an additional factor of $\exp(\beta_{r,q})$. Individual fault objects can be defined independently by means of simple pdfs on their parametric descriptions such as fault length, azimuth, dip, surface roughness, and so on. More details can be found in *Aydin and Caers* [2017].

The main issue here is not, so much, this specific mathematical description but the fact that "a" mathematical description helps control what kind of fault networks are generated, how these hierarchies work, what length and dips should be generated, as opposed to generating them, ad hoc, from direct interpretation on the data (without any explicit prior). This prior is what allows the generation of multiple interpretation, constrained to existing data, without any artifacts of manual interpretation, and the inclusion of geological ideas about faulting.

The parameters of the Strauss model are estimated from analog areas (Nankai Trough, see *Aydin* [2017] for details). An analogy can be made here to classical geostatistics: the variogram is estimated from data (possibly analog areas if not enough samples are available) and then any posterior solution reflects also the spatial variability of the variogram model. Sampling from an MPP is not trivial and needs to proceed by McMC. Important in the chain is the reversibility. This makes the posterior model independent of the initial model. A particularity of McMC with an MPP model is that the initial model needs to match the data, but it need not be a sample from the prior. In other words, any McMC sampling with these kinds of objects requires models that match (or almost match) data. Indeed, an ad-hoc matching of faults structured by interpretation need not be a sample from the geological prior; in fact, it may be inconsistent with the overall geological understanding. To generate initial models for the Markov chain sampler, one can do a simple rejection sampler to find fault parameters that match the data. There would be as many faults as partial fault interpretations; hence, the initial models likely contain too few faults, and thus are biased.

Without going into the details of the McMC sampling [*Aydin*, 2017], a general overview is provided in Figure 6.15 (Metropolis sampling). The likelihood function is a simple mismatch function between the model and the data; hence, this methods falls under approximate Bayesian computation (see Chapter 3, no full likelihood model is specified). Because we are dealing with objects, the perturbations (e.g., move, remove, or modify) need to be reversible such that posterior models end up being consistent with the geologically informed prior. This involves several complex acceptance probabilities that have been developed for marked Strauss point process (MSPP) [*Illian et al.*, 2008], such as a move operation of a randomly selected object residing at $\mathbf{x}_k$ to a new location $\mathbf{x}_k^*$ with acceptance probability:

$$\alpha_{\text{move}} = \exp\left(\sum_{r=1}^{N_h}\sum_{q=1}^{N_h}\beta_{r,q}\sum_{i:h_r=r}\sum_{k:h_q=q}\left(\mathbf{1}\left(\|\mathbf{x}_i-\mathbf{x}_k\|\le\tau_{r,q}\right)\right.\right.$$
$$\left.\left.-\mathbf{1}\left(\|\mathbf{x}_i-\mathbf{x}_k^*\|\le\tau_{r,q}\right)\right)\right) \tag{6.75}$$

Notice how the acceptance probability depends on the prior pdf. Figure 6.16 shows some prior model realizations, some initial models, and some posterior model realizations. The data are the four partial fault interpretations.

6.5.1.2.3. Why Is the Geological Prior so Important?. Why bother with these complex mathematical descriptions? It seems that a simple rejection sampler creates (initial) models that match data that appear to look reasonable. To understand the effect of various modeling decision (Bayesian vs. non-Bayesian), we calculate a global statistic on the generated models. Often faults are important because they disconnect the subsurface into different volumes. Then these volumes may or may not be in communication depending on the sealing of the fault (often a function of fault throw [*Hollund et al.*, 2002]). This sealing has a considerable impact on fluid flow [*Knipe et al.*, 1998; *Manzocchi et al.*, 2008; *Cherpeau et al.*, 2010; *Rohmer and Bouc*, 2010]. The statistic of interest we chose is the size distribution of connected volumes. This global statistic should be the same for prior and posterior (indeed the very few partial interpretations will not affect this global statistic). Figure 6.16 shows this is the case. However, the size distribution of the initial

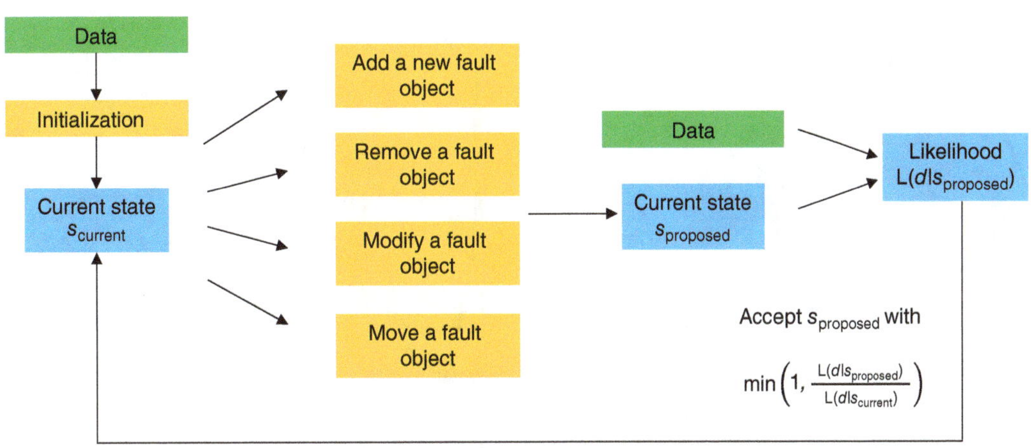

Figure 6.15 Various perturbations in the Metropolis sampling of the Strauss process.

models is quite different, lacking small compartments. This illustrates the bias of ad-hoc fault modeling, generating only as many faults as interpretations. Clearly, the presence of invisible faults cuts the region up, such that small compartments exist. This is critical to any application that involves moving fluids around. Based on the ad-hoc matched models, one may be overconfident about connectivity and hence recovery of fluids from such systems.

6.5.2. Inversion for Grid-Based Geological Structures

6.5.2.1. Introduction. Often grid-based models are built after generating structural models. If the structure is complex then the gridded model needs to reflect this (see Figure 6.3). In many cases, one will need to perform some joint inversion on both structural model uncertainty and property uncertainties. Separating both would ignore the interaction they have on the studied physical/chemical responses, whether fluid flow models or geophysical models. Here we focus specifically on situations where we do not consider any structural elements, just a flat gridded, Cartesian model. In Chapter 8, in the Libya oil field case, we will discuss ways of dealing with structure and properties jointly. Again we focus on the situation where geological information has been provided through a prior movie (see Section 6.3). Similar to the above discussion on object simulation using MSPP, we will sample from the posterior distribution using McMC. The random walk needs to be reversible, meaning (i) posterior samples are independent of the initial models and (ii) the prior information is not changed because of the proposed perturbation (sinuous channels should not become straight, unless that is part of the prior).

6.5.2.2. Perturbation Methods. Much of McMC sampling relies on perturbation mechanisms, changing a model into a new model (in a reversible way), or changing a set of models jointly and then calculating a likelihood. In low dimensions, this can be quite easy to apply (such as the move/remove operations for objects). This is no longer the case with perturbing large gridded model realizations. First, the dimensions are very large, and second geological prior model constraints need to be adhered to. In the multi-Gaussian case (variograms), one way of achieving this perturbation is to use gradual deformation [*Hu*, 2000; *Hu et al.*, 2001]. Consider a current model $\mathbf{m}$ and some other model drawn at random from the multi-Gaussian distribution $\mathbf{m}_{\text{new}}$. Then, a proposal model $\mathbf{m}^*$ models a gradual change from one Gaussian realization $\mathbf{m}$ to any other Gaussian realization $\mathbf{m}_{\text{new}}$

$$\mathbf{m}^* = \mathbf{m}\cos(\theta) + \mathbf{m}_{\text{new}}\sin(\theta) \tag{6.76}$$

The current model $\mathbf{m}$ is multiplied (globally) with a scalar dependent on θ and combined with a new model realization. The value of θ modulates the amount of perturbation

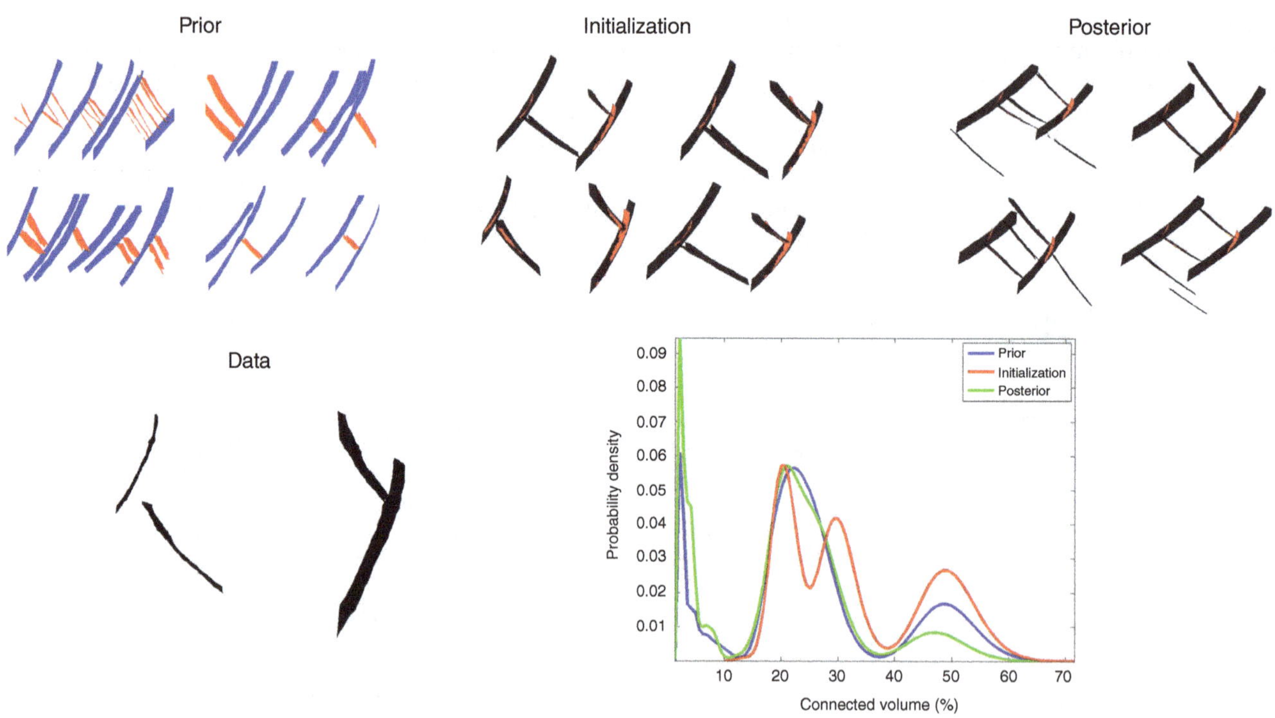

Figure 6.16 Data, prior model realizations, initialization with an ad-hoc matching, and posterior fault networks. Histogram of the size (in percent of total domain size) of the various model sets.

between $\mathbf{m}$ and $\mathbf{m}_{new}$. A simple proof shows that the resulting perturbations preserve the mean and covariance structure [*Reis et al.*, 2000]. Additionally, any perturbed model $\mathbf{m}^*$ can be gradually morphed into another new realization $\mathbf{m}'_{new}$. This allows making small changes (changing θ), or making very large changes by taking a new $\mathbf{m}_{new}$. Gradual deformation can be used in an optimization mode simply by optimizing θ (under a randomized $\mathbf{m}_{new}$); however, such samples are not necessarily samples from a multi-Gaussian posterior distribution [*Mariethoz and Caers*, 2015]. Nevertheless, gradual deformation can be quite useful in generating proposal models that are consistent with the mean and covariance structure. In addition, the θ can be made spatially varying allowing for local perturbation (setting $\theta = 0$) in areas where the model should not be perturbed. The proposal mechanism of gradual deformation is reversible; hence, gradual deformation is a valid proposal mechanism for McMC. It is straightforward to extend gradual deformation to truncated Gaussian or pluri-Gaussian fields. The same principle can also be generalized to combine the uniform random numbers that are underlying most stochastic techniques, for example, to deform object-based simulations of fractures.

The probability perturbation method (PPM) is similar to the gradual deformation in principle but offers a different perspective [*Caers and Hoffman*, 2006]. Instead of combining simulations directly or modifying the underlying random numbers, PPM takes a linear combination of two probability fields to obtain a spatial probability field that is then used as soft data to guide geostatistical simulations. This model perturbation technique is rather general and applicable to object-based, pluri-Gaussian, or training-image models. PPM, as gradual deformation method (GDM), allows for local or global perturbation. In case of global perturbation of a binary variable, a realization of probabilities (on the same grid as $\mathbf{m}$) is defined as

$$p(\mathbf{m},\theta) = (1-\theta)\,\mathbf{m} + \theta p_m \tag{6.77}$$

with p_m a marginal probability of the binary variable. θ modulates the amount of perturbation with $\theta = 0$ entailing no perturbation and $\theta = 1$ equivalent to generating another prior model realization randomly. To achieve a perturbation, the current realization is perturbed using a model of probabilities p defined on the same grid as $\mathbf{m}$. This probability model is then used as a soft probability (in a geostatistical sense, see [*Goovaerts*, 1999]) to generate a new realization. To allow for more flexibility in the perturbation, regions can be introduced, each with a different θ. This achieves a regional perturbation where some regions may change more than others, without creating region border artifacts.

The idea of GDM and PPM is to create dependency in the chain that can be exploited by the sampler. Another way of injecting such dependency is to retain some part of the spatial model realization as conditioning data to generate the next realization in the chain. This can be done in various ways; for example, one can retain a set of discrete locations (iterative spatial resampling, *Mariethoz et al.* [2010]) or one can return a compact block of locations [*Hansen et al.*, 2012]. Other proposals are to retain only the edge of a block [*Fu and Gómez-Hernández*, 2009]. The size of blocks or the amount of resampled points retained allows controlling the amount of perturbation generated.

6.5.2.3. Samplers that Involve Geological Priors

6.5.2.3.1. Sequential Gibbs Sampling. Recall that the Gibbs sampling (Chapter 3) involves sampling from a multivariate distribution by sampling from a full conditional distribution:

$$f(m_n|m_1,m_2,\ldots m_{n-1},m_{n+1},\ldots,m_N) \tag{6.78}$$

The Gibbs sampler requires only for a sample to be generated; the actual specification of the full conditional distribution is not necessary. Second, generating only one single m_n at a time is not very efficient and may lead to long iterations; therefore, it may be more useful to generate a subset $\mathbf{m}_{n\in S}$ of $\mathbf{m}$ at one time (defined either as compact blocks or as a set of points as outlined earlier) requiring sampling from $f(\mathbf{m}_{n\in S}|\,\mathbf{m}_{n\notin S})$. In *Hansen* et al. [2012] the sampling of this distribution is done by sequential simulation (see Chapter 3). This sequential Gibbs sampler is then implemented within the extended Metropolis sampler to generate new samples from the prior. Figure 6.17 shows an example of this re-simulation of blocks to generate a chain of MPS realizations drawn from the prior.

6.5.2.4. Multichain Methods. Traditional McMC methods work well for problems that are not too high dimensional (a few parameters). As was discussed in Section 3.12.7, multichain methods use multiple Markov chains running in parallel and use the information generated by all chains to update models and thereby improve on convergence to the posterior distribution. However, multiple chains have not yet been well adapted to geological priors or any prior involving a spatially distributed property. A neat exception is that of *Lochbuhler et al.* [2015] where prior information enters the chains through so-called summary statistics. Figure 6.18 conveys the idea. A training image is proposed to convey relevant spatial statistics. These images are then compressed, here by means of a discrete cosine transform [DCT; *Jafarpour et al.*, 2009], which allows extracting prior information in terms of histograms. These histograms are used as prior distributions in the DREAM sampler [*Laloy and Vrugt*, 2012].

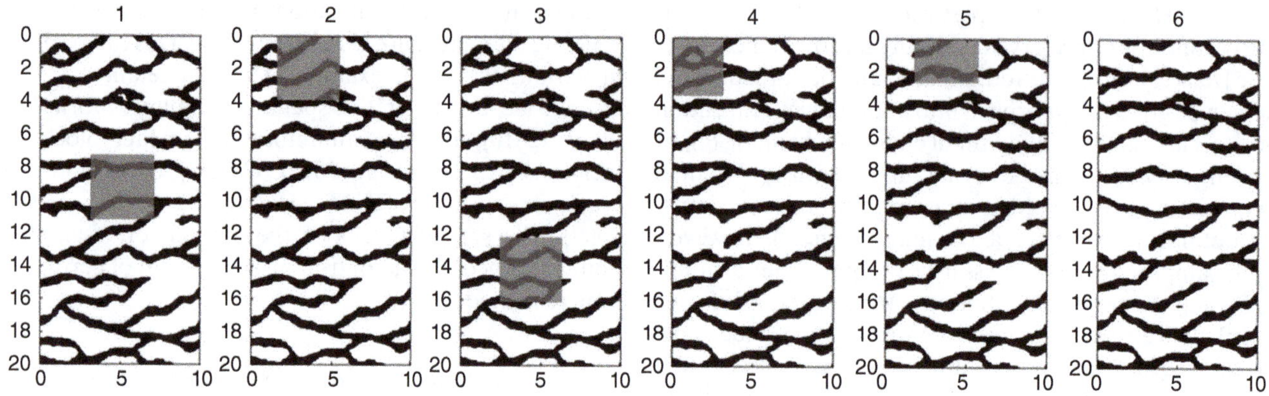

Figure 6.17 Six iterations of the sequential Gibbs. At each iteration a 4 × 4 window is re-simulated [*Hansen et al.*, 2012].

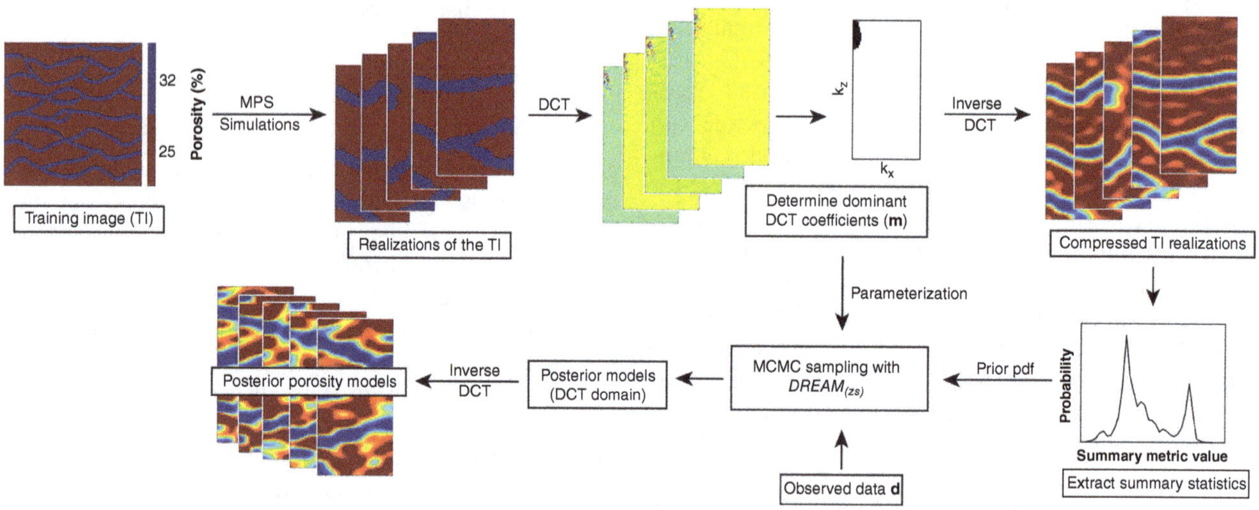

Figure 6.18 Extracting summary statistics from DCT-compressed prior model realizations. These summary statistics are then used as prior model statistics in a multichain sampler [*Lochbuhler et al.*, 2015].

6.6. GEOLOGICAL PRIORS IN GEOPHYSICAL INVERSION

6.6.1. Introduction

Geophysical imaging found its way to major applications in the 1960s, in particular in oil and gas exploration. Many geophysical imaging techniques and ideas have been developed within that industry, simply because of the direct monetary advantage of employing such techniques. At present, the field of hydro-geophysics has emerged as a way to image hydrological processes in the subsurface [*Linde et al.*, 2006; *Hyndman et al.*, 2013; *Binley et al.*, 2015]. In mineral resources assessment, geophysical imaging is increasingly used in exploration and also during the mining phase to assess the nature of the orebody [*Ward et al.*, 1977; *Goodway*, 2012; *Hatherly*, 2013; *Smith*, 2014]. While the purpose may be different, the main principles of geophysical imaging in these various areas are the same. This section is not intended to provide an overview of geophysical imaging and inversion methods, but it is intended to discuss the use of geological priors in the inversion of geophysical data and how such methods integrate into UQ.

Many types of geophysical imaging methods are available (see Table 6.1). The basic principle is to use some physical phenomenon (acoustic waves, electricity, and magnetism), a source for that physical phenomenon (explosions, current electrodes), and a receiver to record what happens when the subsurface is subjected to this physical "agitation." The idea is to infer 3D properties of the subsurface by untangling the received signals in terms of these properties. This is the "inversion." However, this "geophysical inversion" really consists of three different types of inversion.

1. *Migration:* This step is often termed "data processing." It requires putting the signals in the right place and removing noise as much as possible [*Yilmaz*, 2001;

Table 6.1 Overview of geophysical methods, geophysical variable or property, physical variables (or the variable of interest), and their area of application.

Geophysical method	Geophysical variable	Physical variable	Main application domain
Seismic	Elastic modulus, bulk density, attenuation, dispersion	Facies, pore fluid, cementation	All
Seismo-electric	Electrical current density	Voltage coupling coefficient and permeability	Mostly oil/gas
Magnetic	Magnetic susceptibility	Magnetic permeability (metals)	Mining, oil/gas
Gravity	Bulk density	Bulk density	All
GPR	Permittivity, electrical conductivity	Water content, porosity	Mostly hydro
Resistivity	Electrical conductivity	Water content and conductivity, clay content	All
Self-potential	Electrical conductivity	Permeability voltage coupling coefficient	All
NMR	Proton density	Water content, permeability	Hydro, oil/gas
Induced polarization	Electrical conductivity, chargeability	Water content and conductivity, clay content, surface area, permeability	Hydro, oil/gas
Time lapse	Common are seismic, resistivity, induced polarization	Changes in pressure, saturation, subsidence, temperature	Hydro, oil/gas

Berkhout, 2012]. The receivers do not see an image, but they see signals varying over time. To turn these signals into an image of the medium, we therefore need to anchor the signal to the location (to migrate) it came from in the subsurface (the subsurface source). Since the subsurface is unknown, an inversion is needed to do this. This inversion may be the most CPU demanding of all inversions simply because some geophysical methods, for example, seismic rely on a lot of redundancy in the signal to make sure the noise is sufficiently suppressed. For example, the SEAM model [*Fehler and Larner*, 2008], a constructed survey (see Figure 6.20), has 65,000 shots with a total of 200 TB of data.

2. *Geophysical parameter inversion*. Table 6.1 links the geophysical method to the subsurface geophysical parameters (properties or variables) that are being explored. Obviously, when using a seismic method, one is exploring the subsurface variations in velocity or density, while with electrical methods, one is exploring resistivity. This second inversion, therefore, turns the geophysical image into an image of a geophysical parameter.

3. *Rock or soil physical parameter inversion*. The ultimate interest does not lie in the geophysical parameters but in some other physical parameter of interest, as listed in Table 6.1, for example, a mineralization, a saturation, porosity, clay content, lithology, soil type, and so on. In the oil industry, this is termed a "petrophysical inversion" (also recently adopted by the hydrological community). Such inversion requires understanding the influence of the physical or chemical parameters of the medium on the geophysical parameters [*Mavko et al.*, 2009].

In what comes next, we will briefly touch on these various inversions, with an eye on the various sources of uncertainty involved and how geological priors come into the picture.

6.6.2. Creating the Geophysical Image

The geophysical image produced is the outcome of a complex chain of geophysical processing. Here we discuss uncertainty related to "migration." This inversion problem is large because of the large amount of data gathered and the large area covered (10–100 km area, 0–4 km depth). Hence, this (single deterministic) migration may take days to weeks of computing time (and this after making several linearization assumptions in the inversion), see Figure 6.19.

Uncertainty in velocity is particularly prevalent in situations where the subsurface has complex heterogeneity, for example, due to extensive faulting. We will illustrate ways of addressing such uncertainty in the context of subsalt exploration for oil and gas. Salt basins such as those located in the Gulf of Mexico and in the Brazilian offshore have been the site of extensive discoveries. However, because these reservoirs lie below the salt, the seismic image is very poor. The main issue is the shape of the salt body and its higher velocity compared to surrounding sediments. The salt body acts as a lens for such waves, creating poorly imaged regions or gaps in the data due to scattering and refraction. Figure 6.20b shows the illumination, which quantifies which areas are well covered by seismic rays (the red areas) and which areas are not (the blue areas).

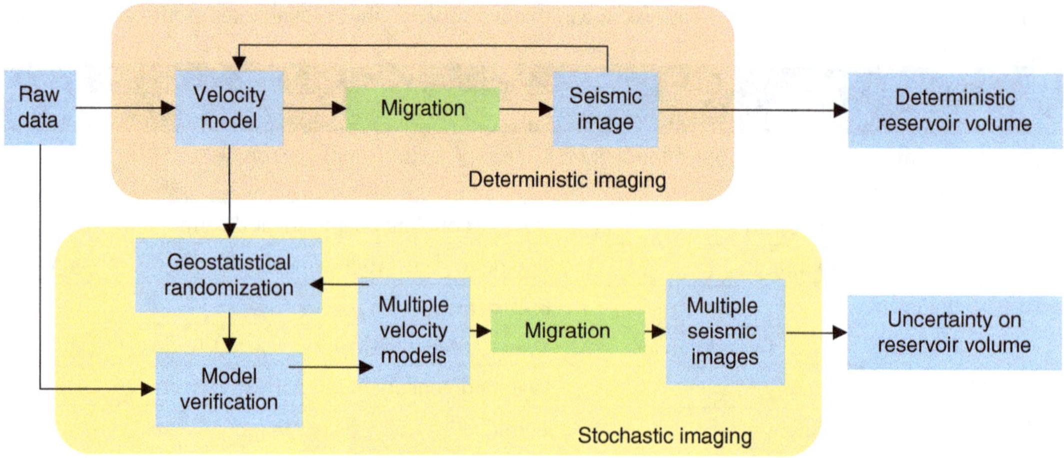

Figure 6.19 Deterministic versus stochastic imaging (adapted from *Li et al.* [2015]).

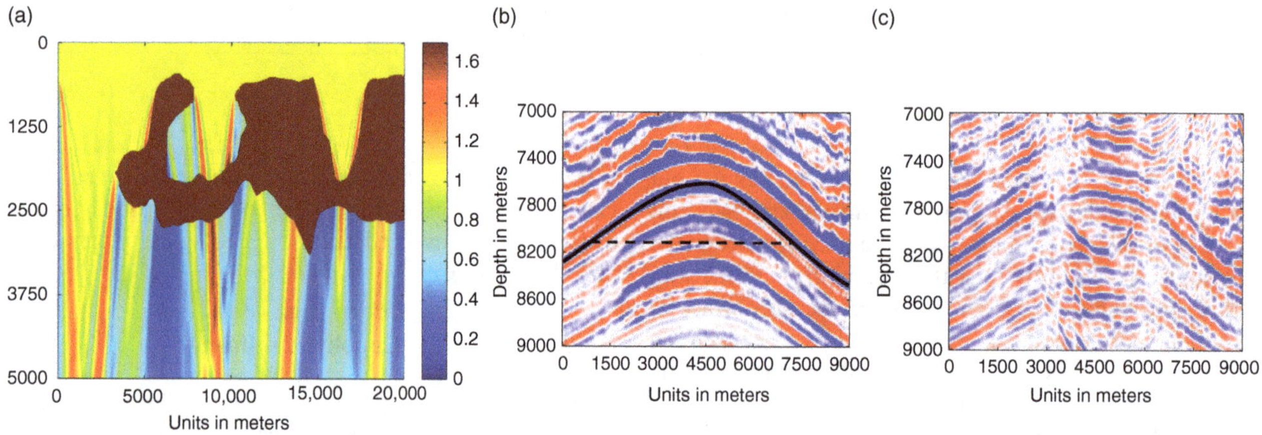

Figure 6.20 (a) A cross-section of the salt body with illumination (SEAM Model [*Fehler and Larner*, 2008]). (b, c) Migrated images of an anticline from the location beneath the salt body structure. (b) is the true image of the structure, while (c) indicates a typical image that can be reasonably obtained from an estimate of the velocity model. The black dotted line indicate the volume V of interest.

If illumination were perfect, then we would obtain a nicely imaged anticline (see Figure 6.20). However, poor illumination creates artifacts depicting structures that do not actually exist. Interpreters may then interpret faults that do not exist, or the structure may be determined at a depth very different from the actual depth.

The deterministic form of imaging may, therefore, lead to a biased estimate of, for example, reservoir volume. Interpreters are also left in the dark as to what the uncertainty of the reservoir is, since basically only one realization of the image is available.

A stochastic form of imaging (Figure 6.19) recognizes that the velocity model is uncertain, in the salt-dome case mostly due the unknown shape of the salt-body, in particular the bottom boundary (the top boundary tends to be flat and hence well imaged). This means that several perturbations of the base velocity model can be made that (i) still match the raw data equally as well as the base model and (ii) create different seismic images. Figure 6.21 shows four such perturbations to the salt-body roughness, using correlated Gaussian noise [*Li et al.*, 2015].

Does this matter? Imaging uncertainty in these kinds of system can produce a zero-th order uncertainty on prediction variables of interest. In oil field exploration, one such variable is the OOIP (see Chapter 1), depending, amongst others, on the total volume of rock in the reservoir system (a steep anticline makes for a smaller volume). Figure 6.21 shows the difference between such volume extracted from a deterministic inversion (the usual single migration) and a stochastic versus deterministic migration. The deterministic inversion significantly overestimates volume and hence OOIP. Given the billions of dollars involved in offshore production, this is not an error without consequence! ("Shell Cuts Reserve Estimate 20% As SEC Scrutinizes Oil Industry", *WSJ*, 12 January 2004].

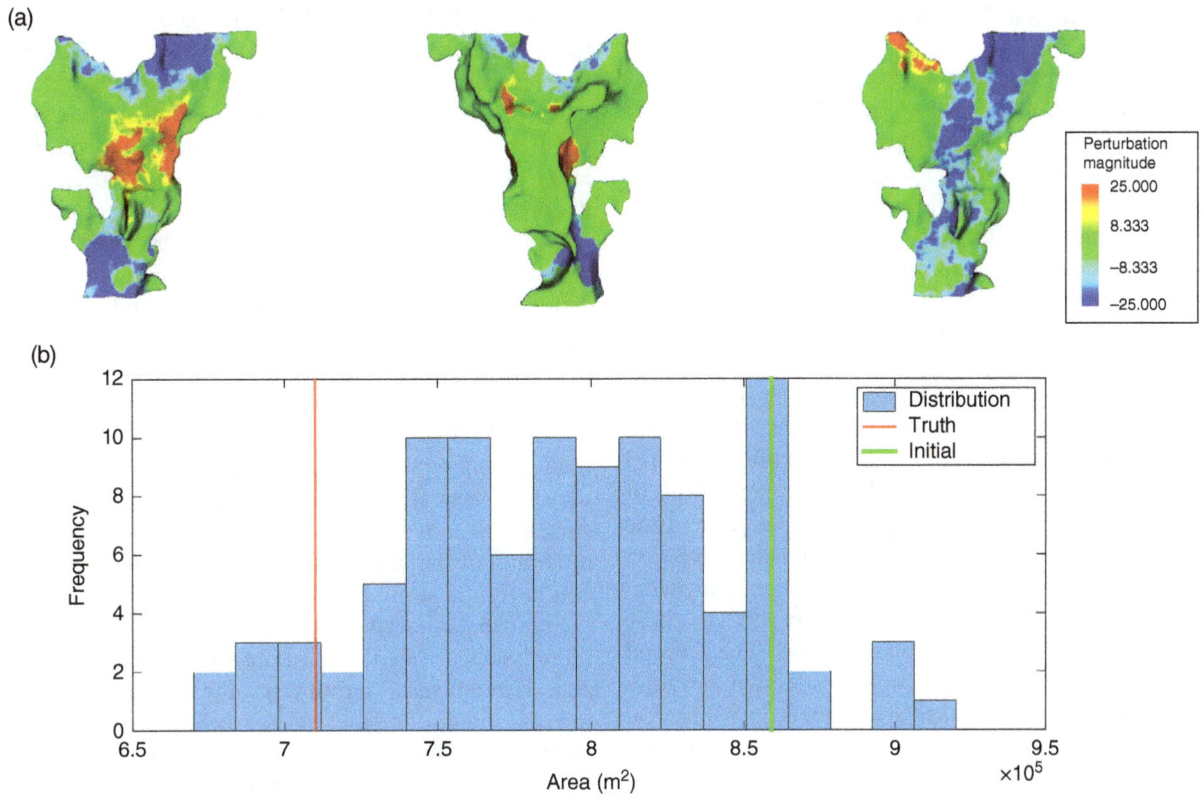

Figure 6.21 (a) Three realizations of the salt body viewed from the top. The green color refers to the base salt body obtained from deterministic inversion, while the red/blue color indicates perturbation of the body. The perturbations are such that they preserve the body roughness. (b) Uncertainty on reservoir volume due to imaging uncertainty.

6.6.3. Rock Physics: Linking the Image with Properties

Any obtained geophysical image can be regarded as observed data to infer either geophysical parameters or rock/soil physics parameters. For this step, one can use deterministic or stochastic inversion methods. The inversion problem is formulated using a combined model parameterization

$$\mathbf{m} = \left(\mathbf{m}_{\text{geoph}}, \mathbf{m}_{\text{phys}}\right) \tag{6.79}$$

Here we differentiate the geophysical parameters as geoph and the target physical parameters as phys. There is a hierarchy in the model parameters, however: it is the combination of physical parameters and the physics of the material studied that will determine the geophysical parameters; hence, the prior distribution of the model is often written as [*Bosch et al.*, 2010]:

$$f_{\text{prior}}\left(\mathbf{m}_{\text{geoph}}, \mathbf{m}_{\text{phys}}\right) = f_{\text{prior}}\left(\mathbf{m}_{\text{geoph}} | \mathbf{m}_{\text{phys}}\right) f_{\text{prior}}\left(\mathbf{m}_{\text{phys}}\right) \tag{6.80}$$

The Bayesian posterior formulation is then often taken as follows:

$$f_{\text{posterior}}\left(\mathbf{m}_{\text{geoph}}, \mathbf{m}_{\text{phys}} | \mathbf{d}_{\text{obs}}\right) \simeq f\left(\mathbf{d} = \mathbf{d}_{\text{obs}} | \mathbf{m}_{\text{geoph}}\right) f_{\text{prior}}\left(\mathbf{m}_{\text{geoph}} | \mathbf{m}_{\text{phys}}\right) f_{\text{prior}}\left(\mathbf{m}_{\text{phys}}\right) \tag{6.81}$$

$f_{\text{prior}}(\mathbf{m}_{\text{phys}})$ models the prior distribution of physical parameters. For example, porosity can be modeled using any of the above geostatistical methods: variogram-based, Boolean, or MPS, based on geological understanding (the geological prior). $f_{\text{prior}}(\mathbf{m}_{\text{geoph}} | \mathbf{m}_{\text{phys}})$ models the uncertain relationship between physical and geophysical parameters. This uncertainty is modeled in a field of science termed "rock physics."

Rock physics (soil physics, petrophysics) studies by means of theoretical or empirical models the behavior of geophysical parameters (e.g., bulk modulus) as a function of physical parameters (factors) such as water content, mineralogy, pore structure, cementation, and so on [*Avseth et al.*, 2005; *Mavko et al.*, 2009]. These models, therefore, play an important role in establishing relationships that can be used when limited sampling (in particular, the limited amount of wells) in a specific site or reservoir is available. However, rock physics models need to be calibrated to the specific site using geophysical data, and possibly also well data (such as core samples, logging). The area of statistical rock physics brings all these uncertainties together: limited wells, uncertain calibration, or difference in scale between the geophysical data, the well data, and any empirically derived petrophysical relationships.

6.6.4. Workflows and Role in UQ

Within UQ geophysical imaging and inversion are not an end-goal. It is but a means to eventually make better decisions based on realistic uncertainty assessment of key variables. These variables may not be directly imaged by geophysical methods. While developing better methods for geophysical imaging, better understanding of the petrophysical relationships is evidently of enormous use. The question of how this data is integrated in workflows for UQ and decision making is equally important.

Figure 6.22 provides an overview of the most common and traditional workflow [*Linde et al.*, 2006; *Bosch et al.*, 2010; *Linde et al.*, 2015]. In this workflow, the various disciplines involved are kept separated and often executed by different modelers. First, some kind of image of one or two geophysical parameters is generated (e.g., acoustic and elastic impedance). The deterministic inversion results in a smooth image, at least smoother than what is often required for applications, in particular those applications involving multiphase flow. The second portion consists of mostly petrophysics-related activities of converting the geophysical parameters into physical parameters, for example, using well data and/or rock physics relationships. The problem now is that the obtained physical parameters are too smooth; hence, some form of "downscaling" is required. This involves a third series of activities typically termed "geostatistical modeling." Here some form of finer scale variability model is formulated whether using variograms, transition probabilities, objects, or training images. Various methods exist that allow integrating this information with the geophysical image, usually using some form of co-kriging or block-kriging [*Goovaerts*, 1997; *Dubrule*, 2003]. This leads to models that use the geophysical image as "soft data," whether for facies or petrophysical properties. These models are constrained also to well information and reflect globally some model of spatial variability that was adapted in the geostatistical simulations. The advantage of dividing modeling activities into three pieces is also a disadvantage. Each piece comes with its own assumptions, estimates, and uncertainties. Because one mostly passes around deterministic images, most of these uncertainties are lost. Therefore, it is not uncommon that future data, in particular data from a completely different nature, cannot be predicted with the resulting models that have too small uncertainties. This has been documented in many case studies [e.g. *Hoffman and Caers*, 2007; *Park et al.*, 2013]. Hence, often ad hoc modifications to the models are made, leading to even smaller uncertainties. One of the reasons (but not the only one as we will see) is that the resulting geostatistical models only represent the statistical relationship between the geophysical image and the target petrophysical variable. It is not necessarily the case that any forward modeling of the data on the posterior geostatistical realization results in models that match the geophysical data.

Figure 6.23 describes a workflow that addresses some of these issues. Instead of working with deterministic inversion of geophysical parameters, the geostatistical modeling is integrated into the workflow. Geostatistical models of physical properties are generated, and these are converted into geophysical parameters. Here any uncertainty on the rock physics models can be integrated as well. Then, any form of Bayesian inversion, for example using McMC, can be executed to generate posterior models of both physical and geophysical variables. The loop can even be extended to the imaging part if the forward model related to the imaging is not too computationally demanding. The workflow in Figure 6.23 is more cumbersome because of the iteration that requires integrating several pieces of code that often have been developed independently.

While this new workflow improves on the original, it still avoids addressing a very important question: What is the geostatistical prior? In the above two workflows,

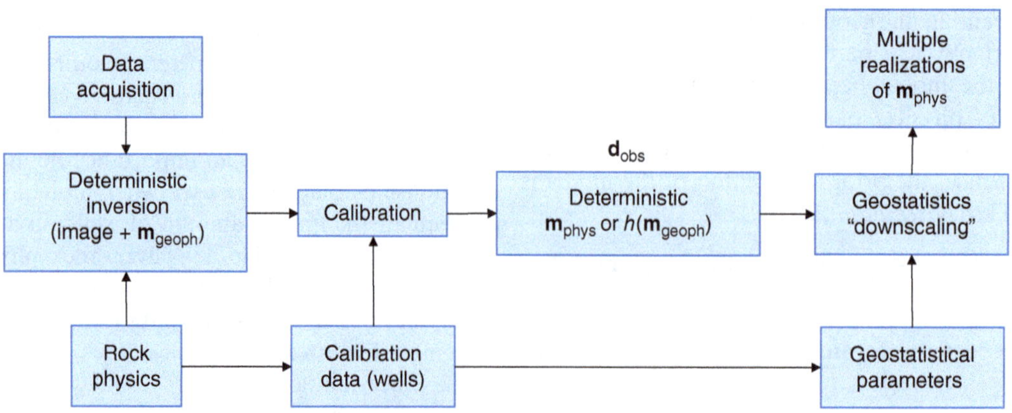

Figure 6.22 Most common workflow from geophysical data acquisition to subsurface geostatistical models.

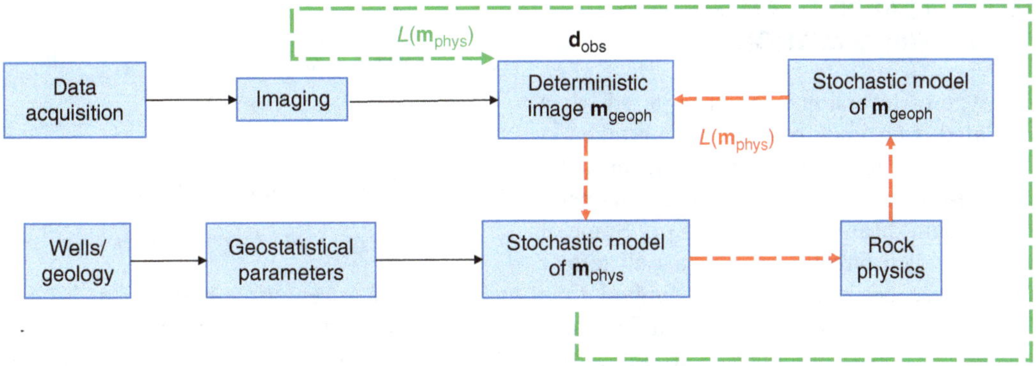

Figure 6.23 Closing the loop between geophysical inversion and geostatistical modeling.

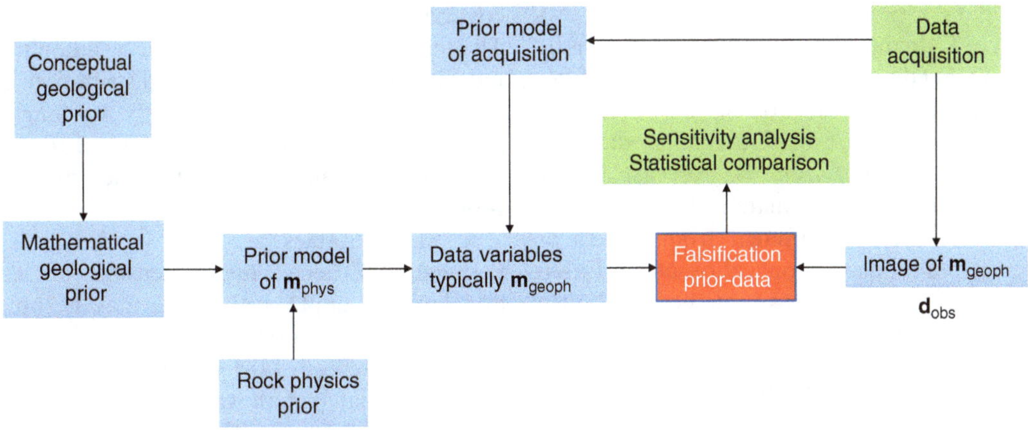

Figure 6.24 A workflow for falsification of the prior, before doing any actual inversion.

the geostatistical prior is typically estimated from wells, in particular when dealing with variogram-based methods, or just assumed, if limited information is available. However, as noted above, such prior would ignore important geological information. It also avoids the question of falsification of the prior: is the assumed prior actually consistent with the data? Neither workflow addressed this; it just assumes that prior and data are consistent. The two workflows also do not diagnose the problem of a too narrow prior, leading to a too narrow posterior, even when matching the data nicely. This is actually quite common and leads to ad-hoc model modifications later on.

Therefore, it is necessary to first attempt to falsify the prior. This would require using a very informative prior. Note that an informed prior may actually represent a more realistic uncertainty than the usual non-informative prior, such a simple uniform or Gaussian or other simple distribution on parameters. Figure 6.24 describes such workflow. Here all uncertainties are formulated and generated: in the acquisition model (e.g., the source wavelet), the rock physical models and the mathematical parameters representing the conceptual geological uncertainties. For example, one may assume a large variety of training images, or variogram models or a diverse set of Boolean models, fault network models, and so on. Then, forward models of the data are generated and a comparison is made with the actual data. This comparison need not simply be a mismatch (a likelihood or objective function). Instead, one may opt to run a global sensitivity analysis (see Chapter 4) and test which of the uncertainties impact the data variables, or one may perform a statistical comparison (e.g., using summary statistics [see *Lochbuhler et al.*, 2015; *Scheidt et al.*, 2015] between the data variables and the observed data. The latter statistical comparison actually allows for possibly rejecting the null-hypothesis that the data can be generated from the prior (a requirement for Bayesianism). In Chapter 7, we present one possible statistical method to achieve this when the data are seismic amplitudes. One advantage of this form of statistical falsification is that is does not require any inversion of data (hence can be very efficient). In Chapter 8, we present several real field applications of the workflow in Figure 6.24.

6.7. GEOLOGICAL PRIORS IN ENSEMBLE FILTERING METHODS

A particular type of inversion is often required when data is accumulated over time, instead of being taken at one time-instance. Examples of these are tracer data, pressure data, well-test data, production data, flow-meter data, stream-flow data, heat-tracers, and so on. Some of this data can be handled directly, such as a well test or pump test since these occur only over a limited time period. In Chapter 3, we discussed methods that filter such data over time into an update of model uncertainty. Here we discuss one such method, termed the "ensemble Kalman filter" (EnKf) [*Houtekamer and Mitchell*, 1998; *Evensen*, 2003; *Aanonsen et al.*, 2009], which has gained popularity in data assimilation. The method borrows the idea of the Kalman filter (see Section 6.4.4) by iteratively calculating the covariances of the Kalman filter empirically from model realizations. However, the question in this book is not how to assimilate data but how to quantify uncertainty on the forecast in the presence of prior geological information. In that regard, the problem is considerably different from oceans or atmospheres, the main area of application of the EnKf. Therefore, we cover first the basic method and then provide a survey of those extensions of the EnKf to integrate geological priors.

In its most basic form, the EnKf is a recursive filter operation where a mismatch in the data is used to adjust the model by a linear update operation. The recursive part refers to its consecutive application in time, as more data becomes available. The theoretical formulation can be derived directly from Eq. (6.21) as well as any Bayesian formulations of the inverse problems. The main assumptions underlying the basic EnKf derivation are to assume

1. a multi-Gaussian distribution on the model and data variables $(\mathbf{d}, \mathbf{m})$
2. a linear relationship between all the variables $(\mathbf{d}, \mathbf{m})$
3. a metric density for $(\mathbf{d}, \mathbf{m})$ that is uniform

In EnKf, the model variables are split into two parts: the static variables $\mathbf{m}_{\text{stat}}$ (not changing in time) and the dynamic variables $\mathbf{m}_{\text{dyn}}$. In the subsurface context, the static variables are typically porosity and permeability, and the dynamic variables are pressures and saturations. The triplet of variables $(\mathbf{d}, \mathbf{m}_{\text{stat}}, \mathbf{m}_{\text{dyn}})$ is now termed the "state vector." In EnKf, an initial ensemble of various $(\mathbf{m}_{\text{stat}}, \mathbf{m}_{\text{dyn}})$ is generated. Consider this initial ensemble at time $t = 0$. Next a forecast step is used to predict the dynamic variable, at the next time $t + \Delta t$ as well as to provide a forecast of the observable variables $\mathbf{d}$ for all models in the ensemble. In the assimilation step, an update of all variables is made based on the difference between the forecasts and the actually observed $\mathbf{d}$ at that time $t + \Delta t$ using a linear filter as follows:

$$\mathbf{y}_{t+\Delta t} = \mathbf{y}_t + K_{t+\Delta t}(\mathbf{d}_{\text{obs},t+\Delta t} - \mathbf{d}_{t+\Delta t}) \quad \text{with} \quad \mathbf{y} = \begin{bmatrix} \mathbf{d} \\ \mathbf{m}_{\text{stat}} \\ \mathbf{m}_{\text{dyn}} \end{bmatrix} \tag{6.82}$$

with the Kalman gain expressed as

$$K = C_{\mathbf{y}_{t+\Delta t}} H^T \left(H C_{\mathbf{y}_{t+\Delta t}} H^T + C_d \right)^{-1} \tag{6.83}$$

where H is a centering matrix (see Chapter 3). The covariance matrix $C_{\mathbf{y}_{t+\Delta t}}$ is calculated from the ensemble. The matrix C_d represents the error covariance on the data variables. Given the assumptions made, the resulting posterior distribution in this linear case is also Gaussian and fully known through the specifications of the various covariance matrices (the original derivation of Kalman). The linear model can be replaced by a nonlinear forward model, but then the posterior distribution is no longer known analytically, and evidently, no longer Gaussian. Nevertheless, the filter can be applied to variables that are non-Gaussian. However, in such extension, several problems may occur:

1. The variables are no longer Cartesian (facies, permeability); hence, metric densities are not uniform, and possibly unphysical values may be obtained (e.g., negative saturation or porosity; there is no preservation of the discreteness of the variables).
2. The updated static variables (such as porosity) may no longer be within the prior formulated on such variables. In fact, the progressive updates of the static model realizations become more Gaussian, even when the prior is distinctly non-Gaussian. As a result, the UQ becomes inconsistent with Bayes' rule. This problem is well recognized, and iterative solutions have been developed to address the issue of UQ [*Evensen and van Leeuwen*, 2000; *Emerick and Reynolds*, 2013].

The leading ideas on addressing these issues is not to apply EnKf directly on gridded models but on some form of re-parametrization that allows preserving the prior model statistics. Such parameterization may address the issue of an artificially dissipating (washing out) prior model as the EnKf assimilates non-Gaussian and nonlinear data. The usefulness of any transformation should also be assessed in how easily (and unique) the back-transform is. For example, taking a logarithm and then an exponential is easier than transforming in feature space (kernels) and then back to physical space (the transformation is difficult and non-unique).

Table 6.2 provides an overview of some of the main ideas and references. These consists of parameterization based on the Gaussian model, meaning transforming the problem such that one can assimilate based on Gaussian random variables, instead of original random variables. Another set of approaches rely on changing the parameterization

Table 6.2 Overview of various parameterization to preserve prior geological information in EnKf.

Method	Geostatistical prior model	References
Truncated (pluri)-Gaussian	Variogram/categorical	*Liu and Oliver* [2005]
Gradual deformation	Variogram/object/MPS	*Heidari et al.* [2013] and *Fienen et al.* [2009]
Pilot point	Variogram	*Heidari et al.* [2013]
Discrete cosine transform	Built from prior realizations	*Jafarpour and McLaughlin* [2008]
Level sets	Any discrete models	*Iglesias and McLaughlin,* [2011] and *Chang et al.* [2010]
PCA/KPCA	Built from prior realizations	*Sarma and Chen* [2009]
Local distributions	Variogram/MPS	*Zhou et al.* [2011] and *Jafarpour and Khodabakhshi* [2011]

MPS = multiple-point geostatistics.

of models to focus on certain properties, such as channels or boundaries. Another set consist of generating a set of prior models from whatever method and then parameterizing them by means of kernel methods.

REFERENCES

Aanonsen, S. I., G. Nævdal, D. S. Oliver, A. C. Reynolds, and B. Vallès (2009), The ensemble Kalman filter in reservoir engineering: A review, *SPE J.*, *14*(3), 393–412, doi:10.2118/117274-PA.

Andrews, D. J. (1980), A stochastic fault model: 1. Static case, *J. Geophys. Res.*, *85*(B7), 3867, doi:10.1029/JB085iB07p03867.

Aoki, Y, S. Shimizu, T. Yamane, T. Tanaka, K. Nakayama, T. Hayashi, and Y. Okuda (2000), Methane hydrate accumulation along the western Nankai Trough, *Gas Hydrates Challenges Future*, *912*, 136–145, doi:10.1111/j.1749-6632.2000.tb06767.x.

Arpat, G. B, and J. Caers (2007), Conditional simulation with patterns, *Math. Geol.*, *39*(2), 177–203, doi:10.1007/s11004-006-9075-3.

Avseth, P., T. Mukerji, and G. Mavko (2005), *Quantitative Seismic Interpretation: Applying Rock Physics to Reduce Interpretation Risk*, vol. *53*, Cambridge University Press, Cambridge, doi:10.1017/CBO9781107415324.004.

Aydin, O. (2017), Quantifying structural uncertainty on fault networks using a marked point process within a bayesian framework, PhD dissertation, Stanford University.

Aydin, O, and J. K. Caers (2017), Quantifying structural uncertainty on fault networks using a marked point process within a bayesian framework, *Tectonophysics*, *712–713*, 101–124.

Aydin, A., and R. A. Schultz (1990), Effect of mechanical interaction on the development of strike-slip faults with echelon patterns, *J. Struct. Geol.*, *12*(1), 123–129.

Backus, G., and F. Gilbert (1970), Uniqueness in the inversion of inaccurate gross earth data, *Philos. Trans. R. Soc. A Math. Phys. Eng. Sci.*, *266*(March), 123–192, doi:10.1098/rsta.1970.0005.

Berkhout, A. J. (2012), *Seismic Migration: Imaging of Acoustic Energy by Wave Field Extrapolation*, vol. *12*, Elsevier, Amsterdam.

Bertoncello, A., T. Sun, H. Li, G. Mariethoz, and J. Caers (2013), Conditioning surface-based geological models to well and thickness data, *Math. Geosci.*, *45*, 873–893.

Binley, A., S. S. Hubbard, J. A. Huisman, A. Revil, D. A. Robinson, K. Singha, and L. D. Slater (2015), The emergence of hydrogeophysics for improved understanding of subsurface processes over multiple scales. *Water Resour. Res.*, *51*(6), 3837–3866, doi:10.1002/2015WR017016.

Borghi, A., P. Renard, and S. Jenni (2012), A pseudo-genetic stochastic model to generate karstic networks, *J. Hydrol.*, *414–415*, 516–529, doi:10.1016/j.jhydrol.2011.11.032.

Bosch, M., T. Mukerji, and E. F. Gonzalez (2010), Seismic inversion for reservoir properties combining statistical rock physics and geostatistics: A review, *Geophysics*, *75*(5), 75A165, doi:10.1190/1.3478209.

Caers, J. (2011), Modeling structural uncertainty, in *Modeling Uncertainty in the Earth Sciences*, 133–151, doi:10.1002/9781119995920.ch8.

Caers, J., and T. Hoffman (2006), The probability perturbation method: A new look at Bayesian inverse modeling, *Math. Geol.*, *38*(1), 81–100, doi:10.1007/s11004-005-9005-9.

Carle, S. F., and G. E. Fogg (1996), Transition probability-based indicator geostatistics, *Math. Geol.*, *28*(4), 453–476, doi:10.1007/BF02083656.

Caumon, G., P. Collon-Drouaillet, C. Le Carlier De Veslud, S. Viseur, and J. Sausse (2009), Surface-based 3D modeling of geological structures, *Math. Geosci.*, *41*(8), 927–945, doi:10.1007/s11004-009-9244-2.

Caumon, G., F. Lepage, C. H. Sword, and J. L. Mallet (2004), Building and editing a sealed geological model, *Math. Geol.*, *36*(4), 405–424, doi:10.1023/B:MATG.0000029297.18098.8a.

Chang, H., D. Zhang, and Z. Lu (2010), History matching of facies distribution with the enKF and level set parameterization, *J. Comput. Phys.*, *229*(20), 8011–8030, doi:10.1016/j.jcp.2010.07.005.

Cherpeau, N., G. Caumon, J. Caers, and B. Lévy (2012), Method for stochastic inverse modeling of fault geometry and connectivity using flow data, *Math. Geosci.*, *44*(2), 147–168, doi:10.1007/s11004-012-9389-2.

Cherpeau, N., G. Caumon, and B. Lévy (2010), Stochastic simulations of fault networks in 3D structural modeling, *C. R. Geosci.*, *342*(9), 687–694, doi:10.1016/j.crte.2010.04.008.

Chilès, J. P. (1988), Fractal and geostatistical methods for modeling of a fracture network, *Math. Geol.*, *20*(6), 631–654, doi:10.1007/BF00890581.

Chugunova, T. L., and L. Y. Hu (2008), Multiple-point simulations constrained by continuous auxiliary data, *Math. Geosci.*, *40*(2), 133–146, doi:10.1007/s11004-007-9142-4.

Comunian, A., P. Renard, and J. Straubhaar (2012), 3D Multiple-point statistics simulation using 2D training images, *Comput. Geosci.*, *40*, 49–65, doi:10.1016/j.cageo.2011.07.009.

Comunian, A., S. K. Jha, B. M. S. Giambastiani, G. Mariethoz, and B. F.J. Kelly (2014), Training images from process-imitating methods an application to the lower namoi aquifer, Murray-Darling Basin, Australia, *Math. Geosci.*, *46*(2), 241–260, doi:10.1007/s11004-013-9505-y.

Davy, P., R. Le Goc, and C. Darcel (2013), A model of fracture nucleation, growth and arrest, and consequences for fracture density and scaling, *J. Geophys. Res. Solid Earth*, *118*(4), 1393–1407, doi:10.1002/jgrb.50120.

Deutsch, C. V., and T. T. Tran (2002), FLUVSIM: A program for object-based stochastic modeling of fluvial depositional systems, *Comput. Geosci.*, *28*(4), 525–535, doi:10.1016/S0098-3004(01)00075-9.

Dorn, O., and D. Lesselier (2006), Level set methods for inverse scattering, *Inverse Prob.*, *22*(4), R67, doi:10.1088/0266-5611/22/4/R01.

Dorn, O., E. L Miller, and C. M. Rappaport (2000), A shape reconstruction method for electromagnetic tomography using adjoint fields and level sets, *Inverse Prob.*, *16*(5), 1119–1156, doi:10.1088/0266-5611/16/5/303.

Dowd, P. A., E. Pardo-Igúzquiza, and C. Xu (2003), Plurigau: A computer program for simulating spatial facies using the truncated plurigaussian method, *Comput. Geosci.*, *29*(2), 123–141, doi:10.1016/S0098-3004(02)00070-5.

Dubrule, O. (2003), *Geostatistics for Seismic Data Integration in Earth Models*, Society of Exploration Geophysicists, London.

Emerick, A. A., and A. C. Reynolds (2013), Ensemble smoother with multiple data assimilation, *Comput. Geosci.*, *55*, 3–15, doi:10.1016/j.cageo.2012.03.011.

Evensen, G. (2003), The ensemble Kalman filter: Theoretical formulation and practical implementation, *Ocean Dyn.*, *53* (4), 343–367, doi:10.1007/s10236-003-0036-9.

Evensen, G., and P. J. van Leeuwen (2000), An ensemble Kalman smoother for nonlinear dynamics, *Mon. Weather Rev.*, *128*(6), 1852–1867, doi:10.1175/1520-0493(2000)128<1852:AEKSFN>2.0.CO;2.

Fehler, M., and K. Larner (2008), SEG Advanced modeling (SEAM): Phase I first year update, *Lead. Edge*, *27*(8), 1006, doi:10.1190/1.2967551.

Fienen, M. N., C. T. Muffels, and R. J. Hunt (2009), On constraining pilot point calibration with regularization in PEST, *Ground Water 47*(6), 835–844, doi:10.1111/j.1745-6584.2009.00579.x.

Frank, T., A. L. Tertois, and J. L. Mallet (2007), 3D-reconstruction of complex geological interfaces from irregularly distributed and noisy point data, *Comput. Geosci.*, *33*(7), 932–943, doi:10.1016/j.cageo.2006.11.014.

Fu, J., and J. J. Gómez-Hernández (2009a), A blocking Markov chain Monte Carlo method for inverse stochastic hydrogeological modeling, *Math. Geosci.*, *41*(2), 105–128, http://dx.doi.org/10.1007/s11004-008-9206-0.

Fu, J., and J.J. Gómez-Hernández (2009b), Uncertainty assessment and data worth in groundwater flow and mass transport modeling using a blocking Markov chain Monte Carlo method, *J. Hydrol.*, *364*(3–4), 328–341, doi:10.1016/j.jhydrol.2008.11.014.

Gabrovsek, F., and W. Dreybrodt (2010), Karstification in unconfined limestone aquifers by mixing of phreatic water with surface water from a local input: A model, *J. Hydrol.*, *386*(1–4), 130–141, doi:10.1016/j.jhydrol.2010.03.015.

Gelman, A., and C. R. Shalizi (2013), Philosophy and the practice of {Bayesian} statistics, *Br. J. Math. Stat. Psychol.*, *66* (1996), 8–38. doi:10.1111/j.2044-8317.2011.02037.x.

Goodway, B. (2012), Introduction to this special section: Mining geophysics, *Lead. Edge*, *31*(3), 288–290, doi:10.1190/1.3694894.

Goovaerts, P. (1997), *Geostatistics for Natural Resources Evaluation*, Oxford University Press, Oxford.

Goovaerts, P. (1999), Geostatistics in soil science: State-of-the-art and perspectives, *Geoderma 89*(1–2), 1–45, doi:10.1016/S0016-7061(98)00078-0.

Groshong, R. H. (2006), *3-D Structural Geology: A Practical Guide to Quantitative Surface and Subsurface Map Interpretation*, Springer, Berlin.

Guardino, F., and M. Srivastava (1993), Multivariate geostatistics: Beyond bivariate moments, in *Geostatistics Troia*, edited by A. Soares, vol. *1*, pp. 133–144, Kluwer Academic Publishers, Dordrecht, The Netherlands.

Guin, A., R. Ramanathan, R. W. Ritzi, D. F. Dominic, I. A. Lunt, T. D. Scheibe, and V. L. Freedman (2014), Simulating the heterogeneity in braided channel belt deposits: 2. Examples of results and comparison to natural deposits, *Water Resour. Res.*, *46*(4), W04516, doi:10.1029/2009WR008112.

Gulick, S. P. S, N. L. B. Bangs, G. F. Moore, J. Ashi, K. M. Martin, D. S. Sawyer, H. J. Tobin, Shin'ichi Kuramoto, and A. Taira (2010), Rapid forearc basin uplift and megasplay fault development from 3D seismic images of Nankai Margin off Kii Peninsula, Japan, *Earth Planet. Sci. Let.*, *300*(1–2), 55–62, doi:10.1016/j.epsl.2010.09.034.

Hansen, T. M., K. S. Cordua, and K. Mosegaard (2012), Inverse problems with non-trivial priors: Efficient solution through sequential gibbs sampling, *Comput. Geosci.*, *16*(3), 593–611, doi:10.1007/s10596-011-9271-1.

Hatherly, P. (2013), Overview on the application of geophysics in coal mining, *Int. J. Coal Geol.*, *114*, 74–84, doi:10.1016/j.coal.2013.02.006.

Heidari, L., V. Gervais, M. Le Ravalec, and H. Wackernagel (2013), History matching of petroleum reservoir models by the ensemble Kalman filter and parameterization methods, *Comput. Geosci.*, *55*, 84–95, doi:10.1016/j.cageo.2012.06.006.

Hermans, T., A. Vandenbohede, L. Lebbe, R. Martin, A. Kemna, J. Beaujean, and F. Nguyen (2012), Imaging artificial salt water infiltration using electrical resistivity tomography constrained by geostatistical data, *J. Hydrol.*, *438–439*, 168–180, doi:10.1016/j.jhydrol.2012.03.021.

Hesthammer, J, and H. Fossen (2000), Uncertainties associated with fault sealing analysis, *Pet. Geosci.*, *6*(1), 37–45, doi:10.1144/petgeo.6.1.37.

Hoffman, B. T., and J. Caers (2007), History matching by jointly perturbing local facies proportions and their spatial distribution: Application to a north sea reservoir, *J. Pet. Sci. Eng.*, *57* (3–4), 257–272, doi:10.1016/j.petrol.2006.10.011.

Hollund, K., P. Mostad, B. F. Nielsen, L. Holden, J. Gjerde, M. G. Contursi, A. J. McCann, C. Townsend, and E. Sverdrup (2002), Havana: A fault modeling tool, *Norwegian Pet. Soc. Spec. Publ.*, *11*(C), 157–171, doi:10.1016/S0928-8937(02) 80013-3.

Honarkhah, M., and J. Caers (2010), Stochastic simulation of patterns using distance-based pattern modeling, *Math. Geosci.*, *42*(5), 487–517, doi:10.1007/s11004-010-9276-7.

Houtekamer, P. L., and H. L. Mitchell (1998), Data assimilation using an ensemble Kalman filter technique, *Mon. Weather Rev.*, *126*(3), 796–811, doi:10.1175/1520-0493(1998) 126<0796:DAUAEK>2.0.CO;2.

Hu, L. Y. (2000), Gradual deformation and iterative calibration of Gaussian-related stochastic models. *Math. Geol.*, *32*(1), 87–108, doi:10.1023/A:1007506918588.

Hu, L. Y, G. Blanc, and B. Noetinger (2001), Gradual deformaton and iterative calibration of sequential stochastic simulations, *Math. Geol.*, *33*(4), 475–489.

Huang, T., X. Li, T. Zhang, and D. T. Lu (2013), GPU-Accelerated direct sampling method for multiple-point statistical simulation, *Comput. Geosci.*, *57*, 13–23, doi:10.1016/j.cageo.2013.03.020.

Huysmans, M., and A. Dassargues (2009), Application of multiple-point geostatistics on modelling groundwater flow and transport in a cross-bedded aquifer (Belgium), *Hydrogeol. J.*, *17*(8), 1901–1911, doi:10.1007/s10040-009-0495-2.

Hyndman, D. W., F. D. Day-Lewis, and K. Singha (Eds.) (2013), Subsurface Hydrology: Data Integration for Properties and Processes, vol. *171*. John Wiley & Sons, Inc., New York.

Iglesias, M. A., and D. McLaughlin (2011), Level-set techniques for facies identification in reservoir modeling, *Inverse Prob.*, *27*, 36, doi:10.1088/0266-5611/27/3/035008.

Illian, J., A. Penttinen, H. Stoyan, and D. Stoyan (2008), *Statistical Analysis and Modelling of Spatial Point Patterns, International Statistical Review*, vol. *76*, Wiley, New York, doi:10.1002/9780470725160.

Jafarpour, B., V. K. Goyal, D. B. McLaughlin, and W. T. Freeman (2009), Transform-domain sparsity regularization for inverse problems in geosciences, *Geophysics*, *74*(5), R69–R83, doi:10.1190/1.3157250.

Jafarpour, B., and M. Khodabakhshi (2011), A probability conditioning method (PCM) for nonlinear flow data integration into multipoint statistical facies simulation, *Math. Geosci.*, *43*(2), 133–164, doi:10.1007/s11004-011-9316-y.

Jafarpour, B., and D. B. McLaughlin (2008) History matching with an ensemble Kalman filter and discrete cosine parameterization, *Comput. Geosci.*, *12*(2), 227–244, doi:10.1007/s10596-008-9080-3.

Jaynes, E. T. (2003), Probability theory: The logic of science, *Math. Intell.*, *27*(2), 83–83, doi:10.1007/BF02985800.

Jeffreys, H. (1946), An invariant form for the prior probability in estimation problems, *Proc. R. Soc. A Math. Phys. Eng. Sci.*, *186*(1007), 453–461, doi:10.1098/rspa.1946.0056.

Jones, N. L, J. R. Walker, and S. F. Carle (2003), Hydrogeologic unit flow characterization using transition probability geostatistics, *Ground Water 43*(2), 285–289, doi:10.1111/j.1745-6584.2005.0007.x.

Jussel, P., F. Stauffer, and T. Dracos (1994), Transport modeling in heterogeneous aquifers: 1. Statistical description and numerical generation of gravel deposits, *Water Resour. Res.*, *30*(6), 1803–1817, doi:10.1029/94WR00162.

Karssenberg, D., Torbj Rn, E T Rnqvist, and J. S. Bridge (2001), Conditioning a process-based model of sedimentary architecture to well data, *J. Sediment. Res.*, *71*(6), 868–879, doi:1527-1404/01/071-868.

Kemna, A., (2000) *Tomographic Inversion of Complex Resistivity: Theory and Application*, Der Andere Verlag, Osnabrück, Germany, 196 p.

Khaninezhad, M. M., B. Jafarpour, and L. Li (2012), Sparse geologic dictionaries for subsurface flow model calibration: Part II. robustness to uncertainty, *Adv. Water Resour.*, *39*, 122–136, doi:10.1016/j.advwatres.2011.10.005.

Kindermann, R., and J. L. Snell (1980), *Markov Random Fields and Their Applications*, American Mathematical Society, Providence, RI.

Knipe, R. J., G. Jones, and Q. J. Fisher (1998), Faulting, fault sealing and fluid flow in hydrocarbon reservoirs: An introduction, *Geol. Soc. Lond. Spec. Publ.*, *147*(1), vii–xxi, doi:10.1144/GSL.SP.1998.147.01.01.

Koltermann, C. E., and S. M. Gorelick (1992) Paleoclimatic signature in terrestrial flood deposits, *Science*, *256*(5065), 1775–1782, doi:10.1126/science.256.5065.1775.

Koltermann, C. E., and S. M. Gorelick (1996), Heterogeneity in sedimentary deposits: A review of structure-imitating, process-imitating, and descriptive approaches, *Water Resour. Res.*, *32*(9), 2617–2658, doi:10.1029/96WR00025.

Lajaunie, C., G. Courrioux, and L. Manuel (1997), Foliation fields and 3D cartography in geology: Principles of a method based on potential interpolation, *Math. Geol.*, *29*(4), 571–584.

Laloy, E., and J. A. Vrugt (2012), High-dimensional posterior exploration of hydrologic models using multiple-try DREAM (ZS) and high-performance computing, *Water Resour. Res.*, *48*(1), W01526, doi:10.1029/2011WR010608.

Lantuejoul, C. (2013), *Geostatistical Simulation: Models and Algorithms*, Springer, Berlin, Heidelberg.

Levenberg, K. (1944), A method for the solution of certain nonlinear problems in least squares, *Q. J. Appl. Math.*, *2*(2), 164–168, doi:10.1017/CBO9781107415324.004.

Li, L., J. Caers, and P. Sava (2015), Assessing seismic uncertainty via geostatistical velocity-model perturbation and image registration: An application to subsalt imaging, *Lead. Edge*, *34*(9), 1064–1070, doi:10.1190/tle34091064.1.

Linde, N., A. Binley, A. Tryggvason, L. B. Pedersen, and A. Revil (2006), Improved hydrogeophysical characterization using joint inversion of cross-hole electrical resistance and ground-penetrating radar traveltime data, *Water Resour. Res.*, *42*(12), W12404, doi:10.1029/2006WR005131.

Linde, N., P. Renard, T. Mukerji, and J. Caers (2015), Geological realism in hydrogeological and geophysical inverse modeling: A review, *Adv. Water Resou.*, *86*, 86–101, doi:10.1016/j.advwatres.2015.09.019.

Liu, N., and D. Oliver (2005), Critical evaluation of the ensemble Kalman filter on history matching of geologic facies, *SPE Res. Eval. Eng.*, *8*(6), 4704–4777, doi:10.2118/92867-PA.

Lochbuhler, T., J. A. Vrugt, M. Sadegh, and N. Linde (2015), Summary statistics from training images as prior information in probabilistic inversion, *Geophys. J. Int.*, *201*(1), 157–171, doi:10.1093/gji/ggv008.

Maerten, L., and F. Maerten (2006), Chronologic modeling of faulted and fractured reservoirs using geomechanically based restoration: Technique and industry applications, *AAPG Bull.*, *90*(8), 1201–1226, doi:10.1306/02240605116.

Mahmud, K., G. Mariethoz, J. Caers, P. Tahmasebi, and A. Baker (2014), Simulation of earth textures by conditional image quilting, *Water Resour. Res.*, *50*(4), 3088–3107, doi:10.1002/2013WR015069.

Mallet, J. L. (2004), Space-time mathematical framework for sedimentary geology, *Math. Geol.*, *36*(1), 1–32, doi:10.1023/B:MATG.0000016228.75495.7c.

Mallet, J. L., (2014), *Elements of Mathematical Sedimentary Geology (the GeoChron Model)*, EAGE, Houten, The Netherlands.

Manzocchi, T., J. N. Carter, A. Skorstad, B. Fjellvoll, K. D. Stephen, J. A. Howell, J. D. Matthews, J. J. Walsh, M. Nepveu, and C. Bos (2008), Sensitivity of the impact of geological uncertainty on production from faulted and unfaulted Shallow-Marine oil reservoirs: Objectives and methods, *Pet. Geosci.*, *14*(1), 3–15, doi:10.1144/1354-079307-790.

Mariethoz, G., P. Renard, and J. Caers (2010), Bayesian inverse problem and optimization with iterative spatial resampling, *Water Resour. Res.*, *46*(11), 1–17, doi:10.1029/2010WR009274.

Mariethoz, G., and J. K. Caers (2015), *Multiple-Point Geostatistics: Stochastic Modeling with Training Images*, Wiley-Blackwell, Hoboken, NJ.

Mariethoz, G., and S. Lefebvre (2014), Bridges between multiple-point geostatistics and texture synthesis: Review and guidelines for future research, *Comput. Geosci.*, *66*(5), 66–80, doi:10.1016/j.cageo.2014.01.001.

Mariethoz, G., P. Renard, F. Cornaton, and O. Jaquet (2009), Truncated plurigaussian simulations to characterize aquifer heterogeneity, *Ground Water* *47*(1), 13–24, doi:10.1111/j.1745-6584.2008.00489.x.

Marquardt, D. W. (1963), An algorithm for least-squares estimation of nonlinear parameters, *J. Soc. Ind. Appl. Math.*, *11*(2), 431–441, doi:10.1137/0111030.

Mavko, G., T. Mukerji, and J. Dvorkin (2009), *The Rock Physics Handbook: Tools for Seismic Analysis of Porous Media*, Cambridge University Press, New York.

Menke, W. (2012), *Geophysical Data Analysis: Discrete Inverse Theory: MATLAB Edition*, vol. *45*, Academic Press, San Diego, CA.

Michael, H. A., H. Li, A. Boucher, T. Sun, J. Caers, and S. M. Gorelick (2010), Combining geologic-process models and geostatistics for conditional simulation of 3-D subsurface heterogeneity, *Water Resour. Res.*, *46*, W05527.

Moretti, I. (2008), Working in complex areas: New restoration workflow based on quality control, 2D and 3D restorations, *Mar. Pet. Geol.*, *25*(3), 205–218, doi:10.1016/j.marpetgeo.2007.07.001.

Mosegaard, K. (1995), Monte carlo sampling of solutions to inverse problems, *J. Geophys. Res.*, *100*(B7), 12,431–12,447.

Mukerji, T., T. Mukerji, G. Mavko, G. Mavko, and P. Rio (1997), Scales of reservoir heterogeneities and impact of seismic resolution on geostatistical integration, *Math. Geol.*, *29*(7), 933–950.

Nicholas, A. P., P. J. Ashworth, G. H. Sambrook Smith, and S. D. Sandbach (2013), Numerical simulation of bar and island morphodynamics in anabranching megarivers, *J. Geophys. Res. Earth Surf.*, *118*(4), 2019–2044, doi:10.1002/jgrf.20132.

Nieto-Samaniego, A. F., and S. A. Alaniz-Alvarez (1997), Origin and tectonic interpretation of multiple fault patterns, *Tectonophysics*, *270*(3), 197–206.

Osher, S. J., and R. P. Fedkiw, (2002), *Level Set Methods and Dynamic Implicit Surfaces*, Springer, New York.

Paola, C. (2000), Quantitative models of sedimentary basin filling, *Sedimentology*, *47*(s1), 121–178, doi:10.1046/j.1365-3091.2000.00006.x.

Park, H., C. Scheidt, D. Fenwick, A. Boucher, and J. Caers (2013), History matching and uncertainty quantification of facies models with multiple geological interpretations, *Comput. Geosci.*, *17*(4), 609–621, doi:10.1007/s10596-013-9343-5.

Priest, S. D., and J. A. Hudson (1976), Discontinuity spacings in rock, *Int. J. Rock Mech. Min. Sci. Geomech. Abstr.*, *13*(5), 135–148, doi:10.1016/0148-9062(76)90818-4.

Pyrcz, M. J., J. B. Boisvert, and C. V. Deutsch (2009), ALLUVSIM: A program for event-based stochastic modeling of fluvial depositional systems, *Comput. Geosci.*, *35*(8), 1671–1685, doi:10.1016/j.cageo.2008.09.012.

Pyrcz, M. J. (2003), A review of some fluvial styles, *Centre for Computational Geostatistics*. papers2://publication/uuid/31C43A49-DE27-4EA3-A1C4-7DB193817057.

Ramanathan, R., A. Guin, R. W. Ritzi, D. F. Dominic, V. L. Freedman, T. D. Scheibe, and I. A. Lunt (2010), Simulating the heterogeneity in braided channel belt deposits: A geometric-based methodology and code, *Water Resour. Res.*, *46*(4), W04515, doi:10.1029/2009WR008111.

Reis, L. C., L. Y. Hu, G. de Marsily, and R. Eschard (2000), Production data integration using a gradual deformation approach: Application to an oil field (Offshore Brazil), *SPE Annual Technical Conference and Exhibition*, 1–4 October, Dallas, TX, doi:10.2118/63064-MS.

Rohmer, J., and O. Bouc (2010), A response surface methodology to address uncertainties in cap rock failure assessment for CO2 geological storage in deep aquifers, *Int. J. Greenhouse Gas Control* *4*(2), 198–208, doi:10.1016/j.ijggc.2009.12.001.

Rongier, G., P. Collon-Drouaillet, and M. Filipponi (2014), Simulation of 3D karst conduits with an object-distance based method integrating geological knowledge. *Geomorphology*, *217*, 152–164, doi:10.1016/j.geomorph.2014.04.024.

Sarma, P., and W. H. Chen (2009), Generalization of the ensemble Kalman filter using kernels for nongaussian random fields, *SPE Reservoir Simulation Symposium*, 2–4 February, The Woodlands, TX, doi:10.2118/119177-MS.

Sarma, P., L. J. Durlofsky, and K. Aziz (2008), Kernel principal component analysis for efficient, differentiable parameterization of multipoint geostatistics, *Math. Geosci.*, *40*(1), 3–32, doi:10.1007/s11004-007-9131-7.

Scheibe, T. D., and D. L. Freyberg (1995), Use of sedimentological information for geometric simulation of natural porous media structure, *Water Resour. Res.*, *31*(12), 3259–3270, doi:10.1029/95WR02570.

Scheidt, C., C. Jeong, T. Mukerji, and J. Caers (2015), Probabilistic falsification of prior geologic uncertainty with seismic amplitude data: Application to a turbidite reservoir case, *Geophysics*, *80*(5), M89–M12, doi:10.1190/geo2015-0084.1.

Sternbergh, S. (1999), *Lectures on Differential Geometry*, American Mathematical Society, Providence, RI.

Smith, R. (2014), Electromagnetic induction methods in mining geophysics from 2008 to 2012. *Surv. Geophys.*, *35*(1), 123–156, doi:10.1007/s10712-013-9227-1.

Straubhaar, J., A. Walgenwitz, and P. Renard (2013), Parallel multiple-point statistics algorithm based on list and tree structures, *Math. Geosci.*, *45*(2), 131–147, doi:10.1007/s11004-012-9437-y.

Strauss, D. J. (1975), A model for clustering, *Biometrika*, *62*(2), 467–475, doi:10.1093/biomet/62.2.467.

Strebelle, S. (2002), Conditional simulation of complex geological structures using multiple-point statistics, *Math. Geol.*, *34*(1), 1–21, doi:10.1023/A:1014009426274.

Tahmasebi, P., A. Hezarkhani, and M. Sahimi (2012), Multiple-point geostatistical modeling based on the cross-correlation functions, *Compu. Geosci.*, *16*(3), 779–797, doi:10.1007/s10596-012-9287-1.

Tarantola, A., and B. Valette (1982), Inverse problems is quest for information.pdf, *J. Geophy.*, *50*, 159–170.

Tarantola, A. (1987), Inverse problem theory: Methods for data fitting and model parameter estimation, *Inverse Problem Theory: Methods for Data Fitting and Model Parameter Estimation.* http://www.scopus.com/inward/record.url?eid=2-s2.0-0023499373&partnerID=tZOtx3y1.

Tarantola, A. (2006), Popper, bayes and the inverse problem, *Nat. Phys.*, *2*(8), 492–494, doi:10.1038/nphys375.

Thibert, B., J. Pierre Gratier, and J. M. Morvan (2005), A direct method for modeling and unfolding developable surfaces and its application to the Ventura Basin (California), *J. Struct. Geol.*, *27*(2), 303–316, doi:10.1016/j.jsg.2004.08.011.

Thore, P., A. Shtuka, M. Lecour, T. Ait-Ettajer, and R. Cognot (2002), Structural uncertainties: Determination, management, and applications, *Geophysics*, *67*(3), 840–852, doi:10.1190/1.1484528.

Tikhonov, A. N., and V. Y. Arsenin (1977), *Solution of Ill-Posed Problems*, Winston & Sons, Washington, DC.

Tjelmeland, H., and J. Besag (1998), Markov random fields with higher-order interactions, *Scand. J. Stat.*, *25*, 415–433, doi:10.1111/1467-9469.00113.

Tsuji, T., R. Hino, Y. Sanada, K. Yamamoto, J. O. Park, T. No, E. Araki, N. Bangs, R. von Huene, G. Moore, M. Kinoshita (2012), In situ stress state from walkaround VSP anisotropy in the Kumano basin southeast of the Kii Peninsula, Japan, *Geochem. Geophys. Geosyst.*, *12*(9), doi:10.1029/2011GC003583.

Vo, H. X., and L. J. Durlofsky (2014), A new differentiable parameterization based on principal component analysis for the low-dimensional representation of complex geological models, *Math. Geosci.*, *46*(7), 775–813, doi:10.1007/s11004-014-9541-2.

Vo, H. X., and L. J. Durlofsky (2016), Regularized kernel PCA for the efficient parameterization of complex geological models, *J. Comput. Phys.*, *322*, 859–881, doi:10.1016/j.jcp.2016.07.011.

Voelcker, H. B., and A. A. G. Requicha (1977), Geometric modeling of mechanical parts and processes, *Computer 10*(12), 48–57, doi:10.1109/C-M.1977.217601.

Ward, S. H., R. E. Campbell, J. D. Corbett, G. W. Hohmann, C. K. Moss, and P. M. Wright (1977), The frontiers of mining geophysics, *Geophysics*, *42*(4), 878–886, doi:http://dx.doi.org/10.1190/1.1440757.

Yilmaz, Ö. (2001), *Seismic Data Analysis: Processing, Inversion, and Interpretation of Seismic Data*, vol. *10*, Society of Exploration Geophysicists, Houston, TX, doi:10.1190/1.9781560801580.

Zhang, X., M. J. Pyrcz, and C. V. Deutsch (2009), Stochastic surface modeling of deepwater depositional systems for improved reservoir models, *J. Petrol. Sci. Eng.*, *68*(1–2), 118–134, doi:10.1016/j.petrol.2009.06.019.

Zhao, H.-K., S. Osher, and R. Fedkiw (2001), Fast surface reconstruction using the level set method. *IEEE Workshop on Variational and Level Set Methods in Computer Vision*, Vancouver, BC, 13 July 2001, pp. 194–201, doi:10.1109/VLSM.2001.938900.

Zhou, H., J. J. Gómez-Hernández, H. J. Hendricks Franssen, and L. Li (2011), An approach to handling non-Gaussianity of parameters and state variables in ensemble Kalman filtering, *Adv. Water Resour.*, *34*(7), 844–864, doi:10.1016/j.advwatres.2011.04.014.

7

Bayesian Evidential Learning

7.1. THE PREDICTION PROBLEM REVISITED

Decision making does not require perfect information; however, it does require understanding and taking into consideration the uncertainties associated with the decision. In the subsurface, decisions are often dependent on some quantity that cannot be directly observed; hence, it needs to be predicted. Predictions are often about some future event (e.g., a production profile of a yet-to-be drilled well) or a quantity that cannot be directly measured (i.e., map of the current concentration of a contaminant). Instead, one relies on data that can be measured (i.e., groundwater levels or electrical resistivity at selected locations) and build appropriate subsurface models that describe the system in question. Applying forward simulation on these subsurface models then allows estimating any desired prediction variable. Recall from Section 3.1.2 that we consider the data, predictions, and model variables to be random variables whose values are unknown or only partially observed. The actual observed data measurements from the field are regarded as a single realization of the data random variables. Our goal is to use the data and prediction variables generated by applying a forward model to the subsurface models in combination with our observed or "real" data to make estimates of the value of the prediction variable.

To illustrate this, consider a simple example. Suppose we are about to move to a different city for a new job and need to make a decision about how much furniture to purchase. This decision will require making a prediction of how large our new apartment will be (denoted as the prediction variable **h**), as well as quantifying uncertainty on this prediction. We know that the size of an apartment is related to its rent, and therefore salary (base and bonus) of its tenant, which we denote as the data (observable) variable **d**. Both apartment size and salary are functions of the socioeconomic conditions within the city. We will call these conditions the socioeconomic model **m**, and for a given **m** we use deterministic *forward* functions to model salary and the size of an affordable apartment. The socioeconomic model is complex involving many parameters such as housing supply, unemployment rate, types of local industries, economic growth, property taxes, and so on, all of which affect housing costs as well as salaries. Furthermore, each of these parameters may be global (applicable all over the city), or vary spatially from one neighborhood to another. This means that the dimension of **m** (number of parameters and spatially varying components) will be much larger than that of **h** (apartment size in square feet) and **d** (base and bonus in dollars), see Eq. (7.1):

$$\dim(\mathbf{m}) \gg \dim(\mathbf{h}), \dim(\mathbf{d}) \tag{7.1}$$

We will term the approach involving inference on model variables from data to predict **h**, *causal analysis*. Suppose we have signed a job offer, so we know what our actual salary is ($\mathbf{d}_{\text{obs}}$). A causal analysis would then attempt to determine a socioeconomic model that will produce a salary that matches $\mathbf{d}_{\text{obs}}$. We then take this matched **m** and apply a forward function to predict what size of an apartment we will be able to afford under these socioeconomic conditions. However, obtaining this matched **m** will be a challenging task, since this involves solving an inverse problem. In Chapter 6, we discussed solutions to inverse problems and illustrated that under nonlinearity of the forward function or if the system is underdetermined, multiple solutions may exist. A single matched model may not actually describe the true conditions in the city and may yield a prediction of apartment size that is different from reality. An alternative approach to solving such problem is *Bayesian evidential learning* (BEL), since it involves a statistical model. In BEL, we first consider a-priori information about the city's socioeconomic conditions, prior to considering the offered salary $\mathbf{d}_{\text{obs}}$. For instance, while we might not know the exact value of the property tax in the city, we may be

Quantifying Uncertainty in Subsurface Systems, Geophysical Monograph 236,
First Edition. Céline Scheidt, Lewis Li, and Jef Caers.

able to say that it varies between 0 and 3% because of state regulations. For each model variable, we state some $f(m_n)$ that captures that prior information (here assumed independent). A random sampling from each of these prior distributions then generates a set of socioeconomic model realizations $\{\mathbf{m}^{(1)}, \mathbf{m}^{(2)}, \ldots, \mathbf{m}^{(L)}\}$. Given a realization $\mathbf{m}^{(\ell)}$, forward modeling produces a value for salary $\mathbf{d}^{(\ell)}$ as well as apartment size $h^{(\ell)}$. It is likely that very few to no $\mathbf{d}^{(\ell)}$ will match our actual salary $\mathbf{d}_{\text{obs}}$. Instead, we generate a scatter plot between $\mathbf{d}^{(\ell)}$ and $h^{(\ell)}$ and fit a statistical relationship (the learning) between salary and apartment size (see Figure 7.1). This statistical relationship can then be used to predict h when $\mathbf{d} = \mathbf{d}_{\text{obs}}$.

Why is this approach termed "evidential" and what are we "learning"? This is related to the discussion in Chapter 5 on Bayesianism, where the data is used as evidence to infer certain hypotheses. The ultimate goal of modeling here is not the model hypotheses (i.e., fractures occur, the water level is low, the system is fluvial). Even if we can assess the probability of these model hypotheses, we still need to calculate how these model hypotheses *cause* (affect) the prediction variables that ultimately matter in decision making. This is the causal approach. In evidential learning, we aim to learn the relationship between the data variables (the evidence) and the decision variables (the decision hypothesis). How much evidence is there that we need to drill a well at some location (x, y)? Is there evidence that contamination has spread: yes or no? The causal form of learning is using

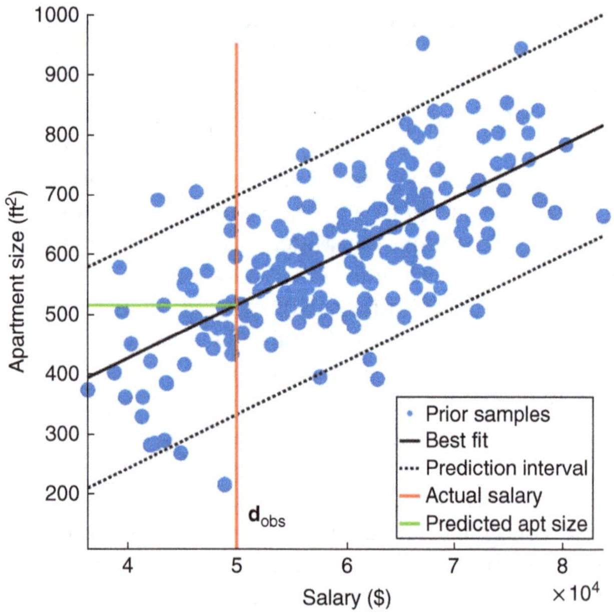

Figure 7.1 A Bayesian evidential analysis uses prior samples to construct a statistical relationship (evidential learning) between salary and apartment size. The actual salary is then used to state uncertainty on apartment size, or the mean is used as an estimate.

evidence as circumstantial: evidence is used for a model, then by circumstance for the prediction. Evidential learning directly models how evidence/data is able to influence decisions. As with any learning, the system requires "examples." These examples are furnished by a-priori model realizations. However, these model variables need not be calibrated (inverted) with data.

This evidential approach is in line with the Bayesianism philosophy of Chapter 5. The prior distribution on the subsurface model parameters leads to stating uncertainty on prior prediction or decision hypothesis $P(\mathbf{h})$. Since this is based on a limited sample, we get an estimate $\hat{P}(\mathbf{h})$. The prior of data variables is estimated from the same prior models as $\hat{P}(\mathbf{d})$. By estimating a statistical relationship, one can then either directly estimate $\hat{P}(\mathbf{h}|\mathbf{d}_{\text{obs}})$ or estimate a likelihood model of the evidence occurring under our current hypothesis $\hat{P}(\mathbf{d}_{\text{obs}}|\mathbf{h})$ to get (Bayes' rule):

$$\hat{P}(\mathbf{h}|\mathbf{d}_{\text{obs}}) = \frac{\hat{P}(\mathbf{d}_{\text{obs}}|\mathbf{h})}{\hat{P}(\mathbf{d}_{\text{obs}})}\hat{P}(\mathbf{h}) \tag{7.2}$$

In the subsurface, the previous socioeconomic model is replaced with a subsurface model $\mathbf{m}$, a numerical model that describes uncertain subsurface properties. The data $\mathbf{d}$ refer to an observable variable, for instance the log of an existing borehole or a geophysical survey of interest, while $\mathbf{d}_{\text{obs}}$ is the field data, that is, the value of the variable $\mathbf{d}$ measured in reality. The prediction $\mathbf{h}$ represents the uncertain quantity we are trying to predict, such as a future production rate, or an exhaustive concentration map. For a given subsurface model, the corresponding data and prediction variable are obtained through forward (often highly nonlinear) physical simulation. This could entail flow simulation, geophysical image migration, reactive transport simulation, and so on. Both data and prediction variables can be static (not changing in time) or dynamic and can manifest as scalars, time series, or spatial maps. Table 7.1 lists some examples of scenarios where decisions depend on data and prediction variables of varying types.

Depending on the discipline, causal analysis (see Figure 7.2) goes by various terms such as, history matching [*Oliver et al.*, 2008], migration velocity analysis [*Sava and Biondi*, 2004], data assimilation [*Houtekamer and Mitchell*, 1998; *Evensen*, 2003; *Reichle*, 2008], dynamic data integration [*Suzuki et al.*, 2008], and so on. However, because inversion is a challenging problem, one frequently invokes ad-hoc modifications to make it work (see Chapters 4 and 6). Furthermore, inverted models are not guaranteed to remain consistent with any additional data that is collected in the future. This necessitates updating or rebuilding the subsurface model each time additional data becomes available. Besides being an expensive and difficult procedure, this has the

Table 7.1 Example decisions and the associated data/prediction variables in subsurface applications.

Decision problem	Data variable	Prediction variable
Should we perform a workover on an existing oil production well?	(Dynamic) Existing well's historical production profile	(Dynamic) Future production rate after performing the workover
Should we remediate a site that has been contaminated?	(Dynamic) Rates of contaminant concentrations at monitoring wells	(Static) Map of containment concentration at a given time
Should we explore a potential mining prospect?	(Static) Magnetic data	(Static) Mineral deposit size
What should the specifications of a geothermal heat pump be?	(Static) Electrical resistivity tomography	(Dynamic) Future temperature decline in extraction well

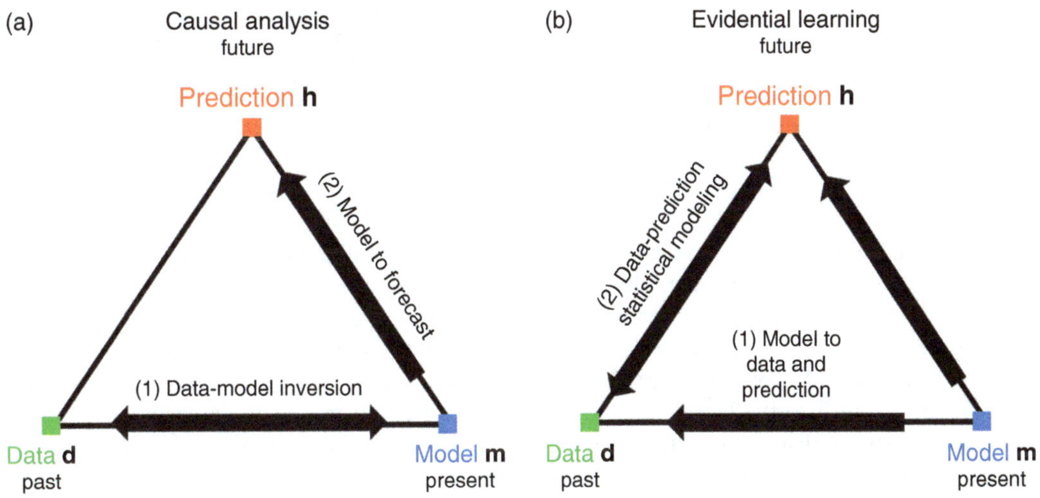

Figure 7.2 Two paradigms for uncertainty quantification. The traditional framework (a) applies causal analysis to match the subsurface models to the data, then use those matched models for predictions. The proposed methodology (b) uses Bayesian evidential learning by which the model is used to construct a statistical relationship between the data and prediction variables.

potential to cause fluctuating or contradictory forecasts over the lifetime of a project.

The BEL approach [*Scheidt et al.*, 2014; *Satija and Caers*, 2015; *Satija et al.*, 2017] is motivated by the recognition that inverted subsurface models are not necessarily the final desired product for decision making but rather the prediction variable and their quantification of uncertainty. In BEL, we reconsider the role of the subsurface model. Rather than serving as a mechanism to match observed data, the subsurface model is used to generate samples to *learn* a statistical model that describes the relationship between the data and prediction variables. This statistical model then serves as the mechanism for predicting $\mathbf{h}$ from $\mathbf{d}_{\text{obs}}$.

In this section, we will examine the elements of statistical learning that are specifically tailored for the kind of problems we typically deal with in the subsurface. Figure 7.3 shows the broad strokes by which this process works and will be further detailed in the next sections. We start by considering the role of the prior in generating the samples used for learning the statistical model. However, statistical learning requires some form of assumption on both the underlying variables and the nature of their relationships. We will discuss these assumptions, tests of their validity, and transformations that can be applied to assuage any deviations. It also requires some form of dimension reduction (feature extraction) for such learning to work, and a way to undo those transformations after modeling. Since the statistical model is not perfect, we need to establish a confidence interval and assess the error of the statistical model in terms of uncertainty quantification (UQ). Finally, we will illustrate the application on several situations that occur in practice.

7.2. COMPONENTS OF STATISTICAL LEARNING

7.2.1. The Statistical Model

The goal of statistical modeling is to characterize the possible outcomes of a random event or quantity and establish how likely each outcome is to occur. Formally, a statistical model is defined as a corresponding pair of the

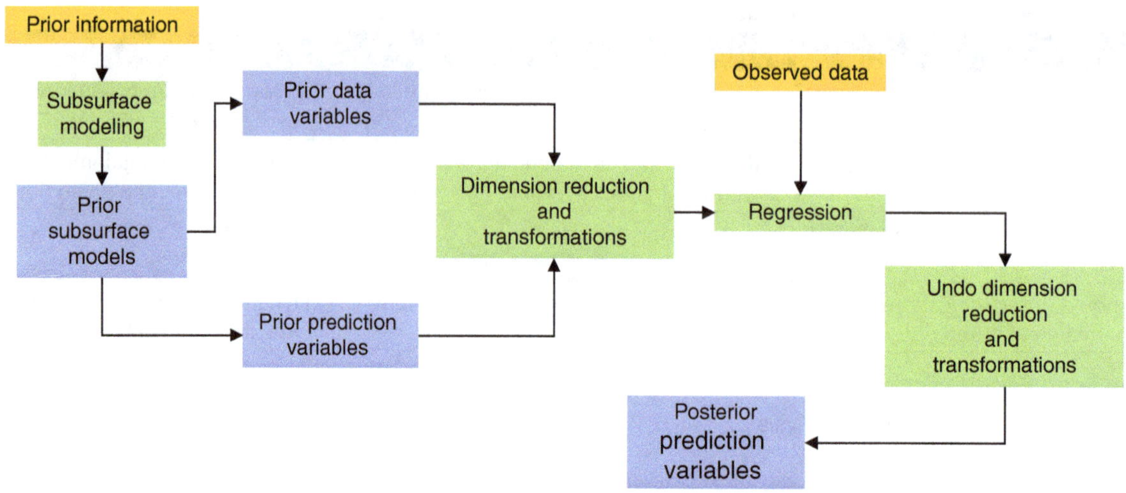

Figure 7.3 Overview of the elements of Bayesian evidential learning and prediction for UQ.

sample space Ω (the set of all possible outcomes), and the set of *probability distribution P* on the sample space. Returning to our apartment salary example, the uncertain quantity is the square footage of our future apartment. The sample space comprises all possible sizes that an apartment can take on (from 0 to the largest apartment in the world). The probability distribution describes the chances of our future apartment being each of the sizes defined in the sample space. As discussed in Chapter 5, the construction of statistical models and UQ is not very meaningful without first collecting some data. Often, the goal is to identify the additional sources of information that narrow the posterior $f(\mathbf{h}|\,\mathbf{d}=\mathbf{d}_{obs})$ which expresses the distribution of the prediction variable **h** (sometimes called the *target* or *dependent* variable), given that the observable variable or data variable **d** has outcome $\mathbf{d}_{obs}$ (also termed *features*, *predictor*, or *independent* variable). The process of uncovering this relationship between the data and prediction variables involves a regression analysis. The choice of the actual statistical model used for carrying out regression analysis will depend on the nature of the variables as well as their relationship.

Regardless of the choice of the statistical model, performing regression requires samples of **h** and **d** to use as the *training set*. The statistical model is fitted or trained in such a way that it estimates the relationship between the variables using the samples in the training set. Once fitted, the statistical model can then be used to make a prediction on the value or distribution of **h** given $\mathbf{d}=\mathbf{d}_{obs}$. The process of obtaining these training samples is called *data collection* (although it entails collecting both samples of data and predictions). Depending on the subject area, data collection can be achieved through surveys, laboratory experiments, or clinical trials, and so on. The essence of data collection is making multiple repeated measurements of the unknown random variables **h** and **d**.

In our apartment–salary case, the training set is a list of apartment sizes and the salaries of their tenant salaries within the city. One could imagine that, to get such samples, we could interview denizens of the city about their salaries and living conditions. However, in the subsurface this is not possible. There is no way for us to directly query the performance of a future well, or take multiple logs of the same well. Instead, we must rely on using subsurface models and forward simulators to generate a set of realizations of **h** and **d**. Another aspect of a statistical model is that it typically comes with a set of underlying assumptions. Assumptions are usually required for any form of statistical modeling, and incorrect assumptions can invalidate any conclusions drawn from the analysis. A common assumption is that the data and prediction random variables are distributed according to some parametric distribution (fully described by a finite number of parameters). The most popular parametric distribution that occurs in statistics is the Gaussian distribution, for the reasons discussed in Section 3.3.5. The advantage of using parametric distributions is twofold. First, it reduces the number of unknowns that we are trying to predict, from the full joint probability distribution to just a few parameters. Second, parametric assumptions provide mathematical convenience as closed-form solutions can often be derived for evaluating the prediction. Another class of assumptions lies in the underlying relationship between the data and prediction variables. Since regression will seek to formulate a function that maps the two variables, structural assumptions on this function can greatly simplify the mathematics. For instance, assuming a linear function between the variables will allow for an analytical solution of the conditional probability (see

Chapters 3 and 6). While these assumptions simplify regression analysis, it is important to verify that they do indeed hold before any results can be taken with confidence.

Another consideration for regression analysis is the dimension of the problem. This refers to both the dimension of the data and prediction variables, and the number of underlying variables. Recall from Section 3.4.1 that the number of required samples increases exponentially with the dimension of the problem. However, in practice, many of the dimensions of the data and prediction variables are redundant. Consider the case where $\mathbf{h}$ is a time-series response; its dimension is equal to the number of time steps by which it has been discretized into. Since physical responses are rarely pure random noise, it suggests that we do not need every single time step to represent the majority of information contained within the signal. For instance, for smoothly varying densely sampled time series, simply discarding every other time step would allow for a dimension reduction without much loss of information. Uncovering such low-dimensional representations has the advantage of reducing the number of samples required for training and simplifying the statistical model. The choice of dimension reduction technique varies depending on the characteristic of the variables being considered. We will discuss some methods as well as strategies for selection in Section 7.2.3.

The choice of the regression approach depends on the validity of the assumptions on the variables and the dimension of the problem. In Section 7.2.4 we will discuss a few types of regression techniques that can be applied when certain assumptions are met. The resulting fitted statistical model is then used with the actual observed value of $\mathbf{d} = \mathbf{d}_{obs}$ to make a probabilistic prediction on $\mathbf{h}$. However, before these predictions can be used for decision-making, it is important to check the validity of the result. In Section 7.2.5 we will develop methods that can falsify the Bayesian evidential learning approach (possibly then in favor of the traditional causal analysis).

7.2.2. Generating the Training Set

The first step of any regression analysis is the collection of samples of the prediction variables and data variables. The goal of sampling is to gather information on the variables of interest in a systematic way, such that they can be used to train or fit the statistical model. Sampling entails first defining the population on which we are trying to establish a statistical model. In our apartment–salary example, the population is defined as the set of all apartments in the city and the salaries of their tenants. It would be impossible to gather information regarding every single apartment and tenant, so we would need to select a smaller and manageable subset of the entire population to study. To this end, we need to define a sampling framework, which is a set in which we can identify every single element and can possibly incorporate each one into our samples. For instance, we could identify all of our friends that live in the city and ask them how much they make and how large their apartment is. These sets of apartment sizes and salaries will then serve as the training set for a statistical model. A fundamentally important characteristic of a good sampling methodology is that it generates samples representative of the population from which it was drawn from. Failure to do so will introduce a sampling bias meaning that certain members of the population are overrepresented in the training set. Accordingly, we would need to query our friends from all walks of life, as statistical modeling using biased samples can result in erroneous conclusions.

We may now be asking ourselves: would it have been easier to call our friends and inquire about their salaries and apartments rather than going through the whole socio-economic modeling process? The answer is, of course, it would have been. However, in order to do so we have assumed that we have multiple friends in the city who we can call. This means we are able to take repeated measurements of the $\mathbf{h}$ and $\mathbf{d}$ random variables. Unfortunately, the same assumption does not hold when dealing with subsurface systems. Recall that a probabilistic approach to UQ perceives reality as a single realization drawn from a random variable. There is no way to draw another realization of reality; once we have drilled a well and logged it to obtain $\mathbf{d}_{obs}$, we cannot re-drill it and measure a second, different instance of $\mathbf{d}_{obs}$. For $\mathbf{h}$, we cannot even measure a single instance, as it refers to a variable that cannot be directly observed at all (i.e., future rates, spatially exhaustive measurements, etc.).

Instead, we must rely on the fact that $\mathbf{d}$ and $\mathbf{h}$ are related to the subsurface model $\mathbf{m}$ through forward functions. The function g_d generates the expected $\mathbf{d}$ for a given $\mathbf{m}$

$$\mathbf{d} = g_d(\mathbf{m}) \tag{7.3}$$

This function tells us deterministically the value $\mathbf{d}$ would take if $\mathbf{m}$ describes the subsurface, and without consideration of noise on $\mathbf{d}$. In practice, g_d is usually a physics-based forward simulation that produces a time-series response (e.g., production rates) or spatial map (e.g., seismic imaging). Likewise, another function g_h applied to $\mathbf{m}$ generates the prediction:

$$\mathbf{h} = g_h(\mathbf{m}) \tag{7.4}$$

To generate a set of subsurface models $\mathbf{m}$, we must first parameterize it such that both the spatially varying and non-gridded components are accounted for. Refer to Section 6.3 for an in-depth discussion of subsurface model parameterization. Since the true subsurface system is unknown, each of these parameters will be uncertain. We may have a-priori beliefs regarding those uncertain parameters. Suppose we parameterized our $\mathbf{m} = (\mathbf{m}_{\text{grid}}, \mathbf{p})$ using N parameters, we then specify the prior parameters distributions as

$$m_i \sim f(m_i) \forall i = 1, \ldots, N \tag{7.5}$$

$f(m_i)$ is the pdf of the i-th model parameter, before any observations of the data and prediction variables have been made. Specifying appropriate priors is perhaps the most challenging aspect of Bayesian analysis, and still an active field of research (see Chapter 5 and Section 6.5).

After stating prior distributions, a set of L samples are generated: $\{\mathbf{m}^{(1)}, \mathbf{m}^{(2)}, \ldots, \mathbf{m}^{(L)}\}$. The sets of $\mathbf{d}$ and $\mathbf{h}$ obtained from applying g_d and g_h on these models serve as the training data for BEL:

$$\left\{\left\{\mathbf{d}^{(1)}, \mathbf{h}^{(1)}\right\}, \left\{\mathbf{d}^{(2)}, \mathbf{h}^{(2)}\right\}, \ldots, \left\{\mathbf{d}^{(L)}, \mathbf{h}^{(L)}\right\}\right\} \tag{7.6}$$

Building a regression model from a limited set of samples comes with an important caveat: extrapolations. Predictions obtained from extrapolation using any regression method should be taken with caution as none of the training samples suggest that such a prediction value is even possible. This phenomena has been evidenced in several practical studies [*Makridakis et al.*, 1982; *Wilkinson*, 1983; *McKenzie et al.*, 1997; *Chiu et al.*, 2007]. If the samples in Eq. (7.6) do not cover or span the observed data, it may be tempting to perform ad-hoc modifications to the prior (such as multiplying a parameter with some ad-hoc value) with the purpose of ensuring that the prior range encompasses the observed data. However, this has been refuted as "ad-hoc" [*Chalmers*, 1999] and can lead to incorrect posteriors. Indeed, any ad-hoc modification of the prior may lead to posteriors to be inconsistent when additional observations are collected in the future. If the prior on $\mathbf{d}$ is realistic, then logically the probability of the observed data to lie outside of the span of the simulated data variables is simply $2/(L+1)$. To see why this is true, consider that each sample drawn from any pdf has equal probability of being the largest drawn so far. Therefore, the probability that $\mathbf{d}_{\text{obs}}$ is larger than all L simulated data variables is $1/(L+1)$. Likewise, the same logic applies for $\mathbf{d}_{\text{obs}}$ being smaller than all L data values, yielding the total probability of $\mathbf{d}_{\text{obs}}$ lying outside the span of $\mathbf{d}$ of $2/(L+1)$. If the simulated data variables do not cover the observations then either (i) one needs more samples or (ii) one needs a different prior. The latter will occur with a probability of $(L-1)/(L+1)$.

7.2.3. Dimension Reduction and Feature Extraction

Traditionally, regression has primarily been applied to cases where the number of samples far exceeds the number of unknown parameters. While there have been recent advancements in high-dimensional regression [*Fodor*, 2002; *Hinze and Volkwein*, 2005], many challenges remain and it remains still an area of active research. The rise in high-dimensional statistical problems is driven by the ability to record and measure large datasets (for instance the resolution at which rates can now be measured). However, increased dimension in data does not necessarily indicate that additional information is obtained, as there could be redundancies within the data. A specific type of redundancy known as *multicollinearity* occurs when dimensions of the variable are highly correlated with each other. Multicollinearity can result in both numerical issues during regression, in addition to degrading the predictive performance of the statistical model [*Graham*, 2003].

In many cases, we may have reasons to believe that while the data and prediction are given in high dimensions, they are actually indirect measurements of a lower-dimensional source that cannot be directly observed. This means that the source of the variability between samples lies on a lower-dimensional manifold. The goal of dimension reduction is, thus, to identify these degrees of freedom that capture the majority of the variance in the data. Performing statistical analysis in this lower-dimensional space not only reduces the required number of samples but can improve the performance of the model due to the removal of multicollinearity.

There has been a recent surge in interest in dimension reduction techniques, and practically some form is usually applied as a preprocessing step for statistical learning procedures. Readers are referred to *Fodor* [2002], *Hinze and Volkwein* [2005], and *van der Maaten et al.* [2009] for a comprehensive survey of existing methodologies. The selection of an algorithm is highly dependent on the nature of the variable. We will next review (see Chapter 2 for the details) three families of methods in the context of BEL.

1. *Principal component analysis* (PCA): The mathematical formulation of PCA was discussed in Section 2.5. Recall that PCA provides a sequence of the best linear approximation to the original high-dimensional variable. By examining the resulting eigenvalues, we can find the number of dimensions that need to be kept capturing the majority of the variability. It should be noted that dimension reduction using PCA is strictly limited to not only sets of vectors but also matrices and spatial maps. A useful application is finding low-dimensional representations of a set of images (e.g., for saturation maps, contaminant maps). This idea was originally developed for purpose of facial recognition and is called eigen-faces [Turk and Pentland, 1991; *Belhumeur et al.*, 1997]. By flattening each image in the training set into a vector, the formulation in Section 3.5.3 can be readily applied. Another important property of PCA is that it is bijective, meaning that the original high-dimensional variable can be recovered by undoing the projection. This is a

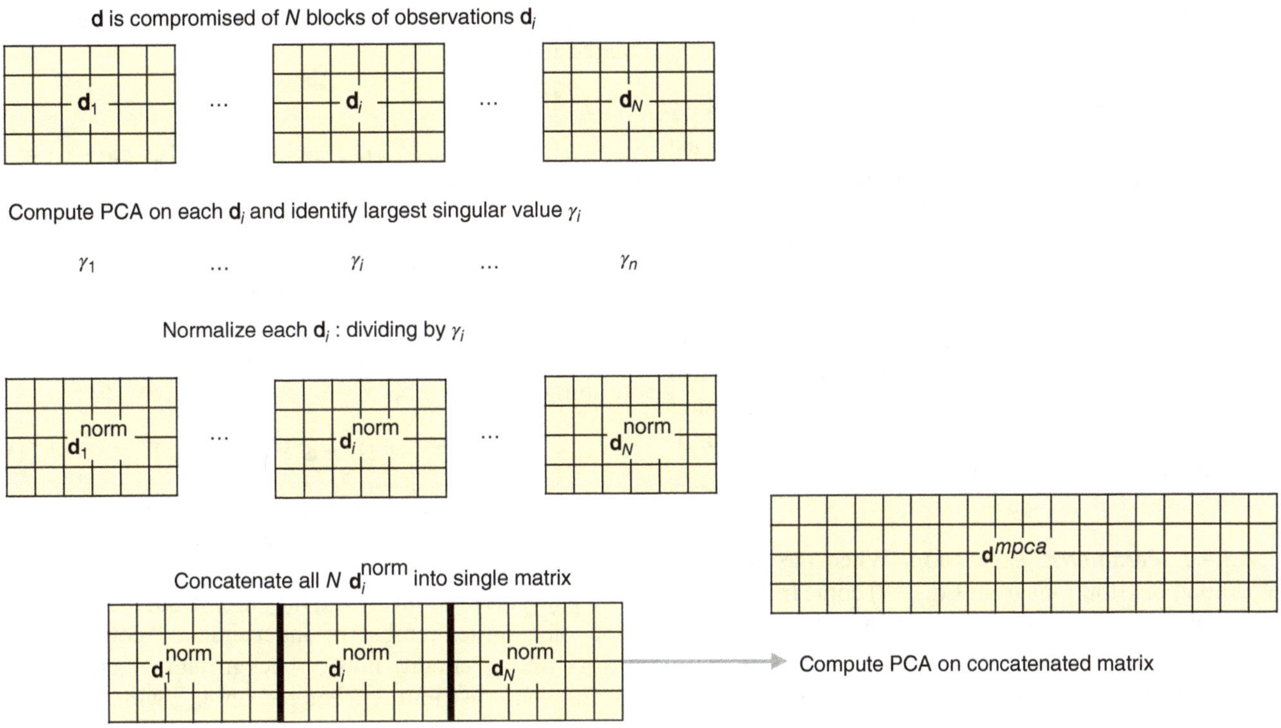

Figure 7.4 Steps in performing mixed PCA.

particularly useful trait in statistical learning, as it allows for regression to be performed in the PC space, and any predictions made in the low-dimensional space can be reconstructed in the original space, easily and uniquely. The precision of the reconstruction depends on the number of components used. For cases where the majority of the variance is captured by the first few eigenvalues, accurate reconstructions can be obtained using only those first few eigen-bases. PCA can also be used to pool data from different sources, identify redundancies, and generate a reduced dimensional projection of the combined data. This is referred to as mixed PCA [*Abdi et al.*, 2013], see Figure 7.4, and could be used, for instance, to combine data from different well locations into a single variable. The procedure is composed of three steps. First, a regular PCA is performed on each of the data sources to obtain the largest singular value. Second, each data source is normalized according to the first singular value; this accounts for any difference in scales amongst the data sources. Third, the normalized data inputs are concatenated and regular PCA is applied to this final matrix.

2. *Multidimensional scaling* (MDS): The mathematical details of multidimensional scaling can be found in Section 3.5.2. Unlike PCA, MDS aims to preserve pairwise distances of high-dimensional samples in low dimensions. This requires computing a distance between samples in the original high-dimensional space, and using that distance matrix to compute a lower-dimensional projection. The advantage of MDS is that a variety of distance metrics can be used, which are suited to describe the dissimilarity at hand. Section 3.2.5 provides a survey of common distances used in this book.

3. *Functional data analysis* (FDA): Functional data analysis is used to identify the dominant modes of variation among data that exhibits some systematic variation. It assumes that the measured variable is being driven by some underlying physical process that is smoothly varying. This allows for the decomposition of the variable into a linear combination of basis function as described in Section 2.6. FDA is usually ideal for dimension reduction on time-series variables that are smoothly varying but complex, in particular, output of computer models of process in the subsurface.

7.2.4. Regression Analysis

In line with a Bayesian philosophy toward UQ, the role of regression analysis in BEL is to estimate the posterior probability distribution of the prediction variable $\mathbf{h}$ given we have observed $\mathbf{d}_{\text{obs}}$. We invoke Bayes' rule to express the posterior as the product between the prior distribution of $\mathbf{h}$: $f(\mathbf{h})$ and the likelihood function $f(\mathbf{d}_{\text{obs}}|\,\mathbf{h})$, normalized by a constant. In general, we first apply dimension reduction and/or other transformations to the prediction and data variables (which we then denote as $\mathbf{d}^*$ and $\mathbf{h}^*$, respectively) and then evaluate the posterior distribution:

$$f\left(\mathbf{h}^* \mid \mathbf{d}^*_{\text{obs}}\right) = \text{const} \times f\left(\mathbf{d}^*_{\text{obs}} \mid \mathbf{h}^*\right) f\left(\mathbf{h}^*\right) \tag{7.7}$$

The process of estimating these distributions will depend on the choice of regression technique. This choice is driven by the nature of the underlying variables as well as their relationship. The literature regarding regression techniques is expansive and very well developed. Interested readers can referred to classical texts such as *Hastieet al.* [2009], *Myers* [1990], and *Smith and Draper* [1998] for an extensive overview of techniques. We will, however, discuss examples of two families of regression, parametric and non-parametric, and their application in evidential analysis and learning.

7.2.4.1. Parametric Regression. When the prior and likelihood distributions from Eq. (7.7) come from certain probability distribution families, an analytical expression can be derived for the posterior distribution $f\left(\mathbf{h}^* \mid \mathbf{d}^*_{\text{obs}}\right)$. This is advantageous as it avoids potentially computationally expensive numerical integrations to obtain the posterior. One such family of probability distributions is the Gaussian distribution. That is, if the likelihood function and prior are Gaussian, the posterior will also be Gaussian. Under such an assumption, Gaussian process regression can be applied [*Tarantola*, 2005; *Satija and Caers*, 2015; *Hermans et al.*, 2016; *Satija et al.*, 2017].

We start by expressing the Gaussian prior $f(\mathbf{h}^*)$ as

$$f(\mathbf{h}^*) = \text{const} \times \exp\left(-\frac{1}{2}\left(\mathbf{h}^* - \bar{\mathbf{h}}^*\right)^T C^{-1}_{h^*h^*}\left(\mathbf{h}^* - \bar{\mathbf{h}}^*\right)\right) \tag{7.8}$$

The mean $\bar{\mathbf{h}}^*$ and covariance $C_{h^*h^*}$ can be readily estimated from the prior samples (after dimension reduction and any transformations) as $\bar{\mathbf{h}}^* = \sum_{\ell=1}^{L} \mathbf{h}^{(\ell)^*}$ and $\hat{C}_{h^*h^*} = \frac{1}{L}\mathbf{h}^{*T}\mathbf{h}^*$. We next consider the likelihood function, which represents a measure of how well a value of $\mathbf{h}^*$ is at explaining $\mathbf{d}^*_{\text{obs}}$. That is to say, what is the probability that a model $\mathbf{m}$ that generates $\mathbf{h}^*$also generates $\mathbf{d}^*_{\text{obs}}$? To express this probability analytically, we need to assume some form of relationship between $\mathbf{d}^*$ and $\mathbf{h}^*$. For instance, if they are linearly related then

$$\mathbf{d}^* = A\mathbf{h}^* + \boldsymbol{\varepsilon}^* \tag{7.9}$$

A is the set of unknown coefficients that map $\mathbf{h}^*$ to $\mathbf{d}^*$, while the error $\boldsymbol{\varepsilon}^*$ is assumed to be a Gaussian with zero mean and covariance C^*_{ε}. To estimate A, we use our training set $\{\{\mathbf{d}^{*(1)}, \mathbf{h}^{*(1)}\}, \{\mathbf{d}^{*(2)}, \mathbf{h}^{*(2)}\}, \ldots, \{\mathbf{d}^{*(L)}, \mathbf{h}^{*(L)}\}\}$ generated from the prior models and apply ordinary least squares to find the solution $\hat{A}$ that minimizes the sum of the squared errors. We can then express the likelihood as a Gaussian centered around $\mathbf{d}^*_{\text{obs}}$ and covariance $C_{d^*d^*}$. For any value of $\mathbf{h}^*$, the likelihood can then be evaluated using

$$f\left(\mathbf{d}^*_{\text{obs}} \mid \mathbf{h}^*\right) = \text{const} \times \exp\left(-\frac{1}{2}\left(\hat{A}\mathbf{h}^* - \mathbf{d}^*_{\text{obs}}\right)^T C^{-1}_{d^*d^*}\left(\hat{A}\mathbf{h}^* - \mathbf{d}^*_{\text{obs}}\right)\right) \tag{7.10}$$

However, what is $C_{d^*d^*}$? This covariance arises due to two sources of error. The first is attributed to the imperfect fitting that occurs in Eq. (7.9). The effect of this can be estimated from the residuals obtained when fitting the training data:

$$\hat{C}_{\varepsilon^*} = \frac{1}{L}\left(\hat{A}\mathbf{h}^* - \mathbf{d}^*\right)^{\mathrm{T}}\left(\hat{A}\mathbf{h}^* - \mathbf{d}^*\right) \tag{7.11}$$

The second source of error arises when measuring the observed data. All scientific measurements are subject to error such that instead of measuring $\mathbf{d}_{\text{obs}}$ we record $\mathbf{d}_{\text{obs}} + \boldsymbol{\eta}$. This random error $\boldsymbol{\eta}$ may arise from the aggregation of a large number of independently contributing errors (e.g., circuit noise, background noise, etc.). A Gaussian distribution centered on $\mathbf{d}_{\text{obs}}$ with a covariance C_{dd} stated in the original high-dimensional space is used to model this error. C_{dd} would then be the error of the sensor used to measure $\mathbf{d}_{\text{obs}}$ (typically given by the sensor manufacturer). However, to transform C_{dd} into its low-dimensional counterpart C^*_{dd} is not trivial as the transformation from $\mathbf{d}$ to $\mathbf{d}^*$ (and back) may consist of a complex series of dimension reduction methods (and other transformations). Only in the case where the transformation is linear (i.e., canonical correlation or PCA) and described by a linear operator B, then

$$C^*_{dd} = BC_{dd}B^T \tag{7.12}$$

However, if the transformations are not linear (e.g., FDA, normal score transforms, etc.), then Eq. (7.12) cannot be used to calculate statistics such as an empirical covariance $\hat{C}^*_{dd}$.

Instead, we need a Monte Carlo sample of the measurement error based on the observations and Monte Carlo sample of $\boldsymbol{\eta}$. The error (e.g., generated by sampling a zero mean Gaussian distribution with covariance C_{dd}) is added to a selected prior sample $\mathbf{d}^{(\ell)}$ to get

$$\mathbf{d}^{(\ell)}_{\text{perturbed}} = \mathbf{d}^{(\ell)} + \boldsymbol{\eta}, \quad \boldsymbol{\eta} \sim \mathcal{N}(0, C_{dd}) \tag{7.13}$$

We next apply the necessary nonlinear transformations (dimension reductions) to get $\mathbf{d}^{(\ell)^*}$ and $\mathbf{d}^{(\ell)^*}_{\text{perturbed}}$. We can then evaluate the difference between the two sets of transformed variables as

$$\boldsymbol{\eta}^{(\ell)^*} = \mathbf{d}^{(\ell)^*}_{\text{perturbed}} - \mathbf{d}^{(\ell)^*} \tag{7.14}$$

By repeating this for all samples, we obtain a set of errors in the transformed space: $\left\{\boldsymbol{\eta}^{(1)^*}, \boldsymbol{\eta}^{(2)^*}, \ldots, \boldsymbol{\eta}^{(L)^*}\right\}$. From this we can directly estimate the empirical covariance $\hat{C}^*_{dd}$. Since both errors are Gaussian, we can write $\hat{C}_{d^*d^*} = \hat{C}^*_{dd} + \hat{C}_{\epsilon^*}$ and substitute into the likelihood function from Eq. (7.10).

With the Gaussian prior distribution and likelihood function established, the same formulation as in Section 3.7.5 [Eq. (3.176)] is used to evaluate the posterior distribution of the prediction variable. As the posterior is also Gaussian, the mean and covariance $(\tilde{\mathbf{h}}^*, \tilde{C}_{h^*h^*})$ are found using the following analytical expressions:

$$\tilde{\mathbf{h}}^* = \bar{\mathbf{h}}^* + \hat{C}_{h^*h^*}\hat{A}^{\mathrm{T}}\left(\hat{A}\,\hat{C}_{h^*h^*}\hat{A}^{\mathrm{T}} + \hat{C}^*_{dd} + \hat{C}_{\epsilon}\right)^{-1}\left(\mathbf{d}^*_{\mathrm{obs}} - \hat{A}\,\bar{\mathbf{h}}^*\right) \tag{7.15}$$

$$\tilde{C}_{h^*h^*} = \hat{C}_{h^*h^*} - \hat{C}_{h^*h^*}\hat{A}^{\mathrm{T}}\left(\hat{A}\hat{C}_{h^*h^*}\hat{A}^{\mathrm{T}} + \hat{C}^*_{dd} + \hat{C}_{\epsilon}\right)^{-1}\hat{A}\,\hat{C}_{h^*h^*} \tag{7.16}$$

The primary advantage of this approach is that sampling from the posterior $f(\mathbf{h}^* | \mathbf{d}^*_{\mathrm{dobs}})$ is straightforward. By reversing any transformations and/or dimension reductions, each sample drawn from the posterior can be projected back into the original space yielding $f(\mathbf{h}|\,\mathbf{d}_{\mathrm{obs}})$. This yields a set of posterior predictions from which statistics such as quantiles can be computed on the prediction (see Figure 7.3). However, this approach is only appropriate when the underlying variables are Gaussian and are linearly correlated. It is important that these assumptions are verified, or the appropriate transformations are performed before proceeding with regression.

7.2.4.2. Non-Parametric Regression. In contrast to parametric regression, non-parametric regression does not make assumptions about the shape of the function that relates the data to the prediction. Rather, these techniques use the samples themselves to infer the nature of the posterior. This is useful when the relationship is nonlinear, multimodal, or not well understood. While non-parametric approaches can handle more complex distributions than their parametric counterparts, they typically require larger sample sizes, and more computationally intensive training.

In the previous sections, we discussed a methodology for estimating the posterior distribution $f(\mathbf{h}^* | \mathbf{d}^*_{\mathrm{obs}})$ under the assumptions of Gaussianity for the distributions and a linear relationship between $\mathbf{h}^*$ and $\mathbf{d}^*_{\mathrm{obs}}$. However, when Gaussianity does not hold, the estimated posterior mean and covariances no longer adequately describe the distribution. Instead, kernel density estimation (KDE, Section 3.3.3) can be used. Consider an illustrative example in Figure 7.5, where both the data and prediction variables are univariate.

The expression for computing the density is

$$\hat{f}(\mathbf{h},\mathbf{d}) = \frac{1}{L\mathbf{w}_h\mathbf{w}_\mathbf{d}}\sum_{\ell=1}^{L} K\left(\frac{\mathbf{h}^{(\ell)} - \mathbf{h}}{\mathbf{w}_h}\right) K\left(\frac{\mathbf{d}^{(\ell)} - \mathbf{d}}{\mathbf{w}_d}\right) \tag{7.17}$$

Recall that K is the kernel function and $\mathbf{w}_h$ and $\mathbf{w}_d$ are the bandwidths corresponding to the prediction and data variables (assuming a diagonal bandwidth matrix). Refer to Chapter 3 for a discussion of kernel choices and bandwidth selection. This allows estimating the joint density $\hat{f}(\mathbf{h},\mathbf{d})$ for any given combination of $\mathbf{h}$ and $\mathbf{d}$. To estimate the conditional distribution, we use

$$\hat{f}(\mathbf{h}|\,\mathbf{d}_{\mathrm{obs}}) = \frac{\hat{f}(\mathbf{h},\mathbf{d}_{\mathrm{obs}})}{\hat{f}(\mathbf{d}_{\mathrm{obs}})} \tag{7.18}$$

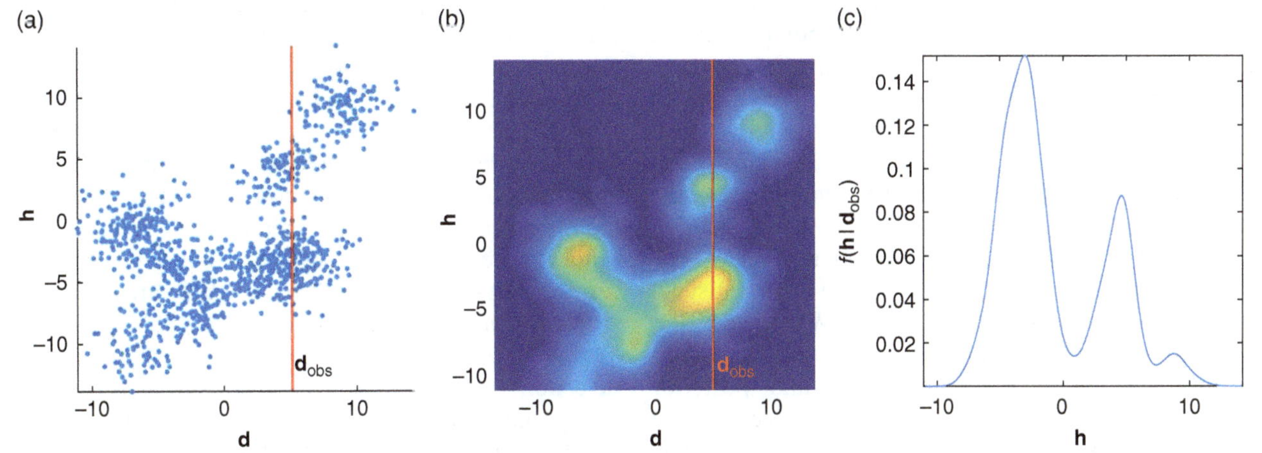

Figure 7.5 Example of density estimation when both the data and prediction variables are univariate. (a) Each blue point represents a prior sample and the desired conditional distribution is $f(\mathbf{h}|\mathbf{d}_{\mathrm{obs}})$ where $\mathbf{d}_{\mathrm{obs}}$ is indicated by the red line. (b) Joint $f(\mathbf{h}, \mathbf{d})$ estimated by performing KDE using a Gaussian kernel. (c) The conditional distribution $f(\mathbf{h}|\mathbf{d}_{\mathrm{obs}})$ obtained using the joint distribution and Eq. (7.19).

To evaluate $\hat{f}(\mathbf{d}_{\text{obs}})$, we simply apply kernel density estimation only on the data variable. Thus, we can rewrite Eq. (7.18) as

$$\hat{f}(\mathbf{h}|\mathbf{d}_{\text{obs}}) = \frac{\frac{1}{\mathbf{w}_h}\sum_{\ell=1}^{L} K\left(\frac{\mathbf{h}^{(\ell)}-\mathbf{h}}{\mathbf{w}_h}\right) K\left(\frac{\mathbf{d}^{(\ell)}-\mathbf{d}_{\text{obs}}}{\mathbf{w}_d}\right)}{\sum_{\ell=1}^{L} K\left(\frac{\mathbf{d}^{(\ell)}-\mathbf{d}_{\text{obs}}}{\mathbf{w}_d}\right)} \quad (7.19)$$

The expected value of Eq. (7.19) is also known as the Nadaraya–Watson kernel estimator [*Härdle et al.*, 2004]. While the example shown in this section was performed on univariate **h** and **d**, it can be readily extended to higher-dimensional data and prediction variables (after potentially applying dimension reduction). However, because of the curse of dimensionality (see Section 3.4.1), the number of samples required to perform KDE increases exponentially as the dimension of the problem increases.

CART (Sections 3.7.4 and 4.4.4) is another family of powerful non-parametric regression tools that can handle both numerical and categorical data variables.

7.2.5. How Much Can We Trust the Statistical Model?

At this point, we have generated samples of data and prediction variables, performed preprocessing, built a statistical model, and then estimated the posterior distribution of the prediction variable. However, before using these predictions in some decision-making context, it is necessary to gauge the quality of the fitted statistical model and assess the confidence in the prediction. A number of scenarios exist that could deleteriously affect the posterior prediction in BEL:

1. *Inconsistent prior distribution*: The data may not be covered by the prior (or, the prior cannot predict the data), which may be due to an inconsistent (e.g., too narrow) prior or the number of samples L from the prior is simply insufficient. We will discuss how this can be detected in Section 7.2.5.1.

2. *Uninformative data variables*: The significance of the posterior prediction depends on the degree to which the data variable is informative about the prediction variable. This significance issue arises because we fit a statistical model based on a limited sample size. We will develop a bootstrap test of significance in Section 7.2.5.2 to quantify this. Note that significance and posterior uncertainty are not necessarily related. One could have a high confidence in a wide uncertainty and a low confidence in a narrow uncertainty. The bootstrap test of significance will allow us to establish this relationship.

3. *Insufficient samples in vicinity of observed data*: The observed data is covered by the prior but an insignificant number of Monte Carlo samples is available to estimate the posterior of **h**. The latter may occur when $\mathbf{d}_{\text{obs}}$ lies in the extremes of the prior. We will develop an importance sampling-based methodology in Section 7.2.5.3 to generate additional subsurface models that are both consistent with $\mathbf{d}_{\text{obs}}$ and the posterior prediction.

7.2.5.1. Inconsistent Prior Distributions. The Bayesian formulation of the prediction problem requires the specification of a subjective belief on a hypothesis. The importance of this subjective prior is well known, and some authors have developed method of falsification of the prior [*Gelman et al.*, 1996; *Gelman and Shalizi*, 2013]. Should the observed data value $\mathbf{d}_{\text{obs}}$ fall outside of the samples of **d** then the probability of the $\mathbf{d}_{\text{obs}}$ under the prior hypothesis will be very small. This in turn results in the posterior having a very small probability as well. Furthermore, as discussed in Section 7.2.2, predictions when $\mathbf{d}_{\text{obs}}$ are outside of the range of samples of **d**, entails extrapolation, which may result in unreliable prediction when using statistical models.

When the data variable is of low dimension, such a test is straightforward by visual inspection. However, when **d** is high-dimensional, a systematic method is required. This task can be viewed as a form of statistical analysis. We assume that the samples within $\{\mathbf{d}^{(1)}, \mathbf{d}^{(2)}, \ldots, \mathbf{d}^{(L)}\}$ were generated from an underlying distribution, and we wish to know if $\mathbf{d}_{\text{obs}}$ deviates from the multivariate distribution modeled by these samples, and thus should be classified as an anomaly or outlier. Owing to the difficulties of high-dimensional outlier detection, it is still recommended that dimension reduction be performed first.

In contrast to conventional classification problems such as those that can be addressed by regression trees (Section 3.7.4) or support vector machines (SVM; Section 3.7.3), applying statistical models for outlier classification may be challenging because the training set may not contain any outliers. This is indeed the case in BEL as the training set contains only samples drawn from the prior. We lack the ability to generate samples that are definitively outside of the prior. In their conventional forms, classification models have difficulties handling these unbalanced training sets; hence, specific outlier detection methods are required. A great deal of research has been conducted in this area; refer to *Aggarwal and Yu* [2001], *Beniger et al.* [1980], and *Hodge and Austin* [2004] for a comprehensive survey of outlier detection methods.

A popular outlier detection algorithm is the one-class SVM (Section 3.7.3). In contrast to conventional SVMs, the one-class SVM aims to fit a minimal volume hypersphere around the samples in $\{\mathbf{d}^{(1)}, \mathbf{d}^{(2)}, \ldots, \mathbf{d}^{(L)}\}$. Any $\mathbf{d}_{\text{obs}}$ that falls outside of this hypersphere is classified as being inconsistent with the prior. Using kernels, this provides a powerful tool for detecting outliers both in high dimensions and with nonlinear decision boundaries. Consider the example in Figure 7.6. The data variable represents a rate from a well. At an initial glance, the observed data does not appear to be an anomaly when viewed in the original time domain. In actuality, unlike any of the prior samples, the observed data starts low and ends up high, and this is inconsistent with the prior. To detect this, we first perform functional principal component analysis (FPCA) which reduces the dimension of the problem by projecting the prior models and observed data into a functional space (here 3D). A one-class SVM using a Gaussian kernel is then trained in functional space, and the decision boundary is identified. If the observed data falls within this boundary, then the prior is classified as being consistent with $\mathbf{d}_{\text{obs}}$, and on the contrary if it is outside the boundary. The prior data variable realizations, observed data, and SVM boundary are plotted in Figure 7.6, which identifies the prior as being inconsistent with $\mathbf{d}_{\text{obs}}$. In this example, the one-class SVM is performed in two dimensions for the sake of illustration; in practice, it can be easily extended to higher dimensions, where visual inspection cannot be performed.

7.2.5.2. Statistical Significance of the Posterior. In BEL, any reduction in uncertainty between the prior and posterior pdfs should be achieved due to the data variable being informative of the prediction variable. While the cause of this informativeness is physical (i.e., geology, physical, and chemical processes), we have modeled it using a statistical model, estimated from samples of **h** and **d** obtained from the sampling the prior. Since this statistical modeling was performed using a finite number of samples, we need to test the statistical significance of the resulting posterior distribution. This is measured using the *p-value* (also called achieved significance level, see Section 3.13), which expresses how *unlikely* an estimated reduction in posterior uncertainty is, if the data variable was not actually informative of the prediction variable. In other words, this measures the probability that the uncertainty reduction we have estimated is an artifact of the random sampling, as opposed to the data actually being informative of the prediction variable. A low $p-$value (i.e., <0.05) would indicate that our posterior is indeed significant. It should be noted that uncertainty and statistical significance are not necessarily correlated. It is possible to generate a narrow posterior uncertainty that is insignificant and vice versa. This depends on a number of factors (the prior uncertainty, the particular data and prediction variable, the forward model, etc.), and thus a method is required to estimate this significance.

In this section, we develop a hypothesis test, which tests whether the data variable is actually informative of the prediction (prediction variables in reduction dimension $\mathbf{h}^*$). The data $\mathbf{d}^*_{\text{obs}}$ is informative when significant difference exists between the prior distribution $f(\mathbf{h}^*)$ and the posterior $f\left(\mathbf{h}^* \mid \mathbf{d}^*_{\text{obs}}\right)$. To test this, we propose the null hypothesis that the data is *not* informative. Under such a

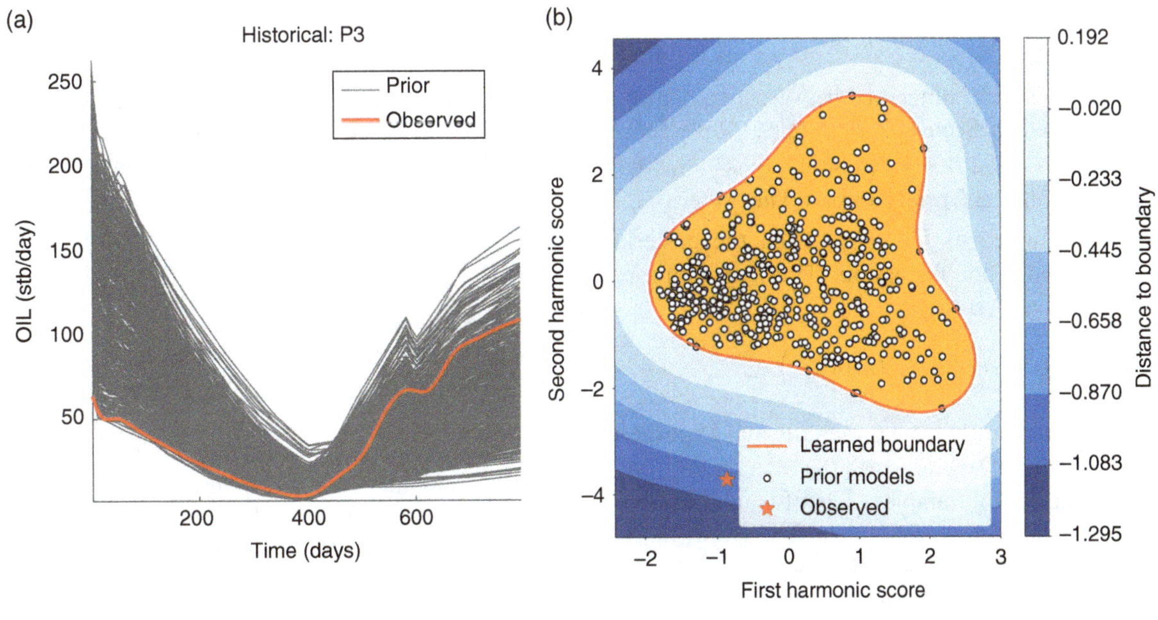

Figure 7.6 (a) Prior and observed data in time domain. (b) Performing outlier detection in 2D functional space. The boundary learned by the one-class SVM is show in orange; observed data falling outside of the boundary suggests that the prior is inconsistent.

hypothesis, the prior and the posterior should exhibit no difference.

The difficulty with directly testing this hypothesis is that we do not have the exact prior and posterior distributions, but rather we only have a limited number of samples drawn from the two distributions. Recall that each sample of $f(\mathbf{h}^*)$ requires a computationally expensive forward model to generate, and thus it is impractical to obtain an exhaustive sample set. Instead, we will rely on bootstrapping (Section 3.13) to test this hypothesis. We start by stating the null hypothesis:

$$H_0 : f(\mathbf{h}^*) = f(\mathbf{h}^* | \mathbf{d}^*_{obs}) \tag{7.20}$$

We next need to define a test statistic that allows us to measure the distance between these two distributions. However, since both of these pdfs are multivariate, it would be difficult to develop a simple measure. There have been attempts to develop such multivariate test statistics [*Holm*, 1979; *Peacock*, 1983; *Fasano and Franceschini*, 1987], but their application is generally limited to two or three dimensions. Instead, we propose the use of the first component of h_1^* as a proxy to evaluate the difference between the two distributions. The first component is assumed to account for most of the variation (as is the case when $\mathbf{h}^*$ is the output of a dimension reduction method such as PCA, MDS, FPCA, etc.). It should be noted that such a test is a necessary but not sufficient test. A sufficient test would need to include all the components.

In the univariate case, the problem of comparing pdfs is simplified. We define the test statistic as a measure of the difference between the two distributions:

$$\theta = \Delta\left(f\left(h_1^*\right), f\left(h_1^* | \mathbf{d}^*_{obs}\right)\right) \tag{7.21}$$

Δ can be any metric of difference between two distributions as discussed in Section 3.5.2.2 (e.g., f-divergence, L1 norm, etc.). The prior distribution is estimated as $\hat{f}\left(h_1^*\right)$ directly from the L prior samples. The posterior distribution is estimated as $\hat{f}\left(h_1^* | \mathbf{d}^*_{obs}\right)$ from the posterior samples obtained from regression. Thus, an empirical measure of the test statistic is

$$\hat{\theta} = \Delta\left(\hat{f}\left(h_1^*\right), \hat{f}\left(h_1^* | \mathbf{d}^*_{obs}\right)\right) \tag{7.22}$$

If $\hat{\theta}$ deviates significantly from 0 then the null hypothesis should be rejected. However, since the calculation of $\hat{\theta}$ is based on limited prior samples, a bootstrap is applied. Specifically, B datasets each of size L are drawn *with* replacement from the original set of samples. For each of these B datasets, a bootstrapped prior is computed and BEL is applied to estimate the bootstrapped posterior distribution (denoted by $\hat{\hat{f}}_b(\mathbf{h})$ and $\hat{\hat{f}}_b(\mathbf{h}|\mathbf{d}_{obs})$). Using these distributions, we can estimate the set of differences:

$$\hat{\hat{\theta}}_b = \Delta\left(\hat{\hat{f}}_b\left(h_1^*\right), \hat{\hat{f}}_b\left(h_1^* | \mathbf{d}^*_{obs}\right)\right),\ b = 1, \ldots, B \tag{7.23}$$

We can use these differences to approximate the achieved significance level, ASL (see Section 3.13). The smaller this number, the stronger evidence is against the null hypothesis. In our case, a small ASL means that the data variable is indeed informative of the prediction variable.

$$\text{ASL} \approx \#\left(\hat{\hat{\theta}}_b \geq \hat{\theta}\right) / B,\ b = 1, \ldots, B \tag{7.24}$$

Conversely, a high ASL is evidence that the null hypothesis is true: the data variable is not informative and not appropriate for prediction. Any estimated posterior uncertainties may be unreliable, and any decisions based on them need to be made with caution. Under such circumstances, traditional inversion is required.

7.2.5.3. Updating by Sequential Importance Resampling. To verify that the posterior prediction is indeed viable, it would be useful to generate the actual subsurface models that are consistent with $\mathbf{d}_{obs}$ and yield prediction that corroborate the posterior uncertainty. This is especially advantageous when few samples from the prior are within the vicinity of $\mathbf{d}_{obs}$. Mathematically, this can be formulated as sampling from the set of subsurface models that are conditioned to the observed data: $f(m_1, m_2, \ldots, m_N | \mathbf{d}_{obs})$. Undoubtedly, this is a daunting task for traditional sampling techniques, such as Monte Carlo, as this target distribution is high dimensional (equal to number of parameters), non-parametric, and difficult to specify. One can imagine that few combinations of model variables will yield a subsurface model that generates a data response that matches $\mathbf{d}_{obs}$ simply by performing Monte Carlo.

Instead, sequential importance resampling (SIR) can be used to obtain models in this target region (see Section 3.10.3). Importance sampling is a well-known statistical method [*Glynn and Iglehart*, 1989] that uses a proposal distribution $q(m_1, m_2, \ldots, m_N | \mathbf{d}_{obs})$ with the aim of making sampling more efficient (e.g., sample more the tail than the center). The proposal distribution can be anything we want as long as it meets the criteria set out in Section 3.10.3.5. By carefully selecting the proposal distribution, based on domain knowledge of the problem, we can emphasize sampling in the regions that we consider to be more important (aka the region of the model space that will generate a data response matching $\mathbf{d}_{obs}$). This increases the efficiency of the sampling procedure and allows obtaining greater precision in the prediction using the same number of samples when compared to regular Monte Carlo.

However, the use of this proposal distribution will introduce a bias (since the proposal distribution is generally not equivalent to the target distribution). Therefore, to compensate, a weight must be assigned to each drawn sample. The weight is equivalent to the likelihood ratio between the target distribution and the proposal distribution. This was illustrated in Section 3.10.3 with the numerical integration of a Gaussian distribution; we refer to the same section for the mathematical details.

Applying SIR to the general subsurface case (making no distribution assumptions, such as multivariate Gaussian) is not trivial. The target distribution is $f(m_1, m_2, \ldots, m_N|\mathbf{d}_{\text{obs}})$, and in the general case, no explicit expression exists. It is, however, more likely that one can determine the conditional distribution for each individual model variable $f(m_n|\mathbf{d}_{\text{obs}})$. Therefore, a possible choice of proposal function is

$$f(m_1, m_2, \ldots, m_N|\mathbf{d}_{\text{obs}}) \approx q(m_1, m_2, \ldots, m_N|\mathbf{d}_{\text{obs}}) = \prod_{n=1}^{N} f(m_n|\mathbf{d}_{\text{obs}}) \quad (7.25)$$

To estimate this proposal distribution q, we proceed as usual with BEL, construct a set of prior subsurface models $\mathbf{m}$, and forward simulate to obtain $\mathbf{d}$ (and $\mathbf{d}^*$ if dimension reduction is required). However, instead of performing regression between $\mathbf{h}^*$ and $\mathbf{d}^*$, we instead apply non-parametric regression (see Section 7.2.4.2) to $\mathbf{d}^*$ and each of the subsurface model parameters m_n to obtain $f(m_n|\,\mathbf{d}_{\text{obs}})$.

Drawing samples from each of these estimated conditional densities is, thus, equivalent to sampling from $q(\mathbf{m}|\,\mathbf{d}_{\text{obs}})$. Consider that we draw L sets of subsurface model parameters:

$$\mathbf{m}_q^{(\ell)} \sim q(\mathbf{m}|\mathbf{d}_{\text{obs}})\ \ell = 1, \ldots, L \quad (7.26)$$

We use the subscript q to denote that these subsurface models are drawn from the proposal distribution rather than the prior. However, because these "proposal models" were drawn from q instead of f, they are biased. Therefore, the predictions generated by these proposal models may not be the same as those of the prior models. To correct this bias, SIR requires a weight to be assigned to each proposal subsurface model:

$$w^{(\ell)} = \frac{f\left(\mathbf{m}_q^{(\ell)}|\mathbf{d}_{\text{obs}}\right)}{q\left(\mathbf{m}_q^{(\ell)}|\mathbf{d}_{\text{obs}}\right)} = \frac{f\left(m_{q,1}^{(\ell)}, m_{q,2}^{(\ell)}, \ldots, m_{q,N}^{(\ell)}|\mathbf{d}_{\text{obs}}\right)}{\prod_{n=1}^{N} f\left(m_{q,n}^{(\ell)}|\mathbf{d}_{\text{obs}}\right)}, \quad \ell = 1, \ldots, L \quad (7.27)$$

To evaluate Eq. (7.27), we need to evaluate $f\left(m_{q,1}^{(\ell)}, m_{q,2}^{(\ell)}, \ldots, m_{q,N}^{(\ell)}|\mathbf{d}_{\text{obs}}\right)$ for each proposal model. Unfortunately, this is not possible, as the numerator represents the high-dimensional target distribution that we cannot specify.

Instead, we will need some way of approximating the ratio in Eq. (7.27). To do this, we use a proxy for $\mathbf{m}$ itself rather the high-dimensional subsurface model itself. Such a proxy would need a function of $\mathbf{m}$. One such candidate is the prediction variable $\mathbf{h}$, since it is a function of $\mathbf{m}$ through the forward model g_h. We can then approximate Eq. (7.27) using the low-dimensional projection of the data and prediction variable as

$$w^{(\ell)} = \frac{f\left(\mathbf{m}_q^{(\ell)}|\mathbf{d}_{\text{obs}}\right)}{q\left(\mathbf{m}_q^{(\ell)}|\mathbf{d}_{\text{obs}}\right)} \approx \frac{\hat{f}\left(g_h\left(\mathbf{m}_q^{(\ell)}\right)|\,\mathbf{d}_{\text{obs}}^*\right)}{\hat{f}_q\left(g_h\left(\mathbf{m}_q^{(\ell)}\right)|\,\mathbf{d}_{\text{obs}}^*\right)} \approx \frac{\hat{f}\left(\mathbf{h}^{(\ell)*}|\,\mathbf{d}_{\text{obs}}^*\right)}{\hat{f}_q\left(\mathbf{h}^{(\ell)*}|\,\mathbf{d}_{\text{obs}}^*\right)} \quad (7.28)$$

The term $\hat{f}\left(\mathbf{h}^{(\ell)*}|\mathbf{d}_{\text{obs}}^*\right)$ represents the conditional distribution of the prediction variable as estimated from the unbiased prior subsurface models. This is computed by performing BEL on the pairs of $\mathbf{h}$ and $\mathbf{d}$ generated by the prior subsurface models. The denominator of Eq. (7.28), $\hat{f}_q\left(\mathbf{h}^{(\ell)*}|\,\mathbf{d}_{\text{obs}}^*\right)$, represents the conditional distribution of the prediction variable as estimated from the biased proposal subsurface models. To evaluate this, we use the samples $\mathbf{m}_q^{(\ell)}$ and compute

$$\mathbf{h}_q^{(\ell)} = g_h\left(\mathbf{m}_q^{(\ell)}\right); \mathbf{d}_q^{(\ell)} = g_d\left(\mathbf{m}_q^{(\ell)}\right),\ \ell = 1, \ldots, L \quad (7.29)$$

where each pair $\left\{\mathbf{h}_q^{(\ell)}, \mathbf{d}_q^{(\ell)}\right\}$ is reduced in dimension to $\left\{\mathbf{h}_q^{(\ell)*}, \mathbf{d}_q^{(\ell)*}\right\}$. This set of samples allows estimating $\hat{f}_q\left(\mathbf{h}^{(\ell)*}|\,\mathbf{d}_{\text{obs}}^*\right)$ as before. Using these two estimated densities, we can evaluate the weight $w^{(\ell)}$ for each proposal model using Eq. (7.28).

SIR attempts to address the inefficiency of the prior distribution in generating samples in regions where it matters: near observed data and where it is important for prediction. The proposal distribution is more efficient in drawing samples that aid prediction as it does take into account $\mathbf{d}_{\text{obs}}^*$. However, the independence assumption in Eq. (7.25) ignores the dependency created by the conditioning to observed data (even if prior model variables are independent). These weights represent how much each proposal model is in accordance with the data–forecast statistical relationship established from the original prior distribution. This means that proposal models that do not contribute to prediction uncertainty will get low weight. The posterior of the prediction variables is then computed using these weights and the proposal model prediction. For instance, the mean posterior prediction in reduced dimensional space is

$$E[\mathbf{h}^*] = \frac{1}{L}\sum_{\ell=1}^{L} w^{(\ell)}\mathbf{h}_q^{(\ell)*} \quad (7.30)$$

Undoing dimension reduction will then yield the posterior mean. Similarly, quantiles such as the P10, P50, and P90 can be taken as weighted quantiles, where the quantile percentage is computed using the fraction of the sum of the total weight rather than the number of samples.

SIR retains the Bayesian formulation of BEL, as additional information is collected (e.g., $\mathbf{d}_{\text{obs}}$ is measured over a longer period of time), the posterior can be iteratively updated. Moreover, the proposal distribution $q(\mathbf{m}|\,\mathbf{d}_{\text{obs}})$ can also be constructed sequentially using the proposal models from the previous iteration to estimate $f(m_n|\,\mathbf{d}_{\text{obs}})$. This allows the SIR algorithm to narrow down the sampling region each time more observations are made.

The sequential repetition of SIR may result in the so-called sample degeneracy. Degeneracy is defined as an increasing skewness of the SIR weights in Eq. (7.27) with each iteration [*Fox*, 2003]. This means that after a few iterations, a few of the proposal samples will have weights that are much larger than the others. This is not very useful, as the goal was to generate multiple samples that are in the vicinity of $\mathbf{d}_{\text{obs}}$. The solution to this is to apply a resampling step. Resampling works by first normalizing each of the computed proposal weights, and then interpreting each normalized weight as the probability of each sample:

$$w_{\text{norm}}^{(\ell)} = \frac{w^{(\ell)}}{\sum_{\ell=1}^{L} w^{(\ell)}} \tag{7.31}$$

We then redraw L samples according to this discrete probability distribution. This removes the low-weight samples and increases the number of high-weight samples. The rationale behind this is that if a sample has low weight at iteration t, it will still have low weight at time $t+1$. A variety of other resampling techniques have been proposed to accomplish this in a systematic manner. An overview of such techniques can be found in *Douc and Cappe* [2005] and *Moral et al.* [2012].

SIR serves two primary purposes within the Bayesian evidential learning framework. The first is to address the situation where the prior is insufficiently sampled within the vicinity of the observed data $\mathbf{d}_{\text{obs}}$. By targeting the sampling procedure, SIR is able to generate additional subsurface models in those areas, allowing for a more refined prediction of the posterior forecast. The second purpose is to generate corresponding subsurface models that contribute to the posterior of prediction assessed by BEL.

7.3. BAYESIAN EVIDENTIAL LEARNING IN PRACTICE

Data and prediction variables in the subsurface may be of different type, ranging from static scalars to spatial maps to dynamic time series. Accordingly, a variety of statistical techniques are needed to account for the diverse nature of these variables. In this section, we will illustrate the application of the techniques outlined in this chapter to a series of situations in which both the prediction and data variables are static or dynamic, meaning not varying in time and varying, so "predicting dynamic from dynamic" means a problem where the goal is to predict a time-varying prediction variable based on a time-varying observation. Several of these applications can be expanded to multiple data sources and hence multiple data variables. Chapter 8 will consider the full application of Bayesian evidential learning within a UQ and decision-making context. In particular, the Libyan oil field case and the Belgian geothermal case use this form of modeling. The reader is, therefore, referred to Chapters 1 and 8 for the context of these applications.

7.3.1. Predicting Dynamic from Dynamic

7.3.1.1. Problem Setup. We first consider the situation where both data and prediction variables are dynamic time series. Typically, this occurs when historical data have been observed over a certain duration, and a prediction regarding a future time series is required. Consider, for instance, an oil field that has been in production for 10 years from five wells (see Section 8.2 for the details); we wish to know, under the current operating scenario, how long the field will be economically profitable. This entails predicting the future field production rate over a given time horizon (30 years) and predicting when it will decline under a given threshold. The data variables constitute a set of five time series describing the historical production rates from the five existing producers over the past 10 years. While both variables are infinite-dimensional temporal responses, in reality they are recorded or simulated as discrete points in time. Here, both variables are discretized at 3-month intervals. A causal analysis would entail inverting subsurface models from the historical rate data of five producers ($\mathbf{d}_{\text{obs}}$) (history matching); then using the inverted models to predict future production, for the same five wells.

In BEL, we first construct 500 prior subsurface models that are not inverted for the production rate. The construction of these prior models is detailed in Section 8.2. The forward model functions, g_d and g_h, are reservoir simulators that simulate the historical production rates and future production rates given a subsurface model and a preset well schedule. By forward simulating each of the prior models, we obtain prior realization of $\mathbf{d}$ and $\mathbf{h}$ (see Figure 7.7).

As an exercise to illustrate the goal of BEL, we will set aside one of the prior models and use its data variable as $\mathbf{d}_{\text{obs}}$ (indicated by red in Figure 7.7). The corresponding prediction variable from this prior realization can later be used to verify the performance of BEL.

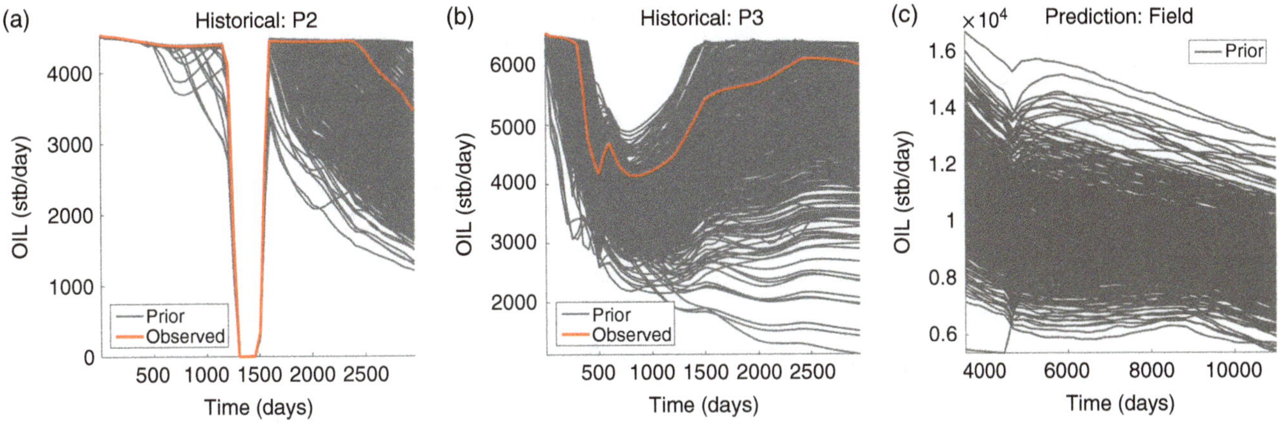

Figure 7.7 (a and b) Prior realizations of data variables for two of the five existing producers (gray). These responses were obtained by flow simulating each of the 500 prior reservoir models. The observed $\mathbf{d}_{\text{obs}}$ production rate is shown in red, as expected very few of the prior production rates match $\mathbf{d}_{\text{obs}}$. (c) Prior realizations of production rates from 10 to 30 years.

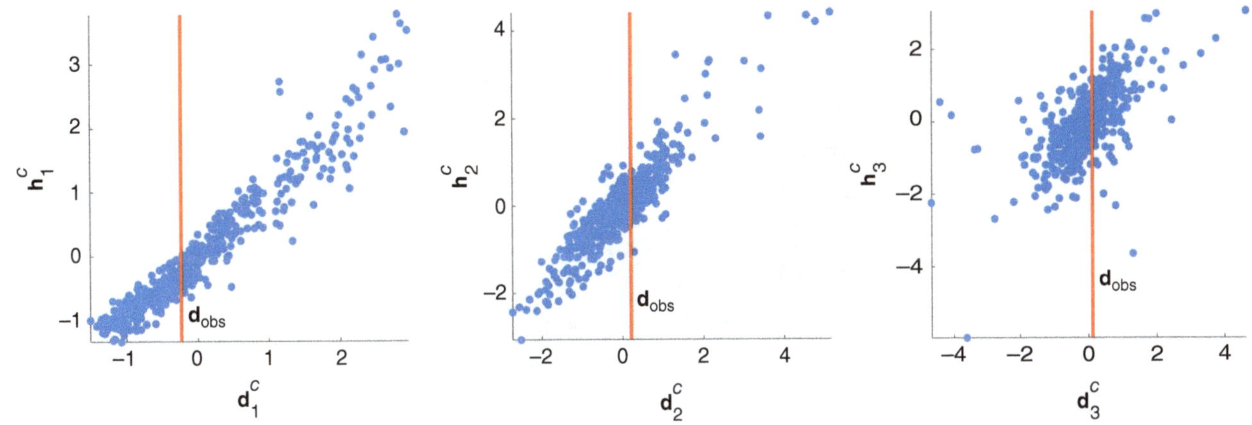

Figure 7.8 Correlation between **d** and **h** in canonical space for first three (of four) components. The canonical projection of $\mathbf{d}_{\text{obs}}$ is indicated by the red line. A sufficiently strong linear correlation between data and prediction variables can be observed.

7.3.1.2. Dimension Reduction. The data variable consists of five production rates over 10 years at 3-month intervals for a total of 200 (5 wells × 40 time-steps) data variables per prior model. Since the data is a time series, a natural choice for dimension reduction would be FPCA (see Section 3.6). Using B-splines as basis functions, FPCA identifies that only four eigenvalues are required to represent over 95% of the variance for production rates of each well. Therefore, we arrive at a functional representation of **d** denoted as $\mathbf{d}^{\text{fpca}}$, a 20-dimensional variable (five wells × four functional components).

Since each of the five production rates are functions of the same underlying subsurface model, it is to be expected that some degree of redundancy or multicollinearity exists between them. To account for this, a mixed PCA (Section 7.2.3) is subsequently applied to $\mathbf{d}^{\text{fpca}}$, which identifies that the first seven eigenvalues account for the majority of the variance between the five producers. The resulting projection of the data variable $\mathbf{d}^{\text{mfpca}}$ has seven dimensions, which is much more manageable for regression. Similarly, applying FPCA to the prediction variable results in the reduction dimension projection $\mathbf{h}^{\text{fpca}}$ that is four dimensional. Since the forecast variable is a single time series (field production rate), there is no need for mixed PCA.

7.3.1.3. Regression. After reducing dimensions of both **d** and **h** to manageable sizes, we can proceed with regression. To linearize the problem further, we apply canonical correlation analysis (CCA) followed by a normal score transform to obtain the normal score canonical components denoted as $\mathbf{d}^c$and $\mathbf{h}^c$, respectively; applying the same transformations to $\mathbf{d}_{\text{obs}}$ yields $\mathbf{d}^c_{\text{obs}}$. Figure 7.8 shows considerable linear correlation between the canonical

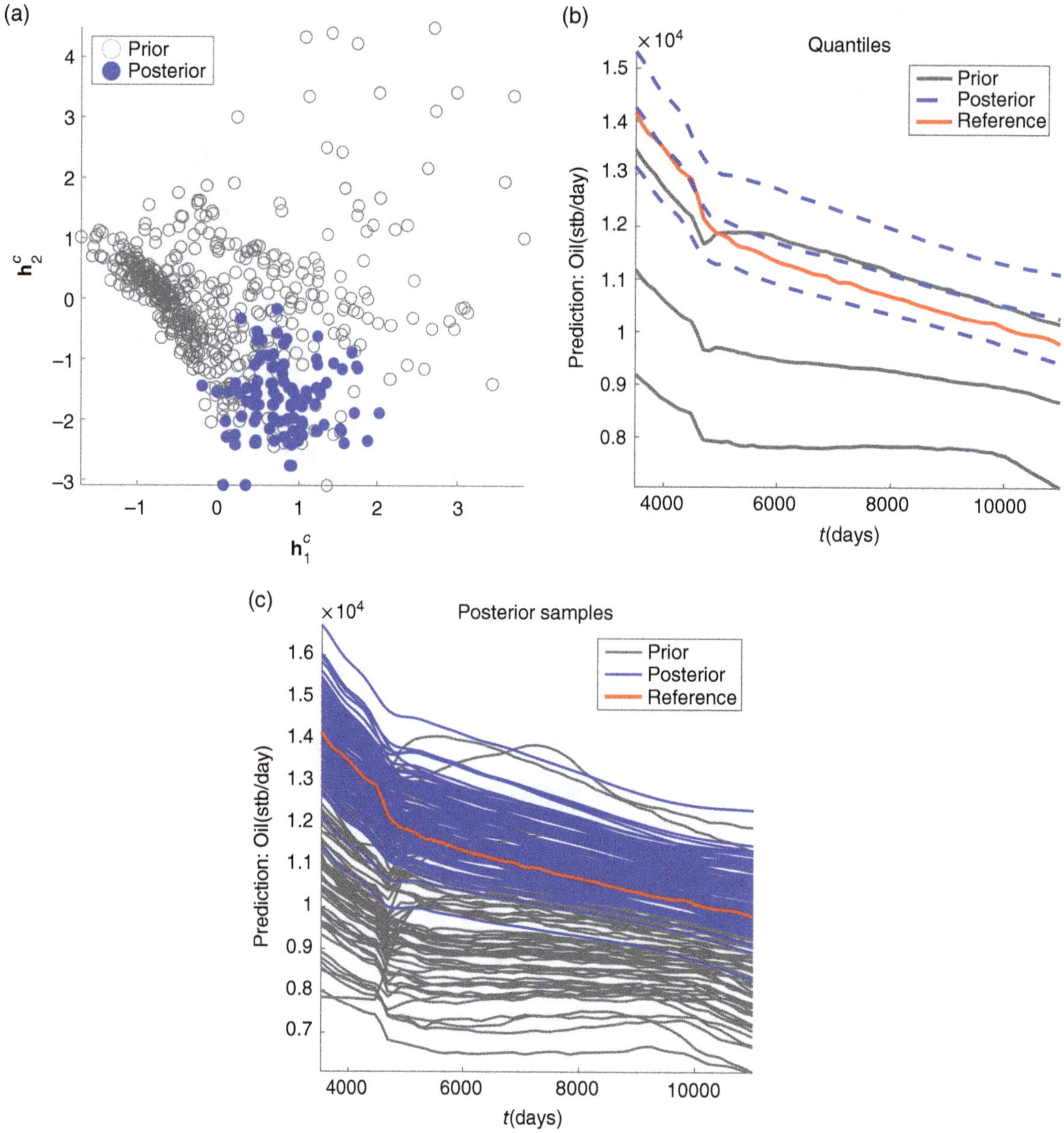

Figure 7.9 (a) Prior (gray) versus posterior (blue) samples plotted in the first two dimensions in canonical space. Note that the posterior covariance is smaller than that of the prior: Bayesian evidential learning reduced uncertainty. (b and c) Posterior of prediction variables in the time domain obtained by undoing each of the previously applied transformations (normal score, CCA, FPCA) on the canonical posterior samples. Quantiles of both prior and posterior forecasts are shown in (b).

data and prediction variables; hence, Gaussian process regression can be applied to estimate $f\left(\mathbf{h}^c|\mathbf{d}^c_{\text{obs}}\right)$ (see Eqs. (7.15) and (7.16)). Moreover, we can also visually confirm that $\mathbf{d}^c_{\text{obs}}$ does fall within the ranges of the prior realizations $\mathbf{d}^c$; hence, the statistical model is interpolating and not extrapolating.

Figure 7.9 shows posterior samples (of the multivariate Gaussian) plotted in the first two canonical dimensions The posterior samples in this space (blue) cover a smaller area than the prior samples (gray), indicating that uncertainty is reduced. These canonical scores can be back-transformed into actual time series by "undoing" the series of transformations (normal score, CCA, FPCA). Statistics such as quantiles can now be computed in the original time domain and be used for subsequent decision making. For verification, the "true" future field-production rate corresponding to the realization that was set aside and used to generate $\mathbf{d}_{\text{obs}}$ is shown in red.

7.3.2. Predicting Static from Dynamic

7.3.2.1. Problem Setup. We next consider a situation in the same Libyan oil field where the data variable remains a time series, but the prediction variable is a static scalar property. This scenario often occurs when the prediction variable is a property that cannot be directly measured, and hence needs to be modeled. Now three injectors are used for water-flooding over a period of 10 years. Water-flooding is simply an enhanced oil-recovery method whereby brine fluids are injected in the reservoir to drive oil to producers. The task is to maximize production by modifying the injection rates. To do so, one often uses the concept of injector efficiency, which measures the incremental oil that is produced per unit of water that is injected into the given injector. In other words, an efficient injector is one that produces the most incremental oil for as little injected water as possible. Intuitively, we would reallocate water from inefficient injectors in the field at the present time to the efficient ones. Injector efficiency is not a quantity that can be directly measured in the field. Rather, it is an uncertain prediction variable that is determined from a reservoir model which itself is uncertain. Injector efficiency can be computer for each generated model realization. We will again set aside one of the prior realizations as "truth," and use the corresponding $\mathbf{d}$ as $\mathbf{d}_{obs}$ (red in Figure 7.9 to predict the efficiency of injector 1 after the 10th year of production. The "true" $\mathbf{h}$ in this case is 35.41%, which will be used later to evaluate the performance of BEL.

7.3.2.2. Dimension Reduction and Regression. The data variable remains the same as in Section 7.3.2 and is a 200-dimensional time series of the producer historical rates profile. The prediction variable (injector 1 efficiency) is a scalar and does not require dimension reduction. The goal of regression is to build a predictor for an injector's efficiency using the seven mixed PCA components of the data variables, then use it to predict the corresponding efficiency given the historical field observations $\mathbf{d}_{obs}$. Figure 7.10 shows the relationship between prediction variable and reduced dimension data variables. Visually, we can see that $\mathbf{d}_{obs}$ falls within the range of the prior data forecasts, allowing us to apply BEL.

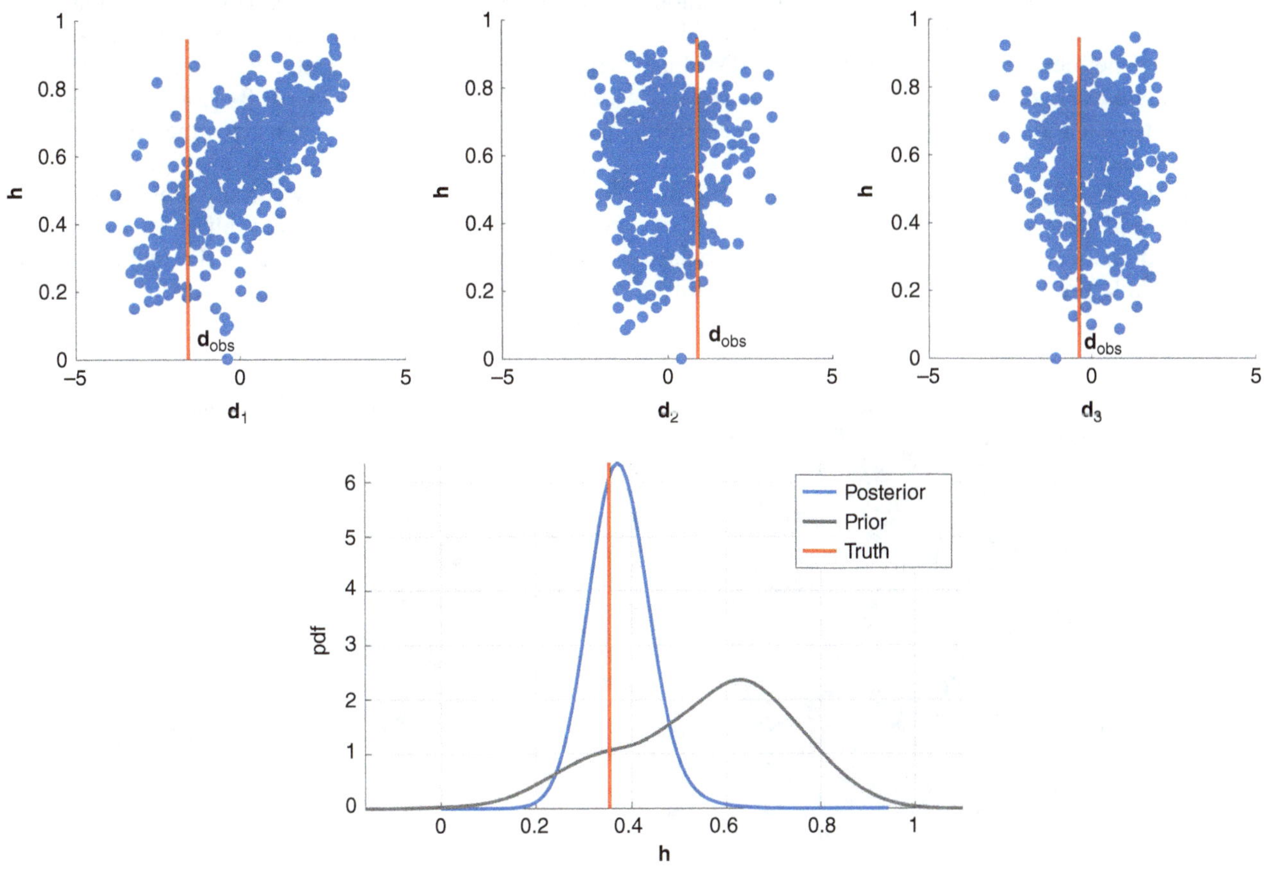

Figure 7.10 Scatterplot comparing the top three mixed PCA components of the data variable with the forecast variable. The projected value of $\mathbf{d}_{obs}$ is shown by the red line. Prior distribution of the prediction variable "injector efficiency" with posterior distribution estimated by means of KDE.

To estimate the posterior distribution $f(h|\boldsymbol{d}_{obs})$, we apply KDE (see Section 3.3.2). In this example, the joint space $f(h, \mathbf{d})$, contains eight dimensions (seven from the data components and one from forecast). Following the same procedure as in Section 7.2.4, we arrive at the posterior distribution shown in Figure 7.10. We observe that BEL does provide a reduction in uncertainty when compared to the prior, and the value of h corresponding to the highest density is 38.91% which is in accordance with the "truth."

7.3.3. Predicting Dynamic from Static

7.3.3.1. Problem Setup. To illustrate this type of situation, we will consider a simple 2D synthetic reservoir model undergoing water flooding and predict the future production rate of both an existing well and a planned producer. The setup is shown in Figure 7.11. The model domain is 200 × 200 cells with each cell having dimensions 50 ft × 50 ft; the thickness of reservoir is 100 ft. The initial oil saturation is 100% throughout the reservoir. A variety of uncertainties are present in this reservoir model, including depositional uncertainty, spatial compaction distribution, in addition to spatial variability describing porosity and permeability heterogeneity. Prior distributions are assumed regarding each of these uncertainties, and geostatistical simulation is used to generate a prior set of 1000 reservoir models.

During the first 5 years of production, water is continuously injected into a well located at the lower left of the model domain at a surface flow rate of 1000 stb/day with a top-hole pressure limit of 12,000 psi. An existing production well is located at the center of the model domain with a bottom hole pressure limit set to 3000 psi. The goal is to predict the next 10 years of production for both the existing and new producers (from 5 to 15 years). Using the prior reservoir models and a reservoir simulator, we obtain the prior field production rates for each model discretized with 50 time steps (see Figure 7.11). We consider as data variables 4D seismic data that informs the spatial distribution of oil saturation. The modeling of such data itself is complex, with many uncertainties. Here we simply mimic the forward modeling of this 4D data by taking the oil saturation of the simulator at 5 years and applying Gaussian filter to smooth the saturation map. This mimics simply the lack of resolution. The set of 1000 of these smoothed spatial maps constitutes the prior data variables (see Figure 7.12).

7.3.3.2. Dimension Reduction. The data variable under consideration is the saturation map of size 200 × 200 cells yielding a 40,000-dimensional variable. This is clearly too large for regression to be effective, and it necessitates a dimension reduction. As described in Section 7.2.3, eigen-images may be an effective parameterization for smoothly varying spatial data. Recall that eigen-image analysis is equivalent to PCA. By concatenating each map into vector, conventional eigenvalue decomposition is used to compute the underlying eigenvectors and corresponding eigenvalues. By keeping those eigenvectors corresponding to the largest eigenvalues, we can identify the orthogonal modes that capture the majority of the variance in saturation maps.

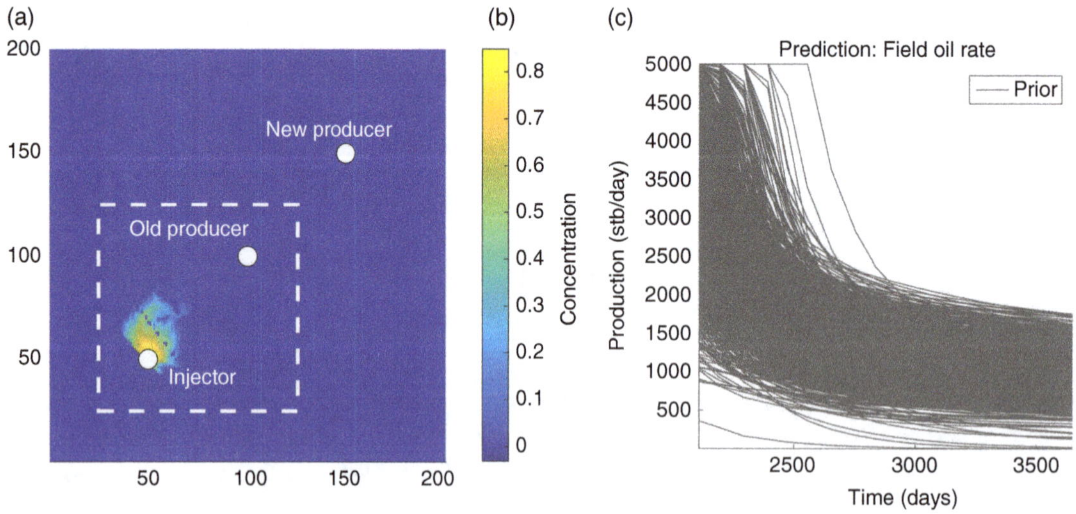

Figure 7.11 (a) Setup of a situation involving injecting water into oil and aiming to track what happens at two producers, an existing and a new one. (b) Color indicates water saturation at 5 years. (c) The data variable is a set of seismic maps (4D seismic) over 5 years, which represents the change in oil saturation.

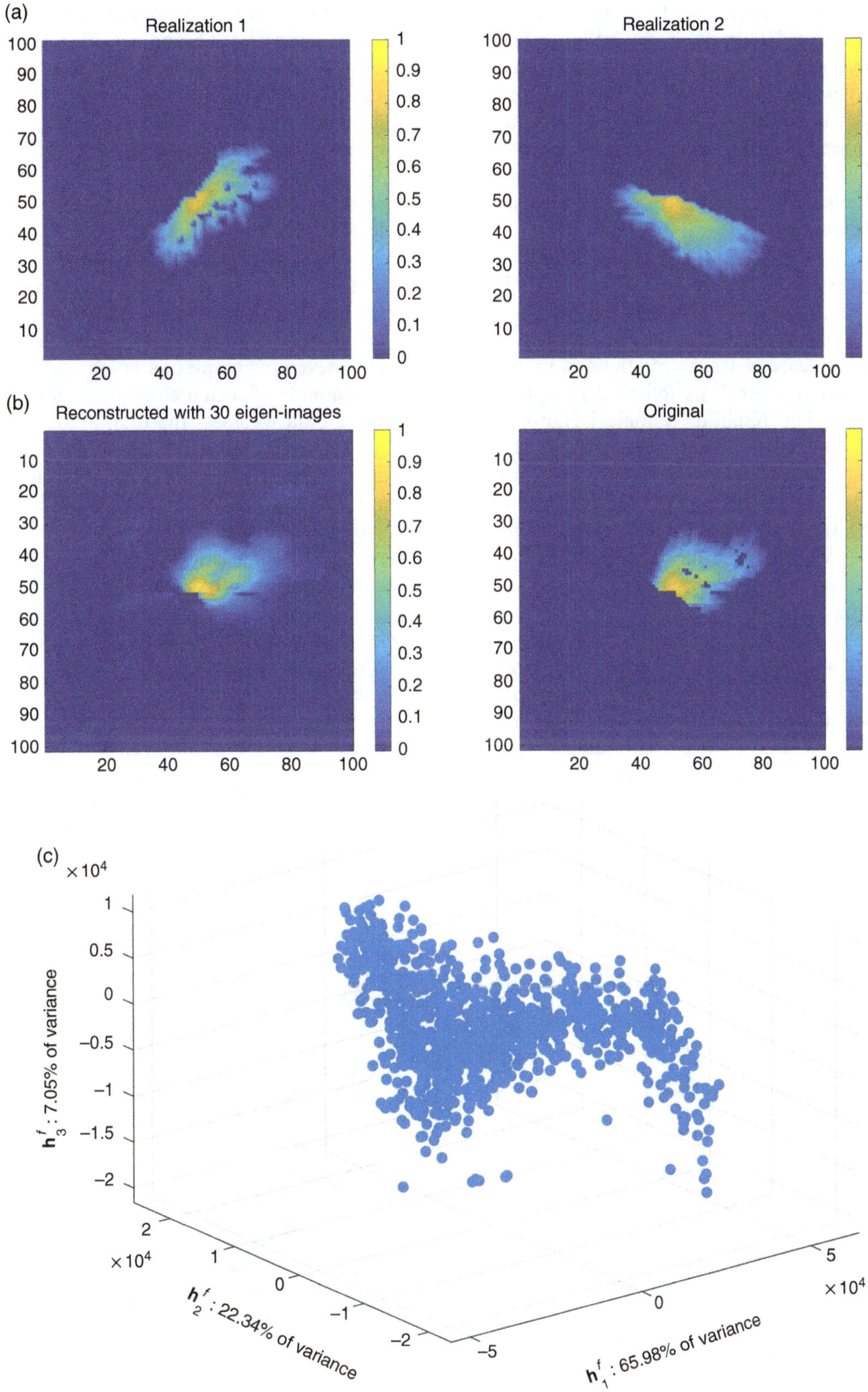

Figure 7.12 (a) Two realizations of the data variables. (b) An example of data realizations and its reconstruction using the first 30 eigen-images. (c) Low-dimensional projection of prior forecast variable using functional data analysis. First three dimensions are shown.

In this example, we were able to retain 90% of the variance in the prior data variables by keeping the first 30 eigen-images (see Figure 7.12). This decreases the dimensionality of the data variable without much loss in variance. We will denote this reduced dimension projection as $\mathbf{d}^e$. The prediction variable is a time-series response. Hence, we will apply FPCA with B-splines to reduce its dimension $\mathbf{h}^{\text{fpca}}$ from 50 to 5 while retaining 99% of the variance.

7.3.3.3. ***Regression.*** Eigen-image analysis and FPCA reduces **d** and **h** to manageable dimensions for regression. We will once again apply Gaussian process regression after the same sequence of transformations as in 7.3.3.1 (Canonical Correlation Analysis followed by a normal score transform). The resulting canonical components are denoted $\mathbf{d}^c$ and h^c, while the observed data $\mathbf{d}_{\text{obs}}$ is transformed to $\mathbf{d}^c_{\text{obs}}$. Posterior samples are then drawn, and the transformations undone to generate posterior forecast in the time domain. These posterior samples are shown along with the quantiles in Figure 7.13.

7.3.4. Predicting Static from Static

7.3.4.1. ***Problem Setup.*** In the absence of any dynamic data or needing to learn a dynamic response, the problem of deriving static prediction from static data often arises. Example of static data is any well measurement, such as cores or logging, 2D, or 3D geophysical data. Target predictions are proportions, volumes, or any model parameter. These problems often arise when appraising a subsurface system. For example, in reservoir modeling, one needs to appraise the original oil in place based on wells and seismic. In groundwater management, one may want to determine drawdown of a well or total amount of contaminant in a given zone.

The causal inversion approach would be to build multiple model realizations using a Bayesian-type inversion and use the posterior models to calculate empirical distribution of the uncertainty of some target variable. This approach allows calculating posterior uncertainty of any variable once the models are derived. But this comes at the cost of a possibly difficult and CPU-demanding approach. BEL can be used as an alternative, even in complex situation.

As an example of such a situation, consider the use of 3D seismic data to assess the total amount of reservoir rock in a subsurface system. We consider here a real field case of estimating proportions in a turbidite reservoir system in West Africa [*Scheidt et al.*, 2015]. The seismic data over the reservoir zone is shown in Figure 7.14. BEL requires stating a prior probability model on all uncertain model variables. It also requires forward modeling of the geophysical data and such forward modeling relies on rock physics as well as other parameters (e.g., seismic wavelet) that are uncertain. Hence, the following uncertainties are included:

1. *Rock physics uncertainty:* Since only few wells are drilled, the relationships between various rock physics properties are uncertain. Additionally, with only few wells it is possible that certain type of rocks, known to be present in these systems, may not be hit by the well path. As a

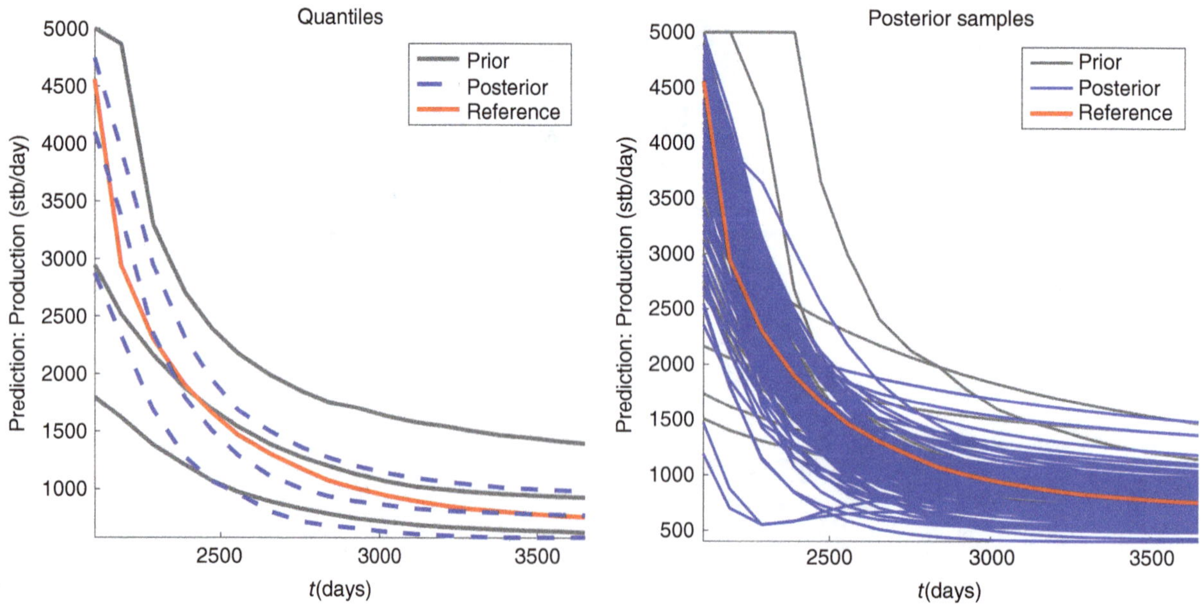

Figure 7.13 Posterior forecasts in the time domain obtained by undoing each of the previously applied transformations (normal score, CCA, FPCA) on the samples drawn in canonical space. Quantiles of both prior and posterior forecasts are shown on the left.

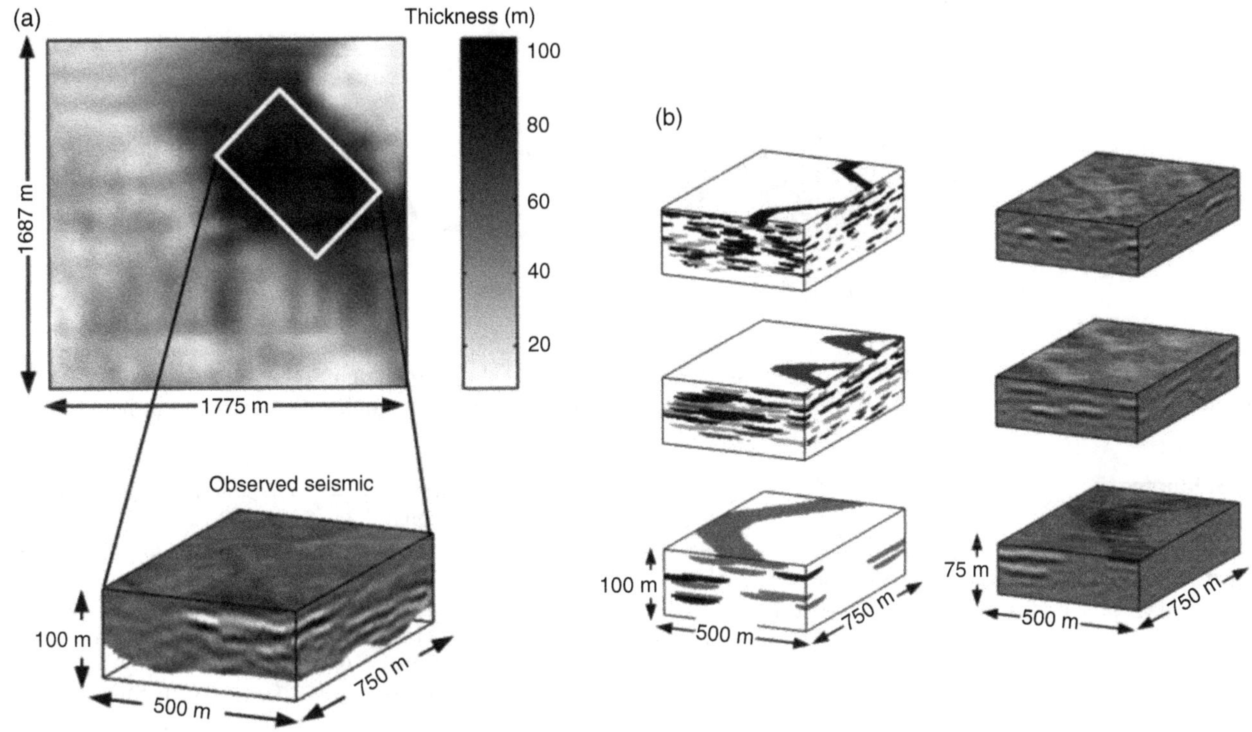

Figure 7.14 (a) 3D seismic survey of a reservoir zone in West Africa. (b) Prior model realizations and forward simulated 3D seismic.

consequence, several alternative rock physics models can be postulated.

2. *Wavelet uncertainty:* The wavelet [often a Ricker wavelet, see *Mavko et al.*, 2009] is used to forward model the seismic data on a given reservoir model. This wavelet is calibrated from wells, but because of measurement error and limited amount of information, this wavelet has an uncertain frequency or bandwidth.

3. *Geological uncertainty:* Turbidite systems in this area are well studied. Turbidites are created by avalanches on the seafloor. They either form channel-like or lobe-like structures. Object-based models can be used to generate geologically realistic reservoir models. The various parameters involved in such modeling (dimensions, spatial occurrence, relationships) are uncertain (see Section 6.6).

7.3.4.2. Dimension Reduction and Regression. BEL consists of first drawing parameters from all prior distributions by means of Monte Carlo, including all parameters involved in forward modeling. Then using that set of parameters to forward model the seismic data (see Figure 7.14) here, we consider seismic amplitude data. The aim is now to build a statistical relationship between the 3D seismic amplitude data and a target parameter of interest. Let us consider as target prediction the total proportion of sand that contains oil. In other words, what we are after is $f(h|\mathbf{d} = \mathbf{d}_{obs})$ with h the proportion and $\mathbf{d}_{obs}$ the observed seismic amplitude data. We need to estimate $f(h|\mathbf{d})$, then evaluate it in $\mathbf{d} = \mathbf{d}_{obs}$. Because the seismic data is of high dimension, we first need to create a space of much lower dimension within which the modeling of $f(h|\mathbf{d})$ can be easily done. To understand how this can be achieved, we need to understand that what we are interested in is not so much the seismic data variables themselves but how they vary when models variable change. In other words, if we can quantify how salient features in the seismic vary, we can achieve a meaningful dimension reduction. Since seismic data variables are essentially waveforms, a way to quantify the global variation of the amplitudes is by means of wavelet analysis [*Debauchies*, 1993]. Wavelets are essentially a form of dimension reduction. Changes in the seismic data variables can be quantified by changes in the wavelet coefficients. Such changes in wavelet coefficients can be quantified by means of distance in the histogram of such wavelet coefficients [see *Scheidt et al.* 2015 for details]. Once a distance is defined, density estimation can proceed by means of kernel density estimation in metric space (see Section 3.3.2). Figure 7.15 shows an MDS plot based on such distance calculations. This plot also contains the actual observed data $\mathbf{d}_{obs}$ and is colored with a model parameter, here turbidite channel width. From such plots, one can easily calculate posterior parameter distribution, including those key target variables, such as the proportion of oil sand (see Figure 7.15).

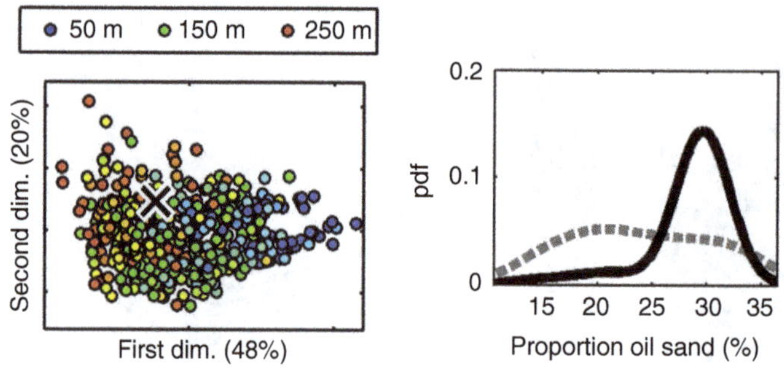

Figure 7.15 MDS plot based on the distance between wavelet coefficients of any two forward simulated 3D seismic models. The color is the width of the turbidite channels. The cross indicates the actual field seismic (right) Using KDE and Bayes' rule the uncertainty of the proportion of sand containing oil is significantly reduced, indicating significant potential in this field.

7.3.5. Summary

So far in this section, we have applied BEL for prediction in four cases where the data and forecast variables take on a variety of forms. Despite these differences in variables, we see the themes outlined in this section are all prevalent. The first step of BEL always consists of constructing a set of prior subsurface model using a-priori information, Monte Carlo experiments, and geostatistical simulation. These prior models are then used to construct the corresponding set of data and prediction variables. In practice, the forward functions used to create these variables can consist of reservoir simulators, geophysical simulators, or simpler static operators.

The next common theme is that of preprocessing by which the prediction and data variables are transformed into lower dimensions or appropriate distributions that satisfy assumptions that may be needed before regression can be applied. The choice of dimension reduction method is dependent on the nature of the variable itself. We saw in these examples that FDA was effective for smoothly varying time-series data, while eigen-image analysis was useful for reducing the dimension of spatial maps. The practitioner is encouraged to explore different multiple dimension reduction techniques. The choice of regression technique depends on the type, dimension, and relationship of the data and forecast variables. For high-dimensional problems, parametric regression is generally preferred over non-parametric techniques unless a large number of prior samples are available, or if the underlying assumptions of parametric regression cannot be met.

In Chapter 8, we will present a few applications of BEL in real-world decision making.

REFERENCES

Abdi, H., L. J. Williams, and D. Valentin (2013), Multiple factor analysis: Principal component analysis for multitable and multiblock data sets. *Wiley Interdiscip. Rev. Comput. Stat.*, *5*(2), 149–179, doi: https://doi.org/10.1002/wics.1246.

Aggarwal, C. C., and P. S. Yu (2001), Outlier detection for high dimensional data, *SIGMOD Conference'01*, Santa Barbara, CA, pp. 37–46, doi: https://doi.org/10.1145/376284.375668.

Belhumeur, P. N., J. P. Hespanha, and D. J. Kriegman (1997), Eigenfaces vs.~{Fisherfaces}: Recognition using class specific linear projection. *Pami*, *19*(7), 711–720.

Beniger, J. R., V. Barnett, and T. Lewis (1980), Outliers in statistical data. *Contemp. Sociol.*, *9*(4), 560, doi: https://doi.org/10.2307/2066277.

Chalmers, A. F. (1999), *What Is this Thing Called Science?* Open University Press. Available from: https://books.google.co.uk/books?id=s13tBAAAQBAJ.

Chiu, W. A., C. Chen, K. Hogan, J. C. Lipscomb, C. S. Scott, and R. Subramaniam (2007), High-to-low dose extrapolation: Issues and approaches. *Hum. Ecol. Risk Assess. Int. J.*, *13*(1), 46–51, doi: https://doi.org/10.1080/10807030601107544.

Debauchies, I. (1993) Ten lectures on wavelets. *SIAM Rev*, doi: https://doi.org/10.1137/1035160.

Del Moral, P., A. Doucet, and A. Jasra (2012), On adaptive resampling strategies for sequential Monte Carlo methods. *Bernoulli*, *18*(1), 252–278, doi: https://doi.org/10.3150/10-BEJ335.

Douc, R., and O. Cappe (2005), Comparison of resampling schemes for particle filtering, ISPA 2005, *Proceedings of the 4th International Symposium on Image and Signal Processing and Analysis*, pp. 64–69. IEEE, doi: https://doi.org/10.1109/ISPA.2005.195385.

Evensen, G. (2003), The Ensemble Kalman Filter: Theoretical formulation and practical implementation. *Ocean Dyn.*, *53*(4), 343–367, doi: https://doi.org/10.1007/s10236-003-0036-9.

Fasano, G., and A. Franceschini (1987), A multidimensional version of the Kolmogorov-Smirnov test. *Mon. Not. R. Astron. Soc.*, *225*, 155–170, doi: https://doi.org/10.1007/s10342-011-0499-z.

Fodor, I. K. (2002), *A Survey of Dimension Reduction Techniques*. Lawrence Livermore National Laboratory, Livermore, CA, doi: https://doi.org/10.2172/15002155.

Fox, D. (2003), Adapting the sample size in particle filters through KLD sampling. *Int. J. Rob. Res.*, *22*(12), 985–1004.

Gelman, A., and C. R. Shalizi (2013), Philosophy and the practice of bayesian statistics. *Br. J. Math. Stat. Psychol.*, *66*(1), 8–38, doi: https://doi.org/10.1111/j.2044-8317.2011.02037.x.

Gelman, A, X.-L. Meng, and H. Stern (1996), Posterior predictive assessment of model fitness via realized discrepancies. *Stat. Sin.*, *6*(4), 733–807, doi: https://doi.org/10.1.1.142.9951.

Graham, M. H. (2003), Confronting multicollinearity in ecological multiple regression. *Ecology*, *84*(11), 2809–2815, doi: https://doi.org/10.1890/02-3114

Glynn, P. W., and D. L. Iglehart (1989), Importance sampling for stochastic simulations. *Manag. Sci.* *35*(11), 1367–1392.

Härdle, W., A. Werwatz, M. Müller, and S. Sperlich (2004), *Nonparametric and Semiparametric Models*, vol. *47*, Springer, Berlin/Heidelberg, doi: https://doi.org/10.1007/978-3-642-17146-8.

Hastie, T., R. Tibshirani, and J. Friedman (2009), *The Elements of Statistical Learning. Bayesian Forecasting and Dynamic Models*, vol. *2*, Springer, New York, doi: https://doi.org/10.1007/b94608.

Hermans, T., E. Oware, and J. Caers (2016), Direct prediction of spatially and temporally varying physical properties from time-lapse electrical resistance data. *Water Resour. Res.*, *52*(9), 7262–7283, doi: https://doi.org/10.1002/2016WR019126.

Hinze, M., and S. Volkwein (2005), Proper orthogonal decomposition surrogate models for nonlinear dynamical systems: Error estimates and suboptimal control, in *Dimension Reduction of Large-Scale Systems*, vol. *45*, edited by P. Benner, D. C. Sorensen, and V. Mehrmann, Springer, Berlin/Heidelberg, doi: https://doi.org/10.1007/3-540-27909-1.

Hodge, V. J., and J. Austin (2004), A survey of outlier detection methodologies. *Artif. Intell. Rev.*, *22*(2), 85–126, doi: https://doi.org/10.1023/B:AIRE.0000045502.10941.a9.

Holm, S. (1979), A simple sequentially rejective multiple test procedure. *Scand. J. Stat.*, *6*(2), 65–70, doi: https://doi.org/10.2307/4615733.

Houtekamer, P. L., and H. L. Mitchell (1998), Data assimilation using an ensemble Kalman filter technique. *Mon. Weather Rev.*, *126*(3), 796–811, doi: https://doi.org/10.1175/1520-0493(1998)126<0796:DAUAEK>2.0.CO;2.

van der Maaten, L., E. Postma, and J. van den Herik (2009), Dimensionality reduction: A comparative review. *J. Mach. Learn. Res.*, *10*(January), 1–41, doi: https://doi.org/10.1080/13506280444000102.

Makridakis, S., A. Andersen, R. Carbone, R. Fildes, M. Hibon, R. Lewandowski, et al. (1982), The accuracy of extrapolation (time series) methods: Results of a forecasting competition. *J. Forecast.*, *1*(2), 111–153, doi: https://doi.org/10.1002/for.3980010202.

Mavko, G., T. Mukerji, and J. Dvorkin (2009), *The Rock Physics Handbook*, Second Edition, doi: https://doi.org/http://dx.doi.org/10.1017/CBO9780511626753.

McKenzie, D., D. Peterson, and E. Alvarado (1997), Extrapolation problems in modeling fire effects at large spatial scales: A review. *Int. J. Wildland Fire*, *6*(4), 165–176, doi: https://doi.org/doi:10.1071/WF9960165.

Myers, R. H. (1990), *Classical and Modern Regression with Applications*, The Duxbury Advanced Series in Statistics and Decision Sciences, vol. *2*, doi: https://doi.org/10.2307/1269347.

Oliver, D. S., A. C. Reynolds, and N. Liu (2008), *Inverse Theory for Petroleum Reservoir Characterization and History Matching*, doi: https://doi.org/10.1017/CBO9780511535642.

Peacock, J.A. (1983), Two-dimensional goodness-of-fit testing in astronomy. *Mon. Not. R. Astron. Soc.*, *202*, 615, doi: https://doi.org/10.1093/mnras/202.3.615.

Reichle, R. H. (2008), Data assimilation methods in the Earth sciences. *Adv. Water Resour.*, *31*(11), 1411–1418, doi: https://doi.org/10.1016/j.advwatres.2008.01.001.

Satija, A., and J. Caers (2015), Direct forecasting of subsurface flow response from non-linear dynamic data by linear least-squares in canonical functional principal component space. *Adv. Water Resour.*, *77*, 69–81, doi: https://doi.org/10.1016/j.advwatres.2015.01.002.

Satija, A., C. Scheidt, L. Li, and J. Caers (2017), Direct forecasting of reservoir performance using production data without history matching. *Comput. Geosci.*, *21*(2), 315–333, doi: https://doi.org/10.1007/s10596-017-9614-7.

Sava, P., and B. Biondi (2004), Wave-equation migration velocity analysis. I. Theory. *Geophys. Prospect.*, *52*(6), 593–606, doi: https://doi.org/10.1111/j.1365-2478.2004.00447.x.

Scheidt, C., P. Renard, and J. Caers (2014), Prediction-focused subsurface modeling: Investigating the need for accuracy in flow-based inverse modeling. *Math. Geosci.*, *1–19*, doi: https://doi.org/10.1007/s11004-014-9521-6.

Scheidt, C., C. Jeong, T. Mukerji, and J. Caers (2015), Probabilistic falsification of prior geologic uncertainty with seismic amplitude data: Application to a turbidite reservoir case. *Geophysics*, *80*(5), M89–M12, doi: https://doi.org/10.1190/geo2015-0084.1.

Smith, H., and N. R. Draper (1998), *Applied Regression Analysis*, Third Edition. Available from: http://www.amazon.co.uk/Applied-Regression-Analysis-Probability-Statistics/dp/0471170828.

Suzuki, S., G. Caumon, and J. Caers (2008), Dynamic data integration for structural modeling: Model screening approach using a distance-based model parameterization. *Comput. Geosci.*, *12*(1), 105–119, doi: https://doi.org/10.1007/s10596-007-9063-9.

Tarantola, A. (2005), *Inverse Problem Theory and Methods for Model Parameter Estimation*, Other Titles in Applied Mathematics, doi: https://doi.org/doi:10.1137/1.9780898717921.

Turk, M., and A. Pentland (1991), Face recognition using eigenfaces. *IEEE Comput. Soc. Conf. Comput. Vis. Pattern Recognit.*, *76*(8), 586–591, doi: https://doi.org/10.1109/CVPR.1991.139758.

Wilkinson, G. G. (1983), The dangers of extrapolation from temperature record statistics. *Weather*, *38*(10), 312–313, doi: https://doi.org/10.1002/j.1477-8696.1983.tb04812.x

8

Quantifying Uncertainty in Subsurface Systems

Co-Authored by: Troels Norvin Vilhelmsen[1], Kate Maher[2], Carla Da Silva[3], Thomas Hermans[4], Ognjen Grujic[5], Jihoon Park[5], and Guang Yang[5]

8.1. INTRODUCTION

We now arrive at the point where things really matter: the applications. Here is where elegant theories, smart workflow, clever mathematics, the rubber meets the road: real (messy!) data, real decisions. The "solutions" we will be presenting to the question posed in Chapter 1, using the materials from previous chapters, are but one approach, one strategy. No single solution fits all problems of uncertainty quantification (UQ), or decision making. The choices we made are our personal preferences. However, a common thread of reasoning occurs in all these cases.

Bayesian philosophy. In Chapter 5, we outlined a view on how Bayesianism applies to UQ in subsurface systems. We reflected on the importance of a proper representation and prior uncertainty of the geological variability/heterogeneity in the subsurface. We relied on building falsifiable prior distributions that are wide and reflect realistically the large set of hypothesis and possibilities that exist.

Monte Carlo. The power of Monte Carlo (and quasi-Monte Carlo) lies in its simplicity and in its ability to be parallelized CPU or GPU-wise. Monte Carlo can be used for many purposes: to perform global sensitivity analysis, to generate geostatistical models, or to produce posterior uncertainty on target prediction variables. It is unlikely that a "one-shot" Monte Carlo will always work. Often, after a first Monte Carlo experiment, prior distributions are falsified by some data and hence these prior distributions need revision, resulting in a second Monte Carlo. Sensitivity analysis may indicate that certain variables can be set to fixed values (or reduced ranges) and this may invoke yet another Monte Carlo.

Global sensitivity analysis. Perhaps of all techniques, global SA may be the most relevant for UQ. Global sensitivity analysis has many purposes; it allows

1. understanding the complexity of the problem;
2. understanding what model variables impact data, decisions, forecasts;
3. simplifying the model;
4. understanding how data informs decisions;
5. targeting the Monte Carlo, thereby reducing variance.

Falsification. Bayesianism is inherently inductionist. Both prior and posterior should be tested (deduction). It is a healthy practice to try and falsify all modeling assumption and hypothesis to the extent that this is possible.

Avoiding complex model inversions. Inverse modeling by deriving model variables from data is a powerful idea. The current practice, however, is to use increasingly complex models and increasingly sophisticated data and to couple many types of physical and chemical models. The entire world may soon be instrumented. Inversion in high dimensions with nonlinear forward models, coupled physics, and non-Gaussian model variables accounting for all uncertainties remains an illusion. Inversion is also sequential, mostly Markovian, meaning it is computationally less appealing than Monte Carlo.

Statistical surrogate models. Doing Monte Carlo and building smart statistical emulators of the various input and output uncertainties is a strategy that will be used

[1]*Department of Geoscience, Aarhus University, Aarhus, Denmark*

[2]*Department of Geological Sciences, Stanford University, Stanford, CA, USA*

[3]*Anadarko, The Woodlands, TX, USA*

[4]*University of Liege, Liege, Belgium*

[5]*Department of Energy Resources Engineering, Stanford University, Stanford, CA, USA*

Quantifying Uncertainty in Subsurface Systems, Geophysical Monograph 236, First Edition. Céline Scheidt, Lewis Li, and Jef Caers.

frequently used to overcome the computational burden of CPU demanding forward models.

Dimension reduction. The subsurface is complex, describing it requires high-dimensional parameterizations. These high dimensions are problematic in applying most statistical methods. A carefully planned dimension reduction on either model, data, or prediction variables will be key to rendering standard statistical methodologies applicable.

Linear models. It may be tempting to use sophisticated nonlinear modeling to address problems with complex interacting components. Many of these methods (e.g., ANN, additive models) are powerful, but they often require substantial expertise and tuning to make them work. Linear models are well understood and robust in high dimensions. The chance of overfitting (and thereby artificially reducing uncertainty) is much more favorable than with nonlinear models. Linear models are not as accurate as nonlinear models in terms of performance, but the idea of weak classifiers, bagging, and boosting may improve accuracy. The trick, therefore, is to turn the problem linear by means of various transformation and to further develop linear modeling, such as kriging in higher dimensions.

In each case, we will first outline and motivate our strategy. We also list the various techniques used, so readers can refer and read-up on those sections. We will not be presenting all the details of these cases. Much more work went into them than perhaps the write-up may suggest. We will instead focus on the key elements of the approaches that are unique to each case.

8.2. PRODUCTION PLANNING AND DEVELOPMENT FOR AN OIL FIELD IN LIBYA

8.2.1. Three Decision Scenarios and a Strategy for UQ

As discussed in Chapter 1, reservoir management decisions made over the lifetime of an oil reservoir are subject to a large set of uncertain variables. We considered three decision scenarios (see Figure 1.4) over the lifecycle of a large oil field of a Wintershall asset in Libya. We will first review these three decision scenarios within the context of the current approaches and then present a strategy that can be used for all the three scenarios.

Decision scenario 1. This decision presents itself when the field has been in production with five producers for 800 days. Waterflooding was commenced at the 400-day mark using three injectors to preserve reservoir pressure. A decision is required regarding the modification of well controls for the three injectors such that the field oil production/NPV is maximized. In practice, waterflooding optimization is performed using a reactive control approach [*Nigel et al.*, 2003], in which nonprofitable producers are shut down. Closed-loop optimization techniques [*Brouwer and Jansen*, 2002; *Aitokhuehi and Durlofsky*, 2005; *Wang et al.*, 2009] have been developed to predictively address problems such as early water breakthroughs. These approaches consist of generating a single history matched reservoir model (deterministic inversion, see Section 6.4), and then applying a rate optimization step on that model. Both steps are time-consuming. Also, such rate optimization has to be done repeatedly in time as the reservoir conditions change. As an alternative to numerical optimization, *Thiele and Batycky* [2006] take advantage of streamline simulations and their ability to quantify the so-called well allocation factors (WAFs) between injector and producers. These WAFs allow calculating injector efficiencies (IEs) that model how effective each injector is at producing offset oil. A heuristic-based decision model is then built that calculates new optimal rates for each injector, with the purpose of re-allocating water from inefficient to efficient injectors. While this approach has been successfully applied to large field cases [*Ghori et al.*, 2007; *Muhammad et al.*, 2007], it still requires the specification of a history-matched models. The challenge, therefore, is to deal with the uncertainty in the reservoir model and the time-dependent nature of the IEs.

Decision scenario 2. The decision question in this scenario is where to drill an infill well after the field has been in production for 3000 days. The aim is to maximize future oil recovery. Optimal well placement has been an intensive area of research both in oil and gas [*Aanonsen et al.*, 1995; *Güyagüler and Horne*, 2001; *Nakajima*, 2003; *Ozdogan and Horne*, 2006; *Wang et al.*, 2012] and in hydrology [*Gorelick*, 1983; *Park and Aral*, 2004; *Maskey et al.*, 2007]. Numerical optimization remains a popular approach for addressing such well placement problems, techniques such as simulated annealing [*Beckner and Song*, 1995], genetic algorithms [*Montes et al.*, 2001; *Emerick et al.*, 2009], or particle swarm optimization (PSO) [*Onwunalu and Durlofsky*, 2009; *Jesmani et al.*, 2015] determine the optimal well placement to maximize recovery. In each approach, the objective function is evaluated by running a flow simulation for a proposed well location and computing the production rate or NPV. As an alternative to full numerical optimization, *da Cruz et al.* [2004] introduce the concept of a quality map, which serves as a proxy measure of NPV by drilling at each grid cell in the reservoir model. By using kriging to interpolate this quality variable from a few locations, the well placement problem becomes much more tractable. The problem is that each of such well placement approaches requires a history-matched reservoir model, and such model is subject to uncertainty. In well placement, the risk associated with infill drilling is much higher than simply adjusting a few well controls. We will need to address the question of how to optimize well-locations subject to the various reservoir model uncertainties.

Decision scenario 3. Here we need to decide when the field is to be plugged and abandoned. Since substantial capital costs is incurred when abandoning a field, it is important for the operator to have an estimate of when this occurs. Such decision requires a long-term prediction of the field production rate and corresponding UQ. A decision corollary to the long-term production is the length of the concession contract. Since petroleum licenses are negotiated for a limited amount of time, it is important for the decision maker to know how long the field can be profitable for. We will consider a situation where the reservoir has been in production for 4000 days. The abandonment decision will depend on the prediction for the field production rate in the future.

Despite the different nature of the three above described decision problems, a common strategy for UQ can be followed (Figure 8.1). In each scenario, only limited spatial information is available (well logs at wells), as well as historical rates from the existing producers.

A causal analysis (inversion or history matching, Chapter 6) of each of the three decision scenarios would require constructing multiple history-matched models. Owing to the complexities of the reservoir and corresponding difficulties associated with history matching, where all uncertainties (structure, petrophysical, and fluid) need to be considered, a Bayesian evidential learning (Chapter 7) appears appealing. Figure 8.1 provides an overview. The workflow consists of generating prior *pdf*s on all uncertain modeling components based on some of the available data. This in turn allows generating multiple reservoir models. These models are constrained to well information but not to any production data. The reservoirs are forward modeled for each of the three situations described above to obtain data variables as well as the prediction variables. The data variables (production at multiple wells) as well as some of the prediction variables (the quality map, the rate decline) are reduced in dimension. The latter will make the building of a statistical relationship between data and prediction variable easier. This statistical model is then used to make predictions in each scenario.

Why this strategy?. *Various forms of uncertainty*. Uncertainty due to a lack of spatial information (limited boreholes), as well as uncertainty in the fluid characteristics, relative permeability, fault transmissibility, and so on, requires a strategy that can properly deal with all forms of uncertainty at the same time.

Difficulties of traditional inversion when dealing with production data. In each of the three decisions, the available data are production profiles from existing producers. Constructing inverted reservoir models (history matching) that match production data is challenging and time-consuming, especially in the presence of complex geological heterogeneity and multiple types of uncertainty.

Rapid decision making required. Certain decisions such as well control optimization need to be made rapidly and

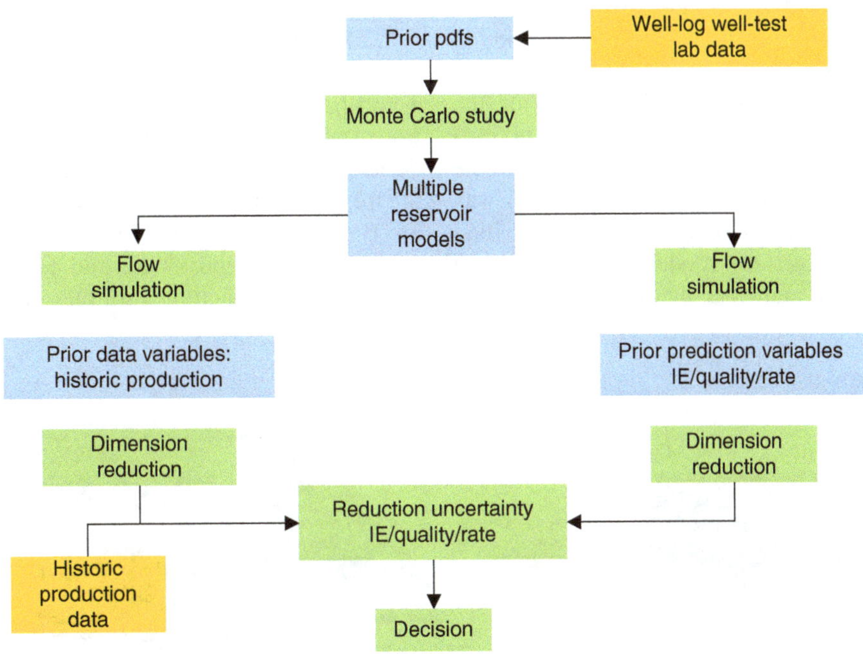

Figure 8.1 A common strategy for UQ in all three decision scenarios.

frequently to maximize productivity. It would be impractical to history match every time such a decision is required.

Because of the different nature of the prediction variables, a variety of statistical tools will be used to account for them. Readers can review the following methodological sections:

1. Bayesian Evidential Learning (Chapter 7)
2. Mixed PCA (Section 7.2.3)
3. Eigen-image Analysis (Section 3.5.1)
4. Canonical Correlation Analysis (Section 3.5.3)
5. Functional Data Analysis (Section 3.6)
6. Gaussian Process Regression (Section 3.7.5)
7. Kernel Density Estimation (Section 3.3.2)

8.2.2. The Prior

8.2.2.1. Prior Reservoir Models. The authors were not involved in the actual geological modeling of this reservoir. Hence, prior distributions were mimicked after what could reasonably be expected for these cases and through discussions with geologists and engineers of the operating company in Kassel, Germany. This also allows us to generate a "truth" case simply by picking a prior reservoir model to generate all the data and predictions (hence we "know" the actual future). In the next section, on the Danish groundwater case (as well as the Belgian and US contamination case), we illustrate how such detailed modeling using geostatistical methods and how such prior distributions are established in real field situations.

Recall from Chapter 1 that a single reservoir model is composed of both gridded (spatial) and non-gridded components (fluid properties). The spatial component includes the structural and stratigraphic model of the reservoir describing the location and geometry of bounding layers and horizons in addition to the faults in the reservoir. In the Libyan case, the structural model was constructed according to the geological setting described in Section 1.2. The structural model (see Figure 8.2) consists of four faults and two deformable horizons. In this case, a brittle deformation is modeled with no smearing. The stratigraphic model is shown in Figure 8.2. In this example, fault locations and displacements are fixed.

Depositional uncertainty is represented by means of training images (see Section 6.3.4). Figure 8.3 shows that the geological conceptual model (as expressed by the training image) consists of three facies: low-quality sand (blue), high-quality sand (green), and diatomite (red). To generate multiple lithological model realizations that reflect the geometries in this training image and are at the same time constrained to lithological interpretation from well data (here five wells), we use a multiple-point geostatistical (MPS) algorithm termed Image Quilting [*Mariethoz and Caers*, 2014] with the well logs as conditioning data. Realizations of 500 facies were generated of which a few are shown in Figure 8.3.

To model porosity, porosity data at the existing wells are used as conditioning data as well as to estimate trends and variogram ranges. Porosity realizations are then generated using sequential Gaussian simulation (Section 2.3) within each lithology. Uncertainty on non-gridded parameters can be expressed using simple prior probability distributions. The names and pdfs of the uncertain non-gridded parameters for the Libyan case are given in Table 8.1. We perform Monte Carlo experiment and generate 500 samples from each pdf. For illustration, we set aside one realization from the prior and use its data variable as $\mathbf{d}_{\text{obs}}$. Its prediction variable is used as the truth and serves as a mechanism for verifying the results of evidential analysis.

8.2.2.2. The Prior for Data and Prediction Variables. Prior realizations of data and prediction variables, $\mathbf{d}$ and $\mathbf{h}$, can now be generated based on the generated prior reservoir model realizations. In each of the three decision scenarios, a reservoir flow simulator is used as the forward functions for both $\mathbf{d}$ and $\mathbf{h}$. The data variables are the production rates for the five existing producers over periods of 800, 3000, and 6000 days respectively. In terms of prediction variables, we need to consider

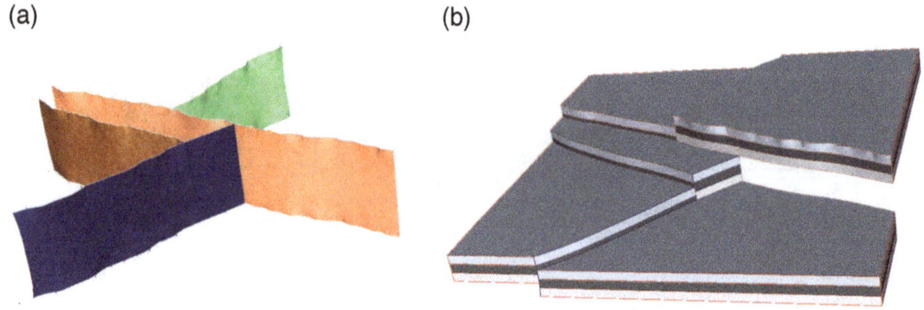

Figure 8.2 Structural framework (a) and stratigraphic (b) grids for the Libyan case. The reservoir model consists of four faults and two deformable horizons.

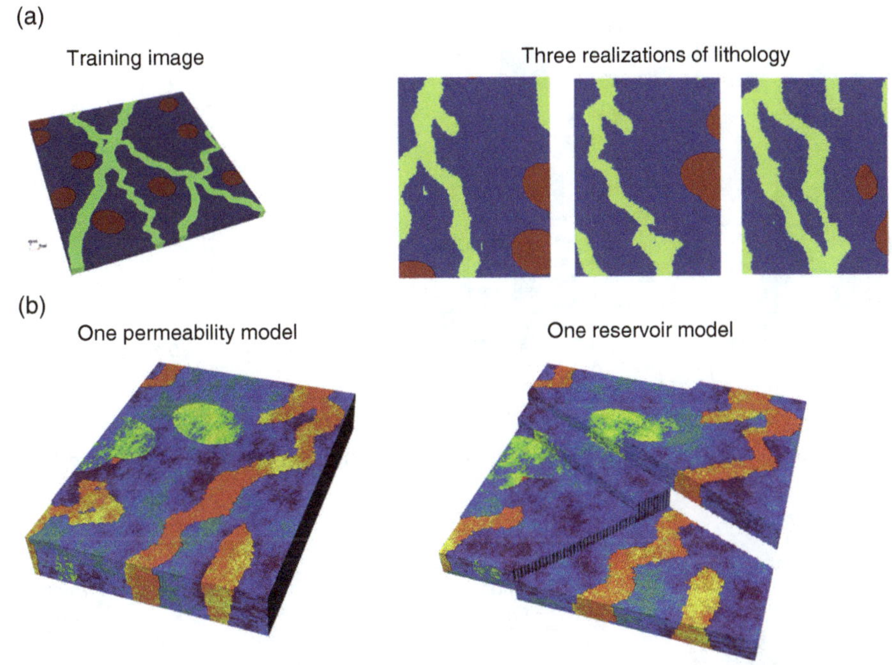

Figure 8.3 (a) A training image with channels of high-quality sand and diatomite bodies is used to construct facies realizations. (b) Permeability realization shown in depositional domain (left) and gridded physical domain (right).

Table 8.1 Prior uncertainties on non-gridded reservoir model variable.

Parameter name	Prior pdf
Oil Water Contact	U[1061,1076]
Fault 1 Transmissibility Multiplier	U[0.2,0.8]
Fault 2 Transmissibility Multiplier	U[0.2,0.8]
Fault 3 Transmissibility Multiplier	U[0.2,0.8]
Fault 4 Transmissibility Multiplier	U[0.2,0.8]
Oil Viscosity	N(4,0.5)
Irreducible Water Saturation	N(0.2,0.05)
Irreducible Oil Saturation	N(0.2,0.05)
End-point rel K Water	U[0.1,0.4]
End-point rel K Oil	U[0.1,0.4]
Corey Coefficient Oil	U[2,3]
Corey Coefficient Water	U[4,6]

(i) injector efficiencies, (ii) quality map, and (iii) future field production rates.

Injector efficiencies. Three injectors are present in the field and have been injecting for the past 800 days. The prediction variable is the efficiency of each injector at the present time. We use the definition of injector efficiency from *Thiele and Batycky* [2006] which is the ratio between offset oil production and water injection. This is a scalar for each injector, for a total of three injectors in this study. To determine the efficiency of an injector for a given prior reservoir model realization, we use a streamline simulator [*Thiele et al.*, 2010] as the forward model. A streamline simulator allows quantifying how much fluid is allocated between each injector and producer by solving a transport problem along each streamline. By running the streamline simulator for each prior reservoir model realization, we obtain realizations (samples) of the prior distributions of each of the injector's efficiencies shown in Figure 8.4.

The data variables are composed of the historical production rates from the five producers over the 800-day duration of the field. These are obtained by performing reservoir simulation on the prior reservoir models. The prior data variables now consist of a set of producer rates discretized at 10-day intervals for a total of 400 dimensions (five producers, each containing 80 time-steps). These are shown in Figure 8.5 along with the actual observed rates $\mathbf{d}_{obs}$ indicated in red.

Quality map. In decision scenario 2, the field has been in production for 3000 days. The data variables are the production profiles of the five existing wells over the past 3000 days discretized at 50-day intervals for a total of 300 dimensions. The prediction variable is the quality map, which is a 2D map the size of the reservoir grid (127 × 165) for a total dimension of 20,095. The quality map aims to reflect, at each location, the expected reservoir performance when opting to drill a production well at that location. Ideally, to obtain this, we would require a

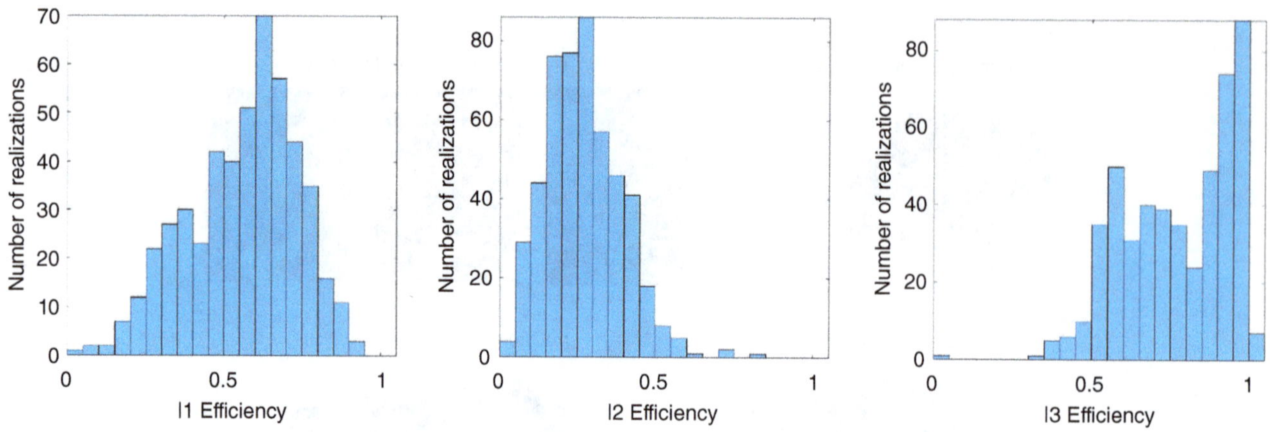

Figure 8.4 Histogram of the prior efficiencies realizations for each of the three existing injectors after 800 days of injection.

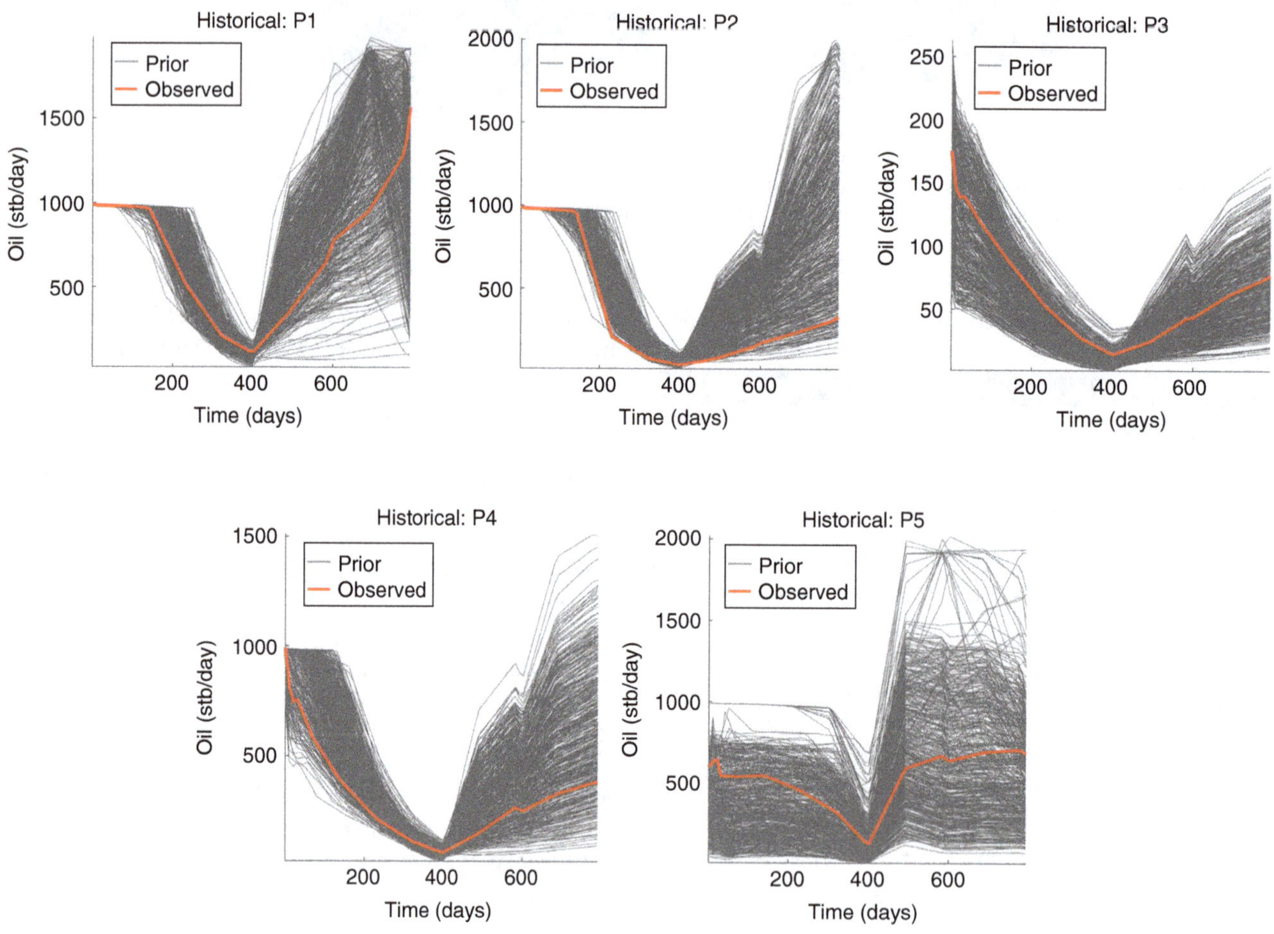

Figure 8.5 Historical production profiles for each of the 500 prior reservoir models (gray). The observed historical production is indicated by the red line.

separate simulation for each possible well location. However, this would be computationally infeasible when dealing with large reservoirs and large number of realizations. Instead, we follow the approach proposed by *da Cruz et al.* [2004], by which only a limited number of potential locations are selected for forward simulation. The simulated incremental oil production over the next 1000 days, for each infill well, is used as spatial data for interpolation with kriging (Figure 8.6). One such quality map is constructed for each prior realization.

Field production rates. The field has now been under production for 4000 days, and a prediction is desired on

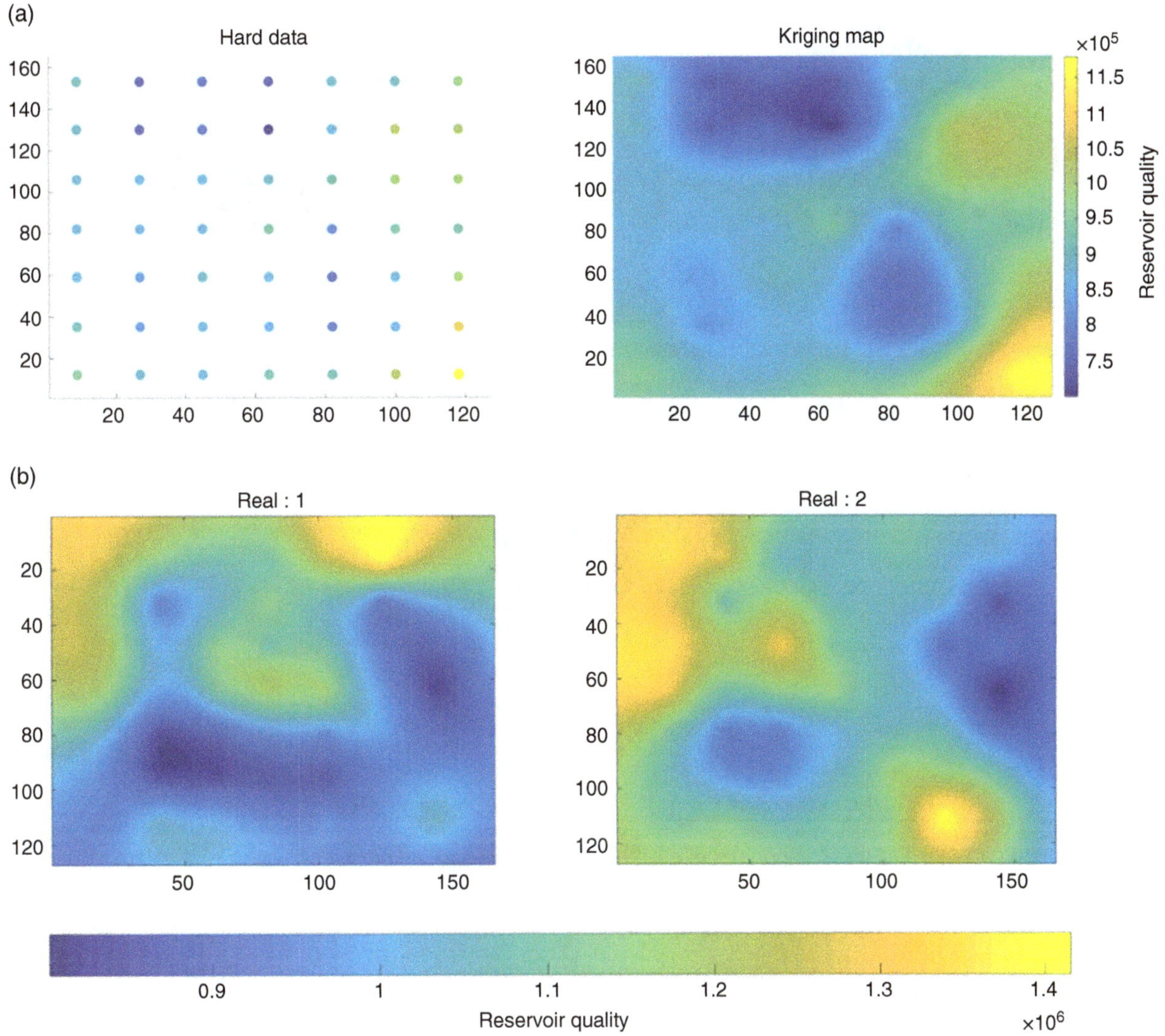

Figure 8.6 (a) Incremental oil production for an infill well drilled at 49 potential well locations are used as data for interpolation (left). Kriging map of reservoir quality using those data points (right). (b) Two realizations of reservoir quality (blue is low quality) drawn from the prior.

the future field production rate. The prediction variable is the field's total production profile from day 4000 to 9000 days assuming current operating conditions. The data variables **d** are the five production profiles of the existing producers over 4000 days, also discretized at 50-day intervals. **d** has a total of 400 dimensions (five producers, each contains 80 time-steps).

8.2.3. Uncertainty Quantification

8.2.3.1. Decision Scenario 1: Well Control. As outlined in Figure 8.1, a dimension reduction on the data variables is needed to make statistical modeling feasible. Since the production profiles in Figure 8.5 are smoothly varying and exhibit systematic variation, functional principal component analysis (FPCA) is an appropriate choice for dimension reduction. A third-order B-spline with 40 knots was selected as the basis function for each of the five data responses. This choice of basis allows for 99% of the variance to be expressed by the largest five eigenvalues, effectively reducing the dimension of **d** from 500 to $\mathbf{d}^{fpca}$, or 25 dimensions. An additional dimension-reduction step uses mixed PCA (MPCA) to account for any possible multi-collinearity of rates between the five producers. MPCA identifies that the top seven components, $\mathbf{d}^{mpca}$, account for the majority (>99%) of the variance between the five producers. Figure 8.7 depicts scatter plots of the first two MPCA components versus the prior injector efficiencies **h**. Since each injector's efficiency is a scalar, there is no need for dimension

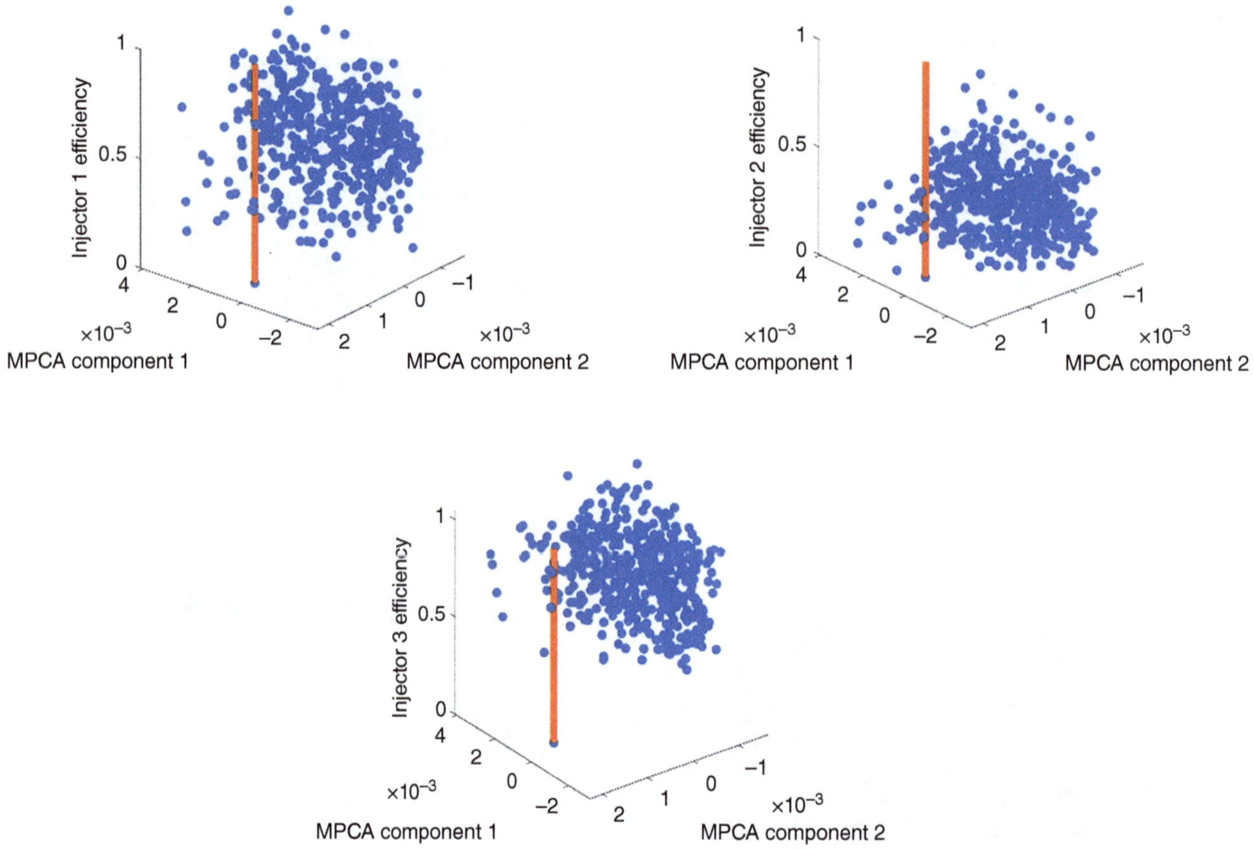

Figure 8.7 Scatter plots of the first two components of $\mathbf{d}^{mpca}$ vs. the prior injector efficiencies $\boldsymbol{h}_{\text{prior}}$ for each injector. The red line indicates the locations of $\mathbf{d}_{\text{obs}}^{mpca}$.

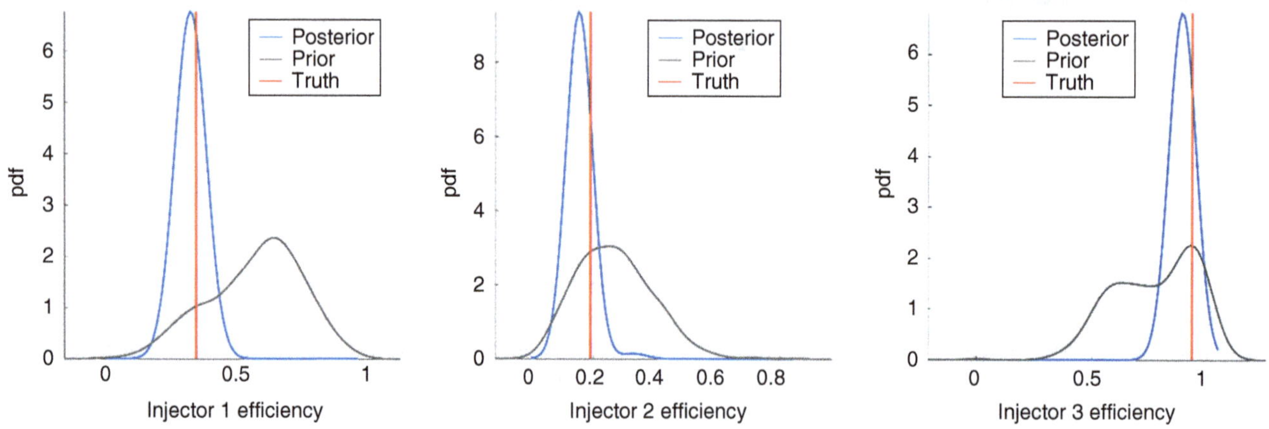

Figure 8.8 Prior (gray) and posterior (blue) distributions obtained from Bayesian evidential learning (by KDE) for each of the three injector efficiencies. The "truth" is indicated by the red line.

reduction on **h**. The red line indicates the actual field observations $\mathbf{d}_{\text{obs}}^{mpca}$.

Since the prediction variable is a scalar, the posterior conditional distribution $f(\mathbf{h}|\,\mathbf{d}_{\text{obs}})$ is univariate. Kernel density estimation (Section 3.3.2) will be used to estimate this posterior distribution. For each of the three injectors, Figure 8.8 shows the posterior distribution estimated by KDE (blue) in comparison with the prior distribution (gray).

8.2.3.2. Decision Scenario 2: Well Placement. To predict a static spatial variable (quality map) from a dynamic data, readers are referred to Sections 7.3.2 and 7.3.3. The main challenge in dealing with spatial is the large dimension of the variable, equivalent to the number of discretized cells. We will use eigen-image analysis (Section 7.2.4) to identify basis images (eigen-images) that represent the underlying modes of variation among the prior quality maps. We next need to determine how many eigen-images should be retained. By examining the eigenvalues, one generally finds that eigen-image analysis requires much more eigenvalues (up to an order of magnitude more) to represent 95% of the variance, in comparison with FPCA. This is due to the underlying variable (spatial map vs. a time series) being more complex. However, retaining a large number of eigen-images required to express 95% of the variance will not reduce the dimension of the variable sufficiently to effectively apply regression. Instead, we need to consider the level of granularity that is required to make the decision. Since the decision is where to place a new infill well, we only need to identify the regions of high reservoir quality, rather than generating a prediction that is accurate to within each grid cell (Figure 8.9).

A number of eigen-images is required that allows expressing the large-scale features within the quality maps. In this instance, 10 eigen-images was deemed (qualitatively) sufficient (38.8% of the total variance). This reduced representation of $\mathbf{h}$, denoted $\mathbf{h}^{\text{eig}}$ is 10 dimensional. We next need to perform FPCA and MPCA on $\mathbf{d}$; the same procedure as in the previous scenario yields $\mathbf{d}^{mpca}$ of seven dimensions.

We now perform regression on $\mathbf{d}^{mpca}$ and $\mathbf{h}^{\text{eig}}$ to evaluate the posterior distribution of the quality maps. Because of the dimensionality of the problem, we apply CCA and normal score transforms, and finally Gaussian process regression to obtain $f\left(\mathbf{h}^c|\mathbf{d}^c_{\text{obs}}\right)$. As done in Section 7.3.3, we will sample from this posterior and undo (back-transform) each of the previously applied transformations and dimension reductions. The result is a set of posterior quality map realizations. We next compute the mean and variance of the posterior realizations in the spatial domain (Figure 8.10). For verification, the true quality map is also shown. We observe that the posterior mean identifies the regions that contain the highest reservoir quality. These two maps (mean and variance) will be used to guide the decision of where to place the new infill well.

8.2.3.3. Decision Scenario 3: Abandonment. In this decision scenario, we need to predict dynamic (future rate) from dynamic (past rate) (see Section 7.3.1). Since both data and prediction variables are time series, FPCA is performed on both $\mathbf{d}$ and $\mathbf{h}$ to get $\mathbf{d}^{fpca}$ and $\mathbf{h}^{fpca}$(dim $\mathbf{h}^{fpca}$ = 4). A MPCA step is performed on the data variables after FPCA to account for multi-collinearity between the five existing producer rates. After MPCA $\mathbf{d}^{mpca}$ is of seven dimension. Prior to performing regression, we apply CCA followed by a normal score transform. The scatter plots of the top three canonical components are shown in Figure 8.11 and indicate the presence of a strong linear correlation.

Equations (7.13) and (7.14) are used to estimate the posterior distribution $f\left(\mathbf{h}^c|\mathbf{d}^c_{\text{obs}}\right)$ as a multivariate

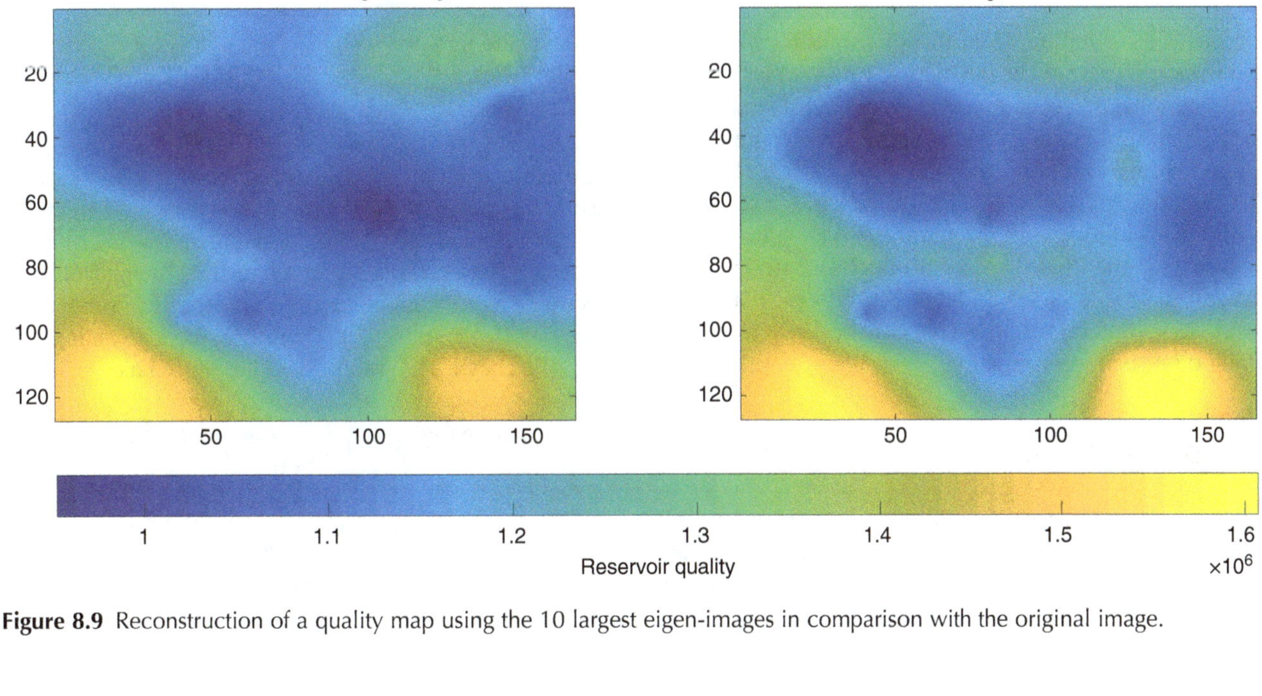

Figure 8.9 Reconstruction of a quality map using the 10 largest eigen-images in comparison with the original image.

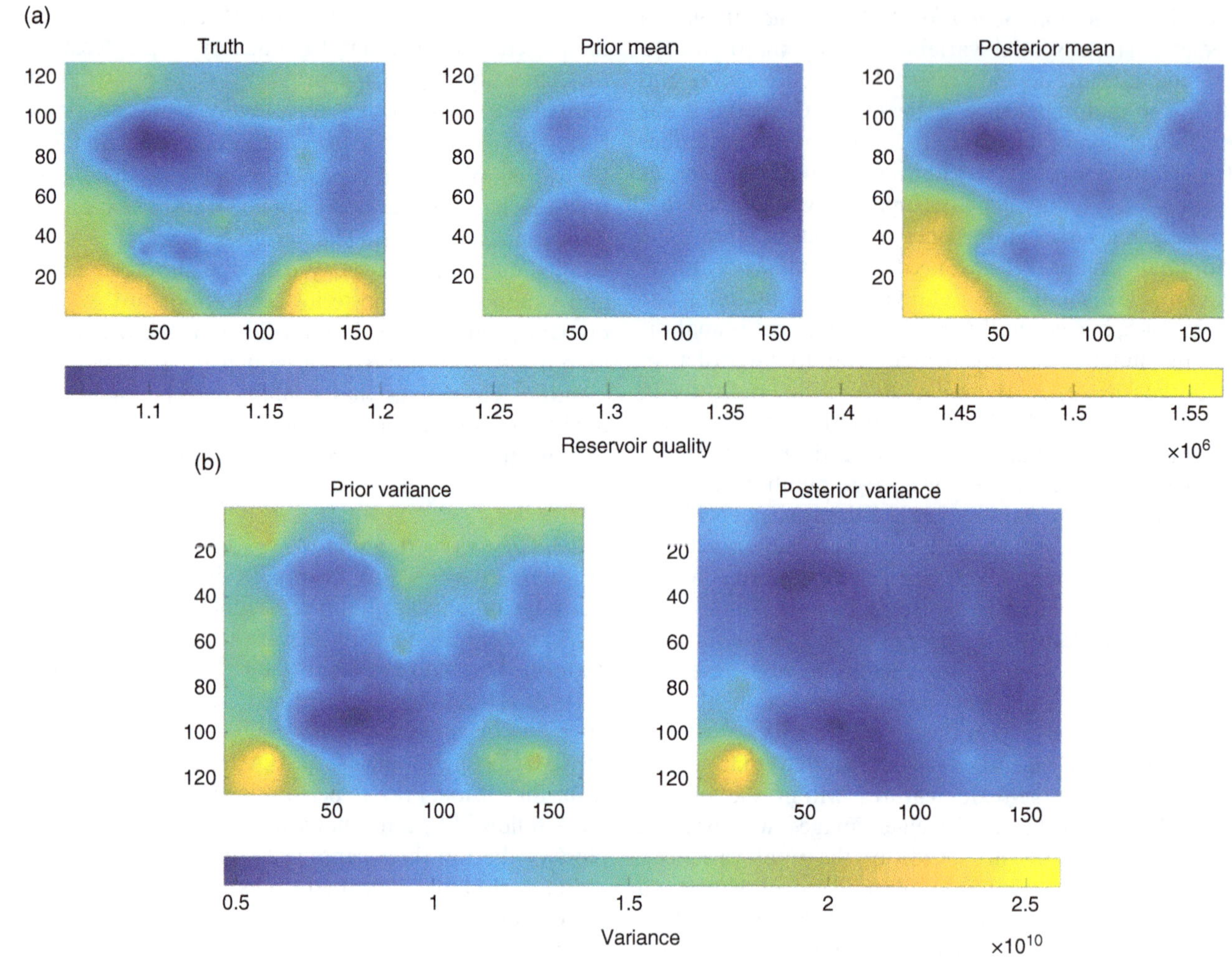

Figure 8.10 (a) True quality map with prior and posterior mean, (b) prior and posterior variance.

Gaussian. By sampling from this distribution, we obtain posterior samples of the prediction in canonical space. These canonical scores are then back-transformed into the original time series by reverting each of the previously applied transformations (normal score, CCA, FPCA). The resulting posterior prediction variables are shown in Figure 8.11 along with the quantiles. As an illustration, we also plot the "true" future production rate from the realization that was set aside as the reference (in red). Given risk preferences, the UQ can be used to decide which date to stop production.

8.2.4. Decision Making

8.2.4.1. Well Control. To address this problem, we use the decision model proposed in *Thiele and Batycky* [2006] in which the new rate of each injector q_i^{new} is computed by multiplying the current injection rate q_i^{old} by a factor of $1 + w_i$, where w_i is the weight of each injector determined from its efficiency.

$$q_i^{\text{new}} = (1 + w_i)q_i^{\text{old}} \tag{8.1}$$

The weight w_i can be positive (increase rate for current injector) or negative (decrease injection rate). *Thiele and Batycky* [2006] propose a functional form for determining this weight based on the average of the efficiencies of all injectors $\bar{e}$ as well as the least and most efficient injectors in the field ($e_{\min}$ and $e_{\max}$). Therefore, for a given injector with efficiency e_i, its weight w_i is computed as

$$w_i = \min\left[w_{\max}, w_{\max}\left(\frac{e_i - \bar{e}}{e_{\max} - \bar{e}}\right)^{\alpha}\right] \text{if } e_i \geq \bar{e} \tag{8.2}$$

$$w_i = \max\left[w_{\min}, w_{\min}\left(\frac{\bar{e} - e_i}{\bar{e} - e_{\min}}\right)^{\alpha}\right] \text{if } e_i < \bar{e} \tag{8.3}$$

$w_{\min}$ and $w_{\max}$ represent the minimum and maximum that the decision maker is willing to modify the injection rate

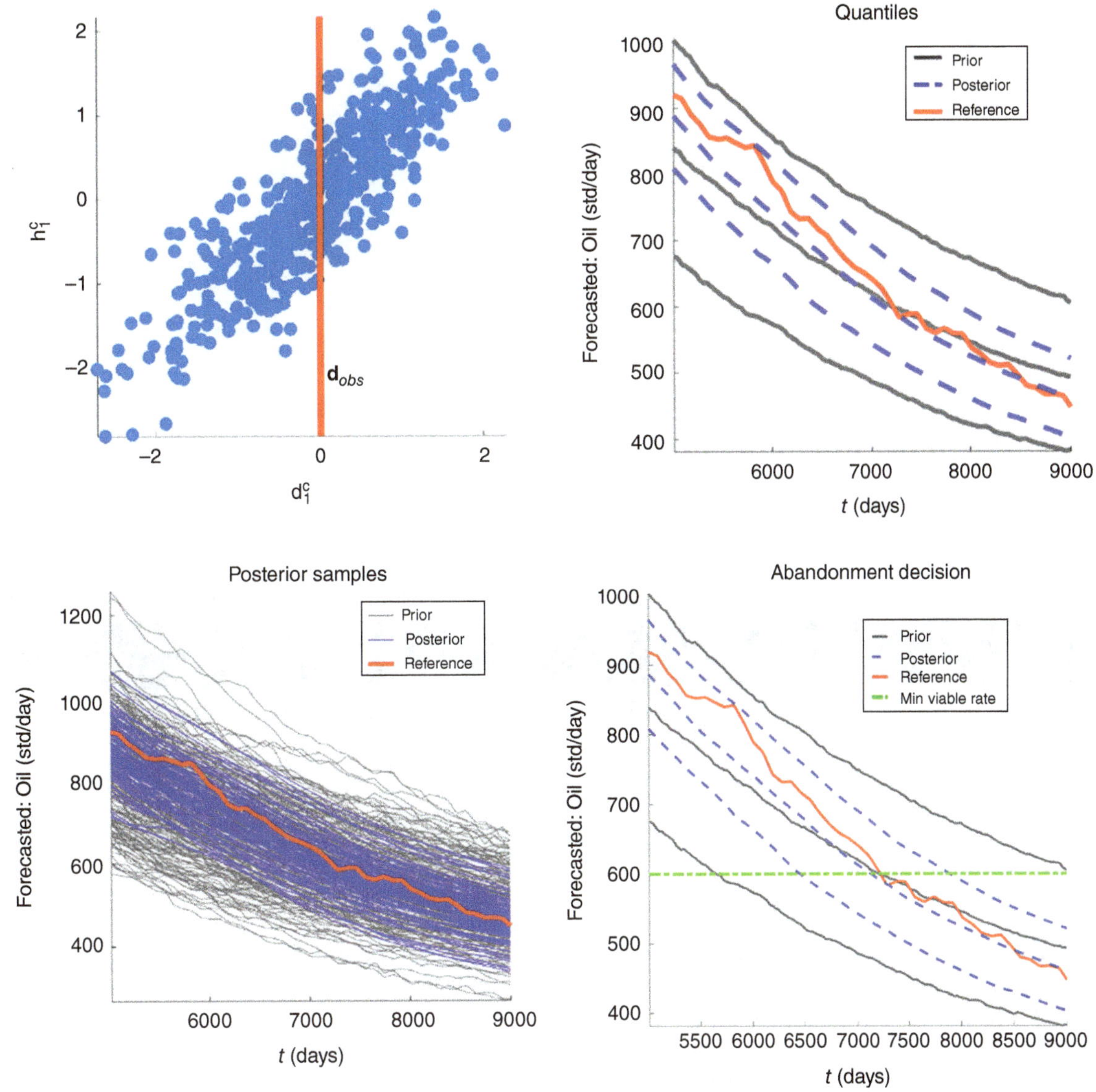

Figure 8.11 Scatterplots of the first canonical components of the data and prediction variables. Posterior field production rates generated by sampling the posterior distribution in canonical space and undoing each of the transformations. Computed quantiles both for the prior and posterior along with the truth is shown in red. For a minimum economically viable rate of 600 stb/day, the posterior predictions indicate that the field should be abandoned at 7897 days, 7136 days, and 6574 days depending on the level of risk the decision maker is willing to accept.

(e.g., $w_{\min} = -1$ would mean that an injector could potentially be shut off completely). The α exponent controls the shape of the weight curve (Figure 8.12) and serves as a mechanism to control how "aggressive" the decision maker wishes to be when re-allocating water from injectors that deviate from the average efficiency $\overline{e}$.

Using this heuristic in lieu of full optimization dramatically speeds up the decision-making process but may result in suboptimal control settings. However, this rapid updating allows decision makers to make informed changes on well controls at a much higher frequency than workflow relying on history matching and rate optimization. For this example, we used $w_{\min} = -0.5$, $w_{\max} = 0.5$, and $\alpha = 0.5$ to compute the optimized injection rates shown in Table 8.2. To verify these updated rates, the "true" realization is flow simulated using both the sets of injection rates for 400 additional days. The cumulative field oil production for the no action and optimized case are shown in Table 8.2; we observe a 3% increase in oil production without increasing the overall field water injection rate.

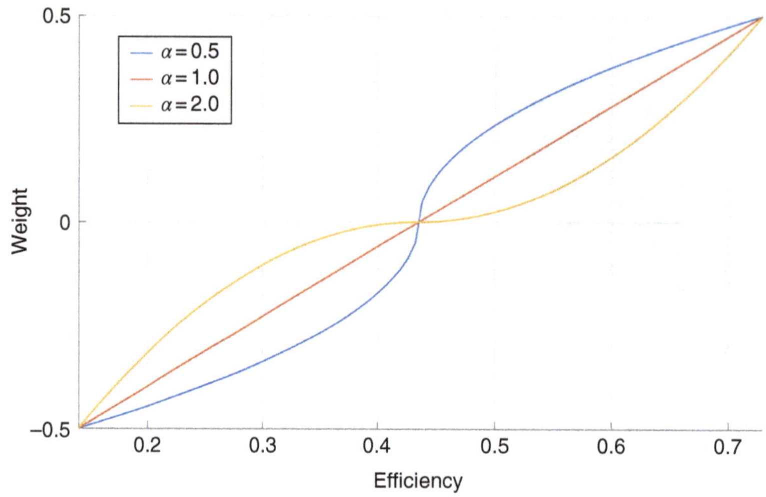

Figure 8.12 Weight functions for different values of the exponent.

Table 8.2 Estimated injection efficiency for "no action" and optimized rate, together with the resulting cumulative oil production.

Rates	Injector 1	Injector 2	Injector 3		Cumulative oil production over next 400 days (stb)
True efficiency	72.15%	13.84%	15.61%	No action	6.968×10^5
Estimated efficiency	72.92%	14.02%	16.20%		
No action injection rate (stb/day)	1460	1830	3788	Optimized injection	7.112×10^5
Optimized injection rate (stb/day)	3004	1273	2772		

*8.2.4.2. **Well Placement.*** Both mean and variance of the posterior quality maps should be taken into consideration when determining the placement of an infill well. The mean quality map is itself a good indicator of the high-quality regions of the reservoir. Intuitively, we would position a new infill well such that it is located as close as possible to as many high-quality cells with smallest uncertainty (as expressed by a posterior variance). Finding the position $\mathbf{x}_{\text{pos}}$ of the optimal infill well location can be expressed as the optimization problem:

$$\max_{\mathbf{x}_{\text{pos}}} \sum_{i=1}^{n_c} \frac{Q_i}{|\mathbf{x}_i - \mathbf{x}_{\text{pos}}|^2} \tag{8.4}$$

n_c is the number of potential well locations, Q_i and $\mathbf{x}_i$ are the quality and position of each potential location. If multiple (n_w) infill well locations $\mathbf{x}_{\text{pos}} = \{x^{(1)}, x^{(2)}, \ldots, x^{(n_w)}\}$ are to be chosen, one can rewrite the optimization problem as

$$\max_{\mathbf{x}_{\text{pos}}} \sum_{n=1}^{n_w} \sum_{i=1}^{n_c} \frac{Q_i}{|x_i - x^{(n)}|^2} \tag{8.5}$$

The variance of the posterior quality maps can be used as an indication of the uncertainty associated with this well placement. A conservative decision maker may choose to avoid placing infill wells in regions where there is significant deviation between posterior realizations. This can be implemented by incorporating additional terms that accounts for this variance into Eqs. (8.4) and (8.5).

For the Libyan case, this methodology was applied to the mean posterior quality map image. The resulting location for a single new infill well is shown in Figure 8.13. By overlaying the selected location with the "true" quality map, we observe that the largest high-quality region was indeed selected as the well location.

*8.2.4.3. **Abandonment.*** For a set of operating expenses and oil prices, an economical model can be used to evaluate the required production of a field for it to remain economically viable. Having computed posterior distributions of field production, we can then evaluate an expected time of when the field should be plugged and abandoned, under that required production rate. Various economic scenarios can be explored by repeating the decision model with different oil prices or operating expenses. For instance, suppose it was determined that the Libyan field requires at least 600 stb/day to be economically viable. According to our prediction, the P10, P50, and P90 of when the field will drop below this rate are 7897, 7136, and 6574 days, respectively (see Figure 8.11).

8.3. DECISION MAKING UNDER UNCERTAINTY FOR GROUNDWATER MANAGEMENT IN DENMARK

8.3.1. A Strategy for UQ

Recall from Chapter 1 the decision problem of re-allocating water extraction from an existing well field to a new well field. Four locations are available as choices for re-allocation. We need to balance return with risk: (return) a successful operation that restores streamflow and wetland, (risk), not achieving the desired re-allocation, and presence of pollution from industrial or farming sources.

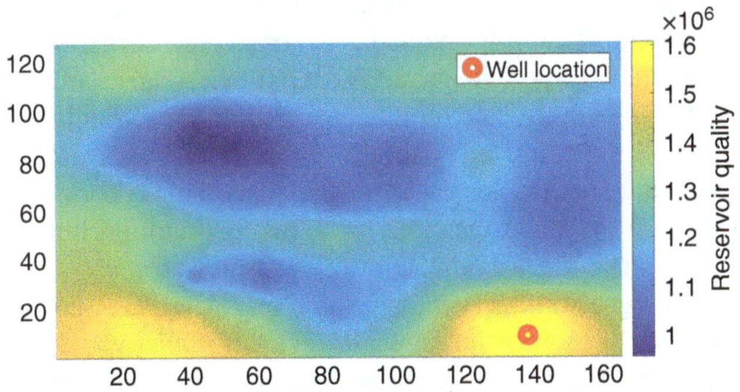

Figure 8.13 Location of infill well selected according to the predicted posterior quality maps and variance, superimposed over the "true" quality map.

The decision model will depend on the prediction of five target variables: (i) caused drawdown, (ii) streamflow reduction potential, (iii) wetland reduction potential, (iv) overlap between well catchment and farmland use, and (v) overlap between well catchment and industry use. To generate the posterior distributions of these five variables, a groundwater model that includes all uncertainties will be built. Model variables are informed by the various data sources available, such as geophysical data, head data, and streamflow data, but such data is subject to error/uncertainty. Our strategy to reach a decision is described in Figure 8.14.

This strategy is summarized as follows. Based on some initial estimates, expert opinions, geological understanding, and geostatistical modeling, we will build a prior model for all uncertainty components in the groundwater model. This prior model will also include any measurement error. This prior model will be used in a Monte Carlo study to generate prior realizations of the data (head and streamflow) and the five prediction variables at the four proposed locations. Next a sensitivity analysis is performed for both data and prediction variables. This sensitivity analysis will provide insight into what

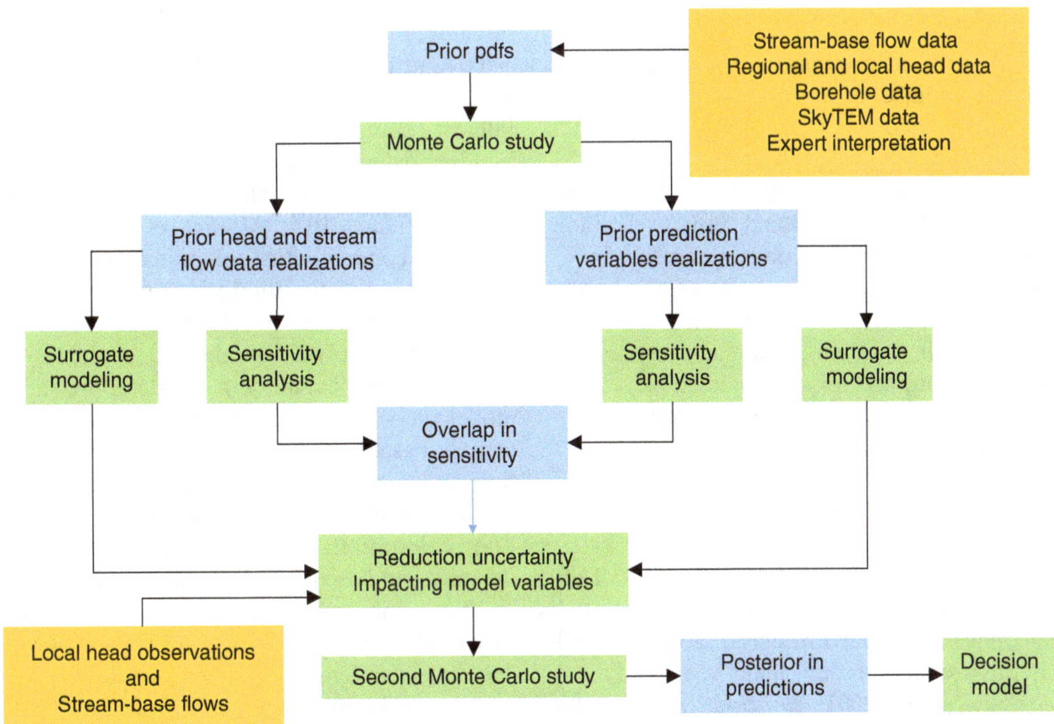

Figure 8.14 A strategy to come to a decision in the Danish case.

modeling components impact both the data and the prediction variables. Only those modeling components that (i) impact the prediction and (ii) whose uncertainty can be reduced with data are of importance in a further analysis. Using the same Monte Carlo study, we build two surrogate models using tree-based regressions. The first regression tree models the relationship between data variables and impacting model components and the second regression model focuses on the relation between the five prediction variables for each location and the model variables. The surrogate models are used to reduce uncertainty on those model components informed by head and streamflow data, the second surrogate model allows performing a second Monte Carlo with the reduced uncertainty model components that then generates the posterior distributions of the prediction variables. The second surrogate model voids any further running of the (CPU-demanding) groundwater models. The resulting posterior distributions on the prediction variables are used in the decision model. Each of these components are elaborated in the next sections.

Why this strategy?. Several reasons have motivated our particular choices:

1. *The SkyTEM data.* This data is of very high quality, certainly for groundwater applications. It is at par with the quality one can expect of seismic data in petroleum reservoir systems. This means that the model architecture (lithology) is reasonably well-informed and hence can be modeled using geostatistics based on a smooth inversion (see Chapter 6).
2. *The complexity of the groundwater model.* Because of the availability of significant amount of data resources spanning multiple decades, we can afford to build a complex model that is well informed. However, complex models come with a large amount of variables. This means that we need to understand how well these variables impact data and prediction responses.
3. *The flow simulation time.* Because of the model complexity and the large simulation times, surrogate models based on full flow simulation are needed for conditioning (inversion) and forward prediction.

The following methods will therefore be used:

1. PCA (Section 3.5.1)
2. MPCA (Section 7.2.3)
3. Deterministic Inversion (Section 6.4)
4. Regression Trees (Section 3.7.4)
5. Approximate Bayesian Computation (Section 3.12.6)

The aim here is not to provide a detailed account of all geological, geophysical, and hydrological modeling work performed in this area which spans many years and many papers [*Henriksen et al.*, 2003; *Jørgensen et al.*, 2003, 2015; *Sandersen and Jørgensen*, 2003; *Thomsen et al.*, 2004; *Jørgensen and Sandersen*, 2006; *Foged et al.*, 2014; *Hoyer et al.*, 2015; *Marker et al.*, 2015]. Instead, we illustrate how such expansive modeling work can be integrated into the workflow of Figure 8.14 to reach a specific local decision.

8.3.2. Designing a Monte Carlo Study

8.3.2.1. The Groundwater Model: Overview. In creating any subsurface model, we need to specify various modeling components largely divided into boundary conditions (BCs) and the geological variability as expressed in lithology, porosity, and hydraulic conductivity. The present groundwater model has common basic elements: recharge, BCs, pumping conditions, stream segments, and surface runoff. Although we are interested in a smaller area near Kasted (see Figure 1.5), this local area cannot simply be isolated from the larger-scale regional variation in groundwater flow. A large model is, thus, built to account for distant outer boundaries and to avoid any adverse effect due to the abstraction from large well fields on water balance. This is certainly the case in heterogeneous systems containing buried valley structures, where hydraulic responses can spread over long distances [*van der Kamp and Maathuis*, 2012]. For that reason, a larger-scale *regional model* was built that can then model the BCs (fluxes) of the *local model*. In terms of modeling effort, there is a fine balance between efforts spent on the local vs. regional model. For this local decision problem, it is likely that local details will matter most, but that the impact of BCs can be modeled through a coarser regional model. This regional structure can then be re-used for several different local decision problems.

To limit the computational burden, and since effects of structural heterogeneity is largest close to the target of interest (pumping wells in this case), it was decided to build a deterministic regional-scale lithological model using the resistivity–clay fraction clustering method [*Foged et al.*, 2014; *Marker et al.*, 2015]. However, hydraulic conductivities are treated as uncertain variables in the deterministic lithology structure. In other words, any spatial uncertainty related to lithology was deemed sufficiently resolved within the regional model (but not within the local model). The deterministic lithological model was obtained in a fashion similar to Section 6.6.3, namely based on rock physics calibration. Such calibration requires the lithological description from boreholes and the geophysical data inverted to a resistivity model. The calibration results in a clay-fraction model. Geophysical-based lithological variation is then created by clustering several properties (in the present case clay-fraction and resistivity) into a set of groups, each denoting a different lithology. A total of 11 lithologies were created (see Figure 8.15).

The local scale and the regional model were simulated jointly using MODFLOW-USG [*Panday et al.*, 2015], to avoid adverse boundary condition effects from the

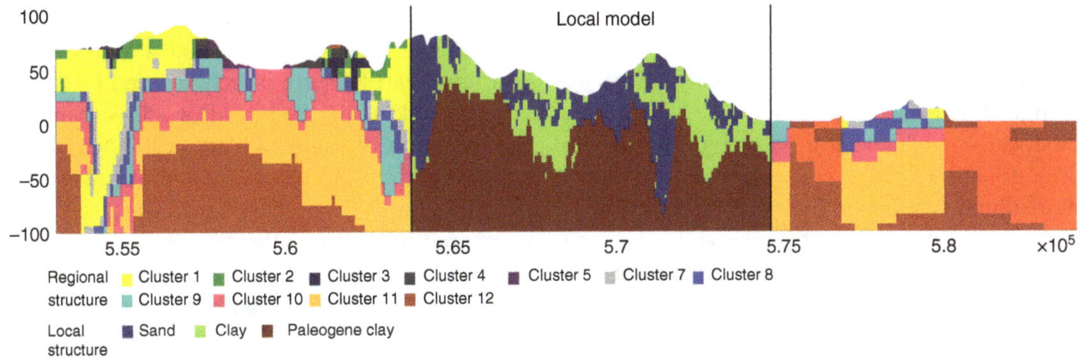

Figure 8.15 Embedding of the local "kasted" model into the regional model. The local model lithologies are stochastically varying, while the regional model is deterministic (with uncertain hydraulic conductivities).

Table 8.3 Summary of parameters and uncertainties.

Parameter name	Parameter code	Amount	Type of uncertainty	Established from
Local model architecture	ma	1	Scenario	Geophysics wells
Regional and local hydraulic conductivity	Kh	22	Log-normal pdfs	Head data Well data
Head boundary conditions	ch	5	Uniform	Experience
River	riv	8	Conductance log-normal DEM: uniform	Experience from previous studies
Drain	drn	8	Conductance log-normal DEM: uniform	Experience from previous studies
Recharge	rch	1	Trapezoidal	Base-flow estimates

interface between the two models. The regional scale model had a refinement of 100 m by 100 m, and the local scale model had a resolution of 50 m by 50 m. Both model areas have 11 numerical layers. Major stream segments were simulated as rivers, and drains were used to simulate surface runoff and runoff from minor trenches and stream segments that may run dry. Besides these, model recharge was simulated using the recharge package, and pumping wells were simulated using the well package in MODFLOW-USG.

Two sources of uncertainty are associated with rivers and drains. Drains are estimated to be located on average at a depth of 1 m below the terrain. The terrain elevation was determined based on a 10-m resolution digital elevation model (DEM), taking the average elevation of the DEM model within the groundwater model cell. The uncertainty of drain depth was modeled using a normal distribution with a standard deviation of 0.3 m. The elevation of the river was determined from the same DEM model, taking the minimum value of the DEM model −1 m within the groundwater model cell. The standard deviation of error on DEM is also 0.3 m. The connection between the rivers/drains and aquifer is modeled using a conductance term. These conductance terms were also treated as uncertain in the analysis.

A spatially variable recharge was estimated based on the land use within both the local and the regional area. This recharge was also treated as an uncertain model variable. It was, however, assumed that the spatial patterns were well known, but not their global level; hence, a constant but uncertain perturbation was applied over the entire area. This uncertain perturbation was modeled using a trapezoid distribution with a lower zero value of 0.6, a lower maximum value of 0.75, a higher maximum value of 0.85, and a maximum zero probability of 1.0.

Approximately half of the outer BCs of the regional model have constant heads, and the remaining are no-flow. This constant head BC is also treated as uncertain, a standard deviation of 0.75 m, was used except for those boundaries at the coastline, where the standard deviation was taken as 0.1 m.

Given this initial setup the following groups of parameters are considered (Table 8.3):

1. *Local model architecture.* This models the uncertainty on the location and the internal structure of the buried valleys. The model architecture will be generated using

geostatistics and constrained to the SkyTEM data. This uncertainty will be discussed in the next section.

2. *Local and regional hydraulic conductivity*. Here we need to distinguish two models:

(i) The local model: each lithology has different hydraulic conductivities.

(ii) The regional model: the hydraulic conductivities assigned to the architecture at the regional scale will influence the fluxes into the local model. The prior distributions include measurement error of the head values.

3. *Rivers*. two groups of model parameters are considered:

(i) River bed conductance: how the river bed is connected to the subsurface.

(ii) River bed elevation: estimated from a DEM, which has its own uncertainties, modeled as constant but unknown perturbation on river bed elevation.

4. *Drains*. It includes simulated surface runoff, runoff from minor trenches and stream segments that may run dry. Drains require conductance and elevation, each subject to uncertainty.

5. *Recharge*. This is water moving from surface water to groundwater and is a function of other processes such as evapotranspiration and infiltration. We assume that the spatial variation of the recharge over this entire area is fixed but multiplied with a constant but unknown scalar.

6. *Regional scale outer BC*. The constant head BCs of the regional scale model are subject to uncertainty. These constant head BCs are largely defined by the location of streams, lakes, or the sea. For the streams and lakes, the elevation of these BCs was estimated similar to the streams and drains, as outlined above. This was also the case for the sea, but here the uncertainty was reduced compared to the two others.

8.3.2.2. Local Model Architecture. Uncertainty in the local model architecture reflects uncertainty on the exact position and crosscutting of buried valleys. Recall that such valleys consist of poorly sorted sediments, some consisting of sand, others of clay. This local heterogeneity is likely affecting the groundwater flow near the four alternative well locations in the decision model. To model spatial uncertainty, we need to opt for a geostatistical method that (i) can be constrained to the geophysical data, (ii) honors the interpreted crosscutting relationship and valley structure, (iii) can be constrained to any borehole data, and (iv) generates multiple realizations relatively fast.

The workflow to achieve this is shown in Figure 8.16. This workflow accounts for both the uncertainty in the geophysical data due to incomplete sampling and the spatial uncertainty of the lithologies. First, one notices that the geophysical data does not cover exhaustively

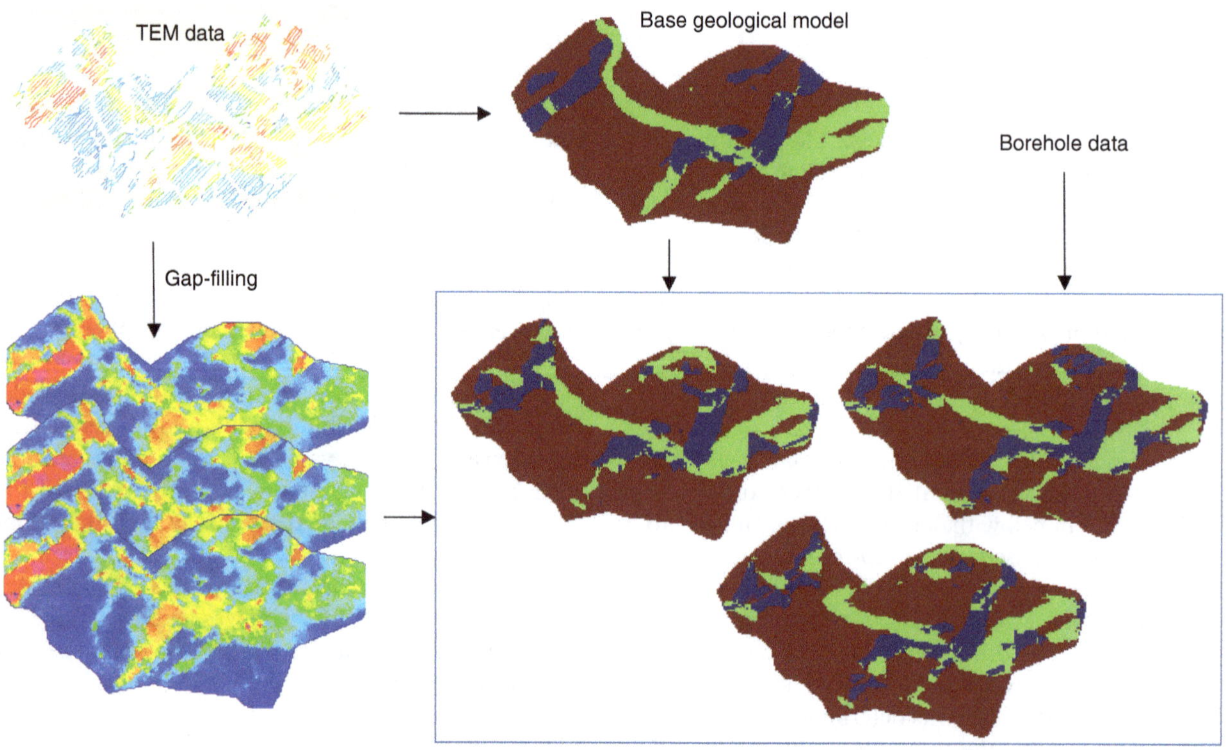

Figure 8.16 Workflow for generating multiple 3D realizations of the local model architecture. Two major uncertainties need to be addressed: (i) gap-filling of the SkyTEM data and (ii) spatial uncertainty related to the cross-cutting of buried valleys of different age.

the area of study because (i) the flight path consists of 1D lines and (ii) the presence of power-lines, cities, and other noise sources produce electromagnetic interference that leaves a gap in this dataset. These gaps therefore need to be filled. A spatial interpolation method such as kriging would fill the gaps but does not deliver an uncertainty model or account for the fact that the geophysical data clearly displays channel-like features. For that reason, we use a MPS method, namely direct sampling [*Mariethoz et al.*, 2010]. Direct sampling preserves the channel-like structures and generates multiple realizations of the gap-filled geophysical data. We refer to *Mariethoz and Caers* [2015] for an example of this type of gap-filling applied to satellite data.

To generate multiple lithological realizations, we again rely on direct sampling. Direct sampling requires the existence of a training image that reflects the geologists' rules and interpretation of the buried valleys. To create such training image, a geological expert [*Hoyer et al.*, 2015] makes a single detailed interpretation of the geophysical data (see Figure 8.16). We retain only three major lithologies from this interpretation: meltwater clay (red), clay till (green), and meltwater sand (blue). Important here is that the buried valleys consist both of a permeability lithology (blue) and a much less permeable lithology (green), whose small-scale variability cannot be deterministically discerned from the TEM data. Direct sampling uses the gap-filled geophysical data (as soft data) and the training image (the spatial variability) to generate lithological model realizations. Each new lithological realization uses a different gap-filled geophysical image realization. A total of 50 local model architecture realizations were generated. Thereby both sources of uncertainties are accounted for.

8.3.2.3. Data and Prediction Variables. The models defined so far account for all data sources except for the 364 head measurements and the three streamflow measurements (see Figure 1.5 for their locations). The latter could possibly further reduce uncertainty on the local model variables. To account for head and streamflow data, inversion methods are often used to determine the model parameters, in particular hydraulic conductivity. Here the problem is much more sophisticated than simply inverting hydraulic conductivity from head data since the local model of uncertainty includes the uncertain lithological model as well as the BCs, which are uncertain because of uncertainties in the regional model. A standard inversion method, deterministic or stochastic, would be difficult to apply.

Therefore, instead of focusing immediately on inversion, we first explore the problem by means of a sensitivity analysis. This analysis will help in understanding not only what model parameters influence the data response (useful for inversion) but also the target prediction variables. This joint sensitivity analysis will aid in understanding to what extent the data are informing the target predictions.

Sensitivity analysis requires defining "input" variables and "output" variables. The input variables are summarized in Table 8.3. Consider as output first the data variables (head and streamflow). Using Monte Carlo simulation, 1000 realizations of all input variables are generated. The groundwater model is executed for each run to record 364 simulated head data and three streamflow data, as well as the 20 prediction variables (five variables, four locations). Since we need to constrain the model to both head and streamflow data, a joint sensitivity is needed on head and streamflow. Therefore, a PCA is performed on the head data, the streamflow data, followed by an MPCA. This MPCA generates (independent) score variables that model joint variation in head and streamflow data. Figure 8.17 shows a score plot of the mixed PCs. Important here is that the actual field observations (the 364 head and the three streamflow measurements) are captured by the prior model.

Consider the following five prediction variables:

1. WDD: water drawdown (minimize). Because we do not know the geological and hydrological conditions at these four locations, it is difficult to evaluate if the desired abstraction rate is possible. Instead, we look at the drawdown caused by pumping: if the drawdown is extremely high, then it is unlikely that the 20% can be extracted (water is available or not). The drawdown should therefore be minimized.

2. SFRP: streamflow reduction potential (maximize). By moving the groundwater extraction from the well field to a new location, the outflow of groundwater to the stream should preferentially increase.

3. WRP: wetland reduction potential (maximize). Groundwater flow to wetlands should increase, if the groundwater level close to the wetland is partially restored by reducing the groundwater abstraction at the well field. However, if the location of the new well is close to wetlands, or the new location is hydraulically well connected to the wetlands, re-allocation will not have the desired effect.

4. FARM: overlap between well catchment and farmland use (minimize). To secure high agricultural production, farmers often use fertilizers and pesticides. These compounds can potentially pose a threat to the quality of the groundwater, and thereby the water extracted. It is, therefore, of interest to minimize the area of the well catchment which is located in an agricultural area.

5. INDU: overlap between well catchment and industry use (minimize). Similar to the agricultural area, industrial areas can potentially pose a threat to the groundwater

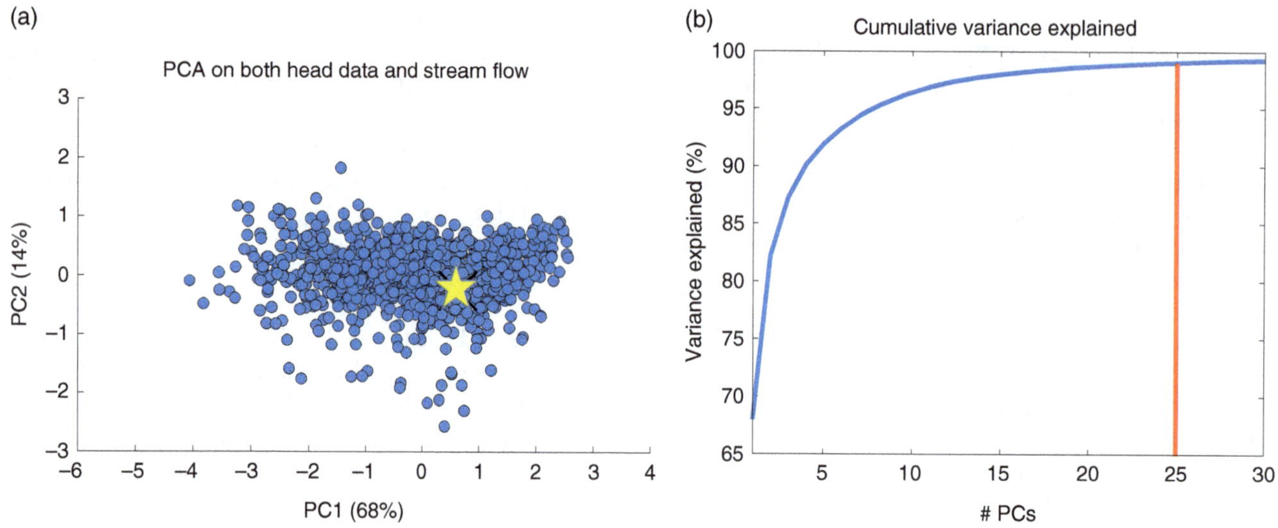

Figure 8.17 (a) Score plot of first two mixed PCs of the head and streamflow data variables. The yellow cross represents the actual field observations. (b) Cumulative variance shows that 25 mixed PCs explain 99% of variation in head and streamflow.

resource. It is, therefore, of interest to minimize the area of the well catchment that is located in industrial areas.

The prediction variables were calculated in two steps. First, the base model scenario was run, using the current groundwater extraction at the well field. This base scenario was used as reference for the predictions. Second, the model was run four times, one for each potential pumping well location. These locations were fixed in the analysis, but the depth of the screening interval was adapted for each new run, such that the pumping well always was screened in the layer of the groundwater model with the highest hydraulic conductivity. For each potential well location, the WDD, was calculated as the drawdown simulated in the cell where the abstraction occurred. The SFRP was calculated as the percentage increase in the outflow to the stream by using the base scenario as reference. The WRP was calculated as the difference in groundwater outflow to the wetland between the base scenario and the outflow caused by moving the abstraction to the new pumping well locations. The FARM and INDU variables were calculated using particle tracking. The catchments for each of the new wells were calculated by backward tracking of particles from the potential new pumping wells to terrain. By extracting the land use from GIS themes, the portion catchment located within farmland and agriculture can be calculated.

To evaluate the effect of the inflow crossing the local model boundary, the water balance of this portion of the model was calculated. Because of the large well fields, whose influence stretches beyond the local model area, the local system cannot be modeled independently from the regional groundwater systems. For each forecast scenario, the *budget* of the local model area (see Figure 1.5) was calculated by determining the inflow and the outflow across the boundary. The influence of the regional water balance can thereby be evaluated in the analysis. The budget variable and its uncertainty, therefore, greatly simplify the representation of uncertainties regarding the connection between the local model and the regional model.

8.3.3. Sensitivity Analysis

To perform sensitivity analysis, we use boosted regression trees (Chapter 4). Trees can be used for two purposes: sensitivity and regression. The regression capability will be used in the next section to generate a surrogate model groundwater flow. One issue that needs to be addressed is the model architecture variable ("ma"), representing the complex spatial variation of buried valley. In order to use this variable in trees (and regression), we need to address the fact that we have 50 different models, hence technically 50 different levels for "ma." Regression trees simply do not work well with that many levels. To address this issue, we first rank the models based on their distance from the base model interpretation (Figure 8.16). The base model is here seen as some "best guess" and the realizations a deviation from it. This makes sense here, because all models are constrained to the high-quality SkyTEM data and so is the base model. Most of the variation in these different model architectures is due to the uncertain spatial interaction between gravel and clay-till (blue and green color in Figure 8.16). As distance, we use the proximity transform distance (see Section 3.5). The distance is used to classify the model architectures

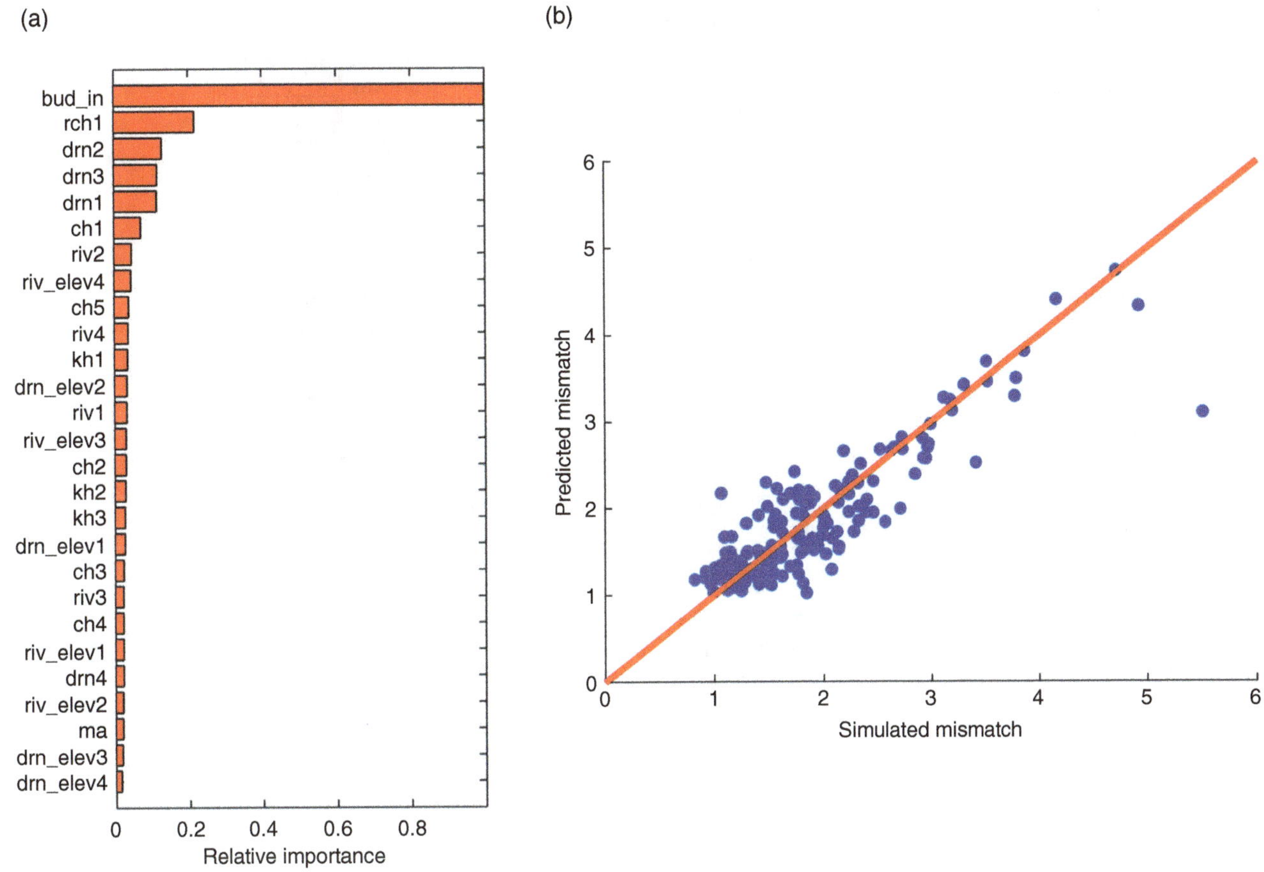

Figure 8.18 (a) Sensitivity analysis of the joint data variables (head and stream). (b) Assessment of how well the tree model predicts the mismatch between simulated data and the observed data.

into five groups. This reduced the problem from 50 model realizations to five model realizations.

We first address sensitivity regarding data variables. More precisely, we calculate sensitivity of model variables with regard to the mismatch between data variables and the actual observed data. In that sense, the regression tree provides insight into how changing certain model variables will affect this mismatch. Second, the regression tree can then serve as surrogate model for the mismatch, which will be useful in inverting the model variables based on the data. Figure 8.18 shows the tree-based sensitivity analysis. An important observation is that the head and streamflow data are dominated by boundary fluxes (budget) and recharge and not by hydraulic conductivity or the lithology model within the local area. Figure 8.19 shows a few of the tree-based sensitivities for the prediction variables. It is encouraging that several prediction variables share sensitivity with the head and flow data, except for the drawdown, which is mostly influenced by hydraulic conductivity and lithological model. Table 8.4 summarizes the sensitivities regarding the 20 prediction variables.

8.3.4. Uncertainty Reduction

The same tree models are now used as surrogate models. The tree for the data variables will serve as surrogate model in approximate Bayesian computation (ABC, Chapter 3), to update the uncertainty on the model variables. ABC becomes feasible because the tree model can be evaluated many thousands of time without much computational effort. In Bayesian computation, we need to define a threshold below which we accept a model realization as matching the data. This threshold here is based on how well the tree fits the actual simulations (see Figure 8.18). Figure 8.20 shows the updated posterior distributions after ABC in comparison with the prior for a few model variables. Logically, sensitive variables are updated, while insensitive variables are not.

The same model realizations accepted by ABC are now used to calculate reduced uncertainty in the prediction variables. Table 8.5 shows the goodness of fit in terms of a correlation coefficient between the simulated prediction and the tree-based prediction. The tree model performs overall very well. A simple Monte Carlo using the posterior

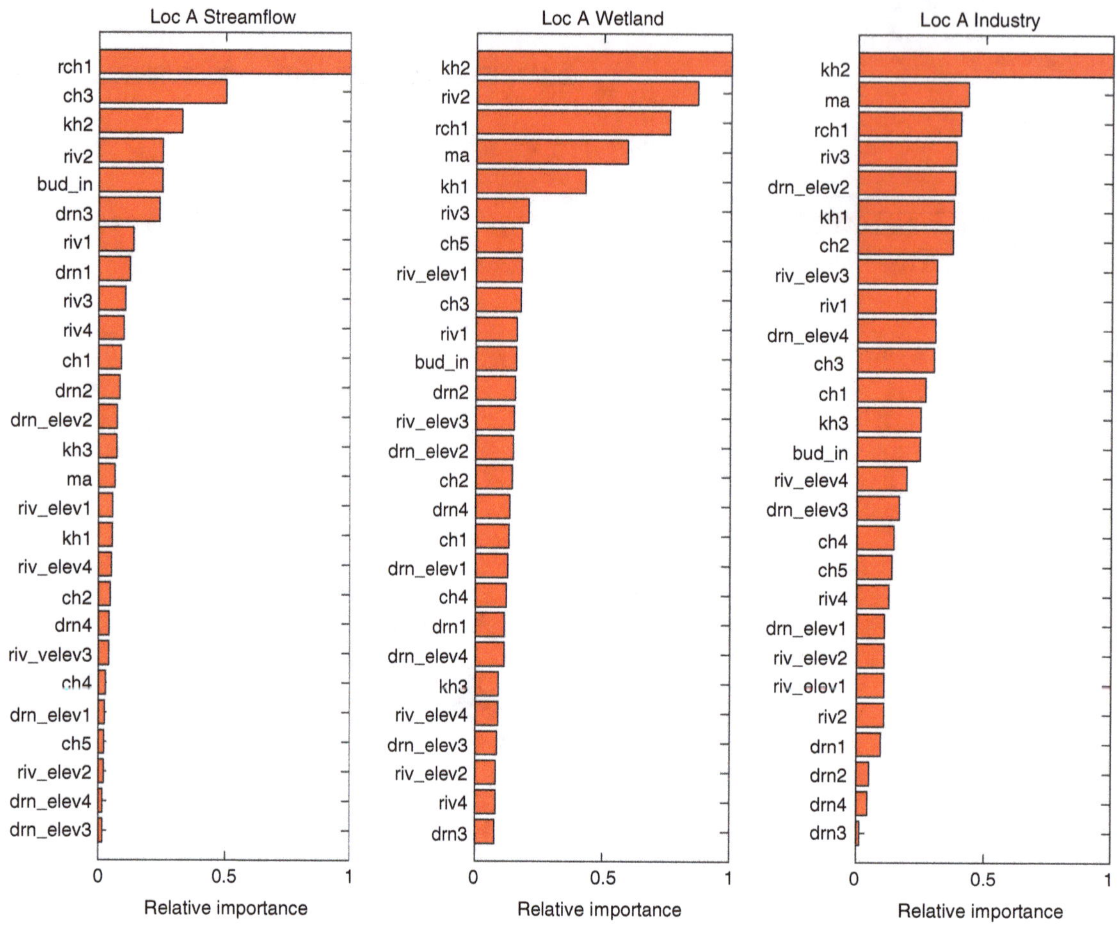

Figure 8.19 Sensitivity analysis for three prediction variables at location A.

Table 8.4 Overview of the most impacting model variables for each location and for each prediction variable.

	Stream SFRP	Wetland WLRP	Drawdown WDD	Farming	Industry
Loc A	Recharge	Recharge	Local K	Local K	Model
	Outer boundary	River	Model	Budget	Local K
	Budget	Local K		Outer boundary	River
	Local K	Model		Recharge	
	Drain			DEM	
Loc B	Local K	Recharge	Local K	Local K	Model
	Recharge	Local K	DEM	Budget	Budget
	River	Model	River		Local K
					Recharge
Loc C	Budget	Recharge	Local K	River	Local K
	Recharge	Local K	Model	Local K	River
	Local K	Model		Model	Model
	River	River		Recharge	
	Drain	Outer boundary		River	
Loc D	Budget	Recharge	Local K	Budget	Budget
	Recharge	Local K	Model	Recharge	Recharge
	River	Model		Local K	Local K
	Local K				

Note: Model = the local lithological model (ma); Recharge = rch1; River = any streamflow-related parameters (riv); Local K = hydraulic conductivity of the local model (kh); Drain = any drain parameter (drn); DEM = digital elevation model (_elev); boundary = outer boundary error (ch parameter).

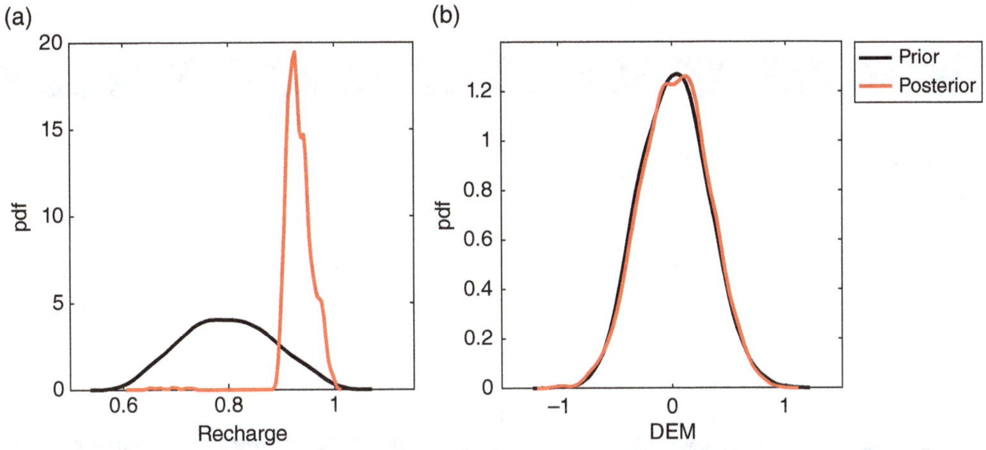

Figure 8.20 Prior and posterior pdfs of recharge (a) and a DEM-related model variable (b).

Table 8.5 Comparison in terms of correlation coefficients between the simulated prediction variables (using MODFLOW) and the tree-based models.

	Stream SFRP	Wetland WLRP	Drawdown WDD	FARM	INDU
Loc A	0.76	0.92	0.98	0.81	0.60
Loc B	0.84	0.89	0.997	0.87	0.83
Loc C	0.78	0.89	0.98	0.84	0.88
Loc D	0.88	0.86	0.97	0.85	0.77

Note: Correlation is between the predictions and the training set for all realizations.

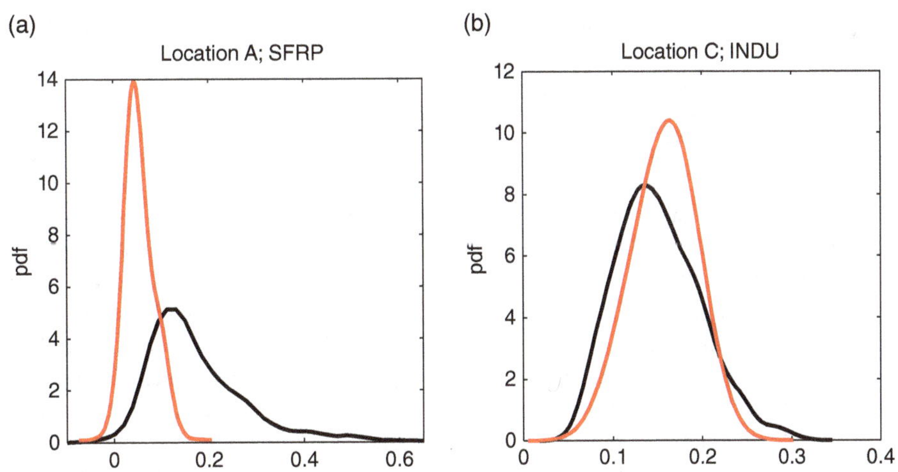

Figure 8.21 Prior and posterior distribution of two prediction variables: (a) SFRP (Streamflow) at location A and (b) industry pollution at location C.

realizations of the previous ABC then results in posterior distributions of these 20 prediction variables (Figure 8.21).

8.3.5. Decision Model

The posterior distributions for the five prediction variables at four locations (a total of 20 posterior pdfs) can now be used in a decision model. Here we follow closely the example case of Section 2.4.5. We need to deal with multiple (conflicting) objectives. As a risk neutral decision maker, we take the P50 of each posterior distribution and construct a table of objectives vs. alternatives (see Table 8.6). From this, we assess which objective best discriminates the alternative to determine the swing weights.

Table 8.6 P50 values of each alternative vs. objective with assignment of the swing rank.

Objectives	Alternatives: Loc A	Loc B	Loc C	Loc D	Best	Worst	Swing rank
Farming pollution Farm (units)	0.895	0.887	0.760	0.616	0.616	0.895	5
Industry pollution Indus (units)	0.066	0.037	0.160	0.227	0.037	0.227	2
Streamflow restoration SFRP (units)	0.048	−0.102	−0.048	0.139	0.139	−0.102	1
Wetland restoration WLR (units)	0.000085	0.00037	0.00037	0.00049	0.00049	0.000085	3
Drawdown WDD (units)	12.40	36.6	15.66	22.74	12.40	36.6	4

Table 8.7 Scoring alternatives vs. objectives, for a risk neutral decision maker.

Objectives	Rank	Weight	Loc A	Loc B	Loc C	Loc D	Type
Farming pollution Farm (units)	5	0.067	0.00	3	48	100	Risk/$ cost
Industry pollution Indus (units)	2	0.267	84	100	35	0	Risk/$ cost
Streamflow restoration SFRP (units)	1	0.333	62	0	22	100	Return/$ benefit
Wetland restoration WLR (units)	3	0.200	0	70	71	100	Return/$ benefit
Drawdown WDD (units)	4	0.133	100	0	87	57	Risk/$ cost
		Total	**56.7**	**40.8**	**42.6**	**61.0**	
		Return-benefit score	20.74	14.08	21.66	53.33	
		Risk-cost score	35.92	26.85	24.16	14.30	

Using linear value functions (a default choice here, but this should be further refined with the decision makers) as a way of stating preference, the various units are transformed into a common scale of 0–100 (see Table 8.7). Based on this information, a total score is calculated for each alternative. Location D receives the highest score.

How can risk be traded off with return? This information is provided through the efficient frontier (see Chapter 2) and calculated by grouping risk (pollution sources and too high drawdown) and returns (wetland & streamflow restoration) (see Table 8.7). Then, we recalculate the scores and plot increasing return vs. decreasing risk (see Figure 8.22). One notices that, although location D has the highest score, it does not score well on risk, as a matter of fact it carries the highest risk (lowest score).

We re-run the same decision model but now for a risk averse decision maker. For example, we take the P10 quantile for returns (to safeguard against not getting as much return) and the P90 quantile for the risks (to safeguard against having more pollution than expected). The risk averse decision maker seems to have more options, as the trade-off between location A and D is more favorable in their case since now location A has more return, and location D has less risk.

8.4. MONITORING SHALLOW GEOTHERMAL SYSTEMS IN BELGIUM

8.4.1. A Strategy for UQ

The previous two cases dealt with geological systems on a 10–100 km with possibly significant amount of information on the subsurface system. We now turn to the use of groundwater as heat exchanger for climatization of building. The problem is now at much smaller scale, 10–100 m, and likely involves much less subsurface information. Here a few boreholes (at best) and time-lapse ERT data. This make sense: because of the small size of the engineering operation, the cost of extensive data acquisition outweighs its benefit. It may, therefore, be tempting to just use a deterministic model to design the system and this is where we start in the next section. However, using a deterministic model leaves many (uneasy) questions. Can we really trust it? What if the chosen parameters differ from the "real" parameters?

A simple deterministic model and some local sensitivity analysis may aid in designing a prior stochastic model with uncertainties hydraulic conductivity and boundary (gradient). Figure 8.23 provides an overview of our strategy. First, a Monte Carlo study and sensitivity analysis on the prediction variable (here the temperature

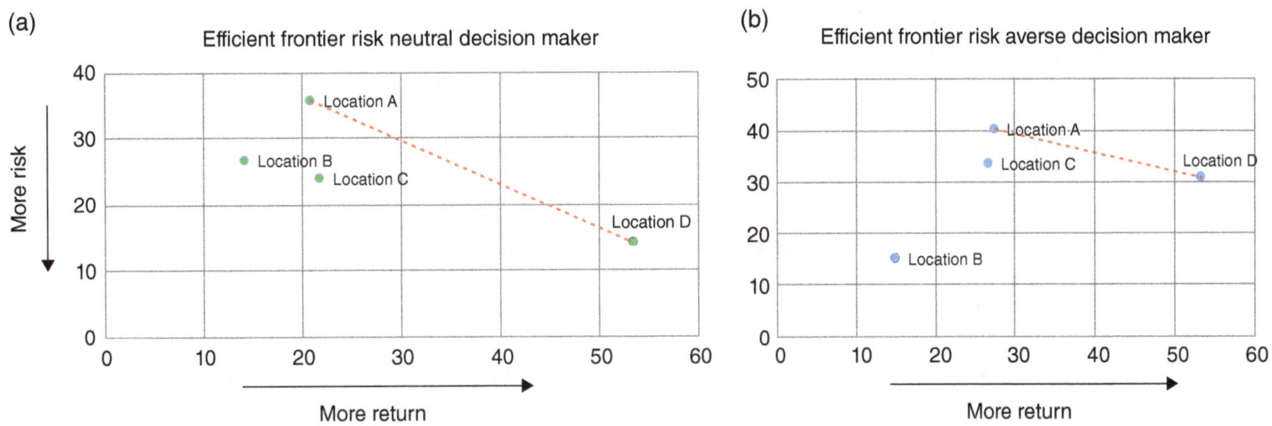

Figure 8.22 Efficient frontiers for a risk neutral (a) and risk averse decision maker (b).

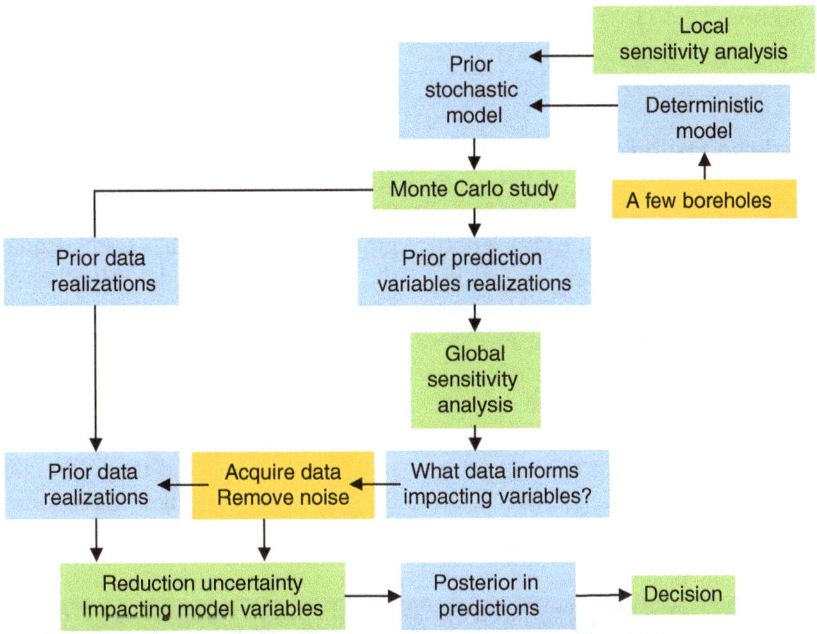

Figure 8.23 Strategy to reach a decision (go ahead with the heat pump) in the shallow geothermal case.

change over time) provides insight into what model components are most impacting the decision variable. This provides insight into what type of geophysical measurements (and its design) inform most impacting model variables. Once that design is known, the actual field data is acquired, and noise removed. Next the data is used to reduce uncertainty on the prediction variables. Here, we use the Bayesian evidential learning (Chapter 7). This requires generating prior realizations of the data variables (ERT) and developing a statistical model that directly predicts the future temperate evolution in the system from the ERT data. The posterior distribution is used to decide whether the groundwater system can be used as a heat exchanger at this site.

Why this strategy?. *Little information on spatial variability*. At best, we get some idea of vertical stratification of the few boreholes available. It would be tempting to simply ignore spatial variability or even variance of hydraulic conductivity and work with a smooth or layered model. However, this would not reflect the impact such heterogeneity may have on ERT measurements and the deduced temperature changes. We, therefore, need to randomize sufficiently the model of spatial variability of hydraulic conductivity. At a minimum, we need to understand its potential impact.

ERT design. Designing time-laps measurements is not a trivial exercise in these kinds of real field applications. To avoid any re-doing because of a failed design, it is important to understand, prior to taking the measurements, whether these measurements will have an impact at all on what predicting temperature change. Part of this design is knowing how long to run the experiment.

Time-lapse inversion. Such inversion is not trivial (see Chapter 6) and involves various steps, such as linking the physical parameters to the measured data, which itself involves modeling. Bayesian evidential learning allows for a simple forward modeling only (no explicit inversion to hydraulic conductivity) and link data directly with prediction, which is ultimately what is needed here.

The following methods will therefore be used:

1. PCA (Section 3.5.1)
2. Deterministic Inversion (Section 6.4)
3. OAT (Section 4.3.1)
4. DGSA (Section 4.4.3.2)
5. CCA (Section 3.5.3)
6. Modeling Noise (Section 7.2.4.1)
7. Bayesian evidential learning (dynamic from dynamic, Section 7.3.3.1)

8.4.2. Deterministic Prediction with Local Sensitivity Analysis

8.4.2.1. A Simple Deterministic Model. UQ does not necessarily require complex modeling from the get-go. Some simple model building may already provide some guidance toward a full UQ and can be useful in cases where very little data is available. A simple deterministic model is not necessarily constructed without a-prior information, but such information is used completely differently from a Bayesian approach. For this case, sites similar to the Meuse river target site are used [*Derouane and Dassargues*, 1998; *Wildemeersch et al.*, 2014; *Hermans et al.*, 2015].

Based on geological information from boreholes [*Wildemeersch et al.*, 2014], a model (the saturated part) is constructed consisting of 14 homogeneous layers each a half meter in thickness (six coarse gravel at the bottom, eight sandy gravel layers at the top). The total model size is 60 m in the direction of flow, 40 m perpendicularly, and 7 m vertically. In that sense, the deterministic model is oriented such that its main axis corresponds to the natural direction of flow which was identified in previous studies. No-flow BCs were assigned to boundaries parallel to this direction. A no-flow BC was also assigned to the bedrock. Between the up and down-gradient boundaries, a natural gradient exists. For the heat transport simulation, a homogeneous initial temperature is assumed equal to the average aquifer temperature encountered in the aquifer. All simulations are done with a control-volume finite element code termed *HydroGeoSphere* [*Therrien et al.*, 2010].

An initial effort consists of performing some model simulations. The base model simulation uses a hydraulic conductivity of 0.05 m/s for the coarse gravel layer and 0.0001 m/s for the sandy gravel layer. The natural gradient on the site has been estimated to be 0.06% from the regional flow model. All other parameters will remain fixed (see Table 8.8).

In an OAT analysis, five simulations with a different gradient are simulated (keep everything else fixed) and four simulations with varying hydraulic conductivity of the coarse gravel layer. Figure 8.24 shows that both the regional gradient and the hydraulic conductivity impact the temperature change. High gradient and high hydraulic

Table 8.8 Modeling parameters for deterministic and stochastic model.

Parameters	Fixed/Variable	Value
Mean of $\log_{10} K$ (m/s)	Variable	U[−4, −1]
Variance $\log_{10} K$ (m/s)	Variable	U[0.05, 1.5]
Range (m)	Variable	U[1, 10]
Anisotropy ratio	Variable	U[0.5, 10]
Orientation	Variable	U[−π/4, −π/4]
Porosity	Variable	U[0.05, 0.40]
Gradient (%)	Variable	U[0, 0.167]
$\log_{10} K$ (m/s) – upper layer	Fixed	10^{-5}
Longitudinal dispersivity (m)	Fixed	1
Transverse dispersivity (m)	Fixed	0.1
Solid thermal conductivity (W/mK)	Fixed	3
Water thermal conductivity (W/mK)	Fixed	0.59
Solid specific heat capacity (J/kgK)	Fixed	1000
Water specific heat capacity (J/kgK)	Fixed	4189

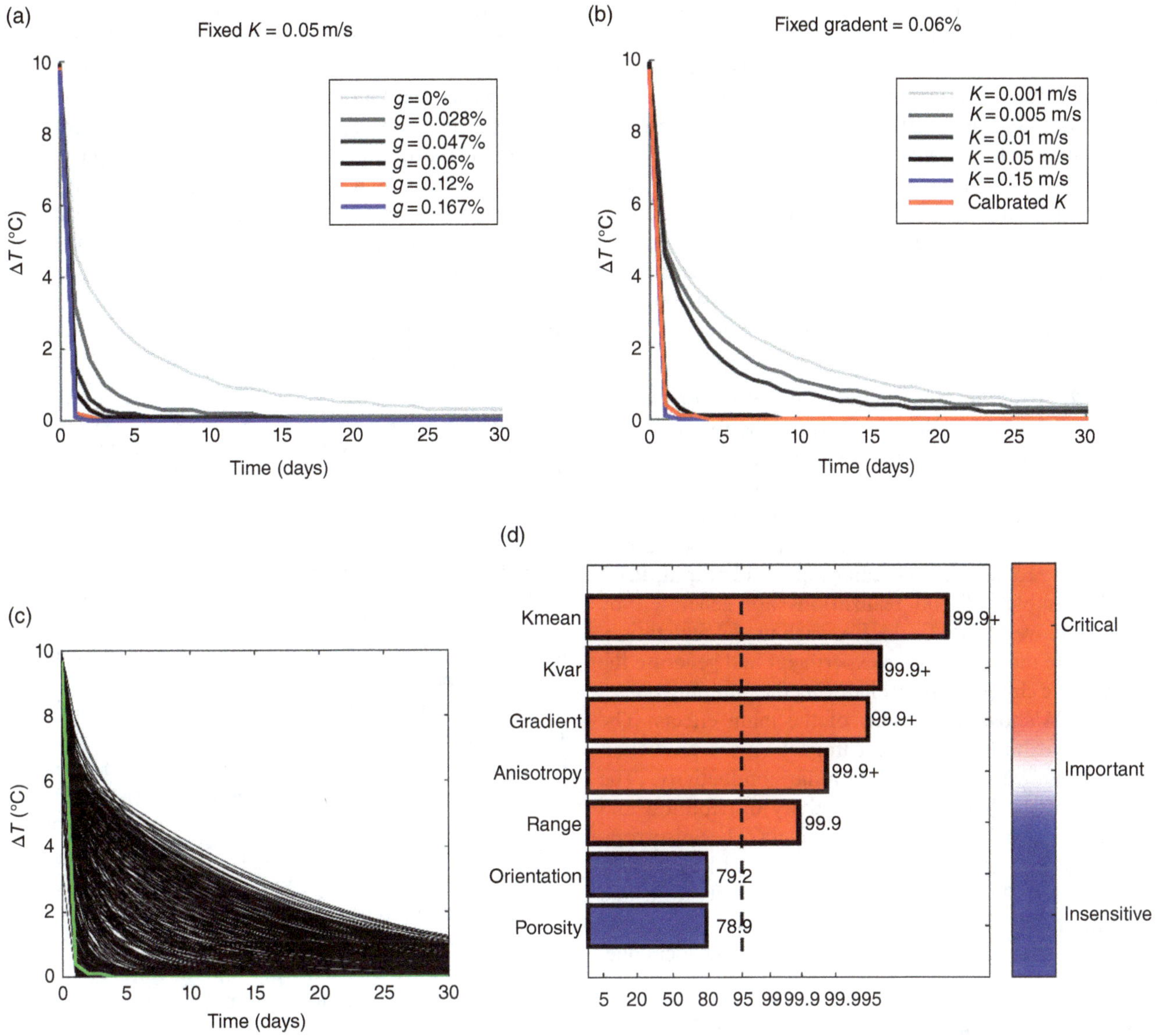

Figure 8.24 Deterministic predictions of the change in temperature during the pumping phase. (a) Local sensitivity of natural hydraulic gradient for a fixed hydraulic conductivity. (b) Local sensitivity of hydraulic conductivity in fixed gradient conditions. (c) Prior samples of the prediction, the green curve is the prediction for a calibrated hydraulic conductivity distribution. (d) Global sensitivity analysis (GSA) of the parameters considered in the prior.

conductivity favor larger groundwater fluxes in the aquifer, moving the heat plume away from the well during the injection phase. Small hydraulic conductivity has an effect similar to the absence of a natural gradient: the thermal energy recovery is much larger.

8.4.3. Bayesian Prediction with Global Sensitivity Analysis

8.4.3.1. Constructing a Prior Model. As discussed in Chapter 5, a Bayesian approach requires a well-thought out prior distribution. In hydrogeological setting with very little data, such as in this particular case (the Danish case is a bit of an outlier), obtaining a meaningful prior may be challenging, but with some research into literature and other resources this is not impossible. The prior model needs to address three main model uncertainties:

1. *Hydrogeological properties.* Uncertainties on hydraulic conductivity and porosity are within ranges found in similar areas of the Meuse site. Note that no direct information on these properties is collected, so the ranges should be quite wide. Uncertainty on porosity may be relevant since thermal properties such as heat capacity and thermal conductivity are a function of porosity.

2. *Spatial variability.* This is modeled using a Gaussian process. This requires a variogram (or spatial covariance).

Very little information is available; hence, we focus mostly on the variogram range which is varied between 1 m (pure nugget) and 10 m (expected radius of influence). Since hydraulic conductivity is anisotropic, we use a scaling factor between vertical and horizontal anisotropy. This scaling factor is uncertain.

3. *Boundary conditions*. the natural gradient was also considered as an uncertain parameter. The gradient was varied between 0 and 0.167%.

In the absence of information, all parameter prior distributions were modeled with uniform distributions, using the principle of indifference (see Chapter 5).

8.4.3.2. Assessing Sensitivity of the Prior Model on Prediction Variables. We now turn to the first question raised in Chapter 1 for this application: Which model parameter is most impacting the prediction of $\Delta T(t)$? This calls for a global sensitivity analysis. Here we use distance-based generalized sensitivity analysis (DGSA) because of its flexibility to handle non-scalar input and output. Prior model variables of 500 realizations are generated by means of naïve Monte Carlo. For each model realization, we simulate the heat storage experiment and generate the temperature change at the well during the pumping phase. Figure 8.24 shows the results of the main effects. The parameter exhibiting the most impact on the prediction is the mean value of the hydraulic conductivity. The variance of the hydraulic conductivity distribution and the gradient are also sensitive parameters. The anisotropy and range also influence the prediction, although to a lesser extent. The influence of the porosity and the orientation on the prediction is limited. The global sensitivity analysis appears to be in line with the local one.

8.4.3.3. Assessing Sensitivity of the Prior Model on Data Variables. Given our insight into sensitivity on the prediction, the evident next question is what data can be used to narrow those model uncertainties that most impact prediction, in particular here hydraulic conductivity and gradient. ERT data is proposed because of its influence on both hydraulic and thermal properties of the aquifer. More specifically, a forced gradient heat tracing experiment monitored with cross-borehole electrical resistivity tomography is considered. Before acquiring such data, the usefulness of the proposed acquisition is investigated. Again, we can use global sensitivity analysis. The reasoning being that a "useful" dataset would at least share parameters sensitivity with the prediction. The proposed lay-out of the experiment is shown in Figure 8.25. The forced gradient lies in the direction of the natural one, so that it enables to fasten groundwater flow without affecting the flow direction encountered in natural conditions.

The heat-tracing experiment was simulated for the same 500 model realizations of the prior set. The temperature distribution was transformed into resistivity variations using a calibrated petrophysical relationship and the resulting change in resistance was computed. The data variables consist of the change in the measured electrical resistance for 410 different electrode configurations at 13 different time-steps; hence, it is of dimension 5330. Because measurements with close electrode spacing are similar and because of similarity in the temporal behavior, we expect a considerable redundancy in this type of data. This becomes clear when performing a PCA, see Figure 8.26, where 10 PC scores explain 98.7% variation in the data. DGSA is, therefore, performed by calculating differences in PC scores (a vector) of various realizations.

The sensitivity results in Figure 8.26 show that the data variables are sensitive to the variance of the hydraulic conductivity distribution and the natural gradient; however, its impact is smaller than for the prediction (heat storage). The reason for this is that the tracing experiment is carried out in forced gradient conditions and not natural gradient.

8.4.3.4. Acquiring Data and Removing Noise. An important issue after acquisition is to deal properly with noise. Overfitting noise data or inverting models without

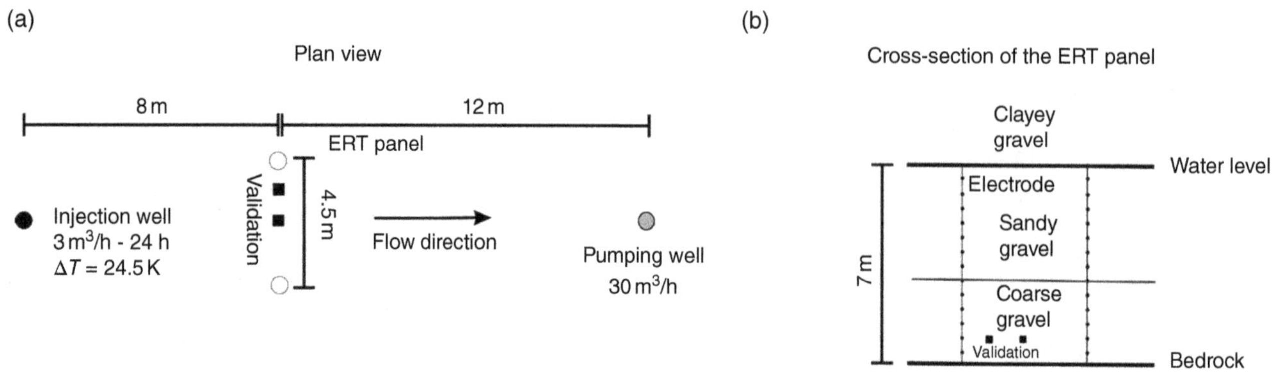

Figure 8.25 (a) Plan view of the experimental setup for the heat tracing experiment. (b) Vertical cross-section of the ERT panel and electrode layout.

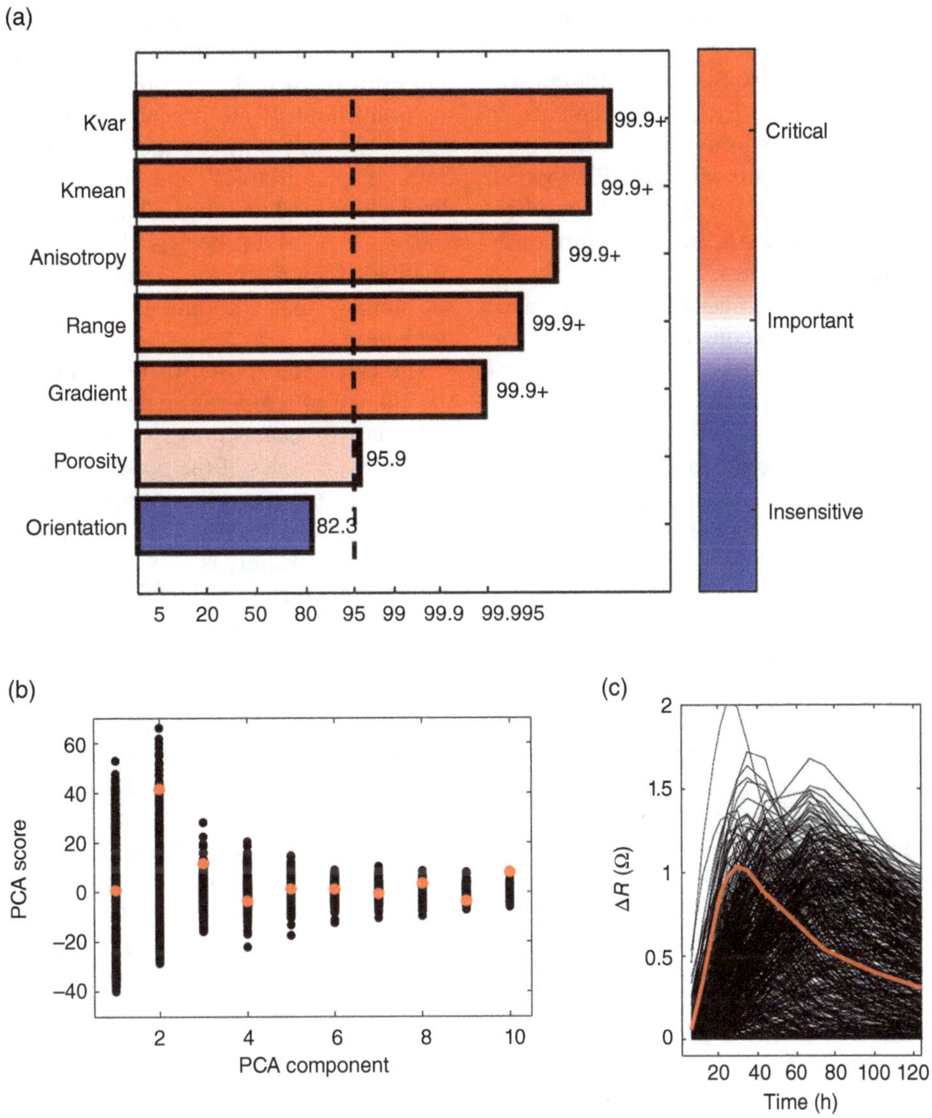

Figure 8.26 (a) Sensitivity analysis of the ERT data on model variables, (b) variation of the PCs, (c) red line corresponds to the field data and black lines correspond to average change in resistance for all data configuration as generated from the prior. The red line and red dot represent the same ERT data.

a proper estimation of the noise component in the data may lead to completely erroneous models [*LaBrecque*, 1996]. Noise is often represented using an error covariance matrix C_d. The issue here is that data is represented in lower-dimensional space using principal component scores. Hence, any noise model assumed on the actual data needs to be translated into a noise model in reduced dimensional space. Because different electrode configurations are affected to various degrees by the noise, depending on the strength of the measured signal, this translation is not trivial and requires a Monte Carlo approach, as outlined in Section 7.3.3.1. This procedure requires first determining some established methods to estimate an error model. In these kinds of experiments, this is done using such as reciprocal measurements in electrical methods [*LaBrecque et al.*, 1996], see Chapter 1. The Monte Carlo study then simply consists of simulating the error and observing how it affects the principal component scores.

8.4.3.5. Attempts to Falsify the Prior. It is critical at this point to attempt to falsify the prior. Just continuing with an untested prior may have disastrous consequences as was discussed in Chapter 5 [*Gelman and Shalizi*, 2013]. Here we can use the acquired data or any other data that is not the target of the study. In fact, using other data can be quite meaningful in both attempting to falsify the prior and the posterior. Bayesian methods require that the posterior solution is part of the span of the prior. The acquired data should, therefore, be predicted by the current prior. We will here verify the prior-data consistency using the entire set of raw ERT measurements, in terms of resistances, and a set in a reduced dimension space. In addition, two piezometers within the ERT panel, located at 1 and 2.25 m from the left ERT borehole (Figure 8.25) were equipped with measuring continuously the temperature within piezometers.

Figure 8.26 shows that the average change in resistance for all electrode configurations is consistent with the prior. We observe that prior models have a similar temporal behavior and similar amplitudes as the data. We now consider the full time-lapse data, but reduced to a few PC score. We can reduce the actual measurements using the same linear (PC) transformation. Figure 8.26 shows how scores of the data lie within the range of the scores produced by the prior model realizations. Again, we cannot falsify the prior.

The attempt to falsify the prior with direct temperature measurements works the same way: data can be compared in physical space with prior or in reduced dimension space. In contrast to ERT measurements, direct temperature measurements are not subject to any integration over a large volume. Figure 8.27 shows the measured temperature variation is captured by the prior, both for the raw data and in the reduced dimension space. These measurements did not falsify the prior. It does not prove the prior "right" or "valid" or "verified," but it makes it a much stronger hypothesis as compared to not doing any attempts of falsification (see our discussion in Chapter 5).

8.4.3.6. Uncertainty Quantification. Since the forced heat-tracing experiment monitored by ERT is informative and prior-consistent, we apply Bayesian evidential learning (see Chapter 7) to predict the temperature at the well during the heat storage experiment. PCA was applied to both data variables and prediction variables. Ten dimensions for **d** and three dimensions for **h** were retained and a CCA established as shown in Figure 8.28. The relatively high correlation confirms the sensitivity analysis, namely that the data is highly informative of the prediction.

The calculated posterior distribution (Figure 8.28) of the change of temperature at the well indicates that the potential for heat storage on the field site is relatively low. Most samples of the posterior have a rapid decrease in temperature once the pumping phase starts. Only a small amount of thermal energy is recovered. The posterior uncertainty is now much smaller than the prior uncertainty, since we established that the data was informative. The geophysical experiment, therefore, suggests foregoing this location as a potential for heat exchange with the subsurface.

One can now go a step further and predict the spatio-temporal distribution of temperature in the saturated part (a 3D = 2D + time variable). The exact same procedure is used. Three different samples are shown at three different time-steps, illustrating the variability of the posterior distribution (see Figure 8.29). They all show a heat plume limited to the bottom part of the aquifer and divided into two distinct parts with a lower temperature in the middle of the section. The confinement of the plume to the

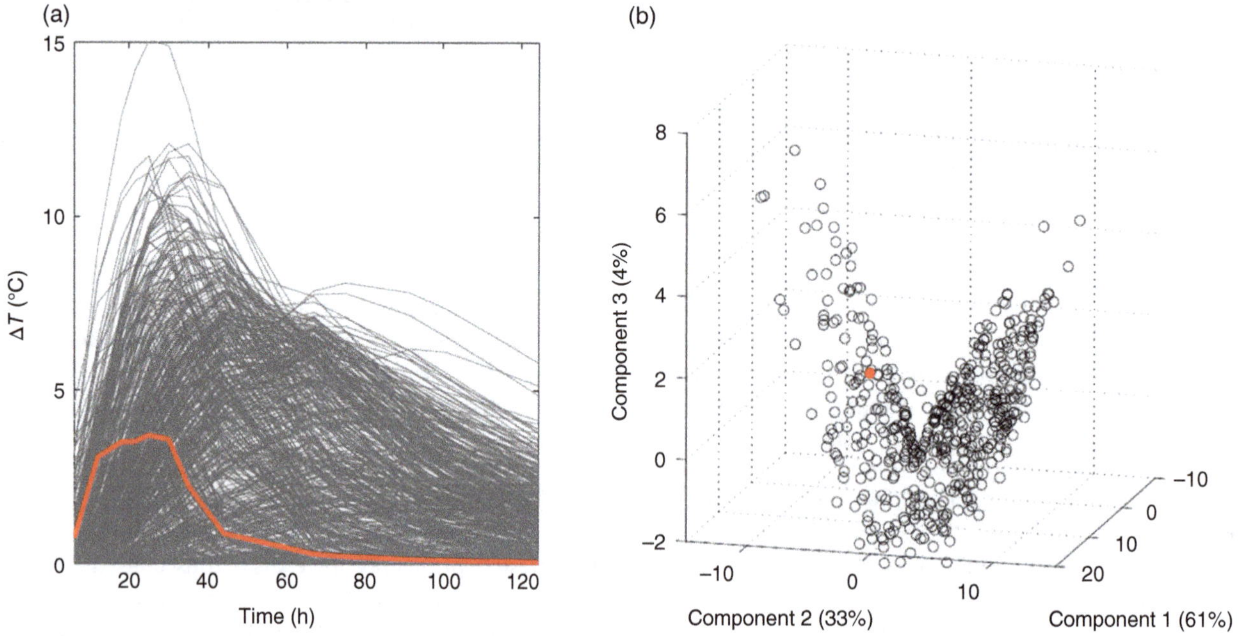

Figure 8.27 (a) Falsification with direct raw and PCA transformed temperature measurements. The red line corresponds to the field data, and the gray lines correspond to the average change in resistance for all data configuration as generated from the prior. (b) The score plot shows that the prior covers the data (red dot).

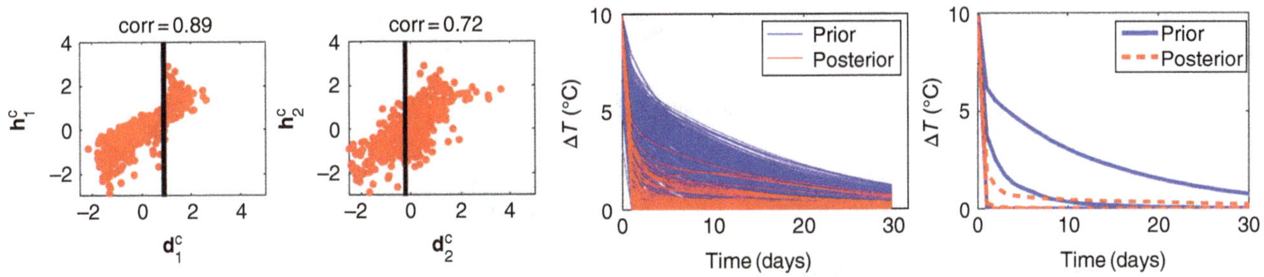

Figure 8.28 First and second dimension of canonical correlation analysis for 5-day experiment. Prior and posterior realizations and quantiles (10, 50, and 90%).

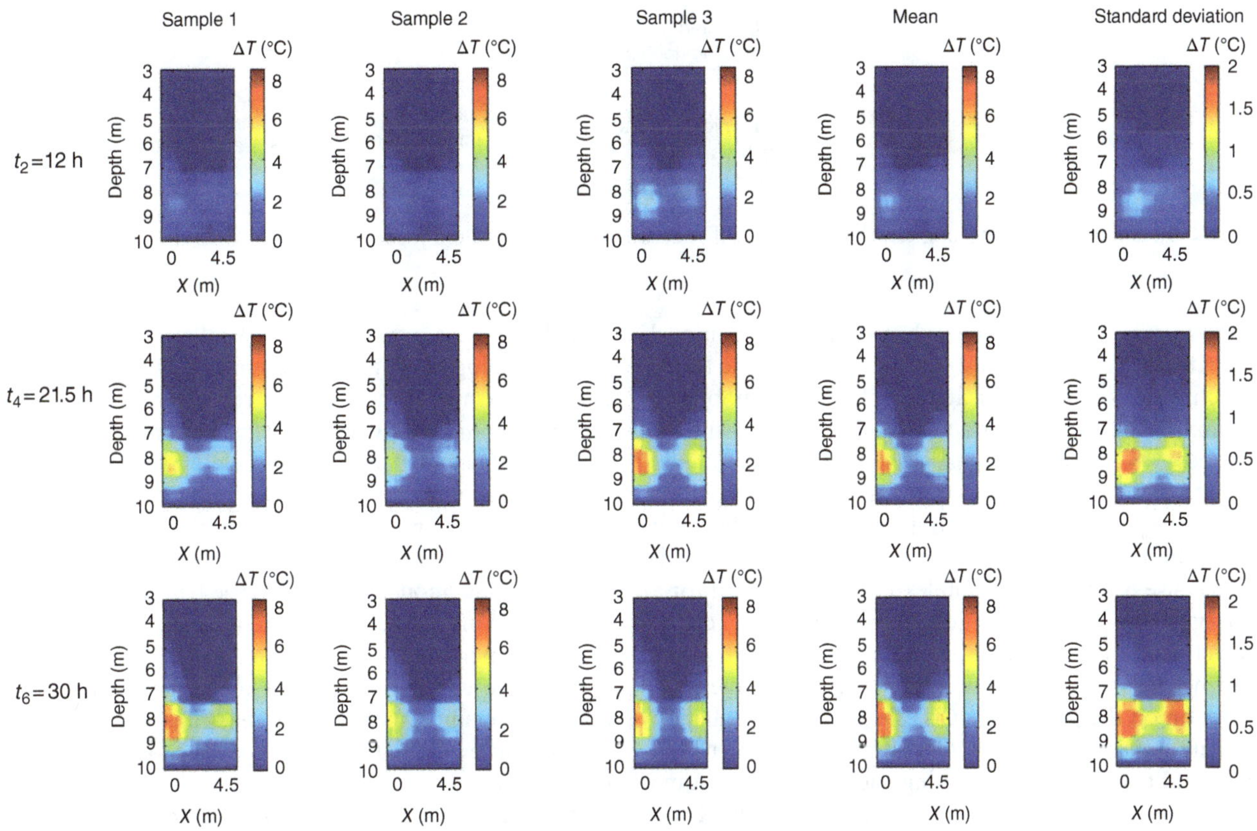

Figure 8.29 Posterior realizations, mean, and standard deviation at three different time-steps.

bottom part of the aquifer is due to the preferential flow in the coarse gravel layer located above the bed rock.

8.5. DESIGNING URANIUM CONTAMINANT REMEDIATION IN THE UNITED STATES

8.5.1. A Strategy for UQ

In this case the aim is to evaluate remediation efficacy of acetate injection to immobilize uranium. Acetate is injected at certain injector locations and the experiment is monitored at several locations by means of tracer concentrations, both reactive and conservative.

This case has elements in common with the shallow geothermal applications except that now the model complexity increases dramatically due to the existence of complex chemical reactions. Many uncertainties (geological, geochemical, and hydrological) need to be addressed. Moreover, it is likely that all these uncertainties interact to create data and prediction outcomes. Figure 8.30 provides an overview of the proposed strategy, summarized as follows.

Based on previous studies, we will formulate prior pdfs on all modeling components, which allows performing a Monte Carlo study. This study will consist of generating multiple model realizations of the joint geological,

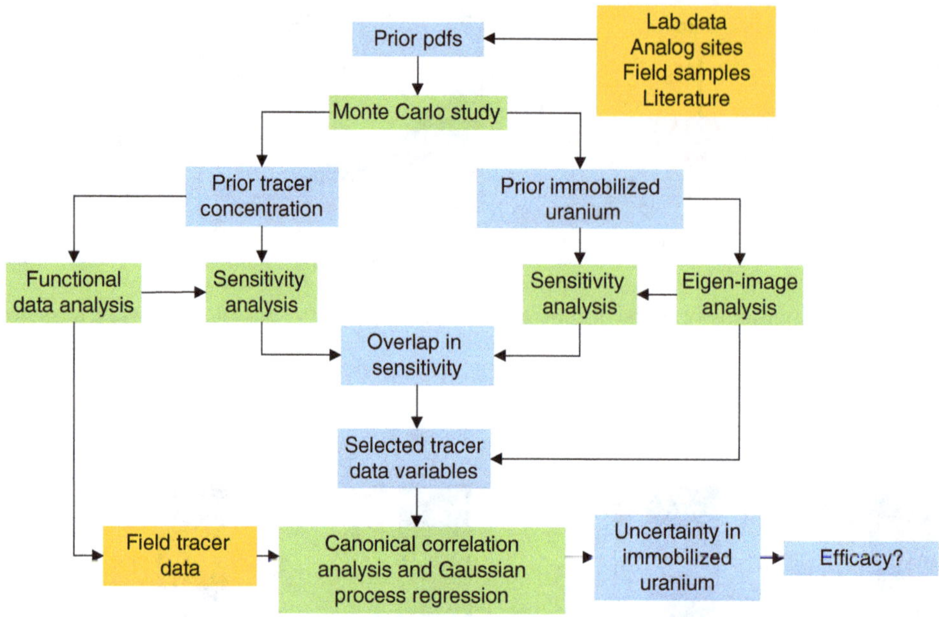

Figure 8.30 Strategy to assess long-term remediation efficacy by means of acetate injection.

geochemical, and hydrological model. These model realizations are then forward simulated to generate tracer concentrations and realizations of the immobilized uranium volume at a certain time (any time could be selected). A sensitivity analysis on both data and prediction variables will provide insight into what data informs best the target prediction: spatial distribution of immobilized uranium. In this case, inversion is impractical because of the model complexity; hence, we opt for a Bayesian evidential learning approach that draws predictions directly based on data. One of the key elements of the workflow is to perform dimension reduction on the prediction variable. Here the prediction variable is a map of immobilized uranium that varies in space and time. Hence, predicting each single space-time location independently is difficult. Instead, we will rely on an eigen-image analysis to reduce dimensionality of images to a limited amount of scores (see Chapter 7). This dimension reduction is needed to perform global sensitivity analysis, in particular DGSA. Directly defining distances between images is non-trivial. Instead, distances are calculated based on the scores after dimension reduction. If a subset of model variables is both sensitive to data and prediction, then a strong canonical correlation is expected. This correlation will be used to reduce uncertainty on the prediction by conditioning to the field tracer data.

Why this strategy?. *The prediction variable is complex.* Successful dimension reduction in the prediction variable is key here. The spatial variation of immobilized uranium is likely to have a non-trivial spatial variability due to existing geological heterogeneity and how that heterogeneity interacts with the chemical reactions.

The forward models are complex and time-consuming. Many of the model components will be interacting to create a response. It would be a poor choice to select one data type (e.g., tracer) and calibrate one subset of model variables while all other remain fixed. Instead, all (impacting model) variables need to be considered jointly in all inversions and predictions. A classical model inversion approach appears impractical, simply because of the large computation times and complexity of the model involving a network of reactions (Figure 8.31). A Bayesian evidential learning approach (Chapter 7) is more suitable.

The following methods will therefore be used:

1. Functional data analysis (Section 3.6)
2. Canonical correlation analysis (Section 3.5.3)
3. Gaussian process regression (Section 3.7.5)
4. Eigen-image reconstruction (Section 7.3.3)
5. Global sensitivity analysis (Section 4.4.3)

8.5.2. Prior Distributions

As in other predictive studies, our setup starts by defining the model parameterization (model variables) and their prior uncertainties. This may require some iteration because we will attempt to falsify the prior by means of the tracer data. A sensitivity analysis may indicate which model variables require larger uncertainties.

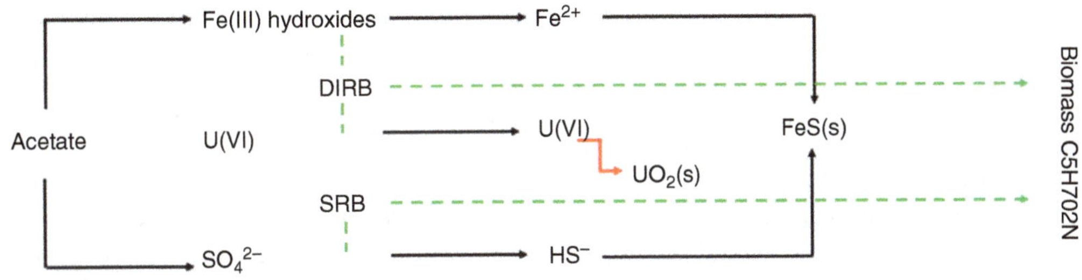

Figure 8.31 Simplified reaction network showing the most important reactions for uranium remediation that takes place when injecting acetate. Modified after *Li et al.* [2011].

Table 8.9 List of uncertain model variables and prior pdfs.

	Parameter name (code)	Parameter type	Parameter distribution
Geological	Mean permeability (Meanlogk)	Continuous	log k~U(−11,−10) m^2
	Variance of permeability(VarioVar)	Continuous	log k~U(0.26,0.69) m^2
	Variogram type of permeability(VarioType)	Discrete	Discrete [1 2 3]
	Variogram correlation length(VarioCorrL)	Continuous	U(3.3,6.6) m
	Variogram azimuth(VarioAzimuth)	Continuous	U(50,90)°
	Mean porosity(MeanPoro)	Continuous	U(0.12,0.17)
	Porosity correlation with permeability(PoroCorr)	Continuous	U(0.5,0.8)
	Method to simulate porosity(MethodSimPoro)	Categorical	Discrete [1 2 3 4]
	Spatial uncertainty(SpatialLogk)	Spatial	500 realizations
Geochemical	Mean Fe(III) mineral content(MeanFerric)	Continuous	U(2.5,10) μmol/g
	Method to simulate Fe(III) mineral content(MethodSimFerric)	Categorical	Discrete [1 2 3 4]
	Fe(III) correlation coefficient with permeability(FerricCorr)	Continuous	U(−0.8,−0.5)
	Mineral surface area(SurfaceArea)	Continuous	U(0.1,2) *base value
	Reaction rate(FerricRate, FerrousRate, UraniumRate, SRBRate)	Continuous	U(0,2) off base reaction rate (varies for different reactions)
	Initial concentrations of different species (ICSulfateJCFerrous, ICUranium)	Spatial	500 conditioned spatial realizations

Note: For reaction rates, see the Reactions (1)–(4) in the text.
*Denotes a reference to that location.

As mentioned in Chapter 1, three groups of uncertainties need specification: geological, biogeochemical, and hydrological. A total of 21 parameters and their pdfs will be specified. Many of these uncertainties are based on sample measurements on site as well as sampling in nearby fields. Table 8.9 lists all parameters considered uncertain for the Rifle case study. Simple continuous parameters follow a uniform distribution (based on the principle of indifference), with bounds collected from previous studies [*Li et al.*, 2011; *Williams et al.*, 2011; *Kowalsky et al.*, 2012]. For discrete/categorical variables or scenarios, each level or scenario is assumed to have equal probability. The forward modeling code for data and prediction variables is *CrunchFlow* [*Steefel et al.*, 2015]. The lateral discretization of the model has cell size of 0.25 m × 0.25 m. The model is 17 m in the direction of flow and 16 m perpendicularly; the model has in total 68 × 64 × 1 cells.

Geological. The area of interest concerns an unconfined aquifer within an alluvial deposit in the floodplain of Colorado River with a thickness around 6–7 m. The water level here is dynamic but located on average at 3.5 m depth. The bedrock is the Wasatch Formation with very low permeability and constitutes the basement of the alluvial aquifer. In this study, the saturated part (2.5 m thick) of the aquifer is modeled as one layer with heterogeneous hydraulic conductivities similar to studies of *Li et al.* [2011] and *Kowalsky et al.* [2012]. We consider five parameters to parameterize spatial variability of permeability (k): the mean of the logarithmic distribution of k and its variance, variogram type, range, and azimuth representing spatial continuity of k. The spatial heterogeneity of permeability is described by one of the following choices for variogram type: Gaussian, spherical, or exponential. The variogram range is varied between 3.3 and 6.6 m following previous study [*Kowalsky et al.*, 2012]. The orientation of the variogram main axis is allowed to deviate from the main flow direction (see below) up to 40°, which is in accordance with the historical flow direction variations. These parameters are used to generate realizations of the permeability field using sequential Gaussian simulation (Section 3.3.5).

Geochemical. The geochemical uncertainties considered here consist of (i) solid-phase mineral content and surface areas, (ii) kinetic rates describing microbial reactions, and (iii) initial concentrations of different ions. Because the Fe(III) mineral is an important electron acceptor at the site that competes with SO_4^{2-},iron-reducing bacteria are thought to be the most important community for mediating reductive immobilization of U(VI). Therefore, spatial variation of Fe(III) mineral content is included. Together with the volumes of solid-phase bioavailable $Fe(OH)_3$, porosity models are also spatially variable. Following the approach of *Li et al.* [2011], a negative correlation between permeability and solid-phase Fe(III) mineral content is enforced when creating the solid-phase models. This negative correlation is based on observations from field samples. The solid-phase Fe(III) content is varied from 2.5 to 10 μmol/g [*Yabusaki et al.*, 2007]. A positive but uncertain correlation between porosity and permeability is assumed. Because of the changes of the variogram type, correlation length, and the method of simulating porosity and ferric solid-phase content, differences in patterns at various scales exist among different porosity and models of solid-phase ferric iron. A complex reaction network following *Li et al.* [2011] is implemented within *CrunchFlow*. The key reactions and species involved in uranium remediation are shown in Figure 8.31. Besides the Fe-reducing microbial reactions (DIRB) that reduce U(VI) to U(IV), the microbial sulfate reduction (SRB) reactions compete with the Fe-reduction pathway to obtain electrons from acetate, due to the high sulfate concentrations in groundwater, and impacts the efficacy of bioremediation. As the initial concentrations of some specific ions (i.e., SO_4^{2-}, Fe^{2+}, UO_2^{2+}) affect the reaction rates, a few heterogeneous realizations of concentration values are imposed. These realizations are conditioned at monitoring wells to initial measurements as reported in *Li et al.* [2011] after the experiment initiated. The in-situ concentrations of uranium range from 0.4 to 1.4 μM (above the standard of 0.18 μM), dissolved oxygen concentration averages less than 0.2 mg/l [see *Williams et al.*, 2011 for details]. Examples of U(VI) realizations are shown in Figure 8.32 reflecting that measurements from samples at wells within the model domain vary spatially [*Yabusaki et al.*, 2007], and hence U(VI) is not spatially homogeneous. Finally, for the prior uncertainty on mineral surface areas and reaction rates, base values from previous studies are used. The upper and lower bounds for reaction rates are set based on the extreme (low and high) uranium reduction cases considered by *Li et al.* [2011].

In *CrunchFlow*, for interactions that involve solid-aqueous phase (e.g., mineral-water) heterogeneous reactions, two separate entries in the database file exist. (i) A thermodynamic entry which gives the stoichiometry of the reaction. The equilibrium constants are a function of temperature, molar volume of the solid phase, and its molecular weight. (ii) A kinetic entry which gives the rate law and rate coefficients for the reaction. *CrunchFlow* currently assumes in all cases that the reaction involves the dissolution of one mole of the mineral or solid phase in investigation (i.e., the stoichiometric coefficient is −1). In our analysis, the thermodynamic entry is kept fixed as

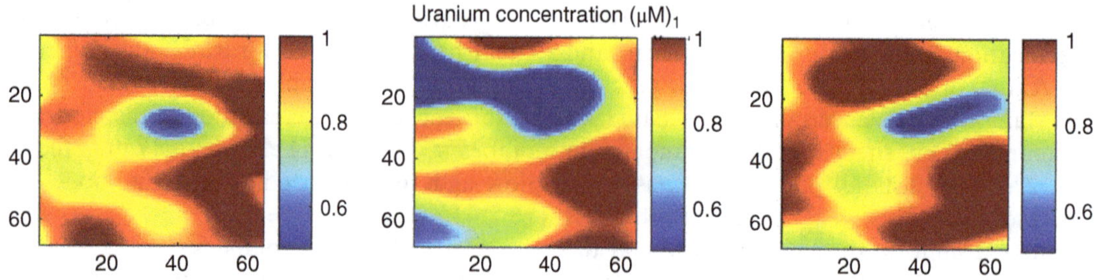

Figure 8.32 Example of heterogeneous realizations of initial U(VI) concentrations conditioned at the monitoring wells. Models are created using sequential Gaussian simulation.

this parameter is normally relatively stable in a shallow subsurface within a short time window. The kinetic entry is varied. For each of the mineral reactions considered, a base rate value and an uncertain offset value vary.

Reactions for which the rates are varied in this study include the following:

$$FeOOH(s) + 1.925H^{+} + 0.033NH_4^{+} + 0.208CH_3COO^{-} \rightarrow Fe^{2+} + 0.033C_5H_7O_2N_{(FeRB)} + 0.25HCO_3^{-} + 1.6H_2O \quad (1)$$

$$Fe^{2+} + H_2S(aq) \Leftrightarrow FeS(am) + 2H^{+} \quad (2)$$

$$UO^{2+} + 0.067NH_4^{+} + 0.417CH_3COO^{-} + 0.8H_2O \rightarrow UO_2(s) + 0.0667C_5H_7O_2N_{(FeRB)} + 0.5HCO_3^{-} + 2.15H^{+} \quad (3)$$

The five reaction types for kinetics in *CrunchFlow* are TST, monod, irreversible, PrecipitationOnly, and DissolutionOnly. For Reactions (1) and (3), monod implementation is used. For Reaction (2), PrecipitationOnly is chosen. For more details, readers are referred to the *CrunchFlow* manual. Monod reactions take inputs of the specification of the activation energy and the various "monod terms." These monod terms indicate the dependence of the reaction rate on electron acceptors and/or electron donors. The relation between reaction rate (R_m) and the monod term is

$$R_m = k_{\max} \prod_i \left(\frac{C_i}{C_i + K_{\text{half}}} \right) \quad (8.6)$$

The quantities in parentheses are the "monod term." Multiple monod terms can be specified, but the most common approach is to provide dual monod form which includes dependences on electron acceptors and donors. In our analysis, the monod terms on acetate is set to be 1e-05. R_m is changed for the three actions with a unit of log(mol)/m^2/s at 25°C.

The base reaction rates for three reactions are (1) −8.2 log(mol)/m^2/s, (2) −6 log(mol)/m^2/s, and (3) −6 log(mol)/m^2/s. An uniformly distributed offset (U(0,2)) for these reactions is added onto the base reaction rates as prior uncertainty model.

In addition, there is an aqueous kinetic reaction considered:

$$SO_4^{2-} + 1.082CH_3COO^{-} + 0.052H^{+} + 0.035NH^{4+} \rightarrow 0.035C_5H_7O_2N_{(SRB)} + 0.104H_2O + 2HCO_3^{-} + HS^{-} \quad (4)$$

A monod reaction type is chosen for Reaction (4), with the monod terms that include acetate (1.0E-4) and sulfate (5.0E-3). Different from the mineral reactions, the unit for aqueous reaction rate is mol/kgw/year. A base reaction rate of 25,000 mol/kgw/year is used with variations being an offset value U(0,2) on the base rate.

Hydrological. The model is oriented such that its main axis (x-axis) corresponds to the overall natural direction of flow, which was identified in previous studies. Therefore, no-flow BCs are assigned to boundaries parallel to this direction. Between the up and down gradient boundaries, a natural gradient is set up as fixed heads at left inlet and right outlet. Fixed gradients can be motivated because of the rather short term of the experiment.

8.5.3. Monte Carlo

Based on the prior model parameters and pdfs described earlier, a Monte Carlo study is conducted. A set of 500 prior model realizations is generated. Each realization constitutes different combinations of all model variables. Figure 8.33 shows some examples of the spatial model realizations of hydraulic conductivity.

Considering that acetate injection for the 2007 experiment only lasted approximately 30 days, the data variables consist of the four tracers measured at 12 monitoring wells over 30 days. The prediction variable is the volume of immobilized uranium at 90 days (longer than the experiment). Figure 8.34 shows prior realizations of tracers for one well and a few realizations of the prediction variable.

8.5.4. Dimension Reduction on Data and Prediction Variables

FPCA was used to reduce dimension on the data variables. The more challenging variable is the prediction variable. Here we use eigen-image analysis (Chapter 7), which is a spatial version of functional data analysis. Figure 8.35 shows the result of such decomposition. The eigen-images exemplify different modes of changes across the ensemble of the mean-subtracted images. For example, the first eigen-image depicts the mean across all the mean-subtracted images; the second eigen-image shows the gradient change across these; higher-order eigen-images reveal higher-order derivatives of the image ensemble. Most changes occurred in the vicinity of the injectors. It also appears that the areal extent of these changes corresponds to the mean correlation length of the permeability field.

Just as with PCA and FPCA, we can use the decomposition to reconstruct the spatial distribution of immobilized uranium from a limited set of scores. Figure 8.35 shows the effect of using different amount of eigen-images in the reconstruction. It appears that with 100 eigen-images (about 80% variance) the difference is acceptable but with some loss of local variations.

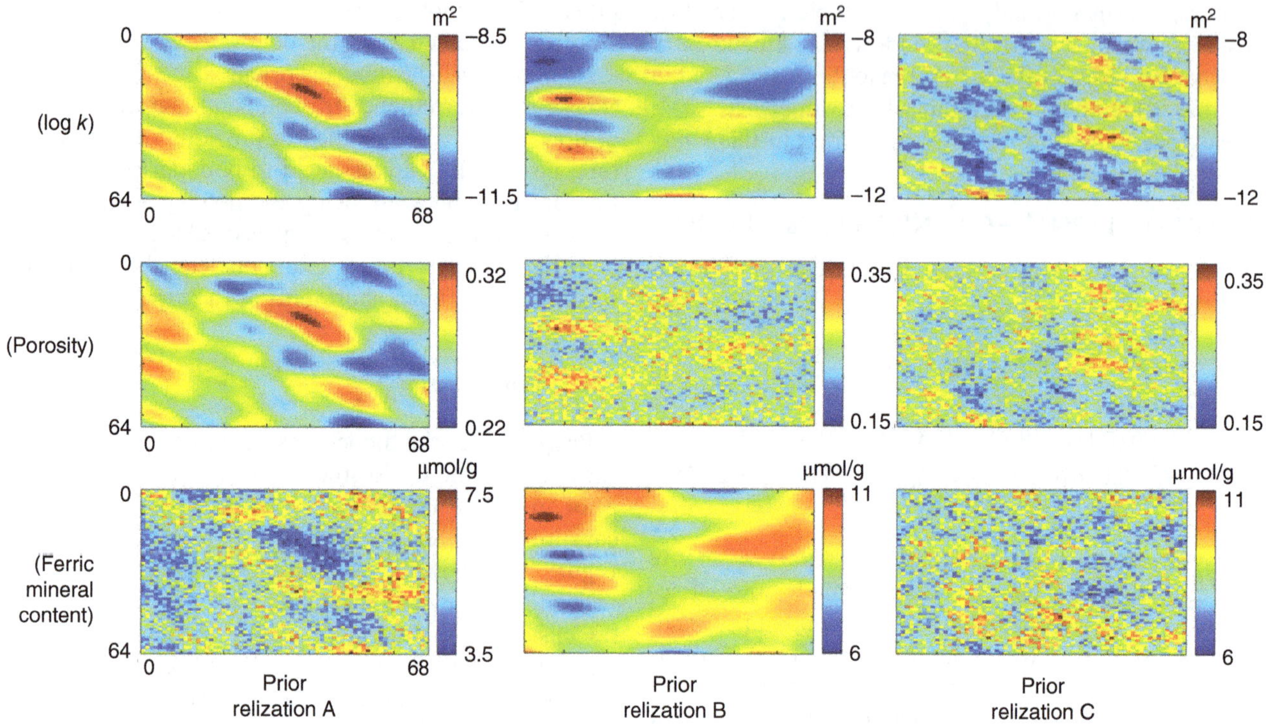

Figure 8.33 Spatial realization of log permeability, porosity, and $Fe(OH)_3$ content.

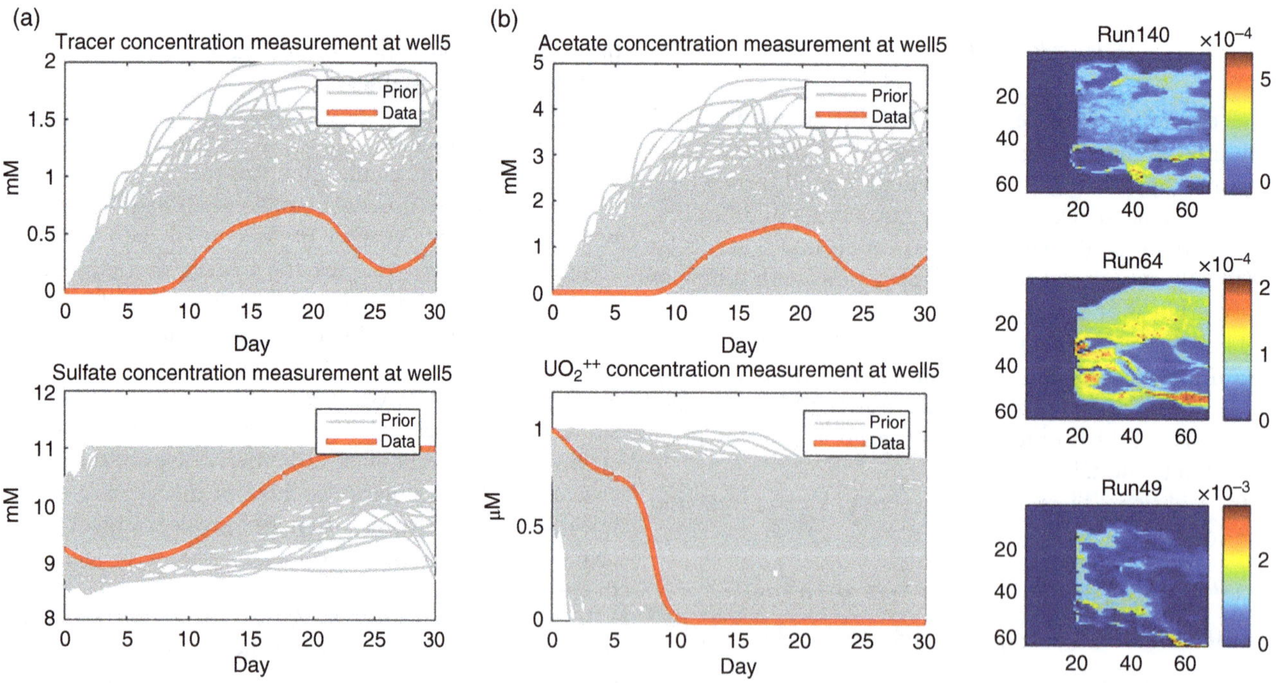

Figure 8.34 Example of realization of the data variables for well5 (a) and the prediction variable, spatial immobilized uranium (b). The gray lines are prior model realization, and the red line is the observed tracer data.

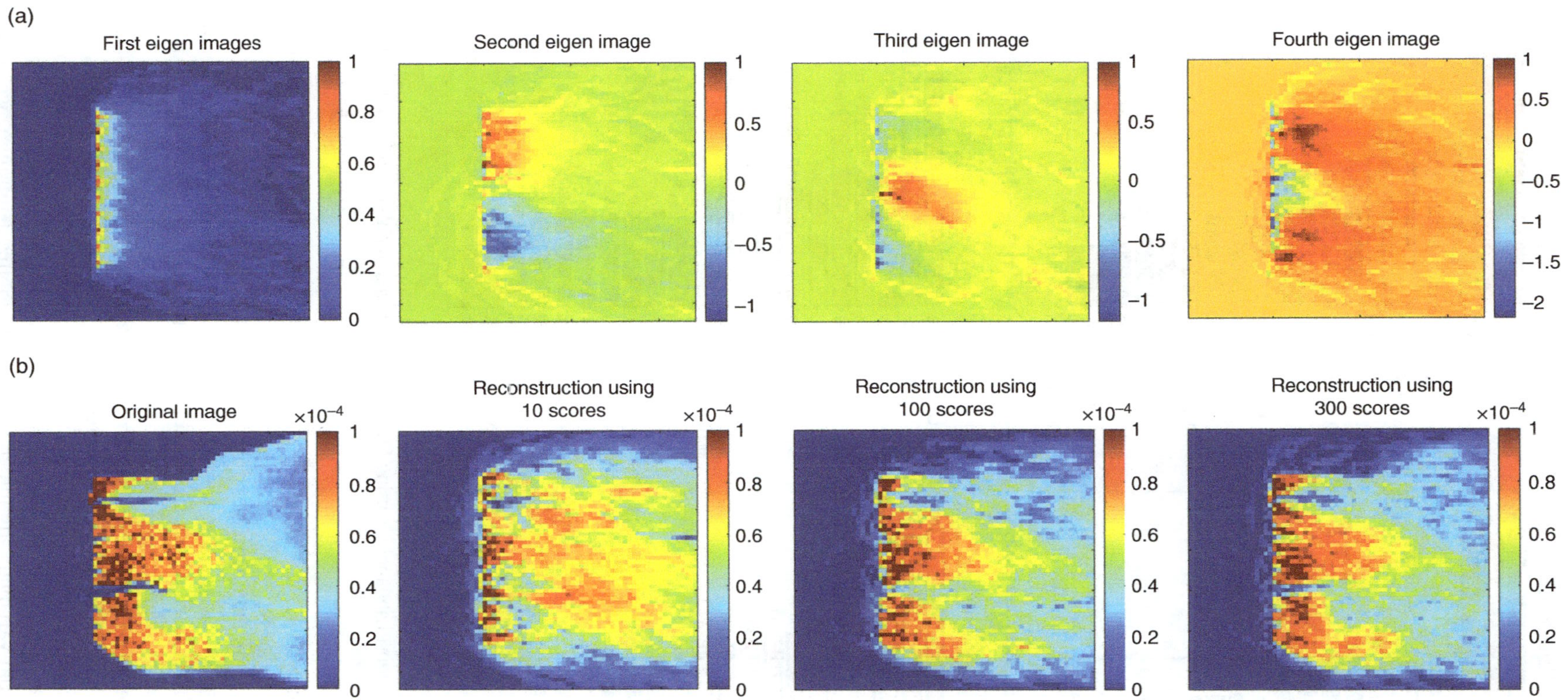

Figure 8.35 Eigen-image decomposition of the ensemble of 500 realizations of immobilized uranium. (a) The first four eigen-images. (b) An example on how to reconstruct images from a limited set of eigen-images.

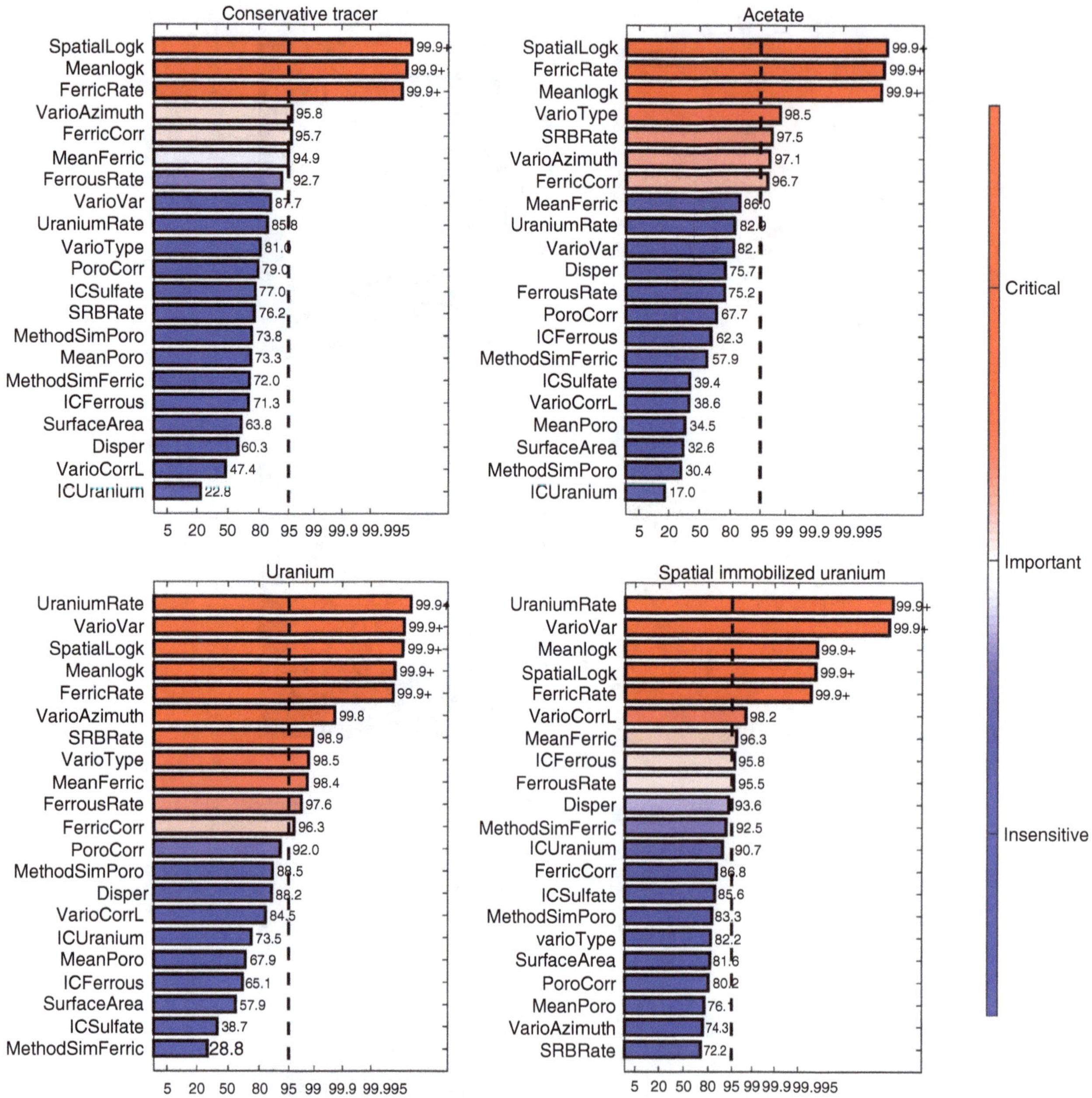

Figure 8.36 Results of a global sensitivity analysis using DGSA on three data variables and one target prediction variable.

8.5.5. Global Sensitivity Analysis

Sensitivity analysis results for data variables for the different species concentration curves are shown in Figure 8.36, together with sensitivity analysis on the prediction variable. In this study, while applying DGSA, Euclidean distance between different curves is used for data variables, while Euclidean distance between scores of images are used for prediction.

For the conservative tracer concentration, we observe that mean log k and its spatial variability are impacting parameters. However, the DIRB reaction rate is perhaps a more surprising impacting factor here. It appears that because the solid-phase Fe(III) is correlated with permeability, its specific surface area and volume, along with the acetate supply, control the rate of Fe reduction. Then Fe reduction is correlated with the travel time and breakthrough curve of the conservative tracers. This illustrates the fact that in this complex system, everything is interacting, creating counterintuitive sensitivities.

For the target prediction, namely the spatial distribution of immobilized uranium, we observe that the uranium reaction rate, the variogram variance of permeability (degree of heterogeneity), mean log k, spatial permeability distribution, and DIRB reaction rate (FerricRate) are the most impacting parameters. This is expected as uranium and DIRB reaction rate control the removal rate of U(VI) from the solute, and other parameters related to the permeability influence the spread of the acetate plume. For example, if the mean log k is too high, the injected acetate is quickly flushed through high-permeability regions (the "channeling effect"), leaving downstream regions of the injectors unexposed.

If we compare the sensitivities of different data sources and prediction, we notice that tracer concentration curves and acetate concentration curves share similar sensitivities, but the most sensitive parameters (e.g., mean permeability, DIRB reaction rate, and spatial permeability) are somewhat different from those of predictions of interest. However, the sensitivity of uranium concentration curves is similar to the prediction variable of interest. This indicates that we would only need to monitor and measure the concentration levels of uranium from the samples, measuring other species will likely not bring additional information as to what we are attempting to predict. A canonical correlation analysis is, therefore, useful to quantify this observation.

8.5.6. Correlation Analysis and UQ

Now that we have established that a correlation exists between the data (uranium concentration at eight locations) and prediction variables (spatial distribution of immobilized Uranium), we generate a posterior distribution of the predictions based on field observations. Because both data and prediction variables are vectors of scores (after dimension reduction), we use CCA. CCA confirms the correlation between data and prediction (see Figure 8.37). Because the correlation is high and linear, we can now proceed with performing a Gaussian process regression to predict the prediction variable scores, and then reconstruct the target saturations (see Figure 8.37). The concentration is not a simple homogenous front, propagating from the injector

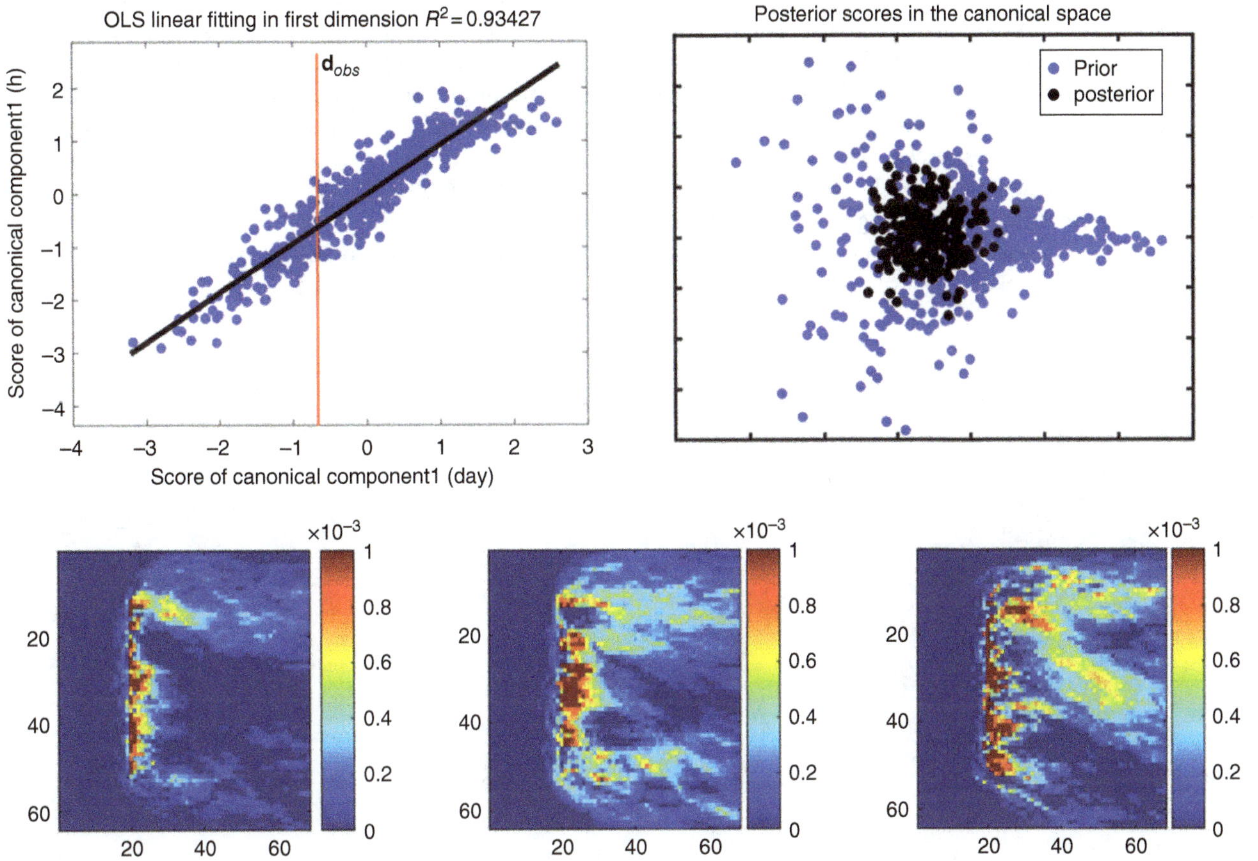

Figure 8.37 CCA analysis between data and prediction variables reveals high correlation in the first canonical score. As a result, the prior uncertainty is reduced considerably into a set of posterior models. This is shown in the score plot as well as with a few posterior realization of the immobilized uranium.

locations, but instead it is influenced by heterogeneity on the geology as well as geochemical reactions. Any future remediation activity will, therefore, need to account for this heterogeneity. For example, geophysical imaging may be of considerable aid in assessing remediation efficacy.

8.6. DEVELOPING SHALE PLAYS IN NORTH AMERICA

8.6.1. Strategy for UQ

Recall from Chapter 1 the three stated questions concerning the development of unconventional resources:

Q1. Which geological and completion parameters impact production most?

Q2. How to predict and quantify uncertainty on production decline in a new well for given geological and completion parameters?

Q3. How many wells need to be in production before a statistical model can confidently estimate production decline in a new location?

The uniqueness of this application of UQ is that these questions will be addressed without any predictive physical models, such as flow simulators, geochemical models, geo-mechanics, or rock mechanics. In theory, prediction based on physics/mechanics is possible, but research in predictive modeling for shales is still in its infancy compared to conventional resources. The current way of hydraulic fracturing shales is merely 10 years old, while modeling of conventional resources goes back more than half a century. Additionally, the rate by which wells are drilled and "fracked" is very high, hence any comprehensive UQ with physical models is challenging from a time framework point of view. Currently, physical models are used in understanding what happens during the fracking process, thereby potentially improving its practice.

The approach will therefore be purely data scientific. We will deal with two real field cases.

Shale case 1 (SC1). A publicly available dataset the Barnett Shale near Dallas, Texas. A large number of wells have already been drilled. The database does not contain geological or completion parameters for each well, because these tend to be propriety. Hence, we will focus mostly on prediction of production in new locations from existing locations and the quantification of uncertainty (see Figure 8.38).

Shale case 2 (SC2). A real field case provided by Anadarko, one of the largest shale producers in the United States. In this case, only 172 wells are available, but now geological and completion parameters are available for each individual well. Here, we will focus on the question of sensitivity as well as the prediction based on the geological and completion covariates and the issue of confidence (uncertainty on the estimate).

The cases we present here could serve as template examples for other data-driven method in subsurface and other Earth Science application. The defining features, distinguishing it from other data science application is the combination of (i) dealing with physical variables, (ii) dealing with spatial/geographical variables and spatiotemporal variables, and (iii) the issue of high dimensions, such as numerous covariates. A general outline of how such methods could proceed is sketched in Figure 8.39. When dealing with real data ("raw data"), a number of

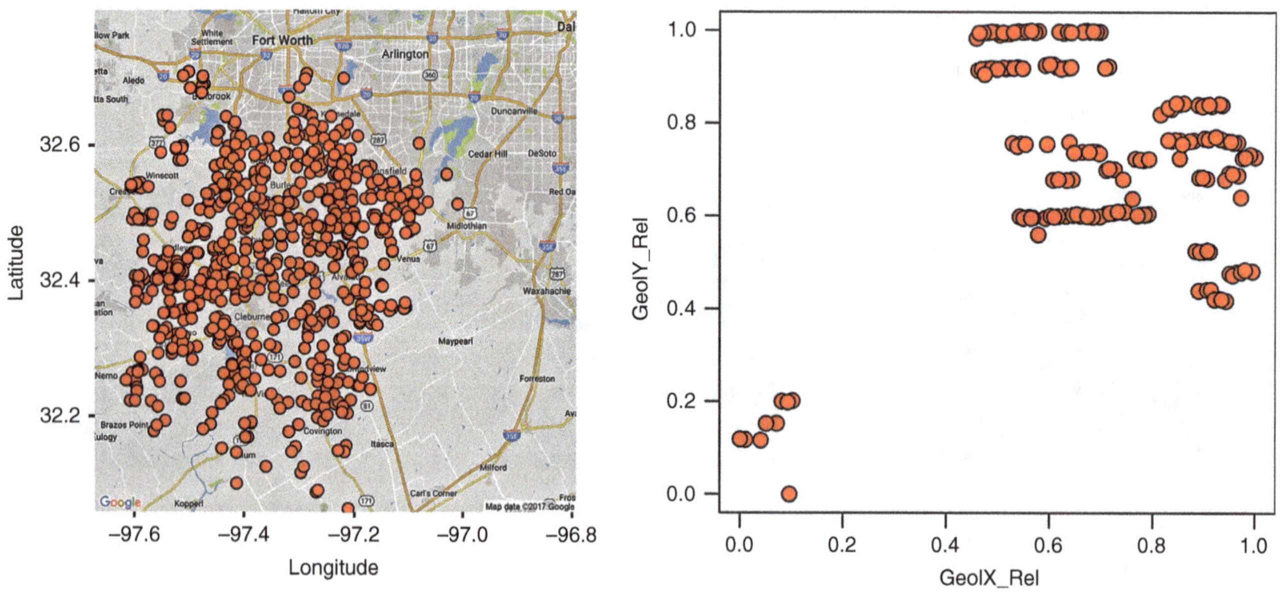

Figure 8.38 Location of wells in the Barnett shale (gas) and the Anadarko dataset (location confidential).

important preprocessing steps will need to be done. First is the definition of "population": which data will one put in the analysis pool and which data is excluded because it is not "representative." This is highly dependent on the application. Real data are subject to noise, and not just random noise. Instrumentation that makes physical measurement may have noise structures. One such structure may simply be human induced by changing settings, turning things off temporarily, and so on. Regression methods, such as those relying on least squares principle, are sensitive to outliers. Regardless of which methods are used, outliers should be dealt with either using robust statistical approaches or simply removing them from the analysis.

To address the above-stated questions, the strategy of Figure 8.40 will be developed. Central to this approach are two methods: CART and kriging (see Section 3.7). Kriging will address the spatial problem of varying production due to spatial geological variation. CART is a nonspatial regression method that addresses the problem of predicting decline rate from a set of covariates (here the geological and completion parameters). CART also allows for calculating sensitivity (relative importance) of parameters on some scalar output (or predictant). Because the output is a decline rate over time, we will use the functional CART method of Chapter 4 (Section 4.4.4). To integrate both the spatial and the covariate aspects of the problem, we will use a functional universal kriging (UK) method that integrates the CART model via a trend model.

Why this strategy?. *Functional data.* The target prediction variable is a function that varies systematically over time; hence, a form of functional data analysis applies well here, such as in the outlier removal, sensitivity analysis, and spatial prediction.

Large set of covariates. The completion variables are likely to have considerable impact on the decline rate. This effect is expected to increase as the operating company learns how to improve their fracking practice.

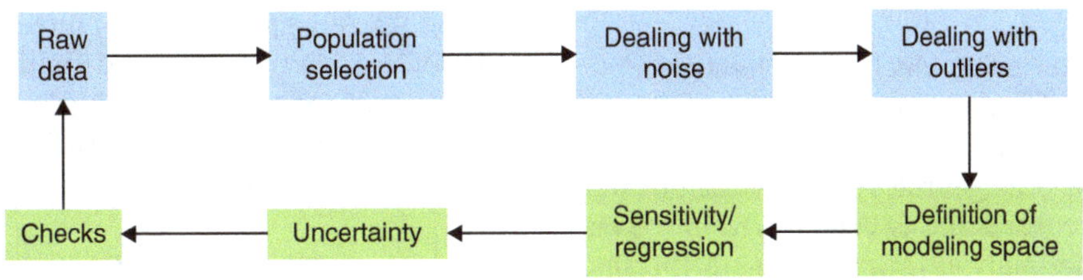

Figure 8.39 A generic workflow for data-based predictions.

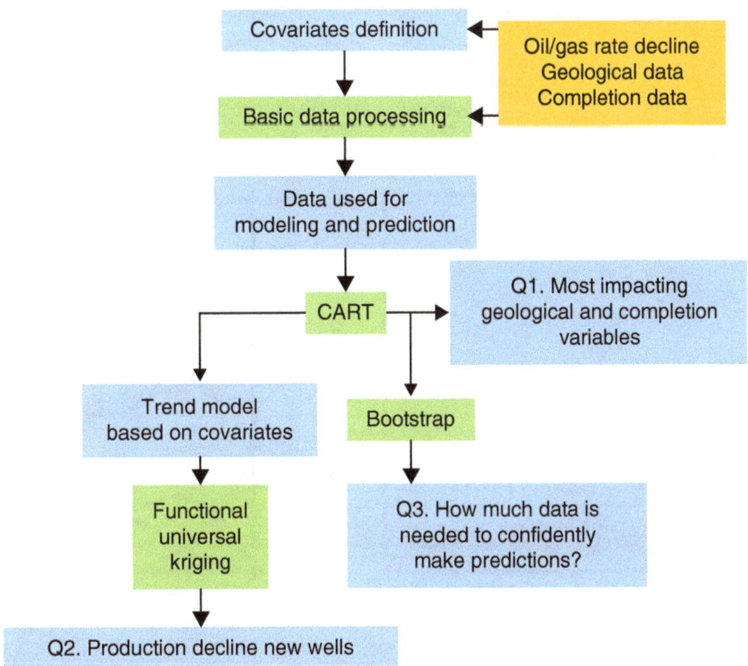

Figure 8.40 Strategy for data-based learning in shale systems (CART = classification and regression trees).

CART methods are ideal to model these complex systems, since CART allows for variables that are discrete, continuous, and by extension (Chapter 4) functional.

8.6.2. Data Preprocessing

In SC1 we use gas production rate curves (GPRC) from 922 wells drilled in the Barnett shale, one of the most prolific and the most developed unconventional gas reservoirs in North America. This dataset was compiled from www.drillinginfo.com, an online oil and gas data repository. A common procedure in data preprocessing is to select a representative set. Shale wells are drilled in all kind of conditions and with a very large variety of well lengths and history. Mixing all data into one group may render prediction impossible if not unrealistic. Therefore, the 922 wells used for the present analysis were selected according to the following criteria:

1. Wells whose lateral length was anywhere between 1800 and 2300 ft and were owned by the same company (operator). This would be an indication that the number of hydraulic fractures was the same or at least very similar across all wells.
2. Wells drilled after 2005, which had at least 5 years of production history (60 months), immediately following the peak gas rate. As a part of this preprocessing, all data entries preceding the peak gas rates (about 3 months) were discarded. Such preprocessing is common in unconventional reservoir data analyses [*Patzek et al.*, 2013], since during that time period wells mostly produce flow-back water that comes as a consequence of hydraulic fracturing.

In both SC1 and SC2 data are quite noisy (see Figure 8.41). SC2 has more noise since the data are recorded daily, while the public domain dataset has only monthly recordings. To smooth out this noise, we use functional data analysis (Section 3.6) and B-spline basis. The number of spline basis functions as well the smoothing penalty on the second derivative are obtained using generalized cross validation [*Ramsay and Silverman*, 2005].

We also need to address the issue of outliers before proceeding with any further statistical analysis. Outlier detection on both the predictors (covariates) and the predictants (production decline curves) needs to be considered. Outlier on covariates can be addressed using robust statistical analysis methods such as the minimum covariance determination, which is a way calculating covariance between covariates by using a part of the dataset not affected by outliers, then using that robust covariance to detect outliers by means of a Mahalanobis distance (see Chapter 3).

Here we focus on outliers on production curves, as this is more unusual than outliers on scalars. One way to address this is to focus on the functional principal component scores. To visualize outliers, we use the extension of the box-plot in 2D: the bagplot [*Rousseeuw and Driessen*, 1999], extended to the "functional" bagplot, see Figure 8.42. A bagplot consist of three regions: bag, fence, and loop (see Rousseeuw for definitions). Basically, the "bag" contains at most 50% of the data while data outside the fence are outliers. We use kernel density estimation to calculate bag and fence.

8.6.3. SC1: Functional Spatial Regression

The aim is to predict decline rate q at locations of future wells, as functions of production decline at previously produced wells. The data is, therefore, a set of production decline rates at existing locations $q(\mathbf{s}_n, t)$, $n = 1, \ldots, N = 922$. This is not a trivial problem since this calls for spatially predicting a function, instead of a scalar, and this

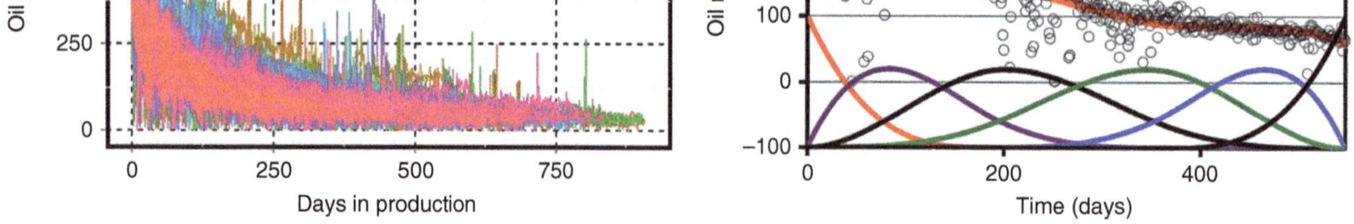

Figure 8.41 Smoothing of oil decline rate in SC2 using functional data analysis.

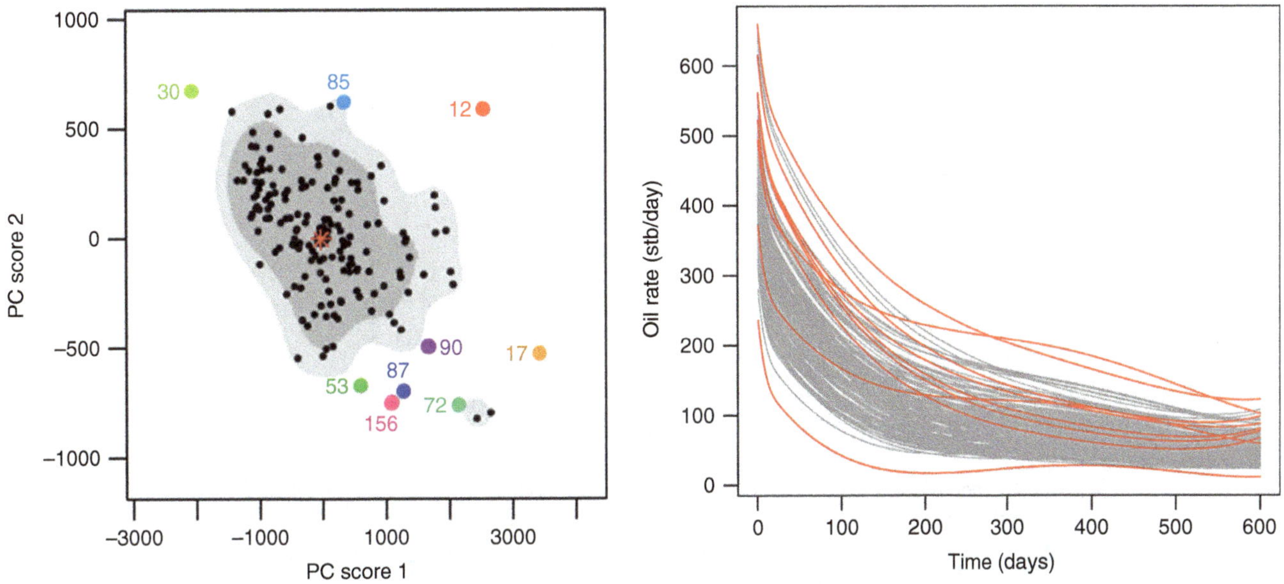

Figure 8.42 Bagplot for outlier detection. Red lines are outliers decline curves that are not used in the training set.

function varies in space. To tackle this problem, we split the prediction problem into two parts.

$$q(\mathbf{s},t) = m(\mathbf{s},t) + r(\mathbf{s},t) \tag{8.7}$$

Production decline $q(t)$ at location $\mathbf{s}$ is modeled with a mean variation and a residual. In geostatistics, this model is used in UK (also known as kriging with a trend model) to make predictions from spatial data. UK extends on Gaussian process regression (ordinary kriging, see Chapter 3) by adding a trend model. In the case of scalar random variables, UK assumes the following trend model:

$$m(\mathbf{s}) = \sum_{k=0}^{K} a_k f_k(\mathbf{s}) \quad f_0(\mathbf{s}) = 1 \tag{8.8}$$

In other words, the trend is a linear combination of user-specific functions. If we consider $\mathbf{s} = (x, y)$, then an example of such trend is as follows:

$$m(\mathbf{s}) = a_0 + a_1 x + a_2 y + a_3 x^2 + a_4 y^2 \tag{8.9}$$

To model the residual, one subtracts the estimated mean from the sample data, and then calculates the spatial covariance or variogram of the residual from the data. Recall from Eq. (3.149) that the spatial covariance can be related to the variogram:

$$\gamma_R(\mathbf{s}_n - \mathbf{s}_{n'}) = \mathrm{Var}(R) - \mathrm{cov}_R(\mathbf{s}_n - \mathbf{s}_{n'}) = \frac{1}{2}\mathrm{Var}(R(\mathbf{s}_n) - R(\mathbf{s}_{n'})) \tag{8.10}$$

To model functions instead of scalars, *Menafoglio and Secchi* [2013] extend the idea of UK to functional UK. To do this, we need to define a space of functions within which a covariance is properly defined. Such generalization is the space L_2 of real-valued, square-integrable functions. In such space, sums, production, and scaling of function, as well as a distance, are defined:

$$\|q(\mathbf{s}_n,t) - q(\mathbf{s}_{n'},t)\|_2^2 = \int (q(\mathbf{s}_n,t) - q(\mathbf{s}_{n'},t))^2 dt \tag{8.11}$$

Once a difference is defined, a covariance or semi-variogram can be defined for functions as follows:

$$\begin{aligned}\gamma(\mathbf{s}_n - \mathbf{s}_{n'}) &= \frac{1}{2}\mathrm{Var}_{L_2}(q(\mathbf{s}_n,t) - q(\mathbf{s}_{n'},t)) \\ &= E\left[\|q(\mathbf{s}_n,t) - q(\mathbf{s}_{n'},t)\|^2\right] - \|E[q(\mathbf{s}_n,t) - q(\mathbf{s}_{n'},t)]\|^2\end{aligned} \tag{8.12}$$

To extend UK of scalars to functions, we rewrite the mean term as follows:

$$m(\mathbf{s},t) = \sum_{k=0}^{K} a_k(t) f_k(\mathbf{s}) \tag{8.13}$$

The empirical estimator of the residual semi-variogram is

$$\begin{aligned}\hat{\gamma}(\mathbf{h}) &= \frac{1}{N(\mathbf{h})}\sum_{n=1}^{N(\mathbf{h})}\left(\|r(\mathbf{s}_n,t) - r(\mathbf{s}_n + \mathbf{h},t)\|^2\right) \\ &= \frac{1}{N(\mathbf{h})}\sum_{n=1}^{N(\mathbf{h})}\left(\int (r(\mathbf{s}_n,t) - r(\mathbf{s}_n + \mathbf{h},t))^2 dt\right)\end{aligned} \tag{8.14}$$

The integral is approximated by a sum. Otherwise, the kriging equations look much like the classical equations of Eq. (3.15). Figure 8.43 shows the variogram for the Barnett-Shale decline data as well as maps of gas rates at certain time, obtained by functional kriging. These

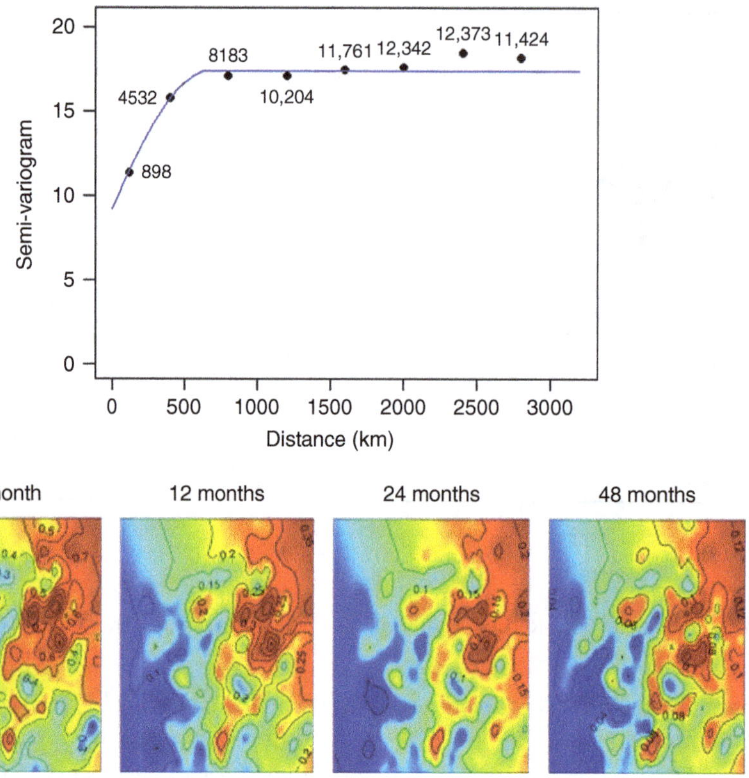

Figure 8.43 Semi-variogram of a function (residual decline rate) with the number of pairs N (h) used for calculation. Gas production (GPRC) at 0, 12, and 24 months estimated by spatial interpolation of existing producers.

maps can be used to make decisions regarding new location to drill, based on existing production declines.

8.6.4. SC2: Functional Spatial Regression with Covariates

In SC2, we need to deal with spatial variation as well as the covariates information, hence in addition to space and time, the decline rates $q(\mathbf{s}, t, \mathbf{c})$ with $\mathbf{c}$ the various covariates (geological and hydraulic fracturing parameters). The data can be represented as follows:

$$(q(\mathbf{s}_1,t),\mathbf{c}_1),(q(\mathbf{s}_2,t),\mathbf{c}_2),\ldots,(q(\mathbf{s}_N,t),\mathbf{c}_N) \tag{8.15}$$

To include this covariate information, we will adapt the following model:

$$q(\mathbf{s},t,\mathbf{c}) = m(\mathbf{s},t,\mathbf{c}) + R(\mathbf{s},t) \tag{8.16}$$

The mean variation is now function of the covariates, while the spatial residual remains as before and will need to be estimated using functional UK. To estimate this mean based on the covariate information, we use classification and regression trees (CART), which naturally perform variable selection, and easily modifiable to deal with functional outputs (see Chapter 4). The main modification to traditional CART is the definition of a cost function for functional variables instead of scalars. For smoothly varying functions such as decline curves, the following cost function is used:

$$\text{cost}(t) = \sum_{m=1}^{M} \sum_{\mathbf{c}_n \in R_m} \int_{t=0}^{t=T} (q(\mathbf{s},t,\mathbf{c}_n) - \mu_m(t))^2 dt \tag{8.17}$$

where R_m is m-th region defined by the tree topology and $\mu_m(t)$ is the mean function associated with m-th region. In CART, we account for the spatial aspect of the problem by adding the $\mathbf{s} = (x, y)$ of the observed decline rates to the covariates. After fitting a functional tree for the trend, the modeling proceeds in the same way as in the case of functional UK. The trend is removed from the training data and the spatial covariance of the functional residual is estimated.

An illustration is given in Figure 8.44. A tree model is fitted with as input the covariate information and output the decline rate. The type of tree used is random forest (see Chapter 2). The variable importance plot shows high ranking of the hydraulic fracturing parameters in this plot, these parameters are also the first to produce a split in the functional regression tree (see Figure 8.44).

The functional random forest approach can also be used to make predictions of decline rate from geological and completion parameters (see Figure 8.45). The advantage of random forest or single tree model is the ability to generate confidence intervals.

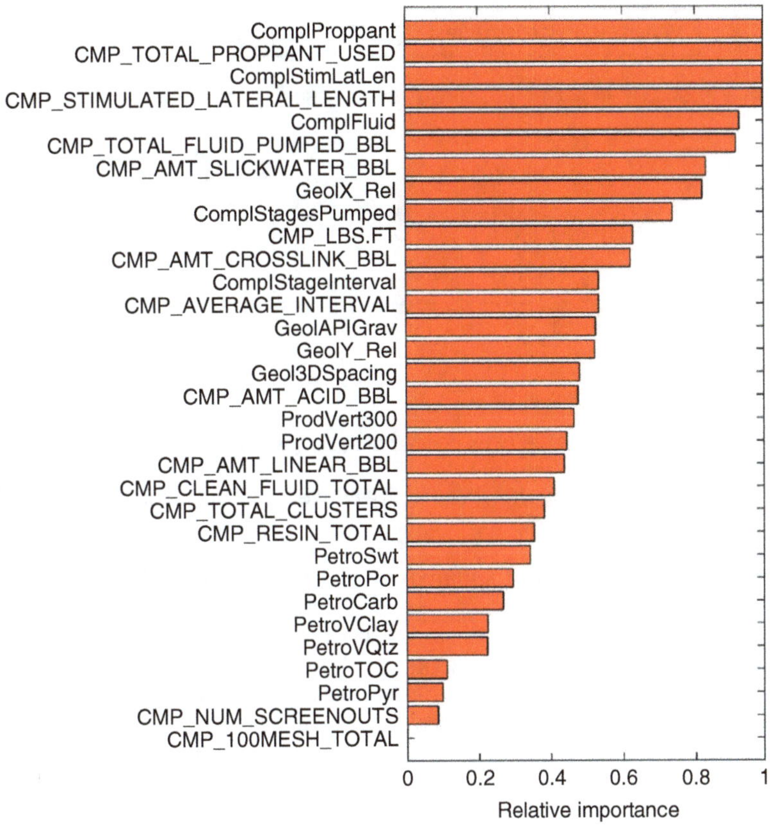

Figure 8.44 Variable importance plot (relative sensitivity) See Table 1.3 for explanation of variables.

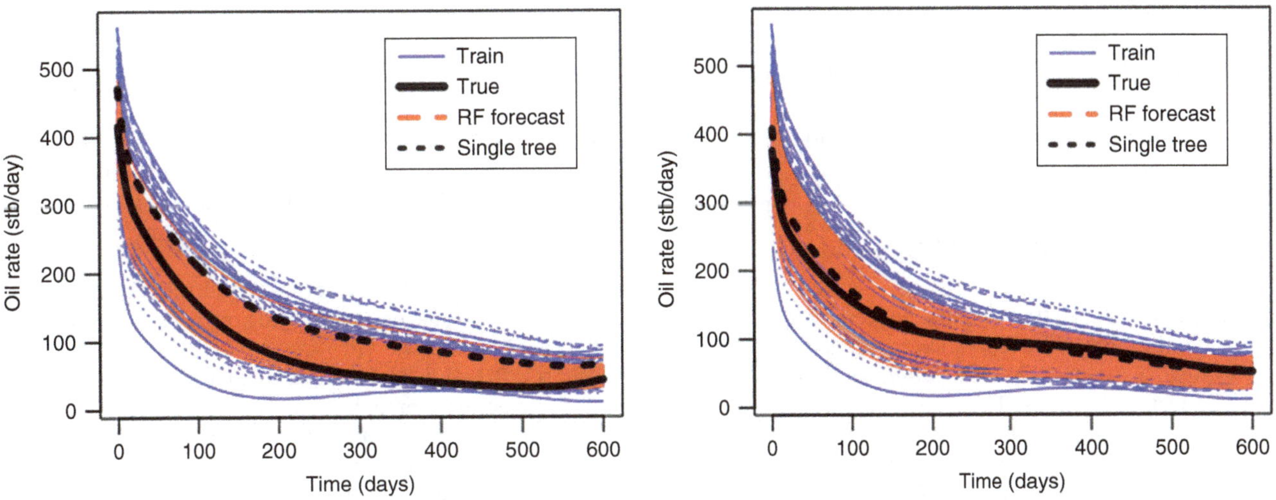

Figure 8.45 An example of tree-based predictions for two wells. Blue curves: training set; red curves: random forest ensemble; thick black line: true well response; dashed black line: forecast of a single tree.

We now visit the question of how many wells need to be drilled for predictions to become reliable. To provide an answer to this question, a small Monte Carlo study was conducted. The size of the training set was varied from 10 to 100 wells. For each training set size 100, training and testing sets were generated by randomly sampling from a pool of 172 original wells. On every iteration, test sets consisted of 71 wells. Three methods [functional random forest (FRF), single tree, and FRF + UK] were fitted to each training set, and then used to predict its

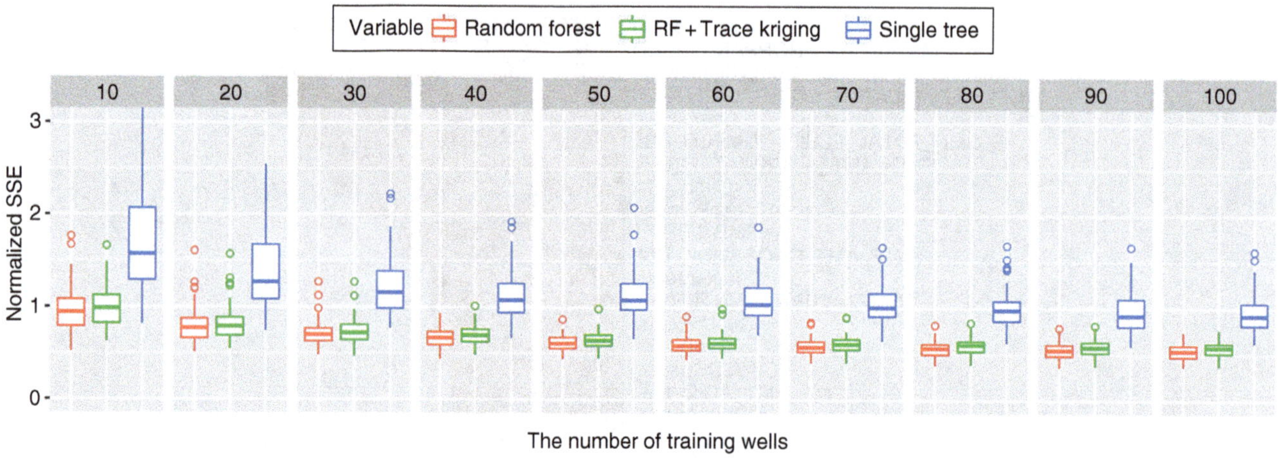

Figure 8.46 The results of the Monte Carlo study showing that 50–60 wells are sufficient for training.

corresponding test set and compute the sum of squared error (SSE):

$$\mathrm{SSE}_i = \| q_i(t) - \hat{q}_i(t) \|^2 \tag{8.18}$$

SSEs are normalized with respect to the averaged squared norm of the entire dataset.

$$\mathrm{SSE}_{av} = \frac{1}{N} \sum_{i=1}^{N} \| q_i(t) - \mu(t) \|^2 \tag{8.19}$$

where $\mu(t)$ is the mean function of all available data (all 172 wells).

Figure 8.46 shows a distribution of the mean of the normalized SSE of 100 test sets for each training set size and each of the three regression methods. What is obvious from this plot is that the error stabilizes around 50–60 wells and that functional random forest had the best forecasting capabilities.

REFERENCES

Aanonsen, S. I., et al. (1995), Optimizing reservoir performance under uncertainty with application to well location, *Proceedings of SPE Annual Technical Conference and Exhibition*, 22–25 October, Dallas, TX, pp. 67–76.

Aitokhuehi, I., and L. J. Durlofsky (2005), Optimizing the performance of smart wells in complex reservoirs using continuously updated geological models, *J. Petrol. Sci. Eng.*, *48*(3–4), 254–264.

Beckner, B. L., and X. Song (1995), Field development planning using simulated annealing: Optimal economic well scheduling and placement, *SPE Annual Technical Conference and Exhibition*, Dallas, October.

Brouwer, D. R., and J. D. Jansen (2002), Dynamic optimization of water flooding with smart wells using optimal control theory, *European Petroleum Conference*, 29–31 October, Aberdeen, (1), pp. 1–14.

da Cruz, P., R. Horne, and C. Deutsch (2004), The quality map: A tool for reservoir uncertainty quantification and decision making, *SPE Reservoir Eval. Eng.*, *7*(December 2003), 3–6.

Derouane, J., and A. Dassargues (1998), Delineation of groundwater protection zones based on tracer tests and transport modeling in alluvial sediments, *Environ. Geol.*, *36*(1–2), 27–36.

Emerick, A. A., et al. (2009), Well placement optimization using a genetic algorithm with nonlinear constraints, *SPE Reservoir Simulation Symposium*, 2–4 February, The Woodlands, Texas, pp. 1–20.

Foged, N., et al. (2014), Large-scale 3-D modeling by integration of resistivity models and borehole data through inversion, *Hydrol. Earth Syst. Sci.*, *18*(11), 4349–4362.

Gelman, A., and C. R. Shalizi (2013), Philosophy and the practice of {Bayesian} statistics, *Br. J. Math. Stat. Psychol.*, *66*(1996), 8–38.

Ghori, S. G., et al. (2007), Improving injector efficiencies using streamline simulation: A case study in a Giant Middle East Field, *SPE Middle East Oil and Gas Show and Conference*, 11–14 March, Manama, Bahrain

Gorelick, S. M. (1983), A review of distributed parameter groundwater management modeling methods, *Water Resour. Res.*, *19*(2), 305–319.

Güyagüler, B. and R. N. Horne (2001), Uncertainty assessment of well placement optimization, *Proceedings of SPE Annual Technical Conference and Exhibition*, 30 September–3 October, New Orleans, LA, pp. 1–13.

Henriksen, H. J., et al. (2003), Methodology for construction, calibration and validation of a national hydrological model for Denmark, *J. Hydrol.*, *280*(1–4), 52–71.

Hermans, T., et al. (2015), Quantitative temperature monitoring of a heat tracing experiment using cross-borehole ERT, *Geothermics*, *53*, 14–26.

Hoyer, A. S., et al., (2015), 3D geological modelling of a complex buried-valley network delineated from borehole and

AEM data, *J. Appl. Geophys.*, *122*, 94–102, doi: http://dx.doi.org/10.1016/j.jappgeo.2015.09.004.

Jesmani, M., et al. (2015), Particle swarm optimization algorithm for optimum well placement subject to realistic field development constraints. *SPE Reservoir Characterisation and Simulation Conference and Exhibition, 2015*, Abu Dhabi, (p. 20).

Jørgensen, F., and P. B. E. Sandersen (2006), Buried and open tunnel valleys in Denmark-erosion beneath multiple ice sheets, *Quat. Sci. Rev.*, *25*(11–12), 1339–1363.

Jørgensen, F., et al. (2003), Geophysical investigations of buried Quaternary valleys in Denmark: An integrated application of transient electromagnetic soundings, reflection seismic surveys and exploratory drillings, *J. Appl. Geophys.*, *53*(4), 215–228.

Jørgensen, F., et al. (2015), Combining 3D geological modelling techniques to address variations in geology, data type and density: An example from Southern Denmark, *Comput. Geosci.*, *81*, 53–63, doi: http://dx.doi.org/10.1016/j.cageo.2015.04.010.

van der Kamp, G., and H. Maathuis (2012), The unusual and large drawdown response of buried-valley aquifers to pumping, *Ground Water*, *50*(2), 207–215.

Kowalsky, M. B., et al. (2012), On parameterization of the inverse problem for estimating aquifer properties using tracer data, *Water Resour. Res.*, *48*(6), W06535.

LaBrecque, D. J. (1996), The effects of noise on Occam's inversion of resistivity tomography data, *Geophysics*, *61*(2), 538.

LaBrecque, D. J., et al. (1996), The effects of noise on Occam's inversion of resistivity tomography data, *Geophysics*, *61*(2), 538–548.

Li, L., et al. (2011), Physicochemical heterogeneity controls on uranium bioreduction rates at the field scale, *Environ. Sci. Technol.*, *45*(23), 9959–9966.

Mariethoz, G., and J. Caers (2014), *Multiple-point Geostatistics: Stochastic Modeling with Training Images*, Wiley Blackwell, Chichester.

Mariethoz, G., and J. K. Caers (2015), *Multiple-point Geostatistics: Stochastic Modeling with Training Images*, Wiley-Blackwell, Chichester.

Mariethoz, G., P. Renard, and J. Straubhaar (2010), The direct sampling method to perform multiple-point geostatistical simulations, *Water Resour. Res.*, *46*(11), W11536.

Marker, P. A., et al. (2015), Performance evaluation of groundwater model hydrostratigraphy from airborne electromagnetic data and lithological borehole logs, *Hydrol. Earth Syst. Sci.*, *19*(9), 3875–3890.

Maskey, S., A. Jonoski, and D. P. Solomatine (2007), Groundwater remediation strategy using global optimization algorithms, *J. Water Resour. Plann. Manage.*, *128*(6), 431–440.

Menafoglio, A., and P. Secchi (2013), A Universal Kriging predictor for spatially dependent functional data of Hilbert Space, *Electron. J. Stat.*, *7*, 2209–2240.

Montes, G., et al. (2001), The use of genetic algorithms in well placement optimization, *SPE Latin American and Caribbean Petroleum Engineering Conference*, 25–28 March, Buenos Aires, Argentina, pp. 1–10.

Muhammad, N., et al. (2007), Streamline simulation for reservoir management of a Super Giant: Sabiriyah field North Kuwait case study streamline potential line, *SPE Middle East Oil and Gas Show and Conference*, 11–14 March, Manama, Bahrain.

Nakajima, L., and D. J. Schiozer (January 2003), Horizontal well placement optimization using quality map definition, *Canadian International Petroleum Conference*, Petroleum Society of Canada.

Nigel, S., et al. (2003), Experience with operation of smart wells to maximize oil recovery from complex reservoirs, *Proceedings of SPE International Improved Oil Recovery Conference in Asia Pacific*, 20–21 October, Kuala Lumpur, Malaysia.

Onwunalu, J. E., and L. J. Durlofsky (2009), Development and application of a new well pattern optimization algorithm for optimizing large-scale field development, Paper 124364 presented at the 2009 SPE Annual Technical Conference and Exhibition, New Orleans, Louisiana, 4–7 October 2009.

Ozdogan, U., and R. Horne (2006), Optimization of well placement under time-dependent uncertainty, *SPE Reservoir Eval. Eng.*, *9*(2), 26–29.

Panday, S., et al. (2015), *MODFLOW-USG version 1.3.00: An unstructured grid version of MODFLOW for simulating groundwater flow and tightly coupled processes using a control volume finite-difference formulation*, U.S. Geological Survey Software Release, 1 December 2015.

Park, C. H., and M. M. Aral (2004), Multi-objective optimization of pumping rates and well placement in coastal aquifers, *J. Hydrol.*, *290*(1–2), 80–99.

Patzek, T. W., F. Male, and M. Marder (2013), Gas production in the Barnett Shale obeys a simple scaling theory, *Proc. Natl. Acad. Sci. USA*, *110*(49), 19731–19736.

Ramsay, J., and B. W. Silverman (2005), *Functional Data Analysis*, Springer series in statistics, Springer, New York.

Rousseeuw, P. J., and K. V. Driessen (1999), A fast algorithm for the minimum covariance determinant estimator. *Technometrics*, *41*(3), 212–223.

Sandersen, P. B. E., and F. Jørgensen (2003), Buried Quaternary valleys in western Denmark-occurrence and inferred implications for groundwater resources and vulnerability, *J. Appl. Geophys.*, *53*(4), 229–248.

Steefel, C. I., et al. (2015), Reactive transport codes for subsurface environmental simulation, *Comput. Geosci.*, *19*(3), 445–478.

Therrien, R., et al. (2010), *HydroGeoSphere: A Three-Dimensional Numerical Model Describing Fully-Integrated Subsurface and Surface Flow and Solute Transport*, Groundwater Simulations group, Canada.

Thiele, M., and R. Batycky (2006), Using streamline-derived injection efficiencies for improved waterflood management, *SPE Reservoir Eval. Eng.*, *9*(2), 5–8.

Thiele, M. R., R. P. Batycky, and D. H. Fenwick (2010), Streamline simulation for modern Reservoir-Engineering workflows, *J. Pet. Tech.*, *62*(1), 64–70.

Thomsen, R., V. H. Søndergaard, and K. I. Sørensen (2004), Hydrogeological mapping as a basis for establishing site-specific groundwater protection zones in Denmark, *Hydrogeol. J.*, *12*(5), 550–562.

Wang, C., G. Li, and A. C. Reynolds (2009), Production optimization in closed-loop reservoir management, *SPE J.*, *14*(3), 506–523.

Wang, H., et al. (2012), Optimal well placement under uncertainty using a retrospective optimization framework, *SPE J.*, *17*(1), 112–121.

Wildemeersch, S., et al. (2014), Coupling heat and chemical tracer experiments for estimating heat transfer parameters in shallow alluvial aquifers, *J. Contam. Hydrol.*, *169*, 90–99.

Williams, K. H., et al. (2011), Acetate availability and its influence on sustainable bioremediation of uranium-contaminated groundwater, *Geomicrobiol. J.*, *28*(5–6), 519–539.

Yabusaki, S. B., et al. (2007), Uranium removal from groundwater via in situ biostimulation: Field-scale modeling of transport and biological processes, *J. Contam. Hydrol.*, *93*(1–4), 216–235.

9

Software and Implementation

9.1. INTRODUCTION

Uncertainty quantification (UQ) in the subsurface would not be possible without the use of modern software and computational resources. Accordingly, the disciplines of UQ and computer science are intrinsically linked, and this relationship is continuously growing stronger as the complexity of the subsurface systems under study increases. The role of software in our Bayesian approach to UQ can be classified into three categories: model generation, forward simulation, and post-processing. In practice, no single software suite may encompass all three components, so implementation of UQ methodologies may require using multiple software packages or codebases.

Model generation refers to the programs used to incorporate prior information and expertise to construct prior subsurface models **m**, usually by means of Monte Carlo. The choice of modeling software is motivated by the type of study (i.e., hydrological, geothermal, oil reservoir, etc.) and is typically specific to the domain of application. The forward simulation component refers to the software package that is used to apply the g_d and g_h operators. As with model generation, the choice of forward simulator is dependent on the domain (i.e., flow, geophysical), as well as the application (i.e., compositional, chemical, etc.), and is at the discretion of the practitioner. The third category corresponds to the methodologies that have been the focus of this book. We refer to this as "post-processing" as it refers to the analysis of modeling parameters and simulation results for quantifying uncertainty. Unlike the former two categories, the same post-processing codebase can be applied to different disciplines, as long as the parameters and simulation results are readily available in an appropriate format.

While practitioners may already have a preferred choice of software for each of the first two stages, in this chapter we will discuss some of the practical considerations that may be ubiquitous to all applications. We next introduce the companion code for this book that implements the post-processing software component. We will elaborate on the technologies that were used as well as provide an overview of executable tutorials based on the examples presented throughout the book. Lastly, we will discuss issues related to modifying and deploying our codebase for analyzing other datasets.

9.2. MODEL GENERATION

The Bayesianism approach to UQ requires the construction of prior models by sampling from the prior distribution $f(\mathbf{m})$. The process of performing a Monte Carlo study is heavily dependent on computer algorithms and their software implementations. Depending on the nature of the prior information, different parameterizations, algorithms, and/or combinations of algorithms will be required. Popular commercial subsurface modeling softwares include, for example, Petrel [*Schlumberger*, 2015], SKUA [*Paradigm*, 2016], and JewelSuite [*Baker-Hughes*, 2017], all of which provide comprehensive modeling capabilities. Other modeling softwares can be used for specific purposes, such as SGEMS [*Remy et al.*, 2009] for geostatistics and FracMan [*Dershowitz et al.*, 1993] for fractures.

9.2.1. Monte Carlo Sampling

Recall that subsurface model consists of both gridded and non-gridded components $\mathbf{m} = (\mathbf{m}_{\text{spatial}}, \mathbf{p})$. For non-gridded model parameters **p** with explicitly specified prior distributions, this requires a Monte Carlo experiment on the prior distributions. For the spatial model parameters $\mathbf{m}_{\text{spatial}}$, Chapter 6 discussed various geostatistical, process-based, level-set, or object-based simulations, or any combination of these. Many of these algorithms have aleatory randomness built in that allows for the

Quantifying Uncertainty in Subsurface Systems, Geophysical Monograph 236,
First Edition. Céline Scheidt, Lewis Li, and Jef Caers.

automated construction of a set of prior model realizations. Any manual operation must be avoided.

9.2.2. The Need for Automation

A major requirement for the practical implementation of Bayesian UQ methodologies is the automation of model generation. Some commercial software suites contain workflow tools that allow users to chain together algorithms to construct complex models. These tools may be graphical in nature such as block diagrams. In other instances, parts of the underlying modeling software may be exposed, such that high-level scripting languages such as Python [*Van Rossum*, 2007], Perl [*Wall*, 1994], and so on may be used to write scripts that automate each of the workflow tasks and connect different software components. By incorporating automation capabilities into model generation software, developers will allow the end user much greater flexibility and efficiency when performing UQ.

9.3. FORWARD SIMULATION

The forward simulation is typically a physical simulation of some phenomena, such as fluid flow or wave propagation, that allows us to obtain the expected data **d** and prediction **h** variables for a given subsurface model realization. As with modeling, forward simulation software is domain specific. For instance, for reservoir simulation, popular commercial packages include ECLIPSE [*Schlumberger*, 2016a], INTERSECT [*Schlumberger*, 2016b], IMEX [*CMG*, 2017], and 3DSL [*Streamsim*, 2017], and academic implementations such as AD-GPRS [*Voskov and Tchelepi*, 2012] and MRST [*Lie et al.*, 2012] are available. Other packages such as TOUGH2 [*Pruess et al.*, 2012], Open-FOAM [*Jasak et al.*, 2007], and MODFLOW [*Harbaugh et al.*, 2000] can be used to simulate fluid flow for a variety of applications. As forward simulation is typically the most computationally expensive component of the UQ workflow, it is the most likely one to invoke the need for high-performance computing, parallelism, or other approaches to improving computational efficiency. The Bayesian Monte Carlo nature of UQ further necessitates this need for acceleration since an ensemble of subsurface models need to be forward simulated.

9.3.1. The Need for Parallelism

Parallelization can be used to accelerate computational performance in two different ways. First, all modern CPUs having multiple cores allow for the distribution of the computational load of a single realization across numerous cores. Many modern simulators have already implemented this parallelization functionality. However, certain forward simulations may be less conducive to this type of parallelization, and increasing the number of cores used may result in diminishing returns. Many factors determine the effectiveness of multi-core simulations, and an optimal choice of cores is perhaps best determined experimentally. Next to CPU, one should note that many computing and scripting languages, such as Python/MATLAB are very easy to run on GPUs making GPU computing more easily accessible.

The second type of parallel acceleration is due to the "embarrassingly parallel" nature of the Bayesian Monte Carlo problem, as each realization can be forward simulated independently. Therefore, the number of realizations that can be forward simulated at a given time is limited only by the number of computers that are available. This allows for maximizing throughput of computer clusters with multiple compute nodes. However, one bottleneck that arises when using commercial simulators is the number of available licenses. Hopefully, vendors will shift toward licensing paradigms that are more conducive for UQ.

9.3.2. Proxy Simulators

In addition to using parallel computing for accelerating the forward simulation software component, we could use proxy simulators. For instance, reduced-order simulators could be used to transform high-dimensional models into lower-dimensional representations to speed up simulation [*Bai*, 2002; *Cardoso et al.*, 2009]. In certain scenarios, reduced physics simulations such as streamline simulations can also be used. For instance, in the oil reservoir undergoing waterflooding in Section 8.2, streamline simulations were used in lieu of finite difference simulators to dramatically decrease computational time. However, use of proxy simulators necessitates understanding of the subsurface system under study, and the applicability of the simulator under these conditions.

9.4. POST-PROCESSING

The final software component is the post-processing component that implements the UQ methodologies presented in the book. Unlike model generation and forward simulation, the post-processing software does not need to be application specific. In fact, despite the diversity of the case studies in Chapter 8, a common codebase was used for the analysis and generating the accompanying illustrations. This codebase was implemented in MATLAB and has been made available online as a git [*Loeliger*, 2009] repository at https://github.com/SCRFPublic/QUSS. We have also included a set of executable tutorials using Jupyter [*Kluyver et al.*, 2016] and sample data sets.

Readers can download this package, and rerun the code to reproduce various examples from the book. In addition, readers can import their own datasets and perform analysis on their own cases. This package may also serve as a starter code for users to develop their own versions. Other packages that implement various UQ methods include DAKOTA [*Adams et al.*, 2014] and PEST [*Doherty*, 2015].

9.4.1. Companion Code Technologies

9.4.1.1. Git. Our codebase is stored as a git repository, which is a versioning system that allows users access to the current software package as well as any future updates. Users can "clone" the repository to make a copy of the codebase, test datasets, and executable tutorials on their local machines. Advanced users can also "fork" the codebase to create their own repository for making modifications to the codebase for catering to their own specific needs. *Blischak et al.*, [2016] provide a succinct introduction to using git for software development. Our repository is hosted using Github, which provides a forum for users to interact with the authors for questions, bug reports, and suggestions.

9.4.1.2. Jupyter Tutorials. In addition to the codebase that implements the approaches in the book, our git repository also contains several executable tutorials under the *tutorials* folder for data science (Chapter 3), sensitivity analysis (Chapter 4), Bayesian evidential learning (BEL), sequential importance resampling (SIR, Chapter 7), and so on. These tutorials are implemented using the Jupyter package [*Kluyver et al.*, 2016], which is a documentation system that is compatible with a variety of programming languages, including MATLAB. It allows to typeset tutorials to contain both code and expected output of the code. The generated documents can not only be viewed within a web browser but can also be downloaded and rerun with modifications, serving as an interactive tutorial.

9.4.2. Deployment Considerations

In our repository, we have provided example datasets that can be used with the tutorials and codebase. To import their own data, users will need to convert their datasets into the appropriate format; that is, we need to ensure that the output of model generation and forward simulation are in a compatible format. This brings us to the notion of bookkeeping and its importance throughout the deployment of all three of the software components required for UQ. A second deployment consideration pertains to the licensing of packages used in the development of post-processing software.

9.4.2.1. The Need for Bookkeeping. Methodologies such as sensitivity analysis (Chapter 4) or Bayesian evidential learning (Chapter 7) require input parameters of each realization and well as its corresponding data and prediction variable. Since the modeling and forward simulation are often performed in separate programs, the bookkeeping of parameters and simulation results can become a non-trivial task. While most modeling and simulation software will have the capability to export parameters and results into text files, the output formats may not be conducive for post-processing, as they contain extraneous information. Additional processing may be required to extract the appropriate data. We have included an example in our companion code (examples/Process3DSLResults.m) of a simulator output log, and the MATLAB script used to marshal the results from a realization, and match it to its corresponding model parameters. Obviously, such a processing script would be modeling/simulator dependent, but it is an important software component in any UQ workflow. Finally, these extracted parameters and data must be stored in a format that can be read by the post-processing software. In our tutorials, the data is organized as MATLAB matrices or structs as explained in detail within the Jupyter documents.

9.4.2.2. Licensing. The final consideration when deploying post-processing software is the licensing of the open source codebases and libraries that may be incorporated. As the post-processing software component is the least computationally expensive aspect of the UQ workflow, it is reasonable to use high-level languages such as R, Python, and MATLAB for their implementation. This opens up the possibility of incorporating high-quality, open-source packages that implement many of the data science tools. For instance, the statistical learning tools used in Chapter 7 are all available within the scikit-learn package [*Pedregosa and Varoquaux*, 2011], which allows us to quickly experiment with different regression methods. However, users will need to be mindful of the open source licenses that such packages are released under. The companion code for this book is released under the MIT license, a permissive license. Other software packages may be released under more prohibitive licenses such as GPL. The reader may refer to *Rosen* [2004] for a discussion of the various types of open source licenses.

REFERENCES

Adams, B. M., L. E. Bauman, W. J. Bohnhoff, K. R. Dalbey, J. P. Eddy, M. S. Ebeida, M. S. Eldred, P. D. Hough, K. T. Hu, J. D. Jakeman, J. A. Stephens, L. P. Swiler, D. M. Vigil, and

T. M. Wildey (2014), DAKOTA, A multilevel parallel Object-oriented framework for design optimization, parameter estimation, uncertainty quantification, and sensitivity analysis: Version 6.0 User's manual, *Rep. SAND2014-4633*, Sandia Natl. Lab., Albuquerque, New Mexico.

Bai, Z. (2002). Krylov subspace techniques for reduced-order modeling of large-scale dynamical systems. *Appl. Numer. Math. 43* (1–2): 9–44. doi: https://doi.org/10.1016/S0168-9274(02)00116-2.

Baker-Hughes. (2017), JewelSuite [Computer Software]. Available from: https://www.bakerhughes.com/products-and-services/reservoir-development-services/reservoir-software/jewelsuite-reservoir-modeling-software.

Blischak, J. D., E. R. Davenport, and G. Wilson (2016), A quick introduction to version control with Git and GitHub. *PLoS Comput. Biol.*, *12*(1), e1004668, doi:https://doi.org/10.1371/journal.pcbi.1004668.

Cardoso, M.A., Durlofsky, L.J., and Sarma, P. (2009). Development and application of reduced-order modeling procedures for subsurface flow simulation. *Internat. J. Numer. Methods Eng. 77* (9): 1322–1350.

CMG (2017). IMEX: Black Oil and Unconventional Simulator. Available from: http://www.cmgl.ca/imex.

Dershowitz, W.S., G. Lee, J. Geier, T. Foxford, P. LaPointe, and A. Thomas (1993). *FracMan, Interactive Discrete Feature Data Analysis, Geometric Modeling, and Exploration Simulation. User Documentation.* Golder Association Inc., Redmond, WA.

Doherty, J. (2015). *Calibration and Uncertainty Analysis for Complex Environmental Models.* Watermark Numerical Computing, Brisbane.

Harbaugh, A. W., E. R. Banta, M. C. Hill, and M. G. Mcdonald (2000), MODFLOW-2000, the US Geological Survey modular ground-water model: User guide to modularization concepts and the ground-water flow process. *U.S. Geological Survey Open-File Report 00-92*, 121.

Jasak, H., Jemcov, A., and Tukovic, Z. (2007). OpenFOAM: A C ++ library for complex physics simulations. *Int. Workshop Coupled Methods Numer. Dyn. m*: 1–20.

Kluyver, T., B. Ragan-kelley, F. Pérez, B. Granger, M. Bussonnier, J. Frederic, K. Kelley, J. Hamrick, J. Grout, S. Corlay, P. Ivanov, D. Avila, S. Abdalla, C. Willing, and Jupyter Development Team. (2016), Jupyter Notebooks: A publishing format for reproducible computational workflows. *Positioning and Power in Academic Publishing: Players, Agents and Agendas, 20th International Conference on Electronic Publishing*, Göttingen, Germany, 87–90, doi: https://doi.org/10.3233/978-1-61499-649-1-87.

Lie, K. A., S. Krogstad, I. S. Ligaarden, J. R. Natvig, H. M. Nilsen, and B. Skaflestad (2012), Open-source MATLAB implementation of consistent discretisations on complex grids. *Comput. Geosci.*, *16*(2), 297–322, doi:https://doi.org/10.1007/s10596-011-9244-4.

Loeliger, J. (2009), *Version Control with Git: Powerful Tools and Techniques for Collaborative Software Development.* Available from: http://it-ebooks.info/book/919/.

Paradigm (2016). SKUA-GOCAD [Computer software]. Available from: http://www.pdgm.com/products/skua-gocad/.

Pedregosa, F. and Varoquaux, G. (2011). Scikit-learn: Machine learning in Python. *J. Mach. Learn. Res. 12*: 2825–2830. doi: https://doi.org/10.1007/s13398-014-0173-7.2.

Pruess, K., C. Oldenburg, and G. Moridis (2012). *TOUGH2 Users Guide, Version 2*, Lawrence Berkeley Natl. Lab., Berkeley, Calif., 197.

Remy, N., A. Boucher, and J. Wu (2009), *Applied Geostatistics with SGeMS: A User's Guide*, Cambridge University Press, Cambridge, doi:https://doi.org/http://0-dx.doi.org.wam.seals.ac.za/10.1017/CBO9781139150019.

Rosen, L. (2004), Open source licensing: Software freedom and intellectual property law, *Open Source Licensing Software Freedom and Intellectual Property Law*, 255–268. Available from: http://www.amazon.com/dp/0131487876.

Schlumberger. (2015), Petrel [Computer Software]. Available from: http://www.software.slb.com/products/petrel.

Schlumberger. (2016a), ECLIPSE Industry-Reference Reservoir Simulator. Available from: https://www.software.slb.com/products/eclipse.

Schlumberger. (2016b). INTERSECT High-Resolution Reservoir Simulator. Available from: https://www.software.slb.com/products/intersect.

Streamsim. (2017), 3DSL Streamline Simulator. Available from: http://www.streamsim.com/technology/streamlines/3dsl-simulator.

Van Rossum, G. (2007), Python Programming Language, *USENIX Annual Technical Conference*, 36, Santa Clara, CA. Available from: http://www.python.org.

Voskov, D.V. and Tchelepi, H.A. (2012). Comparison of nonlinear formulations for two-phase multi-component EoS based simulation. *J. Petrol. Sci. Eng. 82–83*: 101–111. doi: https://doi.org/10.1016/j.petrol.2011.10.012.

Wall (1994). Programming Perl. *Crossroads 1* (2): 10–11. doi: https://doi.org/10.1145/197149.197157.

10

Outlook

10.1. INTRODUCTION

One of the main purposes of this book is to introduce the reader to a variety of real-case problems involving the subsurface, uncertainty, and decision making, as well as its societal importance, in terms of the future of geological resources. We have stated repeatedly that no single best method exists to tackle a problem involving so many complexities as well as fields of science, from fundamental science to applied science, from data to decision. Hopefully, the insight provided through the case studies, how various existing ideas come together, generates a platform for making methodological approaches more applicable in the real world and for the real world to get an appreciation of how principled, science-based, and computational approaches can be powerful in generating informed decisions.

There is little doubt that uncertainty quantification (UQ), as a scientific discipline, will continue to evolve. There are many factors that contribute to this evolution: (i) we need models that predict the future, (ii) more data will be acquired that will need to be "married" with models, and (iii) computation and information technology will only increase. Compare the monochrome, 1984 IBM AT, 16 bit with 6 MHz CPU, 16 MB RAM, 20 MB hard disk, and no sensors to a typical 2017 smartphone with 16 million colors, four times 2.3 GHz CPU (and GPUs), 4 GB RAM, 256 GB hard disk and a proximity sensor, light sensor, barometer, altimeter, magnetometer, accelerometer, and gyroscope. It is estimated that more than three billion people have a smartphone (as many as people living in poverty).

What future holds for the material in this book? As researchers and practitioners of UQ, we are aware of the difficulty of making predictions, and how one's too narrow or cognitive-biased prior may result in an unrealistic UQ. Nevertheless, using seven questions, we would like to speculate about what may come. These are fundamental as well as practical questions that will hopefully outline areas where making progress will have a meaningful impact.

10.2. SEVEN QUESTIONS

10.2.1. Bayes or Not Bayes?

Bayes' is the single most-used principled approach to UQ. By principled we mean referring to Chapter 5 that it is based on a mathematical framework with axioms, definitions, algorithms, and so on. Bayes' is a growing paradigm. Other principled approaches, such as those based on fuzzy logic and possibility theory, are appealing and perhaps complementary to probabilistic approaches but less developed. Non-Bayesian methods in data science (e.g., random forests) are appealing but are they or are they not valid methods for UQ? Should a decision maker care whether the assessment of uncertainty comes from a pure Bayesian method or not? What does such decision entail?

Despite this rigor in the math, it appears that various interpretations and uses of the notion of *prior distribution* exist. Given the nature of subsurface uncertainty, the prior distribution often has considerable impact on the posterior of predictions. This has been shown in the various case studies.

Few standards appear to exist to aid in stating a prior, beyond the usual informative and uninformative prior (see Chapter 5) mostly based on statistical principles. Many within the Bayesian framework use Gaussian and uniform priors, often for mathematical convenience. From our practical experience, we find that the following questions have arisen:

1. *Who states the prior?* This is not a trivial matter. In fact, it refers to the group dynamic that exists when various disciplines (with their own domain experts) come

Quantifying Uncertainty in Subsurface Systems, Geophysical Monograph 236,
First Edition. Céline Scheidt, Lewis Li, and Jef Caers.

together to solve complex problems. If they were left to their own devices, each expert would likely provide a different prior, based on their own different background, probably inspired by the type of "physics" each has expertise on. There seems yet not to be much agreement on how this prior is established within such dynamics (e.g., oil companies).

2. *On what principle does that person select a prior?* A principled approach is important since it leads to some form of reproducibility, a fundamental notion of scientific research and conduct. But what principles? Should that person rely on principle of indifference (a more uninformed prior) or on a physical principle (informed prior)? In question 2, we will address this more deeply since the subsurface system is a natural system, subject to laws of physics that may inform such priors (although those laws may themselves be uncertain, as yet).

3. *How do we know we have a "good" prior? What does "good" mean?* Once a prior is established, do we just move on and assume it is good? How can this be quantified? What if much later we need to completely revise this prior (start from scratch), because some future observation falsifies it? In one interpretation of Bayes, a prior should not need revision; priors are updated into posteriors which then become priors in the next time-stage, when new information arrives. The prior is therefore fundamental to everything that comes next: it is the "base-rate," it informs us how data and prediction variables are related, how much certain data informs model variables, how much we have reduced uncertainty, how decisions are made, and so on. It seems that any complete revision at a later stage should be avoided at all cost. How can this be done?

10.2.2. How to Turn Geological Knowledge into Quantitative Numerical Models?

A vast knowledge of geological understanding exists on the nature of the subsurface. One can take two approaches to summarize such understanding: process and geometry. The subsurface was created by means of physical, chemical, and biological processes operating on a timescale from as few as 100 or 1000 years (e.g., soils) to millions of years (e.g., oil reservoirs). That process has led to a geometry of atoms, molecules, minerals, grains, pores, bedding, layers, unconformities, faults, basins, and so on. This established structure with its established composition can be regarded as a "geometry": a spatial arrangement of "stuff" at the present time. How we got to this geometry is the study of geological processes and much within the realm of geological sciences. Traditionally, this process has been studied descriptively, where geologists use their forensic skills to describe the nature of the process that took place, thereby getting a better understanding on how Earth formed and evolved. They typically use a combination of field work, laboratory experiments, and numerical models to reach such understanding. This can be regarded as pure science, without necessarily any application or engineering goal in mind.

The field of geological sciences is rich since it asks fundamental questions about our planet and the universe. How can this rich knowledge be used beyond the purely scientific approach toward practical applications such as those developed in this book? In terms of applications, we have considered in this book some initial thoughts on linking process to geometry for a delta system (Section 5.9) and an extensional faulting regime (Section 6.3). The traditional way of representing such spatial variation is by means of geostatistics. Geostatistics allows representing spatial variability of geometries through mathematical or computational models. The classical approach is to model these geometries from data (e.g., the variogram), but such models are hard to infer in data sparse environments (like the subsurface) and lack the ability to describe geometries realistically. Such lack of realism may have considerable impact on predictions made. Other geostatistical approaches relying on objects or training images improve on the realism but avoid the fundamental question: how to turn process understanding into a prior distribution of "geometries." The question, therefore, in practice is (i) how to generate understanding of process from limited site-specific data and (ii) how to turn such understanding into a mathematical model that can be used for predicting (and not just understanding). Limited data here encapsulate the combination of boreholes, geophysical data, and expert (a geologist) interpretation, who may rely on data other than those gathered at the site, such as analog data, previous publications, and so on. It is very likely that uncertainty remains on the quantification of such processes, either fundamentally or on the leading physical, chemical, and biological parameters.

Consider the buried valley system in Denmark. Even when the geophysical data is of relatively high quality, it leaves open the sub-resolution uncertainty of the buried valley geometry. How do valleys intersect? What process is at play? What is the process by which clay lenses are created within such valley system? A better understanding of the process generated by means of numerical models for glaciation could generate additional understanding, in the same sense as the flume experiment in Chapter 5 generated better understanding on deltaic deposits. The main question then is how to construct such process models. There appears to be considerable opportunity for research in the areas of numerical process models for geomechanics, rock mechanics, sedimentary processes, diagenetic processes, geochemical processes, and bio-geochemical processes or any combination of these.

For cases with very little data, such as the geothermal case in Belgium, these lofty goals of process understanding may not apply (i) because it may not be needed or (ii) it simply is not feasible, given the lack of information. For such cases, we refer to questions 4 and 5.

10.2.3. What Is a Proper Model Parameterization?

In the univariate case, the definition of a probabilistic model has two components: the variable and the probability measure over its support. Here we consider the former but now extended to high dimensions. What variables are needed to describe the subsurface system? The common approach is to go for certain defaults, for example a regular grid with a number of variables/properties per each grid (porosity/permeability), statistical parameters (also variables) that describe these properties (e.g., a mean, a variogram range). The size of the grid is often taken to be based on computational limitations. For stochastic modeling, this may be several millions, for physical models some order of magnitude less, simply because stochastic simulation is much faster than numerical simulation. Gridded models are usually "upscaled" from the high-resolution stochastic models (the realizations of geometry). Another way to define the model parametrization is based on the nature and amount of data available. The less data, it seems, the coarser the model definition (fewer variables), simply because it is easier to match few data points with coarse models (easier inversion and fast computations) than a heterogeneous set of data with high-resolution models.

This may not always be an efficient or even effective approach. Why should we base such definitions solely on computational needs? In the end, models will be used to make predictions and such predictions may have different sensitivities with respect to the model variables than the data or the computational needs. The problem here is to know this in advance. It seems the only way to know this is by building high-resolution (high-dimensional) model parameterizations and figure this out by means of a sensitivity analysis. This may simply not be feasible, for computational and practical reasons. What can be done? One proposal is to at least consider a few high-resolution model parameterizations (various levels of resolutions) and calculate the error (e.g., on some prediction) statistically. The latter could consist of quantifying the error made based on physical considerations (the numeral discretization error), estimated from the physics of the problem itself. The multilevel approaches outlined in Chapter 3 offer opportunities in this regard.

Another question relates to the very nature of grids. The grid can often be a headache when dealing with considerable structural uncertainty. Such uncertainty changes the topology of the grid. As a result, one needs to rely on sophisticated (e.g., adaptive) gridding. However, such gridding needs to follow basic geological rules (layers cannot end in the middle of nowhere, see Section 6.3) that standard gridding codes are not designed for. All this makes automation very difficult; hence, even a simple Monte Carlo is cumbersome. Implicit methods such as level sets, whose very definition is grid-less, appear to hold promise in including geological rules.

10.2.4. How to Establish a Realistic Prior Distribution?

Once model variables are defined, the probabilistic approaches call for defining the joint distribution of all model variables. A portion of such variables may be independent, and hence the joint distribution may be defined by product of individual distributions. Care must be taken in making such decision. For example, variables may be independently defined (e.g., a slope and intercept of a relationship), but when estimated from the (same) data, they become dependent. This may appear trivial in this simple case, but less obvious when dealing with complex data (from drilling or geophysics) and high-dimensional model variables.

What do we mean by realistic? Largely, this is a subjective notion: what appears realistic for one person does not need to be for another. Practically, it would mean that our choices over the long term appear to generate realistic predictions, for example a 10% percentile reflects a 10% success rate. This is only a qualitative and subjective notion, because each prediction problem has its own unique characteristics, so pooling many different prediction problems may not be meaningful, in terms of making such assessment. A second aspect of such realism is that the prior, in most interpretations of Bayes, needs to "predict" the data. This refers to the prior-data consistency issue discussed at length in Chapters 5, 6, and 7. The most common practical problem is not so much to "select" among alternative prior distributions (model selection) but to establish a prior that does not have an infinitely small probability of predicting the data. When this occurs, it may be very tempting to do two things: (i) change the model parameterization such that data matching becomes easier and (ii) make ad-hoc changes to the prior distributions. An example of such ad-hoc change can be as follows: the first attempt of a prior for the mean proportion of sand fraction is taken as uniform between 15 and 20%. Let us say that it has been established that the current prior (involving many other parameters) cannot predict the data (e.g., geophysical data), but a change (a second attempt) in mean proportion of between 25 and 30% does (fixing all other parameters). The first problem is the ignorance of any interaction that may take place between parameters in predicting data, and the second is that a narrow range is changed into another narrow range, simply anchored around a different center.

Our experience has been that a first attempt at defining a prior, in terms of both model variable definitions and their (joint) distribution, does not perform well at predicting data. For synthetic cases (where we know the future), the same can be said when making predictions. Our experience has been that starting with a wider prior is a good approach and that falsification can be used to narrow this prior, then using a second (or even third) Monte Carlo. Falsification does not require matching data, and it could just be based on comparison between statistics of the data and statistics generated by the prior. However, this would require having quality data. An open question, therefore, exists as to the practice on establishing a realistic prior, without too much tuning on data, yet consistent with data and practical (not far too wide) for solving real cases. A fine line exists between falsification and ad-hoc tuning.

10.2.5. What Combination of Data Is Needed and for What Purpose?

New data acquisition method attempting to resolve various aspects of the subsurface have emerged recently, particularly in geophysics. We mentioned before the employment of sensor technology which has lead, at least in the oil industry, to the deployment of sensors in boreholes for continuous monitoring. Passive seismic, recording the ambient noise field of the subsurface without induced sources, has recently emerged as a way to image subsurface structures.

The main problem here is that no single data source will resolve what is needed to make realistic predictions. Various data sources inform the subsurface in various ways and at various scales. In addition, gathering data can be regarded as a "short-term" experiment that aims to inform longer-term future behaviors. What combination of data sources are needed to address certain decision or prediction problems? The value of information framework introduced in Chapter 2 provides a way forward. However, VOI calculations are cumbersome as they call for model inversion on datasets generated by Monte Carlo from the prior distribution. To that extent, statistical learning and regression may aid in alleviating this issue.

10.2.6. How to Deal with the Computational and Software Issues?

Each year for the past 50 years, we have been able to pack double the number of transistors on a CPU wafer as the year before. This trend, known as Moore's Law, has resulted in the rapid increase of not only computation power but also memory and disk space. In terms of subsurface engineering, this means that each year we have been able to run a larger number of models which themselves are larger and increasingly complex. In turn, this means that our computational capabilities to perform UQ have also been growing at a rapid pace. However, as of the 2010s this rate of doubling has slowed. Transistor sizes have shrunk to near physical limits (the size of silicon atoms). The end of Moore's Law will undoubtedly impact all disciplines that rely on high-performance computing, including subsurface UQ. What avenues can we explore when we can no longer generate and simulate larger and more models using our current computing architectures?

The first area of improvement is specialized hardware to accelerate computations. There has been a recent surge in popularity in using GPUs for scientific computing. For specialized types of calculations, they can provide orders of magnitudes performance improvement over CPUs. There have been considerable efforts to port CPU codebases to run on GPUs in many disciplines, including subsurface engineering. However, GPUs were designed for processing computer graphics. For a greater level of optimization, hardware should be specifically designed for a particular computational task. This has already happened in machine learning, as Google has designed and deployed tensor processing units to accelerate the algorithms behind Google Search, Translate, Photos, and so on. Could we see specialized hardware for subsurface simulations in the future?

A second avenue of progress is in distributed or cloud computing. Previously, individual computers were limited by their local processors speeds, but with the advent of cloud computing, one can draw upon larger and more powerful computers. Services such as Amazon Web Services allow users to perform computations by renting instead of purchasing additional hardware. For large companies, deploying internal distributed computing can maximize the throughput of available hardware, improving the economics of computation.

The final direction is improving the algorithms and software used for UQ. While accelerating physical simulations through algorithmic optimizations may be difficult, there is still potential for improving performance throughout the rest of the UQ workflow. For instance, could we reduce the number of necessary simulations required through means such as proxy modeling? Can we incorporate recent and ongoing advents in artificial intelligence, such as deep learning and reinforcement learning, to draw additional conclusions from a smaller number of models?

10.2.7. What Educational Means Can We Design to Teach UQ?

At Stanford, this book is used to teach a course on UQ for graduate students in Earth Sciences as well as Engineering. How can students prepare themselves for such a complex topic? One of the main difficulties in teaching this material is that it relies on synthesis of many concepts. Most courses teach technical material, for example course in statistics, machine learning, applied geophysics, inverse modeling, decision analysis, and so on. A large hurdle exists in taking all these ideas and methods out of the narrow domain of each course and then put them all together to solve real problems. The synthesis can be more difficult than the technical details, since it relies on a different kind of intelligence.

Who teaches how to synthesize and how? Often, students self-teach, with the help of their advisor using projects or dissertations. But this remains limited in scope, certainly in terms of audience. In terms of publication writing, it is still easier to publish an advance in an individual technical field (or some limited combination) than a synthesis solution, simply because of the limited scope of journals and the emphasis on citations as a measure of advancement at many institutions. Our experience with teaching this material to industry is similar. Industrial applications face considerable complexity, such as organizational, human resources, or technological hurdles that are not even mentioned in this book. Hence, to organize around a common theme of decision making under uncertainty remains difficult, and often pieces of the puzzle are divided among broad groups (geology and geophysics, engineering, decision analysis, data processors, etc.). It is often unclear how these pieces come together and what domain knowledge is used to create this synthesis. Domain expertise is often how one gets hired.

This suggests a considerable educational challenge. On top of that, because of the growing technologies, the amount of information is exploding. A different paradigm of teaching is needed than what was offered over the last few decades, where emphasis has been on technical material and skills, then only applications. Yet students and instructors alike feel more comfortable focusing on the technical material. Perhaps, we are entering a renaissance time in learning. According to Carl Wieman, Noble Laureate in Physics, former Director of the White House Office of Science and Technology, the goal should not be to teach just skills but to teach how to think like scientists. Teaching ways of thinking, for example Bayesian or other, makes what appears to many irrelevant or uninteresting material suddenly compelling. As Wieman states[1]: "... key is to design tasks where students witness real-world examples of how science works." As such, the encouraging trend of joint inter-departmental seminar series and the emphasis on funding large governmental projects that include significant educational components are likely to only accelerate such renaissance.

"I think that when we know that we actually do live in uncertainty, then we ought to admit it; it is of great value to realize that we do not know the answers to different questions. This attitude of mind – this attitude of uncertainty – is vital to the scientist, and it is this attitude of mind which the student must first acquire". Richard P. Feynman, Noble Laureate in Physics, 1965.

[1] nytimes.com/interactive/2013/09/02/science-education-voices.html

INDEX

Acceptance-rejection, sampling from known distribution with, 89, 89f
Achieved significance level (ASL), 102–3, 118, 145, 204
  sensitivity values based on, 119f, 120–21, 120f
Anisotropy ratio, 240t
ANN. *See* Artificial neural networks
API specific gravity, 5
Approximate Bayesian computation, Denmark groundwater management with, 230, 235
Aquifer thermal energy storage (ATES), 14, 14f
Aristotle, 130
Artificial neural networks (ANN), 70, 72t
ASL. *See* Achieved significance level
ATES. *See* Aquifer thermal energy storage

Bacon, Francis, 130
Barker kernel, 164
Bayes, Thomas, 129, 142–43, 267–68
Bayesian evidential learning (BEL)
  components of
    dimension reduction, 198–99, 199f
    feature extraction, 198–99, 199f
    regression analysis, 199–202, 201f
    statistical modeling, 195–97
    training set, 197–98
  overview of elements of, 196f
  paradigms for UC with, 195f
  in practice, 206–13, 207f–14f
  statistical significance of posterior with, 203–4
  SVM for outlier detection in, 202–3, 203f
  uninformative data variables with, 202
  updating by SIR with, 204–6
Bayesian inversion, 172–79
  grid-based geological structures with, 178–79, 180f
  surface-based geological structures with, 174–78, 175f, 176f, 177f, 178f
Bayesianism
  ad-hoc modifications with, 144
  criticism of, 144–45
  deductive testing of inductive, 145–46
  falsification with
    as reaction to induction, 134–36, 135f
    in statistics, 136–37
  in geological sciences, 146–150
  historical perspective for, 130
  induction *vs.* deduction in, 133–34
  inductive logic as basis for, 133–34
  introduction, 129–30
  objective *vs.* subjective probabilities in, 144
  paradigm with
    Kuhn and, 138
    probability theory as, 138–42, 142t
  rationality for, 143–44
  role of experiments in, 132–33, 133f
  science associated with, 130–32, 131f, 132f
Bayes' rule, 42, 129
BEL. *See* Bayesian evidential learning
Belgium geothermal systems, 14–17, 240–246
Boosted trees, 74–76, 75f
Bootstrap
  bias correction, 101
  confidence intervals with, 102–3
  with correlated data, 101–2
  hypothesis testing with, 102–3
  introduction, 99–100, 100f
  nonparametric, 100–101, 100f
  one-sample, 100, 100f
  regression, 101
  summary of procedure for, 100f
  time series, 101
Borehole thermal energy storage (BTES), 13, 14f
Box-Cox transformation, 54
BTES. *See* Borehole thermal energy storage
Bubble-point pressure, 5

California Department of Toxic Substances Control, 17–18
Canonical correlation analysis (CCA)
  example, 67
  linearizing problem with, 207
  theory, 66–67
CART. *See* Classification and regression trees
CCA. *See* Canonical correlation analysis
CE. *See* Certainty equivalent
Central Valley Regional Water Quality Control Board (RWQB), 17
Certainty equivalent (CE), 31, 31f
Chebyshev distance, 52
Classification and regression trees (CART)
  application on simplified DNAPL example, 122f
  boosted trees, 74–76, 75f
  numerical and categorical data variables handled by, 202
  sensitivity, 76–77, 76f
  sensitivity analysis with, 121–22, 122f
  shale development with UQ using, 254–56, 255f, 258
  single tree methods, 73–74, 74f, 75f
Climate change, 1
Cluster analysis
  application of, 88
  choosing number of clusters for
    Davies-Bouldin
      index, 87–88
    silhouette Index, 87, 87f
  kernel methods for clustering, 87
  k-means, 86, 86f
  k-medoids, 86–87

Quantifying Uncertainty in Subsurface Systems, Geophysical Monograph 236,
First Edition. Céline Scheidt, Lewis Li, and Jef Caers.

Companion code technologies
  bookkeeping needed for, 265
  deployment considerations for, 265
  Git, 265
  Jupyter tutorials, 265
  licensing for, 265
Complexity, decision analysis with, 29
Compositions, sampling from known distribution with, 89
Corey coefficient oil, 221t
Corey coefficient water, 221t
Correlation coefficient, SA based on linear models with, 111
Covariance model, 108
Covariance range, 108
Cox, R. T., 138–39
CrunchFlow, 248–49

DAKOTA, 265
Data collection, training samples obtained with, 196
Data-model-prediction-decision, 24
Data science
  basic algebra for
    distances in, 51–54, 52f, 53f
    matrix algebra notation in, 48–50, 49f
    quadratic forms in, 51
    spectral decomposition in, 50–51, 50f
  bootstrap for, 99–103, 100f
  cluster analysis for, 86–88, 86f, 87f
  functional data analysis for, 67–70
  kernel methods for, 81–86, 82f, 83f, 84f, 85f, 87
  Markov chain MC for, 96–99
  Monte Carlo simulation for, 88–94, 89f, 91f, 92f, 94f
  multivariate probability theory for, 54–57
    characteristic property for, 56
    kernel density estimation in, 54–55
    multivariate distribution properties in, 55–56
    multivariate normal distribution in, 56–57
    univariate transformations in, 54
  regression and classification methods for
    artificial neural networks, 70, 72t
    classification and regression trees, 73–77, 74f, 75f, 76f, 121–22, 122f, 202, 254–56, 255f, 258
    gaussian process regression, 77–81, 78f, 79f
    k-nearest neighbor, 70, 72t
    Kriging, 70, 72t, 77–81, 78f, 79f
    multiple linear regression, 70–71
    support vector machine, 70, 72–73, 72t
  sequential MC for, 94–96
Davies-Bouldin index, 87–88
DCT. *See* Discrete cosine transform
Decision analysis
  basic rules in science of, 34
  efficient frontier with, 38–39, 38t, 39f
  graphical tools for
    decision trees, 30f, 39, 40f, 42f
    influence diagrams, 39–40, 40f, 41f
  hierarchical tree for, 35f
  multiple objectives in, 29
  swing weighting for, 36–38, 38t
  thumbtack game example in, 30–32, 30f, 31f
  time component in, 29
  value of information for, 40–42, 41f, 42f
Decision trees, 30f, 39, 40f, 42f
Deduction, 133–34
DEM. *See* Digital elevation model
DE-MC. *See* Differential Evolution Markov chain
Denmark groundwater management, 9–13, 230–237
Dense non-aqueous phase liquid (DNAPL) test, 108
Deterministic inversion
  Kalman filter in, 169
  linear least squares in, 167–68
  modeling assumptions overview, 169, 170f
  nonlinear inversion in, 169
  regularization in, 168
  regularized inversions in, 170–72, 173f
  resolution in, 168–69
DGSA. *See* Distance-based generalized sensitivity analysis
DiffeRential Evolution Adaptive Metropolis method (DREAM), 99
Differential Evolution Markov chain (DE-MC), 99
Differential hydrodynamic trapping mechanism, 7f
Digital elevation model (DEM), 12
  terrain elevation based on, 231
Dimension reduction
  FDA with BEL in, 199
  FPCA for, 207
  MDS with BEL in, 199
  PCA with BEL in, 198–99, 199f
Discrete cosine transform (DCT), 179, 180f
Distance-based generalized sensitivity analysis (DGSA), 117–21
Distances
  axioms formulated for metric space, 51
  based on transformations, 52–53
  basics, 51
  Chebyshev, 52
  between distributions, 53–54
  Hausdorff, 52, 52f, 53f
  Manhattan, 52
DNAPL test. *See* Dense non-aqueous phase liquid test
DREAM. *See* DiffeRential Evolution Adaptive Metropolis method
DREAM sampler, 179

ECLIPSE, 264
Efficient frontier, 38–39, 38t, 39f
Eigen-image reconstruction, uranium remediation with UQ using, 246, 246f, 249, 251f
Electrical resistivity tomography (ERT), 133
  Belgium geothermal systems monitored by, 15–16, 15f, 16f
    uncertainty quantification with, 238–40, 242, 242f, 243f, 244
Empiricism, 130–31
Ensemble filtering methods, geological priors in, 186–87, 187t
Environmental Protection Agency (EPA), 18
ERT. *See* Electrical resistivity tomography

Falsification
  limitations of, 137–38
  as reaction to induction, 134–36, 135f
  in statistics, 136–37
FDA. *See* Functional data analysis
FLUMY model, 150–51, 150f
Formation volume factor, 5
Forward simulation software
  parallelism needed for, 264
  proxy simulators for, 264
Fossil fuels, 1
  oil field production planning in Libya, 3–9, 4f, 6t, 7f, 8f
  shale developing in North America, 20–24, 21f, 22f, 23t, 254–60, 254f–60f
FPCA. *See* Functional principal component analysis

FRF. *See* Functional random forest
Functional data analysis (FDA), 67–70
  functional basis, 68–69, 69f
  functional PCA, 69–70, 71f
  PCA with harmonics, PC functions, and score plot, 71f
Functional principal component analysis (FPCA), 203
  dimension reduction with, 207
Functional random forest (FRF)
  shale development with UQ using, 259–60, 260f
Fuzzy logic, probability theory with, 139–40

GA. *See* Genetic algorithm
Galileo Galilei, 130
Gaussian model, sampling from known distribution with, 89–90
Gaussian process regression
  defined, 77–78, 78f
  generalized linear regression, 79–80
  history of applications of, 80–81
  prediction with, 78–79, 79f
Gauss-Newton scheme, 170
Generalized likelihood estimation (GLUE), 117
Generalized linear regression, 79–80
Generalized SA. *See* Regionalized sensitivity analysis
Genetic algorithm (GA), 99
Geological priors
  Bayesian inversion with, 172–79
    grid-based geological structures, 178–79, 180f
    surface-based geological structures, 174–78, 175f, 176f, 177f, 178f
  conjunction of information with, 158–160
  deterministic inversion for
    illustration of, 169–72, 173f
    Kalman filter in, 169
    linear least squares in, 167–68
    modeling assumptions overview, 169, 170f
    nonlinear inversion in, 169
    regularization in, 168
    regularized inversions in, 170–72, 173f
    resolution in, 168–69
  ensemble filtering methods with, 186–87, 187t
  general discrete inverse problem with, 155–60, 157f, 159f
  geophysical inversion with, 181–185
  inverse modeling for, 156
  parameterization for, 156
  prior model parameterization with
    geostatistics, 162–64, 163f
    gridding, 160–61, 161f, 162f
    process-based prior model, 162
Geological resources
  fossil fuels, 1
    oil field production planning in Libya, 3–9, 4f, 6t, 7f, 8f
    shale developing in North America, 20–24, 21f, 22f, 23t
  geothermal systems in Belgium, 13–17, 14f, 15f, 16f
  groundwater, 2
    Denmark, management of, 9–13, 10t, 11f, 13f, 229–38, 229f, 231f, 231t, 232f, 234f–37f, 236t–38t
  heat, 1–2
  minerals, 2
  storage of, 2
Geological uncertainty, BEL in practice with, 213
Geophysical surveys, Belgium geothermal systems monitored by, 14–17, 15f, 16f
Geostatistics
  multiple-point, 162–63, 163f
  variogram-based, 163–64
Geothermal systems. *See* Belgium geothermal systems
Gibbs sampler, 97, 179, 180f
Git, 265
Global sensitivity analysis (GSA), 111–121, 241–244
GLUE. *See* Generalized likelihood estimation
Grid-based geological structures
  Bayesian inversion with
    Gibbs sampler in, 179, 180f
    multichain methods in, 179, 180f
    perturbation methods in, 178–79
    samplers involving geological priors in, 179, 180f
Groundwater, 2
  Denmark, management of, 9–13, 10t, 11f, 13f, 229–38, 229f, 231f, 231t, 232f, 234f–37f, 236t–38t
GSA methods. *See* Global sensitivity analysis methods

Handbook of Decision Analysis (Parnell *et al.*), 30
Hausdorff distance, 52, 52f, 53f
Heat, 1–2
HF. *See* Hydraulic fracturing
Hume, David, 130
Hydraulic conductivity, 47, 108
Hydraulic fracturing (HF), 4
Hydrodynamic trapping mechanism, 7f

IEs. *See* Injector efficiencies
Importance sampling, 93–94, 94f. *See also* Sequential importance resampling
  measures of importance, 114–16, 116f
Induction
  Bayesianism using, 133–34
  falsification as reaction to, 134–36, 135f
Influence diagrams, 39–40, 40f, 41f
Injector efficiencies (IEs), 218, 221, 222f, 228t
Intuitionism, probability theory with, 139
Inversion. *See also* Deterministic inversion
  Bayesian inversion with geological priors, 172–79
    grid-based geological structures, 178–79, 180f
    surface-based geological structures, 174–78, 175f, 176f, 177f, 178f
  conditions necessary for formulation of, 155–56
  conjunction of information with
    conditional distributions as conjunctions, 159, 159f
    conjunction, 158–59
    information sources, 158
    limited formulations, 159–60
  dimension reduction from prior movie with
    general purpose of, 166
    kernel methods for, 166–67, 167f
  forward modeling for, 156
  general discrete inverse problem with, 155–60, 157f, 159f
  geological priors in ensemble filtering methods, 186–87, 187t
  geophysical inversion with geological priors
    creating geophysical image, 181–82, 182f, 183f
    linking image with properties, 183
    overview of, 180–81, 181t
    rock physics, 183
    role in UQ, 184–85, 184f, 185f
  introduction, 155
  inverse modeling for, 156
  model expansion from limited prior movie with, 166–67, 167f
  parameterization for, 156
  physical variables in probability theory with
    metric densities, 156–58, 157f
    model and data space, 156

Inversion. *See also* Deterministic inversion (*cont'd*)
prior model parameterization with
geostatistics, 162–64, 163f
gridding, 160–61, 161f, 162f
non-grid-based prior movies, 164–66, 164f, 165f
prior movie, 160
process-based prior model, 162
sampling from known distribution with, 89
subsurface systems with, 217
time-lapse, 240
Irreducible oil saturation, 221t
Irreducible water saturation, 221t

JewelSuite, 263
Jupyter tutorials, 265

Kant, Emanuel, 130
Karhunen-Loeve (KL) expansion, 166
KDE. *See* Kernel density estimation
Kernel density estimation (KDE), 54–55
posterior distribution estimated by, 224, 224f
Kernel methods, 81–87
Kernel PCA (KPCA)
dimension reduction with, 166–67
geological priors in, 187t
method for, 84–85
pre-image problem with, 85–86
KL expansion. *See* Karhunen-Loeve expansion
K-nearest neighbor, 70, 72t
Kolmogoroff, A. N, 138–39
Kolmogorov-Smirnov (KS) statistical test, 117, 117f
KPCA. *See* Kernel PCA
Kriging
regression and classification with, 70, 72t, 77–81, 78f, 79f
shale development with UQ using, 255, 255f, 257, 260f
KS statistical test. *See* Kolmogorov-Smirnov statistical test
Kuhn, Thomas, 130, 138

Latin hypercube sampling (LHS), 91–92, 92f
Level-set representation of surfaces, 165–66, 165f
LHS. *See* Latin hypercube sampling
LNAPL. *See* Light non-aqueous phase liquids
Longitudinal dispersivity, 240t
Lottery, 30
Low-enthalpy geothermal systems, 13–14, 14f

Making Good Decisions (Bratvold and Begg), 30
Manhattan distance, 52
Markov chain MC (McMC), 163
approximate bayesian computation, 98–99
assessing convergence with, 98
defined, 96
Gibbs sampler, 97
Metropolis-Hastings sampler, 97–98
motivation, 96–97
multichain, 99
random walk, 97
Markov random field (MRF) model, 163
Matèrn covariance model, 108
MATLAB, 264–65
Matrix algebra, notation in, 48–50, 49f
McMC. *See* Markov chain MC

MC simulation. *See* Monte Carlo simulation
MDS. *See* Multidimensional scaling
Mean squared error (MSE), 90
Measurement error, 48
Metric densities, inversion with, 156–58, 157f
Metropolis-Hastings sampler, 97–98
Minerals resources, 2
Mineral surface area, 247t
Mixed PCA (MPCA)
Denmark groundwater management with, 230, 233
dimension reduction in context of UC, 223
Modeling noise, Belgium geothermal systems with UQ using, 240, 244
MODFLOW, 264
MODFLOW-USG, 230–31
Monte Carlo (MC) simulation. *See also* Differential Evolution Markov chain; Markov chain MC; Particle Markov chain Monte Carlo; Sequential MC
introduction, 88–89
MC integration, 88–89
naïve, 90
sampling from known distribution, 88–90, 89f
shale development with UQ using, 259–60, 260f
software for, 263–64
stochastic simulation, 88
variance reduction methods, 90–94, 91f, 92f, 94f
Morris method, 109–11, 110f
interactions with, 125t
model assumption with, 125t
results using three different seeds, 124f
sensitivity measure with, 125t
MPCA. *See* Mixed PCA
MPS. *See* Multiple-point statistics
MRF model. *See* Markov random field model
MRST, 264
MSE. *See* Mean squared error
Multichain McMC, 99
Multidimensional scaling (MDS)
dimension reduction technique in context of BEL, 199
example, 63–64, 65f
kernel density estimation in metric space, 65–66, 67f
kernel transformation with, 85f
metric densities in plot using, 156
projection, 64–65, 66f
Multiple objectives, decision analysis with, 29
Multiple-point geostatistics, 162–63, 163f
Multiple-point statistics (MPS), 162, 163f
Multivariate Gaussian, sampling from known distribution with, 89–90
Multivariate probability theory
characteristic property for, 56
kernel density estimation in, 54–55
multivariate distribution properties in, 55–56
multivariate normal distribution in, 56–57
univariate transformations in, 54

Nadaraya-Watson kernel estimator, 202
Naïve Monte Carlo, 90
Non-grid-based prior movies
level-set representation of surfaces, 165–66, 165f
object-based priors, 164, 164f
Nonparametric bootstrap
bias correction, 101
one-sample, 100, 100f

regression, 101
time series, 101
Non-parametric regression, 201–2, 201f
North America shale development, 20–24, 21f, 22f, 23t
uncertainty quantification with, 254–60, 254f–60f
NPV. *See* Net present value

OAT method
Belgium geothermal systems with UQ using, 240
model assumption with, 125t
parameters for, 108–9
sensitivity measure with, 125t
simplicity of, 108
Tornado chart with, 109f
Object-based priors, 164, 164f
Oil viscosity, 221t
Oil water contact, 221t
Open-FOAM, 264
Orthogonal component analysis
canonical correlation analysis, 66–67
multidimensional scaling in, 63–66, 65f, 66f, 67f
principal component analysis in, 60–63, 61f, 62f, 63f, 64f, 68f, 69–70, 71f

PAM. *See* Partitioning around medoids
Paradigm, 195f
Kuhn and, 138
probability theory as
alternative or compliment with, 141–42, 142f
axioms of, 138–39
dogma of precision with, 140–41
fuzzy logic with, 139–40
intuitionism with, 139
Kolmogorov-Cox, 138–39
Parametric regression, 200–201
Pareto plots, t-statistics for regression model in, 114f
Partial correlation coefficient (PCC), 125
Particle Markov chain Monte Carlo (PMCMC), 99
Partitioning around medoids (PAM), 86–87
Payoff, 34
PCA. *See* Principal component analysis
PCC. *See* Partial correlation coefficient
Performance score, 34
Performance score matrix, 36, 36t, 37f
PEST, 265
Petrel, 263
PMCMC. *See* Particle Markov chain Monte Carlo
Porosity, 240t
Predictive models, research required for resource challenge by, 2
Pressure-volume-temperature (PVT) properties, 5
Principal component analysis (PCA)
application to spatial models of, 61–63, 61f, 62f, 63f, 64f
bijective dimension reduction with, 166
dimension reduction technique in context of BEL, 198–99, 199f
functional, 69–70, 71f
kernel, 84–86, 166
mixed, 223
practice, 61
theory, 60–61
Probability theory
alternative or compliment with, 141–42, 142f
axioms of, 138–39
fuzzy logic with, 139–40
intuitionism with, 139
Kolmogorov-Cox, 138–39
Prospect, 30
Proxy simulators, 264
PVT properties. *See* Pressure-volume-temperature properties
Python, 264

Quality map, 221–22, 223f

Random walk, 97
Reaction rate, 247t
Regionalized sensitivity analysis (RSA), 116–17, 117f
Regression. *See also* Gaussian process regression
functional spatial, 256–58, 258f
linear, 111–13, 113t, 114f, 125t
non-parametric, 201–2, 201f
nonparametric bootstrap with, 101
parametric, 200–201
parametric regression, 200–201
Regression trees. *See also* Classification and regression trees
Denmark groundwater management with, 230, 234–35
Reliability, 31
Reservoir charge, 4
Reservoir fluid properties, 5
Reservoir petrophysical properties, 4–5
Reservoir structure, 4
Reservoir volume
equation for, 3
probability distributions for, 4f
Risk preference, 35
River gradient, 108
Rock physics uncertainty, BEL in practice with, 212–13
RSA. *See* Regionalized sensitivity analysis
RWQB. *See* Central Valley Regional Water Quality Control Board

SA. *See* Sensitivity analysis
Sampling from known distribution
acceptance-rejection, 89, 89f
compositions, 89
inversion, 89
MC used to approximate, 90
multivariate Gaussian, 89–90
Scatter plots, SA based on linear models with, 111, 112f
Scenario, 30
Science. *See also* Data science; Geoscience
empiricism in, 130–31
facts as basis for knowledge in, 132
first facts, then theory in, 131–32
role of experiments in, 132–33, 133f
seeing is believing in, 131, 131f, 132f
Screening techniques
Morris method, 109–11, 110f, 124f, 125t
OAT method, 108–9, 109f, 125t
Sensitivity, 76–77, 76f
Sensitivity analysis (SA). *See also* Distance-based generalized sensitivity analysis; Global sensitivity analysis methods; Regionalized sensitivity analysis
distance-based generalized, 117–21, 118f, 119f, 120f, 125t
generalized/regionalized, 116–21, 117f–20f
GSA methods for, 111–22, 112f, 113t, 114f, 116f–20f, 122f
introduction, 107–8
notation and application example of, 108
regionalized, 116–17, 117f

Sensitivity analysis (SA). *See also* Distance-based generalized sensitivity analysis; Global sensitivity analysis methods; Regionalized sensitivity analysis (*cont'd*)
reviews of, 107
screening techniques for
Morris method, 109–11, 110f, 124f, 125t
OAT method, 108–9, 109f, 125t
stochasticity in models' impact for, 122–23, 123f, 124f
summary, 124–25, 125t
terminology, 107
Sequential importance resampling (SIR)
SMC with, 95–96
statistical modeling updating by, 204–6
Sequential MC (SMC)
problem formulation, 94–95
sequential importance resampling, 95–96
SGEMS, 263
Shale development in North America
CART for, 254–56, 255f, 258
data preprocessing for, 256, 256f, 257f
data science in, 23–24
functional data in, 255
functional random forest for, 259–60, 260f
functional spatial regression for, 256–58, 258f
kriging for, 255, 255f, 257, 260f
large set of covariates in, 255–56
Monte Carlo study for, 259–60, 260f
parameters in designing unconventional, 23t
reservoirs and production in, 21–22, 22f, 23t
single tree for, 259–60, 260f
strategy for UQ in, 254–56, 254f, 255f
Silhouette Index, 87, 87f
Silverman's rule of thumb, 55
Single tree
methods, 73–74, 74f, 75f
shale development with UQ using, 259–60, 260f
SIR. *See* Sequential importance resampling
SMC. *See* Sequential MC
Sobol' indices, 116f
interactions with, 125t
sensitivity measure with, 125t
stochasticity with, 125t
Software and implementation
companion code technologies
bookkeeping needed for, 265
deployment considerations for, 265
Git, 265
Jupyter tutorials, 265
licensing for, 265
dealing with software issues, 270
forward simulation
parallelism needed for, 264
proxy simulators for, 264
introduction, 263
model generation
automation needed for, 264
Monte Carlo sampling in, 263–64
post-processing, 264–65
Solid specific heat capacity, 240t
Solid thermal conductivity, 240t
Solution gas-oil ratio, 5
Spatial uncertainty, 247t
Spectral decomposition, 50–51, 50f
SRC. *See* Standardized regression coefficients
SSE. *See* Sum of squared error
Standardized regression coefficients (SRC)
linear regression analysis with, 111–13, 113t
sensitivity measured with, 111
Statistical surrogate models, subsurface systems with, 217–18
Streamflow reduction potential. *See* SFRP
The Structure of Scientific Revolution (Kuhn), 138
Sum of squared error (SSE), 260
Support vector machine (SVM), 70, 72–73, 72t
outlier detection in statistical model with, 202–3, 203f
Surface-based geological structures
Bayesian inversion with, 174–78, 175f, 176f, 177f, 178f
faults networks from partial fault interpretations with
geological priors for faults, 174–76, 175f, 176f
geological prior's importance, 177–78, 178f
from geological prior to marked point process prior, 176–77, 177f
SVM. *See* Support vector machine
Swing weighting, 36–38, 38t

Thermal response test (TRT), 15
Thermal tracer test (TTT), 15
3DSL, 264
Thumbtack game, 30–32, 30f, 31f
TI. *See* Training image
Time component, decision analysis with, 29
Time-lapse inversion, Belgium geothermal systems with UQ using, 240
Time series, nonparametric bootstrap with, 101
TOUGH2, 264
Training image (TI), 148–49, 149f
Transverse dispersivity, 240t
Tree-based SA, 121–22, 122f
discrete parameter with, 125t
high dimensional response with, 125t
input distribution with, 125t
interactions with, 125t
model assumption with, 125t
sensitivity measure with, 125t
stochasticity with, 125t
TRT. *See* Thermal response test
TTT. *See* Thermal tracer test

Uncertainty
decision making under, 29–42
geological, 213
Uncertainty quantification (UQ)
Bayesianism in context of, 129
Belgium geothermal systems with
Bayesian evidential learning for, 240, 244, 246
Bayesian prediction with GSA methods in, 241–45, 241f–45f
CCA for, 240, 244
deterministic inversion for, 240
DGSA for, 240, 242
modeling noise for, 240, 244
OAT method for, 240
PCA for, 240, 242, 243f, 244f
strategy for UQ in, 238–40, 239f
time-lapse inversion in, 240
data-model-prediction-decision for, 24
distances component in, 51–54, 52f, 53f
elements of BEL for, 196f

modeling components and data sources for, 6t
outlook for, 267–71
paradigms using BEL for, 195f
probability theory as paradigm for, 138–42, 142f
role of geophysical inversion in, 184–85, 184f, 185f
seven questions on
Bayes or not Bayes, 267–68
geological knowledge into quantitative numerical models, 268–69
how to deal with software issues, 270
proper model parameterization, 269
realistic prior distribution establishment, 269–70
what combination of data is needed, 270
what educational means for design to teach UQ, 271
statistical model confidence interval with, 195
in subsurface systems
introduction, 217–18
Univariate transformations, 54
UQ. *See* Uncertainty quantification
Uranium remediation in United States, 17–20, 18t, 19f, 20f
uncertainty quantification with, 245–54, 246f, 247f, 247t, 249f, 250f–53f
Utility function, 35
Utility metric, 35

Value, three elements influencing, 31
Value functions, 35
Value metric, 34–35
Value of Information in the Earth Sciences (Eidsvik *et al.*), 30
Value of perfect information (VOPI), 31
Variance of permeability, 247t
Variance reduction methods
control variates and multilevel MC, 92–93
importance sampling, 93–94, 94f
introduction, 90
Latin hypercube sampling, 91–92, 92f
stratified sampling, 91, 91f
Variogram azimuth, 247t
Variogram-based geostatistics, 163–64
Variogram correlation length, 247t
Variogram type of permeability, 247t
VOPI. *See* Value of perfect information

WAFs. *See* Well allocation factors
Water drawdown. *See* WDD
Water specific heat capacity, 240t
Water thermal conductivity, 240t
Wavelet uncertainty, BEL in practice with, 213
WDD (Water drawdown), 233–34, 236t
Well allocation factors (WAFs), 218
WRP (Wetland reduction potential), 233–34, 236t, 237t

www.ingramcontent.com/pod-product-compliance
Lightning Source LLC
LaVergne TN
LVHW082118200926
848484LV00008B/1973